Harnessing Bistable Structural Dynamics

Harnessing Bistable Structural Dynamics

For Vibration Control, Energy Harvesting and Sensing

Ryan L. Harne
The Ohio State University
Columbus, Ohio, USA

K.W. Wang
University of Michigan
Ann Arbor, Michigan, USA

WILEY

Registered Office
John Wiley & Sons Ltd, The Atrium, Southern Gate, Chichester, West Sussex, PO19 8SQ, UK

Editorial Offices
111 River Street, Hoboken, NJ 07030, USA
9600 Garsington Road, Oxford, OX4 2DQ, UK
The Atrium, Southern Gate, Chichester, West Sussex, PO19 8SQ, UK

For details of our global editorial offices, customer services, and more information about Wiley products visit us at www.wiley.com.

Wiley also publishes its books in a variety of electronic formats and by print-on-demand. Some content that appears in standard print versions of this book may not be available in other formats.

Library of Congress Cataloging-in-Publication data applied for

ISBN: 9781119128045

Cover design by Wiley
Cover image: Courtesy of the author

Set in 10/12pt Warnock by SPi Global, Pondicherry, India

10 9 8 7 6 5 4 3 2 1

Contents

Preface

This book describes a recent perspective that seeks to strategically harness the unique dynamics of bistable structural systems for engineering advances, with focus on the three technical areas of vibration control, energy harvesting, and sensing.

When a structural system exhibits two statically-stable configurations, it is said to be *bistable*. This class of structures has been employed in various mechanical, civil, marine, and aerospace engineering applications for many years. The bistable components include arches and post-buckled beams/columns, panels or shells having a shallow curvature, and curved microbeam or dome diaphragm transducers, to name just a few examples. In many historical assessments, it was undesirable for the bistable structure to "snap" to the second state of static equilibrium – a phenomenon referred to as *snap-through* – because the consequence might be unfavorable to the health and performance of the overall engineered system. Thus, the structures or materials were used in ways to avoid static or dynamic loading (e.g., pressure on a shell) that could cause the bistable system to switch from the original, stable configuration to the other stable equilibrium (e.g., an inverted shell).

It is from such a perspective that the focus of this book departs. Recently, researchers have been challenged to reconsider bistable structural systems within a variety of emerging engineering contexts. Many scientists and engineers have discovered and explored the dynamics of bistable structures that may be deliberately exploited to the advantage of certain applications. Innovative ideas have been proposed to intelligently induce snap-through behaviors such that the performance of the overall system is enhanced and/or new functionality is realized. This new spirit of engineering system development is the foundational viewpoint of this book: *harnessing bistable structural dynamics*.

The new ideas have been found to be well-suited for many applications across a variety of engineering disciplines. Among them, researchers have particularly investigated the exploitation of bistable structural dynamics in the areas of (i) *vibration control*, (ii) *vibration energy harvesting*, and (iii) *sensing and detection*. In the first area, the dynamics of bistable devices integrated with the structural system are used so as to isolate, dissipate, or reactively attenuate the input energies that excite the system. Depending on the performance objective, the multi-faceted dynamics of the bistable members are strategically employed to best control the vibrations and improve the operational integrity of the system.

The aim of energy harvesting is to electromechanically convert ambient vibrations into usable electrical power resources. To this end, maintaining persistent snap-through behaviors is a common aim because of the large dynamic mechanical and electrical response amplitudes that they induce in the energy harvesting systems. Hence, the energetics of snap-through promotes a significant potential for energy conversion. In the context of sensing and detection, the ability to recognize small changes in monitored parameters, which represent time-varying system characteristics, is critical in providing the earliest indicator of structural change. Thus, by harnessing the sudden transition between low amplitude oscillations around a stable equilibrium to energetic snap-through vibrations spanning the static equilibria, bistable dynamics-based detection strategies have been found as promising, novel approaches well-suited for a variety of sensing contexts.

The recent interest in harnessing bistable structural dynamics for engineering advances in vibration control, energy harvesting, and sensing has led to a flourishing research development that ranges from rigorous theoretical formulations to exploratory experimental studies. The comprehensive investigations have drawn a variety of conclusions regarding the exploitation of bistable dynamics that may best promote the aims of the respective technical area and applications, and the continued active research engagements suggest that much is left to be discovered.

The insights on the effective exploitation of bistable dynamics to enhance the aims of the aforementioned three technical areas are currently scattered amongst numerous scientific publications and proceedings, including a selection of works by the authors. To derive the greatest benefits from the assorted findings, there is a need to consolidate the results and, with the organized evidence, to derive coherent conclusions that are informed by the broad range of research. This book seeks to meet this need through the presentation of extensive topical reviews and detailed case studies which complement the overall conclusions. For the benefit of the ongoing efforts, this book also seeks to identify the emerging areas and outstanding needs that require future attention before the exploitation of bistable dynamics meets its fullest potential for vibration control, energy harvesting, and sensing and detection.

A large body of scientists and engineers is represented among these technical areas. The approaches and conclusions described in detail in this book will inspire researchers investigating the exploitation of bistability in engineering systems, as well as enlightening a broad range of readers interested in vibration control, energy harvesting, and sensing to the attractive opportunities that may be engendered via harnessing bistable structural dynamics. By reviewing the developments and presenting specific studies as examples, a goal of this book is to provide an accessible avenue for a large readership to an appreciation of the recent findings from the engineering objective lens. We assume that readers have a college level undergraduate education in an engineering curriculum including engineering mathematics and structural dynamics or vibrations. Exposure to more advanced topics such as electromechanical systems, nonlinear dynamics, and stochastic vibrations is beneficial but not essential. Additionally, this book focuses on all-mechanical, electromechanical, or all-electrical bistable systems, which may be realized or adequately modeled as one-dimensional systems; such platforms represent a large number and wide variety of bistable devices that have been investigated. For readers interested in more details on a specific topic or concept, references to many books and papers are provided throughout the book.

We are indebted to financial support from sponsors in order to conduct and document the research included in this book. Parts of the research were supported by grants from the Air Force Office of Scientific Research, the Defense Advanced Research Projects Agency, the National Science Foundation, by the University of Michigan Collegiate Professorship Fund, and by funds from the Department of Mechanical and Aerospace Engineering at The Ohio State University. We also recognize the contributions from our graduate students and co-workers so that the breadth of research topics in this book may be so comprehensively investigated. These individuals include David Johnson, Jinki Kim, Fabio Semperlotti, Manoj Thota, Zhen Wu, and Kai Yang. In addition, we wish to acknowledge our many colleagues within the professional communities of adaptive materials and structures, dynamics and vibration, energy harvesting, structural sensing and health monitoring, and systems and controls who have cultivated a vigorous environment of search and discovery within which we have enthusiastically undertaken and compiled the efforts of this book. Finally, our deepest appreciations go to our families and loved ones for their tremendous support throughout the years.

Ryan L. Harne
Columbus, Ohio, USA

K. W. Wang
Ann Arbor, Michigan, USA

1

Background and Introduction

This chapter provides a background to the common realizations and dynamics of bistable structures which have recently been harnessed to advance the aims of the three technical areas of interest in this book: vibration control, vibration energy harvesting, and sensing and detection. The structural forms and dynamical behaviors of bistable systems are first introduced to demonstrate the broad versatility of design and response which are commonly exploited, and to highlight the dynamics which are enabled via leveraging bistability. Then, two aspects of bistable structural dynamics are identified as trademark features which serve as common rationales for exploiting bistability in engineering applications. These aspects are elaborated upon through concise descriptions of example implementations within the technical areas considered here. Finally, an outline of the remaining chapters is provided.

1.1 Examples of Bistable Structures and Systems

To introduce the essential static characteristics of a bistable structure, a schematic of an example one-dimensional, mechanical bistable system is shown in Figure 1.1. To set a clear convention, hereafter the terms bistable *structures* and *systems* will be used interchangeably, irrespective of whether the object considered is all-mechanical, electromechanical, or all-electrical. For the bistable system shown in Figure 1.1, two identical springs of undeformed lengths l_o connect a lumped mass to a surrounding frame of span $2d$. It is assumed that all displacements of the system are in a horizontal direction, such that the frame motions z and mass displacements X move along one axis. When the undeformed spring length is less than or equal to half of the span of the frame, $l_o \leq d$, the system is monostable and the mass will come to rest at the zero displacement position, $X = 0$. In contrast, when the undeformed spring length is greater than half of the span of the frame, $l_o > d$, the springs exert a force on the mass such that the mass cannot be easily maintained in the central configuration. The zero displacement configuration is unstable, while two stable positions of the mass are adjacent (and, here, symmetric) to the central, unstable state. As a result of the geometric design condition $l_o > d$, the mass-spring and frame system is said

Harnessing Bistable Structural Dynamics: For Vibration Control, Energy Harvesting and Sensing, First Edition. Ryan L. Harne and K.W. Wang.

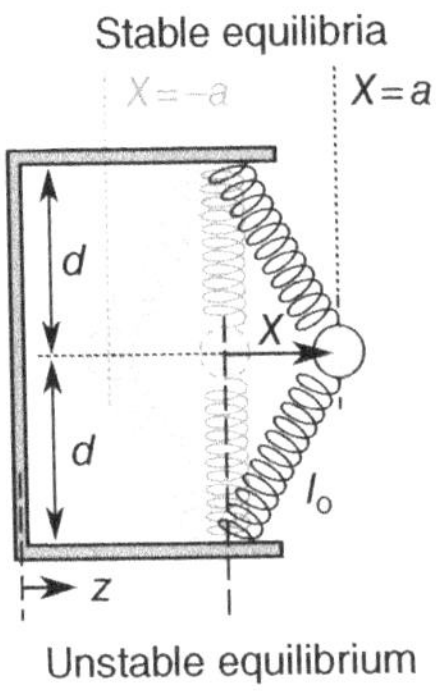

Figure 1.1 Schematic of a bistable system composed of springs, mass, and frame. Here $l_o > d$ which makes the central configuration $X = 0$ an unstable position of static equilibrium.

to be *bistable*. The stable equilibria of the structure are shown to be configurations of the mass such that the displacements are $X = \pm a$.

Figure 1.2a,b illustrates the force, $F(X)$, and potential energy, $U(X)$, of the bistable system, respectively, as functions of the mass displacement. Potential energy is determined by $U = \int F \cdot dX$. In this example, the bistable structure is symmetric and the only forces resisting the horizontal mass displacements are due to the identical springs. Figure 1.2a shows that the total restoring force in the X-axis is zero when the inertial mass is positioned at any of the equilibria. On the other hand, Figure 1.2b shows that the potential energy is locally maximized at the unstable central configuration of the inertial mass $X = 0$, while the adjacent, stable equilibria at $X = \pm a$ locally minimize the potential energy of the system. Therefore, based on the principle of minimum total potential energy, disturbances to the inertial mass when it is originally positioned at the unstable equilibrium will propel the mass towards one of the stable equilibria.

An instructive analogy is that of a ball on a terrain with elevation profile shaped like the potential energy plot in Figure 1.2b. While situated precisely at the peak of the central hill (at $X = 0$), the ball will remain stationary even though the gravitational potential energy is high. But if given a slight perturbation, the ball will roll into the nearest valley where it settles into the displacement position which minimizes potential energy (specifically, gravitational potential energy in this analogy).

By Hooke's law, the stiffness of a spring element is determined by the spatial derivative of the restoring force, dF/dX. Considering the total spring force profile at the unstable equilibrium in Figure 1.2a, it is apparent that the bistable spring is characterized as having a *negative stiffness* for this mass position. In contrast to a spring which *resists* the motion of the mass in a given direction, a spring exhibiting negative stiffness over a range of displacements will *assist* the motion of the mass. As a result, the small perturbation to the inertial mass when precisely positioned at the unstable equilibrium will lead the bistable spring to propel the mass away from the central location to one of the stable system configurations.

All the bistable systems considered in this book share these fundamental, static characteristics illustrated using the mechanical example in Figure 1.1. In fact, the existence of two statically-stable equilibria configurations and one unstable configuration make it straightforward to identify bistable structures or systems. However, the type of geometrical constraints exemplified in the mechanical

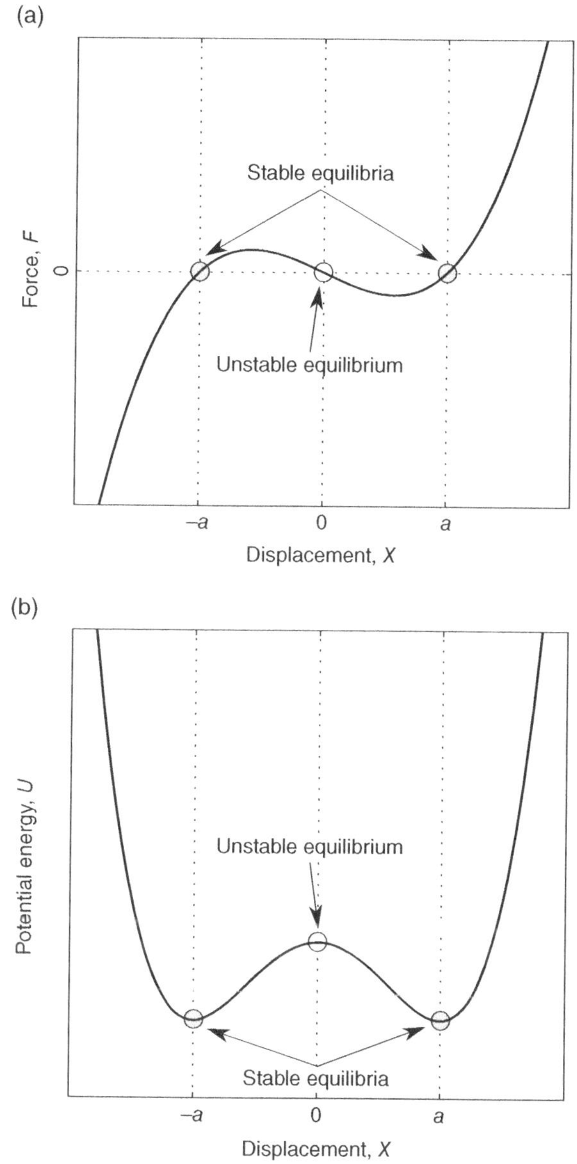

Figure 1.2 Dependence of (a) spring force and (b) stored potential energy on the displacement position of the inertial mass.

device shown in Figure 1.1 are just one possible approach to effect bistability. For the technical areas of vibration control, energy harvesting, or sensing, numerous and diverse methods are employed to realize bistability.

To sample the many approaches, Figure 1.3 shows recently investigated engineering systems that utilize one of the various methods to effect bistability. In Figure 1.3a, modules of series- and parallel-assembled double-beam units are compressed within a housing support near the threshold of buckling. When the

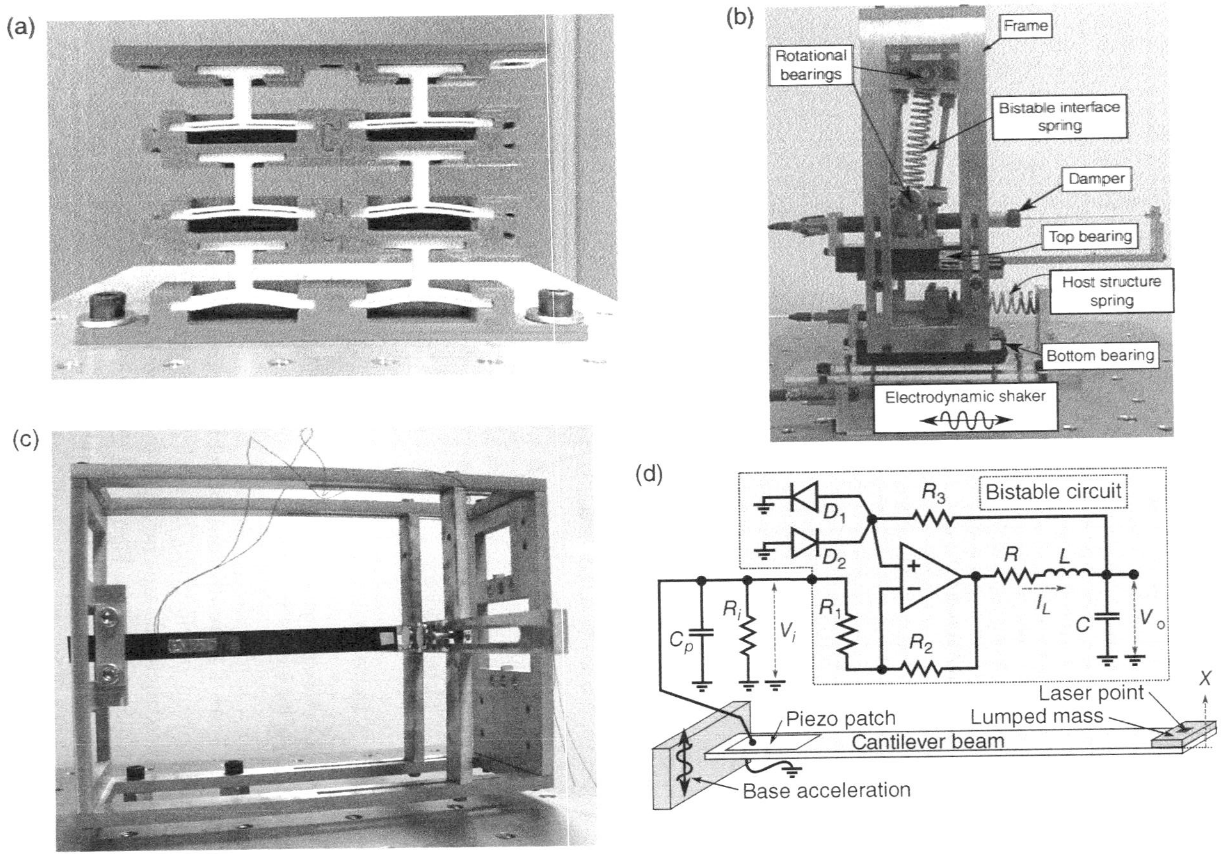

Figure 1.3 (a) Damping module of buckling beams arranged in series and parallel within a housing frame for energy attenuation between the ends of the module. (b) Bistable device for vibration isolation of suspended top bearing mass. (c) Cantilevered beam with piezoelectric PVDF patches and magnetic attraction induced bistability for energy harvesting. (d) Bistable circuitry attached to host beam structure via piezoelectric transducer for sensing structural change.

statically compressed modular structure is excited with periodic dynamic loads, the modules provide high damping due to the transitions among the various stable topologies. Such transition phenomena result in a large dissipation of the input energy according to the number of state changes that occur. In Figure 1.3b, a two degrees-of-freedom (DOF) vibration suspension concept is shown [1]. Here, bistability is effected between the DOF (the moving frame and top bearing mass) via geometric relations, comparing the length of the interfacing spring to the distance between the frame and top bearing mass. Due to the activation of unique bistable dynamics, the investigations of the two DOF suspension system uncover an exceptional reduction of motion transmissibility as compared to the counterpart two DOF linear suspension [1]. Note that the geometrical constraints in the two DOF architecture shown in Figure 1.3b which lead to bistability between the two moving bodies are similar to those constraints employed in the translational single DOF bistable system example shown in Figure 1.1.

In some vibration energy harvesting applications [2], a composite plate with attached piezoelectric patches is made bistable through the generation of a static stress for the flattened plate configuration, such that the plate maintains one of two stable equilibria shapes having finite curvatures. There are numerous design and fabrication parameters which may be adjusted to tailor the two stable plate curvatures, including the composite material layer selection, lay-up order, layer relative rotations, and thermal conditions under which the layers are stacked and cured [2,3]. When the plate is excited at its center by external vibrations, the energetic snap-through actions of the plate from one stable curvature to the next greatly strain piezoelectric materials attached to the plate surface. In consequence, the input vibrational energy is converted by the piezoelectric material to an oscillating flow of current which can be exploited for energy storage purposes (e.g., battery charging) or conditioned for direct utilization as a supply for low-power microelectronics. In another vibration energy harvesting investigation, bistability is realized using a combination of elastic and magnetic restoring forces on a cantilever beam, as shown in Figure 1.3c [4]. Motions of the ferromagnetic cantilever beam tip are resisted via the beam's inherent elasticity but the attractive influences of the magnet pair are tuned so as to draw the beam away from the original cantilevered configuration. The piezoelectric PVDF (polyvinylidene fluoride) patches attached to the beam surface are significantly strained as ground motions excite the cantilever base and cause the beam to oscillate between the stable beam positions. Similar to the case of the bistable plate, the charge generation on the piezoelectric material electrodes may then be harnessed for energy harvesting purposes. The broad frequency bandwidth and high sensitivity at low frequencies of snap-through dynamics set bistable energy harvesters apart from other harvester platform designs and strongly justify the recent attention given to bistable structures for energy harvesting applications [5,6].

Bistability may also be effected via electromechanical or electrical means. For instance, due to electrostatic actuation including bias and oscillating voltages, a microbeam having an initial curvature may be excited to oscillate between two stable beam shapes [7]. In this example, the beam may or may not be inherently bistable due to elastic influences alone. Nevertheless, the electrostatic forces provide a means to ensure bistability and then to actuate the beam between the two stable shapes. The applications for micro- and nanoscale bistable systems are numerous, and include their implementation as electromechanical signal filters, switches, actuators, or as novel mechanical memory elements, to name only a few of the functions for which they have been explored [7–10]. Additionally, the circuit schematic shown in Figure 1.3d realizes bistability simply by the circuit design [11]. In the absence of an input voltage V_i, the output voltage V_o across the capacitor C will settle on a finite positive or negative value. Bistable circuitry is a well-established tool in the study of many physical sciences [12], but the exploitation of such circuit designs to advance the aims in engineering contexts is an emerging area of technical interest. For example, when the circuit input is connected to a structural transducer, such as a piezoelectric patch as shown in Figure 1.3d, the activation of the high amplitude snap-through response of the bistable circuit output voltage V_o can be harnessed as an indicator of change in the structural system. By this strategy, one may monitor shifts in important parameters, such as a reduction in stiffness which could indicate damage. Thus, small variations in structural responses of the beam are tracked using large changes in bistable circuit voltage dynamics. Through a relation between the changing structural model parameters and the critical conditions that activate the circuit snap-through dynamics, one realizes a novel pathway for robust and sensitive detection of change [11].

As exemplified by the cross-disciplinary sample described above, the bistable systems considered in this book represent diverse mechanical, electromechanical, and electrical platforms that span a vast range of length scales. Yet, regardless of the specific manifestation and inducement of bistability, the existence of two statically-stable configurations and one unstable configuration is the common denominator for all bistable systems. However, the mechanics of bistable structures are historically-established science [13]. Thus, following the previous introduction of the essential static features of bistable systems and how they have been realized in structural or system forms, the next section introduces the numerous, characteristic *dynamics* of bistable structures. These are the behaviors which have been recently harnessed to the advantage of the technical areas of interest in this book.

1.2 Characteristics of Bistable Structural Dynamics

Regardless of the method by which bistability is effected, each platform shown in Figure 1.3 exemplifies dynamics common to all bistable systems. Moreover, these dynamics have particularly unique manifestations as compared to the responses exhibited by linear or monostable nonlinear systems, which justifies the recent research attention given to bistable structures in the technical areas of interest in this book. The following sections review the distinct dynamic characteristics of bistable structures and provide illustrative and exemplary plots of the behaviors. Before elaborating on the unique dynamical features, a brief description is provided below regarding the means to model and predict such dynamics and the interpretations of the parameters that have been employed to generate the illustrative results.

The illustrative plots presented in the following sections are computed as the dynamic solutions to the governing equation of motion of a bistable system. The analytical and numerical techniques to solve the equation, so as to generate the following plots, are described in detail in Chapter 2. A conventional governing equation construction is used which is known to accurately model the dynamics of numerous bistable structure realizations that have been examined across a wide range of sciences and engineering fields. The equation may be expressed in a normalized form as

$$\frac{d^2x}{d\tau^2} + \gamma\frac{dx}{d\tau} - x + \beta x^3 = p\cos\omega\tau \tag{1.1}$$

where $x(\tau)$ is considered to be the non-dimensional, generalized displacement of the bistable structural system which is a function of non-dimensional time τ; γ is a damping factor; β is a degree of nonlinearity; p is the excitation level. The term ω is the excitation frequency which is normalized according to the system's natural frequency computed using the magnitude of the linear stiffness and the mass of the non-normalized system [14]; thus, $\omega = 1$ indicates a resonance-like excitation, although the analogy is imperfect. The transient solution to Eq. (1.1) also depends on the initial conditions of displacement and velocity, $x(0)=x_0$ and $\dot{x}(0) = \dot{x}_0$, respectively, where an overdot indicates differentiation with respect to the normalized time, $d/d\tau$.

Although the following examples describe the characteristic dynamics computed from Eq. (1.1) in regards to a bistable *mechanical* system response, in which case terms such as displacement and velocity will be used, it is noted for completeness that Eq. (1.1) also models bistable systems realized in electrical domains. Thus, the coordinate x may represent a normalized voltage or current and the corresponding system and excitation parameters (p,ω,γ,β) may be described in terms of the appropriate counterpart electrical components.

The unforced, static solution to Eq. (1.1) leads to the determination of the equilibria configurations of the bistable system. In this case, one solves an equation expressed by

$$-x + \beta x^3 = 0 \tag{1.2}$$

The solutions to the polynomial in Eq. (1.2) are $x = \pm 1$ and 0. Further mathematical treatment of these results, detailed in Chapter 2, reveals that the equilibria at $x = \pm 1$ are *stable* whereas the equilibrium at $x = 0$ is *unstable*.

Using the analytically or numerically computed solutions to the governing equation and by taking into consideration the bistable system equilibria configurations, the following sections elaborate on the characteristics of bistable structural dynamics. While some of the general dynamical features are exhibited by other types of nonlinear systems, the unique ways in which they are realized by bistable structures are detailed to highlight the important distinctions. For brevity, the following sections describe and elucidate the plotted results while the figure captions detail the specific parameter set combinations and relevant computational aspects to generate the data.

1.2.1 Coexistence of Single-periodic, Steady-state Responses

In contrast to linear systems, when driven by single-frequency excitations many nonlinear systems may potentially exhibit one or more forms of steady-state response that *coexist* as a result of the same excitation conditions. These vibrations occur primarily at the excitation frequency. The coexistence of dynamic responses means that for a set of design parameters and prescribed harmonic excitation level and frequency, the nonlinear system may undergo different dynamics in time, often one of two distinct harmonic response amplitudes. The specific dynamic that occurs is dependent upon the system initial conditions at the starting time of the single-frequency excitation.

In the case of bistable systems, there are numerous types of single-periodic, steady-state dynamics which may coexist. First, one may classify the dynamics into two regimes: *intrawell* oscillations which occur around one of the two stable equilibria, or *interwell* oscillations which vibrate across the unstable equilibrium twice per excitation cycle. Because the potential energy profile of bistable systems, shown in Figure 1.2b, is conventionally referred to as the *double-well* potential, the terms *intrawell* and *interwell* denote that the oscillations of the inertial mass remain confined to one of the local wells of potential energy or cross back-and-forth between them, respectively. One may then separate these two dynamic regimes into low and high amplitude versions of the intra- and interwell dynamics. As a result, there are *four* forms of single-periodic, steady-state dynamics exhibited by a bistable structure, some of that may coexist for the same excitation and system design parameters.

As examples of the two forms of *intrawell* dynamics, Figure 1.4a presents numerically computed displacement responses using Eq. (1.1). Here, the different initial conditions lead to unique steady-state dynamics while all other parameters remain the same. It is clear that the intrawell dynamics, plotted as the black or gray solid curves, oscillate around one of the two stable equilibria shown as thick dotted lines. Although not able to be observed using one example, the low and high amplitude oscillations, as predicted from Eq. (1.1), may occur around either stable equilibrium. In the example presented in Figure 1.4a, the initial conditions are such that the high amplitude oscillations vibrate around the stable equilibrium at $x = -1$ while the low amplitude responses occur around $x = +1$. Figure 1.4b plots the displacement amplitude across all frequencies as a consequence of the excitation at frequency ω. The data are computed using the fast Fourier transform (FFT) of the displacement time series shown in Figure 1.4a; the final 50 of 100 excitation periods are used in the computation. The spectra shown in Figure 1.4b plainly indicate that the different intrawell dynamics are distinct in their amplitude at the excitation frequency, shown as the circle points. The input energy at the frequency of ω is apparently diffused to other harmonics; further discussion of this feature is deferred until Section 1.2.6.

Figure 1.5a plots an example in which the low and high amplitude *interwell* dynamics are found to coexist for the same excitation frequency and level. The interwell behaviors are distinct from the intrawell responses in that the prior oscillate across the unstable equilibrium position, which was identified above to be the normalized position of $x = 0$. Apart from the difference in the amplitudes

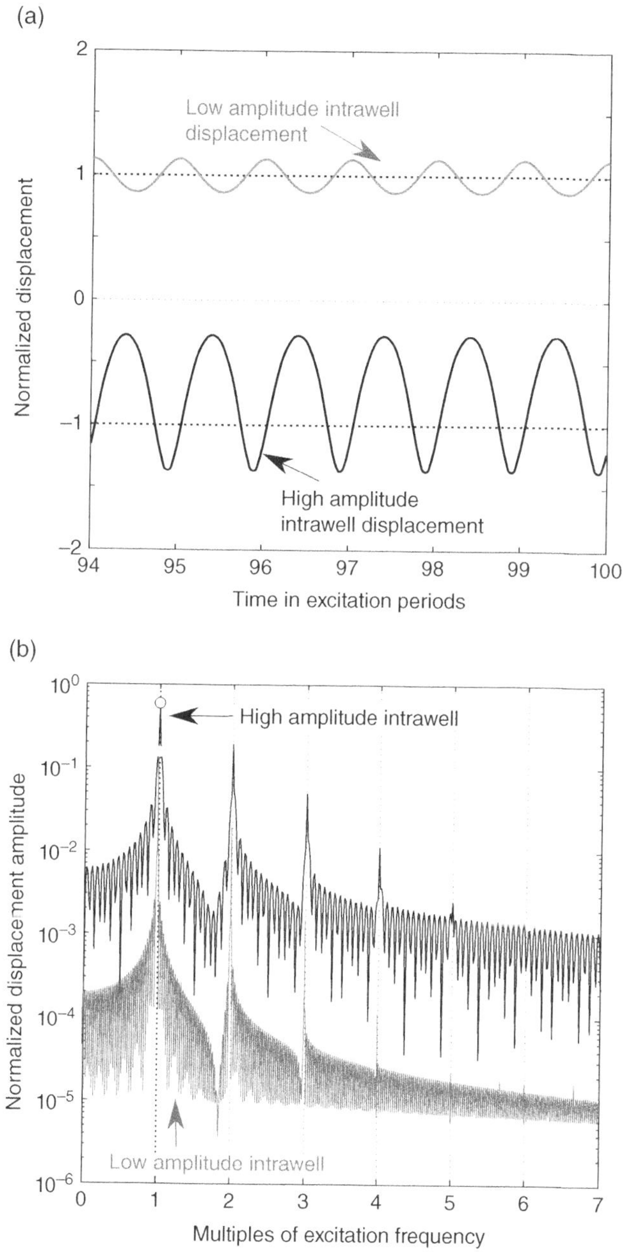

Figure 1.4 (a) For an excitation frequency at which both the low and high intrawell dynamics coexist, the time series of the displacement responses are presented. The thick dotted lines indicate the statically-stable equilibria. (b) The displacement magnitude spectral responses of the low and high amplitude intrawell dynamics are shown as computed from the time series data plotted in (a). The energy diffusion to other harmonics of the excitation frequency is apparent and distinct between the response forms. In (a,b), the parameters are $(p, \omega, \gamma, \beta) = (0.1, 1.1, 0.09, 1)$, with initial conditions $(x_0, \dot{x}_0) = (1.0041, 0.0961)$ for the low amplitude oscillations and (2.2858, −0.9302) for the high amplitude dynamics.

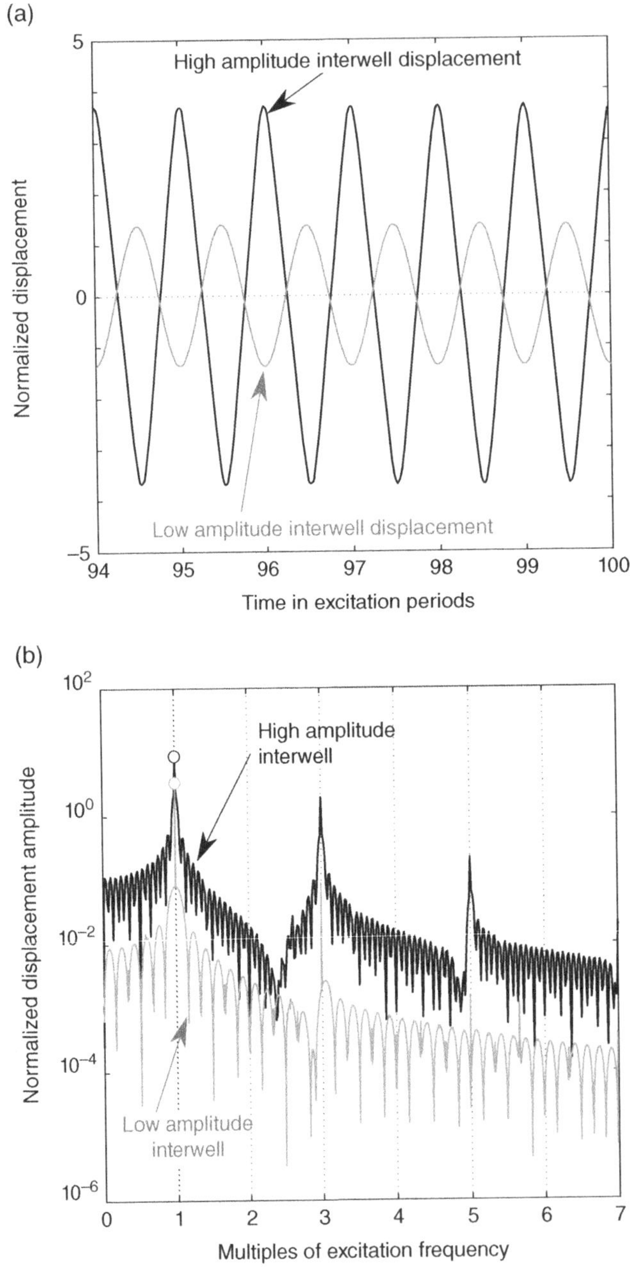

Figure 1.5 (a) For an excitation frequency at which both the low and high interwell dynamics coexist, the time series of the displacement responses is presented. (b) The displacement magnitude spectral responses of the low and high amplitude interwell dynamics are shown as computed from the time series data plotted in (a). The energy diffusion to other harmonics of the excitation frequency is apparent and distinct between the response forms. In (a, b), the parameters are $(p, \omega, \gamma, \beta) = (8, 2.5, 0.09, 1)$, with initial conditions (−1.9297, −0.0944) for the low amplitude interwell responses and $(x_0, \dot{x}_0) = (-2.9443, -0.1791)$ for the high amplitude interwell vibrations.

of the two dynamic forms, it is clear that the responses are approximately 180 degrees out-of-phase. Although not shown in Figure 1.5a, the high amplitude snap-through dynamics occur with phase lags no greater than 90 degrees with respect to the excitation, while the low amplitude interwell oscillations lag the excitation typically by around 180 degrees. The unique phase relationships enable a practical means to distinguish the two forms of interwell dynamics during experimentation in which case they both oscillate back-and-forth across the unstable equilibrium but clearly have different phase relationships with respect to the excitation. Similar to the intrawell dynamic spectral features, Figure 1.5b shows that the two interwell dynamics of bistable systems are different in terms of response amplitude at the excitation frequency of ω (shown as the circle points) as well as in terms of the energy diffusion to other integer multiple harmonics of the excitation.

By evaluating Eq. (1.1) over a broad range of excitation frequencies using an analytical strategy, Figure 1.6a plots the normalized displacement amplitudes of a bistable system (shown as solid curves) as compared to the linear system governed by the corresponding normalized governing equation of motion $\ddot{x}+\gamma\dot{x}+x=p\cos\omega\tau$ (dash-dot curve). Additionally, Figure 1.6a shows thin gray curves which represent *unstable* dynamics of the bistable system. In other words, these are responses which are mathematical solutions to the governing equations but which are not physically realizable dynamics.

The amplitudes of the bistable system intrawell dynamics show resonance-like frequency dependence similar to the linear system, but the prior are "bent" towards lower frequencies. This is an indicator of a "softening" type of nonlinearity, further details of which are provided in later chapters of this book. It is also observed in Figure 1.6a that the high amplitude interwell dynamics exhibit large displacement magnitudes over a broad range of frequencies. In fact, the large amplitude snap-through dynamics are predicted to occur even for a progressively vanishing excitation frequency, in other words as $\omega \to 0$. This prediction may be intuitively appreciated through a basic consideration of the mechanics involved. Note that as the frequency approaches zero, the motions of the linear system mass are confined to how greatly the linear spring may be statically deformed for the applied load. In contrast, as the excitation frequency approaches quasi-static conditions, the displacements of a bistable system mass will still undergo a large stroke from one stable equilibrium to the other so long as the system is "pushed" with a large enough load level to induce a snap-through dynamic. As such, it may be stated that snap-through is a *non-resonant* dynamic because there is no reliance on resonance-like features to excite the energetic motions. The exceptional sensitivity to broadband and low frequency inputs is a repeatedly exploited feature of bistable systems in the technical areas described in this book.

A major difference between the responses of the bistable structure and the counterpart linear oscillator are the *bifurcations* in the bistable system responses, observed in consequence to variation in the excitation frequency, ω. A bifurcation is a large qualitative change in a system state (static and/or dynamic) due to an infinitesimally small change in an individual parameter (whether internal or external to the system) [15,16]. Three bifurcations in the dynamics of the bistable system are apparent in Figure 1.6a. The first is seen to occur where the low amplitude intrawell oscillations destabilize if the excitation frequency exceeds $\omega \approx 1.1$. While the linear system exhibits smooth and continuous variation of displacement amplitudes as the excitation frequency is varied, the bistable system dynamics will undergo a sudden jump from low (point B) to high amplitude

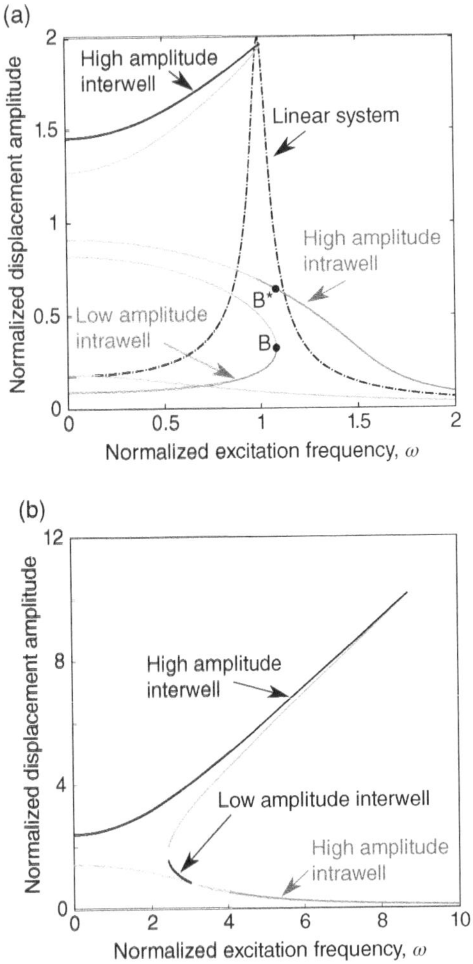

Figure 1.6 Depending on the excitation level, the single-periodic, steady-state response of a bistable system may exhibit non-unique dynamic forms. (a) A high amplitude interwell and both low and high amplitude intrawell responses may occur depending on the excitation frequency. Here, the parameters are $(p, \gamma, \beta) = (0.18, 0.09, 0.7)$. The corresponding linear system stationary response computed using the counterpart equation $\ddot{x} + \gamma\dot{x} + x = p\cos\omega\tau$. (b) A greater excitation level than that used in (a) may induce low or high interwell responses or an intrawell response. Here, the parameters are $(p, \gamma, \beta) = (8, 0.09, 1)$. (c) Responses of softening and hardening Duffing oscillators contrasted against the corresponding linear system. Here, the parameters are $(p, \gamma, \beta) = (0.18, 0.09, 0.7)$. The thin gray curves indicate unstable response forms.

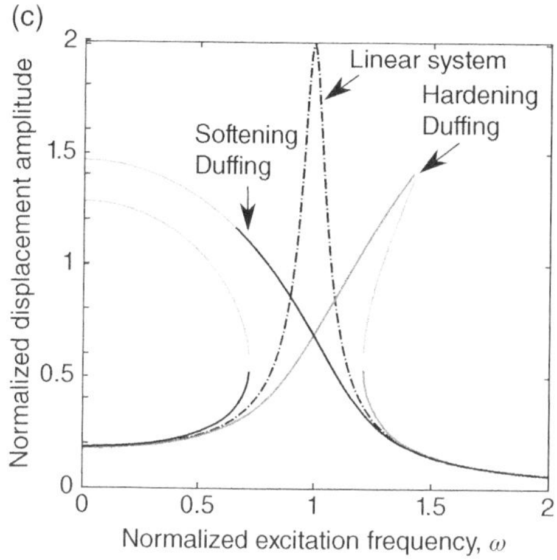

Figure 1.6 (Continued)

(point B*) intrawell oscillations when the frequency exceeds $\omega \approx 1.1$. Another bifurcation is evident in the high amplitude intrawell response if the excitation frequency decreases below $\omega \approx 0.95$; if the frequency is swept below this critical point, the system is predicted to undergo either low amplitude intrawell oscillations or snap-through. The final bifurcation featured in the plot occurs in this bandwidth when the bistable structure is initialized in the high amplitude interwell dynamic state and the excitation frequency exceeds $\omega \approx 1$. This latter bifurcation from an exceptionally large amplitude response (snap-through) to an intrawell dynamic is a distinct characteristic of bistable systems, and the significant response amplitude difference involved can be favorably exploited in several scientific and engineering contexts. For the previous two bifurcations where three, coexistent steady-state responses are reduced to two, the initial conditions at the time of bifurcation activation determine into which dynamic response the bistable system settles.

Figure 1.6b shows results predicted using different system and excitation parameters, and indicates that the low and high *interwell* dynamics may coexist near an excitation frequency of $\omega \approx 2.5$. It is found that the low amplitude interwell oscillations occur across a narrow bandwidth of excitation frequencies. The trend for the high amplitude interwell dynamics is a nearly linear increase in displacement magnitude as the excitation frequency progressively increases. This trend continues up to a critical frequency at which the single-periodic snap-through behaviors become destabilized: a similar bifurcation feature as that observed in Figure 1.6a. Finally, the results plotted in Figure 1.6b show that a high amplitude intrawell dynamic is possible for excitation frequencies greater than $\omega \approx 4$. The phase relationship of these intrawell oscillations with respect to the excitation is the characteristic factor to identify the dynamic regime.

The analytically predicted results in Figure 1.6 provide a clear example of the coexistence of numerous, steady-state dynamic forms of bistable structures as well as the possibility for sudden transitions in the dynamic regime due to small changes in system or excitation parameters. The multiplicity of the response forms, characterized by their unique displacement amplitudes and excitation phase lags, makes bistable structural dynamics particularly distinct from those of other linear and nonlinear systems. Indeed, snap-through is truly a distinguishing feature of bistable systems.

It is worthwhile to compare the steady-state, single-periodic dynamics of bistable structures to those of the traditional hardening and softening Duffing oscillators, which are regularly implemented as archetypal models for engineering systems in the real-world [17,18]. These oscillators are governed by the equations $\ddot{x}+\gamma\dot{x}+x\pm\beta x^3 = p\cos\omega\tau$ where the positive (negative) nonlinear restoring force term denotes hardening (softening) behavior. It is evident that this governing equation is similar to Eq. (1.1) for the bistable system, repeated here for convenience: $\ddot{x}+\gamma\dot{x}-x+\beta x^3 = p\cos\omega\tau$. On the other hand, the resulting static and dynamic behaviors are considerably distinct. Note that the only static solution to the equations $\ddot{x}+\gamma\dot{x}+x\pm\beta x^3 = p\cos\omega\tau$ is $x=0$, indicating that the Duffing systems have just one static equilibrium and are thus termed *monostable*. Using the methods presented in Chapter 2, the steady-state dynamics of the forced Duffing equations may be approximately solved and representative results are shown in Figure 1.6c using parameters of (p, γ, β) identical to those used in Figure 1.6a. Here again, the corresponding linear system dynamics are also shown, and light gray curves denote the unstable behaviors of the Duffing systems. Figure 1.6c illustrates that the hardening Duffing oscillator exhibits a "bending" or "leaning" of the resonance curve to higher frequencies with respect to linear system trends. In contrast, the softening Duffing oscillator shows response amplitude curves that lean to lower frequencies. Figure 1.6c indicates that the Duffing systems may exhibit *dynamic bistability*, in that more than one steady-state oscillation regime may occur for a given excitation frequency (similar results are obtained by varying the other parameters β, γ, and p). In comparing the different responses of bistable and these *monostable* Duffing oscillators between Figures 1.6a and 1.6c, respectively, it is evident that the intrawell oscillation regime of the bistable system is comparable to the softening Duffing response, while the Duffing oscillators exhibit no such dynamic behavior as snap-through since the Duffing systems do not possess *static bistability*. Truly, it is static bistability that enables snap-through and draws a dramatic distinction in the resulting, potential dynamic behaviors of bistable systems which the following sections continue to detail.

1.2.2 Sensitivity to Initial Conditions

It was shown in Figures 1.4 and 1.5 that system and excitation parameters as well as the initial conditions influence the resulting dynamic state of a bistable system. In fact, this is a characteristic of all nonlinear systems which may exhibit multiple coexistent steady-state dynamics. On the other hand, due to the two stable equilibria, there are important implications of the initial condition sensitivity demonstrated by bistable structures. A clear example of this feature is shown in Figure 1.7 where a difference in initial normalized displacement of 10^{-6} leads to completely different results as time elapses. It is seen that, based on the

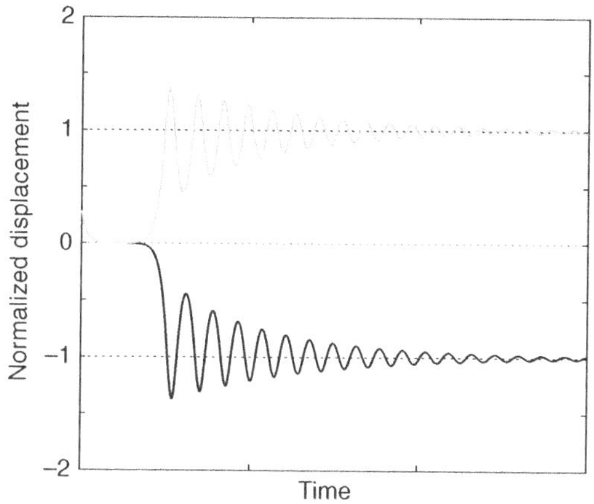

Figure 1.7 The transient responses of bistable systems may exhibit high sensitivity to initial conditions. In this example, the initial conditions are $(x_0, \dot{x}_0)$=(0.292721, −0.3) for the response settling to the normalized displacement stable equilibria position of −1 (black curve), while the small change of initial conditions to $(x_0, \dot{x}_0)$ = (0.292722, −0.3) causes the transient dynamics to tend to the stable equilibrium of normalized displacement +1 (gray curve). Here, the parameters are (γ, β) = (0.09, 1). The thick dotted lines indicate the statically-stable equilibria.

initial displacement, the unforced, transient dynamics of the oscillating mass decay and ultimately come to rest at one of the two stable equilibria. Thus, unlike the initial condition sensitivity of other nonlinear systems which becomes evident due to *steady-state* excitations, the *unforced* dynamics of bistable structures may be strongly sensitive to changes in initial conditions. The unique final value of normalized mass displacement is also relevant in the event that the bistable system is harmonically-excited. As was described in Section 1.2.1, the low or high amplitude intrawell dynamics may be realized as oscillations around either of the stable equilibria, which means that the average value of the steady-state dynamic is unique. All together, the initial condition sensitivities of bistable systems are manifest in distinct ways as compared to other nonlinear systems, because the offset or bias of the inertial mass due to the stable equilibria is a trademark characteristic of bistable structures.

1.2.3 Aperiodic or Chaotic Oscillations

When subjected to single-frequency excitations, certain classes of nonlinear systems may oscillate aperiodically. These motions are not to be confused with noise-induced behaviors and may indeed appear to be completely random. Yet, they can be deterministically predicted using the system governing equation and knowledge of the initial conditions. This phenomenon can be classified and quantified as *chaos*, and it occurs for a variety of physical, fluid, electrical, chemical, optical, and acoustic systems [19]. When it is possible, chaos is not always the dynamic form which is exhibited due to single-frequency excitation. Comparable to the potential for coexistent steady-state dynamic regimes exemplified in Figure 1.6, a specific set of system and excitation parameters can lead to aperiodic responses in a bistable system. However, the analytical tools to predict the steady-state dynamics in Figure 1.6 are incapable of identifying chaotic dynamics because such analytical techniques assume the dynamics of the bistable system are periodic. Consequently, a common strategy is to numerically integrate the governing equations when chaos is anticipated to approximate the parameter dependence of realizing chaotic responses for bistable structures. The manifestation and importance of chaos for the engineering implementation of bistable systems is described below.

The chaotic dynamics exhibited by bistable systems are realized as aperiodic interwell behaviors. As a first look into the features of chaos for a bistable system, Eq. (1.1) is numerically integrated using two slightly different sets of initial conditions and the computed displacement trajectories are plotted in Figure 1.8. There are several aspects to observe. First, the dynamics are clearly aperiodic although the input excitation is a single sinusoidal motion. Even after 500 cycles of excitation, Figure 1.8b shows that the displacement responses maintain an apparently random nature. The behaviors are not to be confused with a prolonged decay of transient dynamics related to initial conditions because the bistable system has appreciable damping. Continued numerical integration of the governing equation shows that the aperiodicity is maintained. In addition, the initial condition sensitivity of the chaotic responses as time progresses is exemplified in Figure 1.8a where two aperiodic displacement trajectories are seen to diverge after about 10 cycles of excitation. Following 500 periods of excitation, Figure 1.8b shows that there is little resemblance between the results. Although not shown in the example in Figure 1.8, a feature may be exhibited where chaos is realized in brief bursts. This *transient chaos* phenomenon occurs when intrawell dynamics of the bistable system appear to be the established steady-state response, but then a sudden and temporary bout of aperiodic interwell responses may be activated. The engineering implications of transient chaos are important because large amplitude chaotic dynamics may suddenly overtake an otherwise predictable intrawell dynamic of the bistable structure. In addition, once the transient chaos passes and the intrawell behaviors re-emerge, the system might in fact oscillate around the other stable equilibrium. Thus, an instance of transient chaos may lead to a change in the static state of the bistable system once the phenomenon has passed (should the excitations be removed). Such potential for state change is unique to the bistable structural systems which exhibit chaos. Although it is important to understand the potential for chaos, to date there have been few practical implementations of chaos for bistable systems in the technical areas of interest in this book. Engineers in vibration control, energy harvesting, and sensing typically avoid design and operational regimes that may induce chaotic dynamics. Interested readers are encouraged to refer to other works for further details regarding the chaotic dynamics of bistable structures [19–23].

1.2.4 Excitation Level Dependence

The forced, steady-state dynamics of nonlinear oscillators often exhibit an intricate dependence on the level of excitation. Whereas the response magnitudes of linear systems change with direct proportionality to the excitation level, the

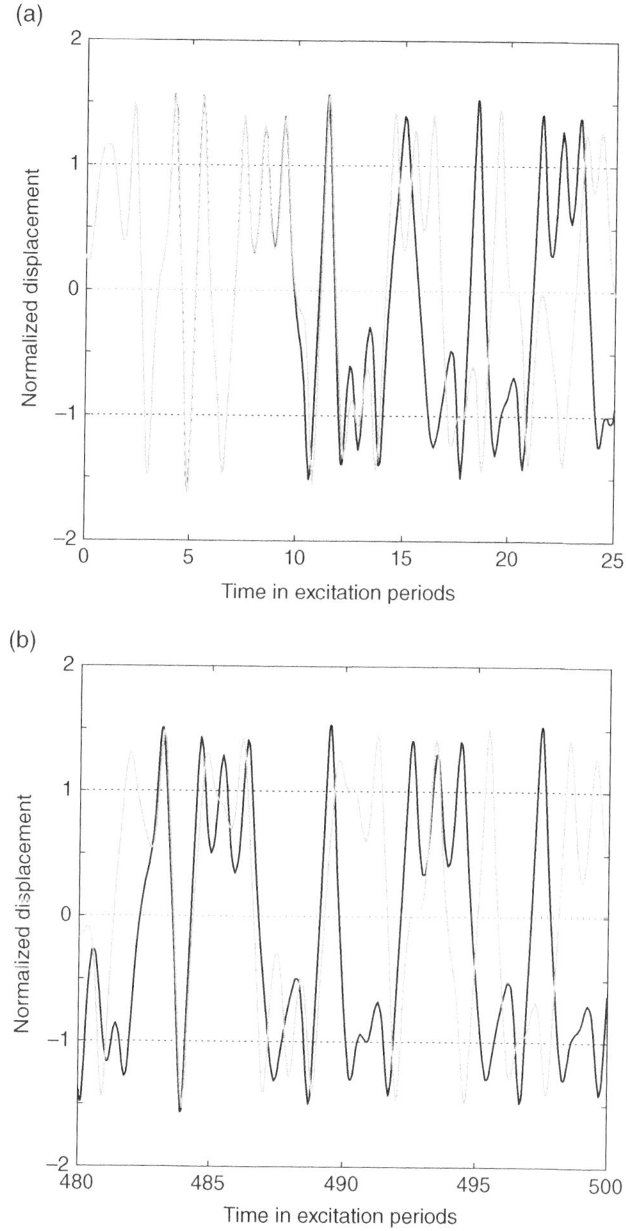

Figure 1.8 (a) An harmonically-excited bistable system may lead to aperiodic dynamics. This characteristic is also strongly sensitive to initial conditions. Thus, after only a small number of excitation periods, two nearly identical initial conditions show great divergence in the transient, aperiodic snap-through dynamics of the bistable systems. (b) After 500 excitation periods, the aperiodic responses bear little resemblance. Here, the parameters are $(p, \omega, \gamma, \beta)$=(0.37, 1.29, 0.1, 1), with initial conditions $(x_0, \dot{x}_0) = (0.292707, -0.3)$ leading to the responses plotted in the black curves and with initial conditions $(x_0, \dot{x}_0) = (0.292706, -0.3)$ leading to the responses plotted in the gray curves. The thick dotted lines indicate the statically-stable equilibria. The normalized sampling frequency in the 4th-order Runge–Kutta numerical integration is 32.

dynamic amplitudes of nonlinear systems as well as the existence of a given dynamic form will, in general, vary non-proportionally due to excitation level change. Because of the multitude of single-periodic behaviors that bistable systems may exhibit, the excitation level dependence is critical to examine for a bistable structure applied in an engineering practice. In other words, it is essential to determine the suitable design and excitation conditions that help to ensure the desired dynamic form is activated and the target performance objectives are achieved.

Figure 1.9 provides an example of the excitation level dependence of a bistable system. The analytically predicted results are computed for three excitation frequencies. Figure 1.9a shows the results for frequency $\omega = 0.8$ while Figure 1.9b,c show results for $\omega = 1$ and $\omega = 1.2$, respectively. The data for the bistable system are shown as solid curves whereas the counterpart linear system responses are identified by the dash-dot curves. Comparable to the excitation *frequency* dependence illustrated in Figure 1.6, the excitation *level* is an important parameter to determine the occurrence of the unique dynamic forms of the bistable structure; in Figure 1.9 these dynamic types are the low and high amplitude intrawell oscillations and the high amplitude interwell, snap-through vibrations. For the case of excitation frequency at the resonance-like condition, $\omega = 1$, shown in Figure 1.9b, the peak amplitudes which may be achieved for the high amplitude intra- and interwell dynamics of the bistable system are nearly identical to the response of the linear system excited at resonance. On the other hand, when excited off-resonance such as $\omega = 0.8$ or 1.2, Figure 1.9a,c, respectively, the dynamics of the bistable structure, and in particular the interwell snap-through vibrations, may achieve substantially greater amplitudes for the same excitation level. For the scientist or engineer who wishes to exploit the large, steady-state amplitudes of displacement enabled by a bistable system's snap-through characteristics, it is clear that an intelligent assessment of the excitation conditions under which the platform will operate – both the level and frequency of the forcing – is required to ensure the activation of the energetic behaviors.

Another informative assessment metric of the excitation level dependence for nonlinear systems is the frequency response function (frf), which is often computed as the ratio of a response magnitude (e.g., displacement, velocity, or acceleration) to the excitation level. As an example, Figure 1.10 plots the displacement frf magnitude of a bistable system (solid and dashed curves) as compared to that determined for a linear system (dash-dot curve). It is clear that the linear system response amplitudes are independent of the excitation level, as anticipated from a fundamental understanding of system dynamics. In contrast, due to the variation in excitation level, the bistable system displacements change in terms of frf amplitude as well as the frequency bandwidths across which each dynamic regime occurs. The arrows on the plot indicate the trend of change as the excitation level is increased. It is seen that increasing excitation level reduces the maximum frequency of existence for the low amplitude intrawell oscillations and increases the minimum frequency of existence for the high amplitude intrawell oscillations: a gap in the frequency bandwidth thus occurs across which only the snap-through dynamics are anticipated. At the same time, as the excitation level increases the snap-through dynamics are predicted to be sustained to higher frequencies. Additionally, the frf magnitude of the snap-through displacement response reduces. Thus, from Figure 1.10 it is clear that the maximum amplitude of the frf of bistable system snap-through dynamics is most nearly equal to that of the linear system only for low excitation levels when considering the resonance-like excitation frequency $\omega = 1$. On the other hand, the realization of the snap-through dynamic is more greatly ensured as the excitation increases because the frequency bandwidth of coexistence with intrawell dynamics is found to reduce. The excitation level dependence characteristics illustrated in Figure 1.9 and Figure 1.10 are important to understand in order to appropriately harness the distinct bistable structural dynamics for engineering applications.

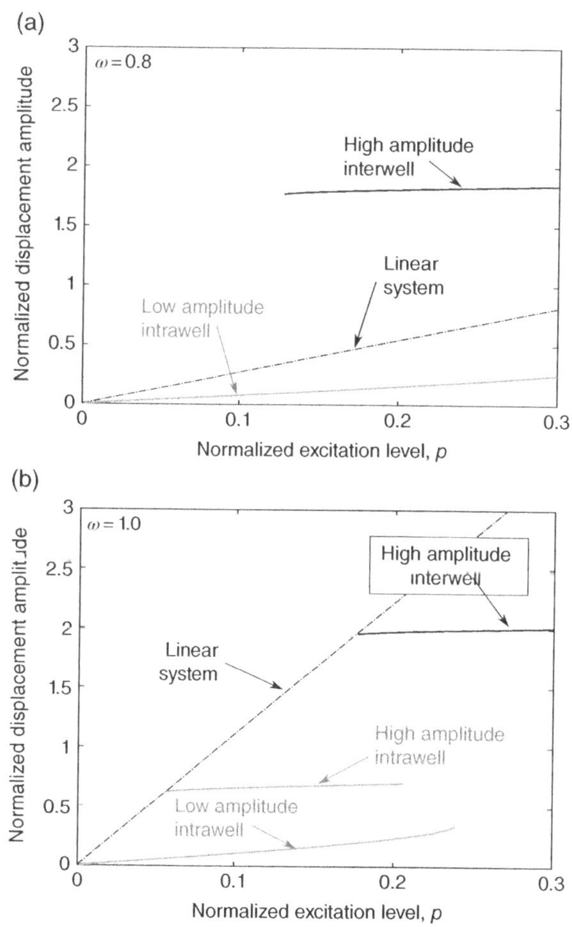

Figure 1.9 Depending on the excitation frequency, the single-periodic, steady-state response of a bistable system may exhibit non-unique dynamic forms. In the following examples, changes in normalized excitation frequency ω lead to variation in the amplitude and existence of the low and high amplitude intrawell dynamics and the high amplitude interwell vibrations of a bistable system, as functions of changing normalized excitation level p. In contrast, only the amplitude of the counterpart linear system responses is seen to vary. Results computed for (a) a normalized excitation frequency $\omega = 0.8$, (b) $\omega = 1.0$, and (c) $\omega = 1.2$. Here, the parameters are $(\gamma,\ \beta) = (0.09, 0.7)$. The corresponding linear system stationary response computed using the counterpart equation $\ddot{x} + \gamma\dot{x} + x = p\cos\omega\tau$.

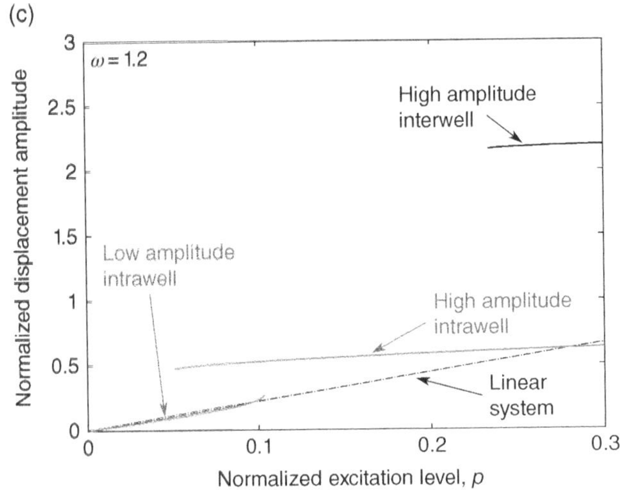

Figure 1.9 (Continued)

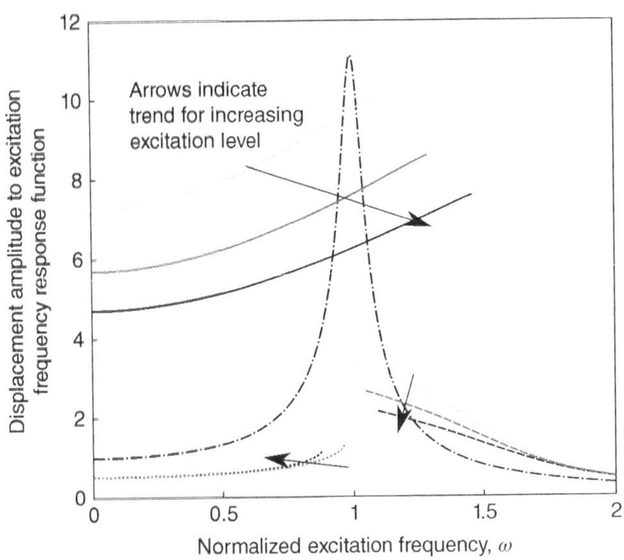

Figure 1.10 Examples of the frequency response function (frf) dependence of a bistable system as compared to a linear system. The frf is the displacement amplitude with respect to the excitation level. The linear system frf (dash-dot curve) is independent of excitation level. For bistable systems, the interwell response frfs (solid curves) reduce in amplitude but extend to higher frequencies as the excitation level increases. The high amplitude intrawell frfs (dashed curves) reduce in amplitude and frequency range as the excitation level increases, while the low amplitude intrawell dynamics (dotted curves) primarily reduce in frequency bandwidth without notable change in frf level. The light to dark curves for the bistable system are determined using p from 0.17, 0.22, to 0.27. Here, the parameters are $(\gamma, \beta) = (0.09, 1)$. The linear system frf is computed using damping factor $\gamma = 0.09$.

1.2.5 Stochastic Resonance

Although Section 1.2.4 indicated that greater excitation levels may help to consistently activate the snap-through dynamics of bistable systems by reducing the likelihood that intrawell dynamics could coexist, there is an additional avenue to realize the energetic interwell behaviors. The phenomenon is called *stochastic resonance* and it is a thoroughly investigated feature of bistable systems, particularly in physics-based research [24,25]. Stochastic resonance refers to the effect of exciting a bistable system with a combination of low-level, single-frequency harmonic input and low-level noise so as to activate snap-through dynamics. Individually, the two excitation inputs are insufficient to induce the interwell behaviors but their combination has dramatic influence.

Figure 1.11 shows an example of this dynamical feature. It is seen that the individual periodic or white Gaussian noise excitations upon the bistable system lead only to low amplitude oscillations around one of the two stable equilibria, whether they are periodic or stochastic in nature. However, in combination, the excitations activate snap-through dynamics. By considering the normalized time axis and the times at which the snap-through transitions occur, it is also apparent that the resulting large amplitude motions occur primarily at the same frequency as the periodic input. As a result, this phenomenon provides a novel avenue to magnify a low-level input by the "injection" of either noise or periodic excitations to activate the high amplitude interwell dynamics in the bistable system with a minimum of extra cost in terms of the input energy. Stochastic resonance is not unique to nonlinear systems exhibiting bistability [24], but the degree of response amplification that bistable structures may realize due to

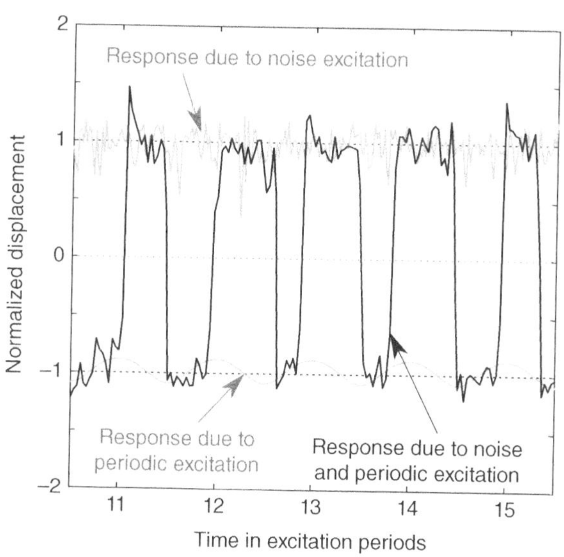

Figure 1.11 The stochastic resonance phenomenon occurs when combined periodic and stochastic excitations, which are individually insufficient to activate snap-through dynamics of a bistable system, lead to large amplitude interwell dynamics which closely follow the excitation period of the periodic excitations. Here, the parameters are $(p, \omega, \gamma, \beta) = (0.2, 0.05, 0.5, 1)$ for the cases involving the periodic excitations. The normalized, additive white noise variance is 0.04. All cases use initial conditions $(x_0, \dot{x}_0) = (0, 0)$.

stochastic resonance is notably greater because of the switching back and forth between the two stable equilibria. This distinction is one reason for the increased attention on stochastic resonance in the context of bistable structures and systems throughout the scientific and engineering fields.

1.2.6 Harmonic Energy Diffusion

The results presented in Figure 1.4b and Figure 1.5b indicate that the four, steady-state types of bistable structural dynamics are distinct in terms of their respective amplitudes. Looking closer at the spectral plots, it is also apparent that the input energy at frequency ω becomes manifest via dynamics of the bistable structure at other frequencies. In other words, the energy is *diffused* from the single-frequency input to other harmonics. Specifically, for the intrawell dynamics Figure 1.4b shows that the input energy becomes concentrated in the system dynamics at integer multiples of the excitation frequency. In this case, the input excitations at a frequency ω are exhibited by bistable system dynamics at frequencies $n\omega$ where $n = 1, 2, 3, \ldots$ It is also evident in Figure 1.4b that the high amplitude intrawell dynamics exhibit a greater degree of such energy diffusion than the low amplitude responses.

Similar results are seen in Figure 1.5b where the coexistent low and high amplitude interwell dynamics diffuse the input energies at frequency ω to responses at frequencies $m\omega$ where $m = 1, 3, 5, \ldots$ In the cases presented in Figures 1.4b and 1.5b, the harmonic energy diffusion is not dramatic because the corresponding time series plots, Figures 1.4a and 1.5a, show that the bulk of the dynamic behaviors is indeed at the same frequency of excitation, albeit with phase shifts. (One may verify this conclusion by comparing the primary periods of the displacement responses to the horizontal axes which are shown as time in normalized excitation periods). Nevertheless, the spectral results reveal that the bistable system redistributes a non-trivial amount of the input energy into dynamics at other frequencies.

Yet, in some cases, the harmonic energy diffusion may become extreme. Figure 1.12 provides clear examples of this possibility such that the single-frequency input at ω is mostly concentrated at other frequencies. Figure 1.12a shows that for some system and excitation parameters, the inputs at frequency ω become primarily concentrated at frequency $\omega/3$ (black curve) or 2ω (gray curve). The time series results plainly reveal that the dynamics which achieve such significant harmonic energy diffusion are interwell behaviors, and specifically large amplitude snap-through effects. In Figure 1.12b, the spectral results computed from the corresponding displacement time series verify the dramatic redistribution of input energy via the particular manifestation of bistable system dynamics. It is seen that in the one case, the energy is predominantly redistributed to frequencies $m\omega/3$ where $m = 1, 3, 5, \ldots$ (black curve), while in the other case the spectral peaks occur at $2m\omega$ (gray curve). No less important than when the energy remains primarily concentrated at the excitation frequency, the extreme harmonic energy diffusion due to bistable system snap-through dynamics has critical engineering implications for their appropriate utilization.

In general, the redistribution of the dynamics from a single-periodic response to a variety of frequencies is not unique to bistable systems and can be exhibited

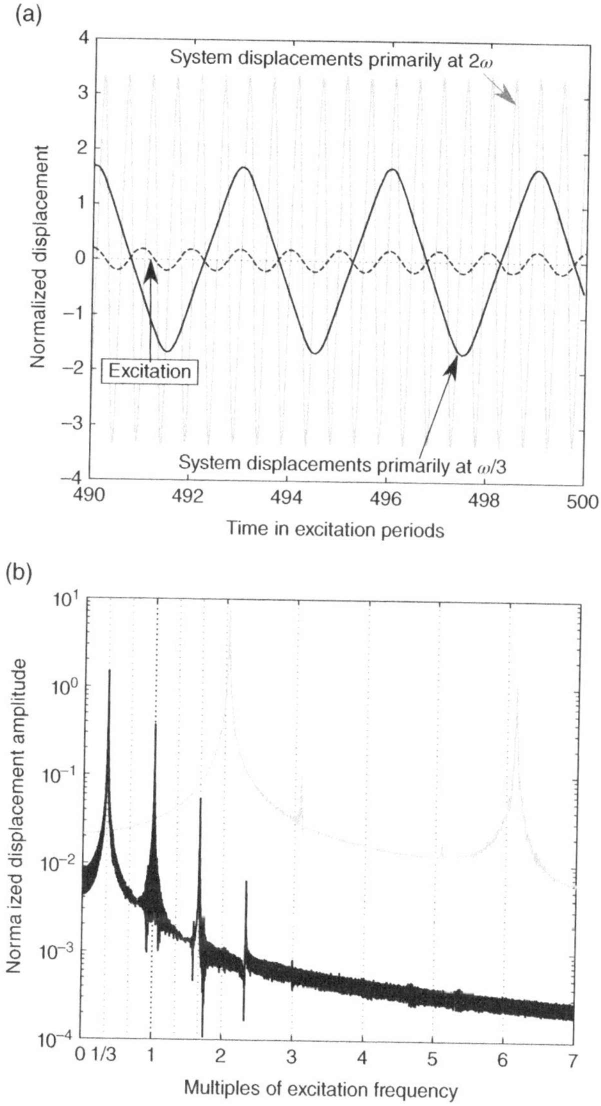

Figure 1.12 (a) Although bistable systems may be excited at a frequency ω, the steady-state dynamics may occur primarily at integer multiples or fractions of the excitation frequency. Two cases are shown for which the system primarily vibrates at a frequency of 2ω (gray curve) or $\omega/3$ (black curve. (b) The displacement magnitude spectral responses of the bistable system dynamics, as computed from the time series data plotted in (a), show clearly the unique spectral concentrations of the input excitation energies. In (a,b), for the dynamics tending to the frequency of 2ω, the parameters are $(p, \omega, \gamma, \beta) = (0.2, 1.3, 0.0001, 1)$, with initial conditions $(x_0, \dot{x}_0) = (-3.2987, 1.5132)$; while for the dynamics tending to the frequency $\omega/3$, the parameters are $(p, \omega, \gamma, \beta) = (0.2, 3, 0.01, 1)$ with initial conditions $(x_0, \dot{x}_0) = (-0.7507, 1.4841)$.

by other nonlinear systems. On the other hand, the multitude of the possible, coexistent steady-state dynamic forms of bistable structures lends itself to an intricate manifestation of harmonic energy diffusion that is not evident in other nonlinear systems. This provides bistable systems with a much greater variety of multiharmonic dynamic characteristics that may be intelligently harnessed.

1.3 The Exploitation of Bistable Structural Dynamics

The review above indicates that the dynamics of bistable systems include a rich assortment of behaviors. All together, the numerous distinct features suggest that an attractive versatility of response may be exploited by intelligently utilizing or switching among the different types of dynamics that are realized by a single degree of freedom (DOF), bistable architecture. As compared to the considerably more uniform response characteristics of linear systems and the fewer opportunities to exploit with other nonlinear platforms, the dynamics of bistable structures and the various means by which to realize bistability motivate the scientific and engineering utilization of bistable system architectures to advance a variety of technical fields.

The rationale for exploiting bistable structural dynamics depends upon the application. In some cases, the reasons that encourage use of certain dynamic forms are the same that could discourage their use for other purposes. Therefore, the perspective of the researcher and the performance metrics of interest are the guiding factors for appropriately harnessing bistable system dynamics. Across the fields of vibration control, vibration energy harvesting, and sensing, there are certain shared or underlying aims that the dynamics of bistable systems have been found to meet effectively in ways and to degrees beyond what is possible when utilizing alternative system platforms.

The shared aims may broadly be described as achieving exceptionally *energetic responses* and enabling *adaptability*. For bistable structures, the basis for realizing notably energetic responses is the opportunity to activate snap-through dynamics, which do not have a counterpart in single DOF linear or other monostable nonlinear systems. The adaptability of bistable system dynamics is due to the potential for multiple, coexistent steady-state responses – whether they are single-periodic, multiharmonic, or aperiodic – such that an intelligent exploitation of system design and excitations can enable one to selectively trigger the dynamics that are preferred for a certain operating condition or application.

In the technical areas of interest in this book, the energetic responses and adaptability of bistable structural dynamics draw the several engineering applications around the general objectives of enhanced performance versatility and functionality. In the following sections, the individual aims of vibration control, energy harvesting, and sensing applications are briefly detailed. Within each context, a first look is taken at how the objectives of the areas have recently been achieved and advanced through the exploitation of bistable system dynamics, particularly as they relate to harnessing the energetic and adaptable responses.

1.3.1 Vibration Control

Dissipating, isolating, or absorbing the vibrations of structures is an historic goal: elastic carriage suspensions were employed during, if not also before, the Iron Age [26], while the first example of a vibration absorber for reactive attenuation purposes is often dated to Hermann Frahm's patent in 1910 for an anti-roll tank for ships [27]. Apart from the assurance of comfort via limiting the jarring of carriage vibrations or alleviating the possibility of sea-sickness, the control of vibrations is more recently a matter of structural necessity.

The use of engineering systems in ways or in environments that cause significant vibration levels, for example space vehicle launches or even the ordinary operation of rotorcraft, requires that vibration control systems operate effectively and with high performance to ensure the success or integrity of the structure in its mission. The numerous costs associated with energy to fuel a wide variety of vehicular systems, including aerospace, marine, and automotive vehicles, have stimulated the use of lightweight structures and materials to minimize mass to allowable safety standards. This means that the inherent energy dissipation characteristics of the system are of great importance. Without sufficient, and oftentimes high, damping capacity in the lightweight systems, the day-to-day operation of many modern engineering structures and materials carries the risk of incurring a catastrophic failure due to working too closely to the limits of system integrity. On the other hand, although high damping is critical for some applications, low damping may be just as vital. For example, the need to obtain accurate wave- or vibration-based signal readouts from a structure to monitor its operating state, also termed structural health monitoring, benefits from a lightly damped structure at the time of data acquisition for the sensors to acquire clear measurements. As a result, modern implementation of engineering systems has led to a set of operational requirements for vibration control performance, some conflicting with each other: for instance, a lightweight structure or material solution that provides high damping during rotorcraft operation but low damping during a rotor blade condition monitoring examination.

From age to age, the achievement of these overall vibration control aims has been limited by the traditional employment of passive materials or devices having tightly prescribed capabilities, often as a result of operating the system within a linear dynamic regime. The introduction of nonlinear components for energy dissipation, such as fluid [28] or friction dampers [29], led to improved performance potential, particularly for high vibration levels. Nevertheless, the more extensive exploration and exploitation of nonlinear vibration control concepts, including the harnessing of bistable elements, is a comparatively recent initiative [30,31].

Indeed, there are a variety of advantages to be gained by employing a bistable device in its own right within the architecture of a vibration damping, isolation, or control system. The energetic snap-through dynamics lead to high velocity motions for the inertial mass, due to the large displacement strokes from one stable equilibrium to the other and associated low dynamic stiffness. In addition, snap-through is activated across a broad bandwidth of frequencies when compared to a linear system and its resonant dynamics, as exemplified by Figure 1.6. Collectively, these characteristics of large kinetic energy, low dynamic stiffness, and potential for adaptively triggering large amplitude, periodic motions provide opportunities to advance aims in dissipating, isolating, and absorbing vibrations in myriad contexts.

One idea recently explored to achieve damping characteristics uncommon in traditional bulk materials is the implementation of bistable elements into structural or material systems to take advantage of the snap-through dynamics.

In such implementations, the bistability of the system is deliberately retained. In these initiatives, by capitalizing on the activation of snap-through dynamics of an inertial mass, a large dissipation of energy is realized: extreme damping. This idea has been theoretically and experimentally explored using numerous incorporations of bistable elements into the systems, such as negative stiffness inclusions within a host material matrix [32] and as an integrated device composed of bistable structures in parallel with dampers [33,34], to sample only a couple of the various system realizations. The broadband spectral characteristics of snap-through motions provide a particular motivation for their exploitation to enable a high damping capacity. As described in Section 1.2.1, snap-through is a non-resonant dynamic which means there is no reliance on resonance-like phenomena to achieve the large amplitude interwell motions: only a sufficient excitation level is required to induce the energetic state change from one stable equilibrium to the other. Thus, incorporating bistable elements into a structural or material system provides the means for cultivating high damping properties across a wide range of excitation frequencies.

As described above, modern engineering systems that require high damping may also benefit from an intrinsic ability to tailor energy dissipation or vibration suppression characteristics. Although active control systems may be used to effect real-time variations in the dynamic characteristics of engineered structures, there are significant advantages to developing and employing passive-adaptive systems that are inherently multifunctional via mechanical properties tuning. In this way, the system weight remains low and complexity is alleviated because the properties of the system may be adapted passively without need for the weight, maintenance, and energy expense of active control hardware. By exploiting the excitation level and frequency dependence of the snap-through dynamics of bistable structures, one may realize adaptable harmonic energy dissipation features [34]. A simple implementation of the concept is described following the basic trends shown in Figure 1.9. Thus, for a structure with bistable damper inclusions a small dissipation of energy is realized when the excitation level is low, but as the level increases the activation of snap-through dynamics can be harnessed for more aggressive levels of damping. Careful investigations of damping and excitation level dependence on the snap-through characteristics of bistable systems [35,36] have provided informative guides to achieve adaptable damping features.

Bistable systems with buckled structural members are also leveraged for vibration isolation purposes. The dynamic stiffness near the point of structural buckling is very low to significantly reduce transmitted force and displacement, while the static stiffness of buckled members can be quite large in order to support high load [37,38]. Indeed, the use of buckled components in their own right is straightforward as well as effective when subjected to the range of operational conditions for which they were developed. On the other hand, many applications of vibration isolation involve significant time-variation in the characteristics of the excitation source, whether in terms of primary frequencies or amplitudes. Consequently, researchers are considering strategies by which to enhance the robustness of performance for vibration isolation systems containing bistable components, such as by using combinations of linear and bistable elastic components [39,40]. In this way, the energetic snap-through dynamics are leveraged in the best ways possible. As revealed in Section 1.2, the dynamics of bistable structures are rich and, as described elsewhere throughout this book, the coupling of bistable components to other linear elastic members may induce even richer opportunity for dynamic response regimes that may be activated according to different operating conditions. Thus, the exploitation of bistable structures for vibration isolation purposes provides a unique high-risk, high-payoff context that researchers are investigating in detail and with enthusiasm.

Another extension upon the idea of bistable structural integrations for high performance vibration control is the idea to employ snap-through dynamics as a mechanism for vibration absorption. This initiative stems from the established understanding of the operating principles of classical vibration absorbers: the attachments suppress the motions of the excited host system to which they are attached via forces the absorbers generate, which are equal in amplitude and opposite in phase with respect to those produced by the harmonic excitations. As the logic then follows, a bistable vibration absorber employing large amplitude and broadband snap-through dynamics may have particular advantages for reactive vibration control purposes if the dynamics are suitably harnessed in such a way as to supply the needed out-of-phase forces upon the host system. Several researchers have begun to explore this new outlook [41–43] and have so far considered bistable device attachments and primary structures consisting of lumped-parameter constituents, in other words lumped masses, coil springs, and so on. The snap-through dynamics of a bistable system thus open a second avenue of research for vibration control studies, wherein the energetic responses can be intelligently tailored to provide precisely the required reactive force upon a host structure for vibration absorption aims.

The strict and sometimes conflicting standards of high vibration control performance in many modern engineering systems indicate the achievement of the numerous aims using passively-adaptable structural members or integrations is of significant value. The selective activation of energetic snap-through behaviors has been found to be a novel approach to meet the various objectives. The encouraging research results to date serve as strong motivation for the ongoing explorations of harnessing bistable system dynamics in vibration control applications.

1.3.2 Vibration Energy Harvesting

The conversion of ambient vibrations, caused by environmental sources or structural motions, into a usable electrical power resource is an emerging technical field. The rapidly growing interest in and attention to vibration energy harvesting opportunities have led to systems and methods that, in the relatively short span of time from the initial explorations, are already at the early stages of commercialization. There are numerous factors that have encouraged the investigations and stimulate the continued efforts.

One motivating factor is to harvest the available vibrational energies to enable low-powered, self-sustainable microelectronics employed in remote or embedded applications where other energy resources are unavailable or inconveniently accessed via line transmission [44–46]. The sources of such vibration considered are numerous but a sample of such environmental oscillations include: bridge vibrations induced by vehicles, wind loading, pedestrian traffic, and so on; human motions such as walking or jogging; transport vehicle vibrations like aircraft wing oscillations or railway car bearing accelerations. The applications for the electrical energy harnessed from such vibrations include the proliferating implementation of distributed, structural health monitoring nodes and the recent interest in implanted devices like automotive tire pressure sensors or even biomedical devices. The architectures utilizing the harvested energy typically have low power requirements (around mW) due to the efficiency of the electronics, yet would have limited operational life-spans (or fixed intervals of manual service) if powered by conventional batteries that are exhausted over time. By employing electromechanical systems which capture and convert sufficient vibrational energy, one may then use the electrical output to charge a storage element or directly power an electronic system, following a power-conditioning phase.

Another motivating factor in vibration-based energy harvesting is to generate large power (W or greater), and potentially to supplement the conventional, continental electric grid [47,48]. A prime example of vibrations considered by this perspective is the heaving motions of ocean waves. For instance, it is estimated that the overall wave energy resource of western European coastal fronts is itself greater than the world's present-day energy demand (~2 TW!) [47], assuming it could be adequately concentrated and harvested. Other vibration sources considered in this perspective of energy harvesting are large-scale structural system motions such as the wind- or seismic-induced sway of skyscrapers and automotive suspension motions [49]. Due to the significant attention to and demand for alternative, sustainable energy resources to reduce civilization's reliance on exhaustible supplies, the conversion of vibrational energy for large-scale use is an established area of active research and commercial interest.

In spite of the two different motivating factors, a common aim is apparent: to capture, or additionally amplify, the ambient vibrational energies within an electromechanical energy harvesting system. Once consolidated in this way, the transduced electrical current is conditioned to charge storage elements (for example, rechargeable batteries or capacitors) or directly power a system. The vibrations considered in these studies are commonly periodic or stochastic in waveform. In addition, the electromechanical transduction approaches are typically those which generate an electrical signal due to the relative velocity of the harvester's moving mass, for example, piezoelectric materials on a vibrating cantilever beam or a magnetic mass-spring oscillator moving through a coil. For periodic vibrations, it is logical to use an energy harvesting system that resonates or similarly amplifies the input motions by matching the sensitivity of the harvester dynamic responses to the input vibrations. As a result, early investigations explored linear energy harvesting architectures that resonate at the predominant frequency of the periodic input vibrations. Thus, the sensitivity of the electromechanical systems is appropriately tuned for the input spectrum, a strategy common to both research perspectives in the energy harvesting field. By capturing and ideally amplifying the input motions via a resonant-type energy harvesting platform, a beneficial electromechanical conversion of the vibrations is achieved.

A serious disadvantage of linear resonant harvester systems is the narrow band of frequencies across which the high amplitude dynamics occur. Because the spectral concentration of periodic vibrations changes regularly, for example, due to change in temperature of the host structural system or different human walking gaits, the reliance on a limited frequency bandwidth of high performance makes the practical implementation of linear resonance-based energy harvesting platforms a less than promising approach. Therefore, harnessing the broadband snap-through behaviors of a bistable energy harvester has clear advantages in terms of performance and robustness to changing excitation conditions. As was shown in Figure 1.6a, the large amplitude interwell dynamics of bistable systems are sensitive to a wide range of excitation frequencies, and notably sensitive in the extremely low frequency bandwidth. The low frequency range is particularly challenging for a linear resonant system to exploit due to stroke motion constraints that can inhibit the practicality of employing a resonant harvester.

The research and development interest in bistable energy harvester architectures and their appropriate dynamical implementation has consequently flourished [5,6]. By strategically exploiting the activation of snap-through dynamics, large energy conversion is realized. The motivation to harness the dynamics of bistable energy harvesting systems is grounded upon the pursuit of greater performance via larger amplitude snap-through dynamics as well as the practical realization that robust solutions are necessary to see the energy harvesting vision to success.

Typically, the objective is to activate continuous or repeated snap-through dynamics in the bistable harvester platform. However, for very low frequencies, the design of a bistable structure sensitive to such frequency spectra is a challenging endeavor. As a result, an alternative idea is considered, whereby occasional snap-through behaviors are exploited. The concept is often termed frequency up-conversion [5,50], in other words utilizing the very low frequency inputs to activate a state change in the bistable harvester that leads to a "ring down" intrawell oscillation. The ring down dynamics are partially dissipated via the electromechanical conversion, which provides a novel power generation strategy for extremely low frequency input vibrations. Thus, whether continuously or intermittently activated, the utilization of snap-through dynamics is a widely explored strategy to harness periodic input vibrations for energy harvesting aims.

On the other hand, not all ambient vibration resources are periodic. In fact, structural vibrations such as bridge/walkway oscillations due to vehicle or human motions are very often stochastic [51,52], and ocean wave heaves, which typically have a predominant spectral peak when evaluated over a given time period, have a large spread of the vibrational energy around the frequency of greatest energy concentration [47]. As a result, studies take care to investigate the stochastic excitation performance of energy harvesters [53]. In this context, bistable systems have unique advantages over alternative energy harvesting platforms when excited by noise. Whether by the stochastic resonance phenomenon using combined noise and periodic excitations or due only to stochastic inputs, an intelligent and effective design of the bistable harvester may activate the favorable snap-through dynamics when the stochastic inputs are otherwise insufficient to greatly excite alternative architectures. The performance advantages of bistable vibration energy harvesters in stochastic excitation environments have been demonstrated for various stochastic spectra: including white Gaussian [54,55], exponentially-correlated [4], and band-limited [56] noises.

On the whole, the capacity to exploit a sensitive and robust array of energetic snap-through dynamics makes bistable energy harvester platforms effective candidates in the ever-growing exploration of novel means to transform the ambient vibrations around us into sustainable electrical energy resources.

1.3.3 Sensing and Detection

There are historic needs to monitor the changing dynamic responses of structures due to variation in the applied loads or the internal composition of the monitored system itself. The needs are numerous and the engineering and scientific applications of sensing structural changes are found on an enormous extent of length scales. Some examples of current research and industrial interest include the detection of miniscule force or mass change on micro- or nanoscale resonators [57], the health condition monitoring of commercial aircraft before, during, and after flights [58], and tracking the long-term wear-and-tear on massive civil infrastructure systems such as bridges [59]. By employing sensors to detect parameter shifts in the observed dynamic responses, the safety, performance, and longevity of numerous structural systems are ensured, which has broad and positive impact on society.

Sensitivity and accuracy are two critical objectives in detection methodologies. How these aims are achieved depends upon the length scale of interest as well as the detection approach which is employed. Vibration-based sensing strategies are commonly used due to an ease of implementation and ability to monitor global features of a structure. Across scales, a conventional implementation of vibration-based sensing is to track shifts in important structural frequency response data. For example, in microscale sensing of gaseous or chemical mass (like hazardous materials or fumes), the primary structural responses to observe are the changing resonant characteristics of the sensor due to the adsorbing mass on the sensor surface. Thus, the resonant frequency or phase shift at resonance is tracked over time and translated through a sensor model to the adsorbed quantity of mass [60]. The same overall detection strategy is used in meso-/macro-scale vibration-based sensing with frequency shifts, although on such scales the shifting dynamic characteristics might be attributed to damage features [58].

A serious difficulty for the otherwise favorable frequency shift-based detection strategies is an inherent limitation in the confident quantification of small shifts in resonance frequency or phase due to structural variation. The difficulty becomes exaggerated when the sensor readout (or structural response) is corrupted by the influences of environmental or sensor noise and system damping. Such concerns challenge the accuracy of detection for all scales of application. Consequently, to overcome the obstacles to accuracy and robustness, an alternative, emerging vibration-based detection strategy has been explored based upon the activation of bifurcations in the response of strongly nonlinear sensors [61–63].

For clarity, representative bifurcations of a bistable system are illustrated in the analytical response prediction plots in Section 1.2. For example, in Figure 1.6a it is seen that a steady-state, low amplitude intrawell dynamic response of a bistable structure will be destabilized if the excitation frequency exceeds a critical value of $\omega \approx 1.08$; at such a time, the bistable system behaviors would suddenly transition to the high amplitude intrawell dynamic. In the context of sensing and detection, the onset of the bifurcation is translated through a sensor model back to the near-infinitesimal structural change which induced the large response amplitude jump. Thus, the limits of parameter shift detection in an ideal bifurcation-based method approach the observation of quantum events [64,65]. This ability sharply contrasts the far more constrained sensitivity bounds imposed on traditional frequency shift-based approaches due to the smallest increment of change that can be resolved in the monitored analog signal based on hardware performance.

Bifurcation-based sensing methods have recently been successfully applied to the microscale, where the nonlinearity required to undergo bifurcations is easily effected, whether by the application of particular excitations or due to the typical, forced oscillation amplitude of the sensors with respect to resting geometry [66,18]. From these studies, evidence of considerable robustness to noise and damping in the proposed sensing strategy has been demonstrated [67,68], which contrasts the limitations encountered by employing the traditional approach of tracking linear resonance features for microscale mass sensing. The underlying requirements in bifurcation-based detection strategies are the exploitation of strong nonlinearity in the sensor/structure and an intelligent methodology to activate a bifurcation in such a way as to denote change in the parameter one wishes to monitor. Bistable sensor platforms thoroughly meet these needs and indeed are introducing opportunities to extend bifurcation-based sensing to new applications beyond the microscale.

Bistable structures have distinct advantages in their exploitation as strongly nonlinear sensor platforms, such as a greater number of bifurcations to exploit due to the unique dynamical forms represented by the intra- and interwell oscillations; a dramatic difference in dynamic response levels may be induced which provides more apparent indicators of change following the bifurcation event; a wide variety of all-mechanical design compositions that could be employed, like pre-curved beams [69], to empower sensing operation free of constant active load requirements to maintain bistability; and bistable sensors may undergo static stable state transitions, for example to enable means for passive detection of thresholds like impact or acceleration events [70]. Thus, bistable sensor architectures are strong candidates in the development and implementation of bifurcation-based detection strategies by the combined exploitation of the energetic behaviors of snap-through and the selective (in other words, adaptable) activation of bifurcations from one dynamic or static state/regime to another.

The ease with which strong nonlinearity is realized makes the microscale a favorable domain to develop and explore bifurcation-based sensing systems and methods to greatly advance detection capabilities beyond traditional frequency shift-based strategies. Yet, there are numerous important sensing and monitoring interests on the macroscale that also suffer from the drawbacks of traditional frequency shift-based sensing approaches. Unfortunately, most macroscale structures do not exhibit strong nonlinearity under ordinary operations, indicating a critical constituent of bifurcation-based sensing is often absent in the inherent structural systems. Recently, to surmount this concern, researchers have explored the integration of bistable electric circuitry with a piezoelectric transducer and the host, monitored structure which is facilitated by a transducer; one such integrated system schematic is shown in Figure 1.3d [11]. Using this architecture, a strong nonlinearity – bistability – is introduced into the structural-electrical system, and is specifically contained only within the circuit due to the one-way current flow provided by the circuit design. The monitoring system is thus capable of harnessing many of the benefits of a bistable sensor for bifurcation-based sensing, and the robustness of detection accuracy to adverse noise and damping is found to be achieved for macroscale structural monitoring [71]. By the strategic exploitation of a bistable circuitry and structural integration, the advantages of bifurcation-based sensing are able to be leveraged on the macroscale. Thus, the energetic responses and dynamic adaptability of bistable sensors helps to justify their recent popular exploration in many sensing applications and research areas [11,72,73].

The deliberate activation of the distinct dynamic responses of bistable sensors finds another useful implementation in the development of microscale electromechanical signal filters or detectors [74]. The overall concept of electromechanical signal filtering is to exploit frequency-dependent response characteristics to induce strongly variable response amplitudes or spectral concentrations in the oscillator dynamics so as to detect or pass certain input frequencies or convert the input frequency to another value [66]. On small scales, all-electrical signal filters are prone to high damping and it has been shown that electromechanical counterparts may empower substantially sharper passbands and cut-off rates for accurate and precise filtering performance. When strongly nonlinear microscale oscillators are employed, their implementation may be as straightforward as the activation of a bifurcation in the sensor response from a low (or no) amplitude oscillation to one of significantly greater amplitude [75], which can be engineered to effect a signal filtering influence due to the on/off-type of dynamic variation. More intricate filtering concepts have been investigated to exploit multiharmonic behaviors in the microscale devices, such as the inducement of additional harmonics to leverage a greater versatility in filtering capability using an individual device architecture [76].

Due to the wide variety of behaviors that may be exploited and the dramatic difference in response amplitude between dynamic regimes for sharp filtering features, the dynamics of bistable oscillators indicate such microscale systems are strong candidates in the development of effective and adaptable filtering platforms [7]. For instance, Figure 1.12 exemplifies the versatility of bistable systems via multiharmonic behaviors produced in consequence to a single frequency of excitation: a frequency ω is input into the system but induces dynamics primarily at $\omega/3$ or 2ω. As a result, a microscale bistable oscillator can be harnessed to monitor particular harmonics of a given input frequency, minimizing the number of filters needed to be employed in series to achieve the same task. Thus, whether used for direct frequency bandpass filtering via the activation of low-to-high response amplitude bifurcations or for the versatile harmonic response modulation features, bistable systems present new opportunities for electromechanical signal filtering and detection. The recent studies exploring such potential suggest there is much left to be uncovered and leveraged [7,70].

1.4 Outline of This Book

In summary, bistable systems can be realized using numerous methods, and the architecture that is selected for an application depends strongly on how it will be leveraged, whether the aims are to control vibration or convert vibrational energies to electrical power or to detect infinitesimal mass particles upon a micromechanical sensor. Although a variety of characteristics make the dynamics of bistable structures unique, the highly energetic behaviors and the ability to intelligently adapt dynamics for specific operational purposes are common rationales for the exploitation of bistable systems. Following this introductory assessment of the features of bistable structural dynamics that make them attractive for a variety of scientific and engineering purposes, the remainder of this book details the recent research achievements that have been made via strategically harnessing the dynamics of bistable systems, specifically within the fields of (i) vibration control; (ii) vibration energy harvesting; and, (iii) sensing and detection.

Prior to elaborating on the state of the art in these areas, Chapter 2 provides an overview of the theoretical foundations which are central to the modeling and analysis of bistable structural responses. The theoretical developments are provided in such a way as to allow readers with sufficient knowledge of vibrations and structural dynamics an accessible introduction to the common approaches which are used to mathematically investigate the strongly nonlinear dynamics of bistable systems. Readers with more advanced theoretical backgrounds may wish to continue directly to the subsequent chapters which contain more detailed mathematical developments specific to the bistable system architectures considered therein.

Chapter 3 describes how researchers are discovering new and high performance means for vibration damping, isolation, and absorption by exploiting the distinct, energetic dynamics of bistable structures. Concepts such as bistable isolators, snap-through absorbers, and adaptive damping elements are detailed from their modeling formulations and treatments through their experimental investigations, showing that effective performance relies on a knowledgeable design and utilization of bistability. Recent explorations of multi-body vibration control architectures are presented, indicating that harnessing the energetic and adaptable behaviors of bistable devices within system architectures can lead to enhanced performance when they are appropriately leveraged.

Chapter 4 describes the popular and successful implementation of bistable dynamics in vibration energy harvesting systems to enable exceptional and robust performance when excited by a variety of ambient vibration sources. The chapter details how researchers are using rigorous mathematical frameworks supplemented with careful laboratory and field examinations to derive generalized insight into the effective exploitation of snap-through dynamics for high and consistent energy harvesting performance. Analytical and experimental developments are presented which clearly show that single and multiple DOF architectures may intelligently harness the energetic dynamics of bistable structures to broadly advance the aims of vibration-based energy harvesting.

Chapter 5 discusses how the adaptive and energetic dynamics of bistable structures have been exploited for the sensing and detection of changing system characteristics. Through research efforts that represent a large range of length scales and sensing applications, it is shown that intelligently leveraged transitions from one dynamic regime to another provide bistable sensor architectures with a detection capacity that is particularly sensitive and versatile. The chapter details analytical and experimental studies that investigate new bifurcation-based sensing strategies which find that the dynamics of bistable sensors may be effectively harnessed for high robustness to inevitable noise and damping influences.

To conclude, Chapter 6 surveys the emerging areas and outstanding issues that remain to be addressed in order to realize the greatest benefits for the three technical fields of interest here via harnessing the dynamics of bistable structures. The issues are described in terms of analytical understanding of the effective strategies for leveraging the dynamical behaviors as well as practical matters involved in the design, fabrication, experimentation, and deployment of bistable systems. Looking further forward, the book closes with a prospective summary of research avenues which may yet emerge, and discusses how the future ventures may be beneficially grounded via the state-of-the-art development and exploitation of bistable structural dynamics.

References

[1] K. Yang, R.L. Harne, K.W. Wang, *et al.*, Dynamic stabilization of a bistable suspension system attached to a flexible host structure for operational safety enhancement, *Journal of Sound and Vibration* **333**, 6651–6661 (2014).

[2] A.F. Arrieta, P. Hagedorn, A. Erturk, *et al.*, A piezoelectric bistable plate for nonlinear broadband energy harvesting, *Applied Physics Letters* **97**, 104102 (2010).

[3] D.N. Betts, H.A. Kim, C.R. Bowen, Modeling and optimization of bistable composite laminates for piezoelectric actuation, *Journal of Intelligent Material Systems and Structures* **22**, 2181–2191 (2011).

[4] R.L. Harne, K.W. Wang, Prospects for nonlinear energy harvesting systems designed near the elastic stability limit when driven by colored noise, *ASME Journal of Vibration and Acoustics* **136**, 021009 (2014).

[5] R.L. Harne, K.W. Wang, A review of the recent research on vibration energy harvesting via bistable systems, *Smart Materials and Structures* **22**, 023001 (2013).

[6] S.P. Pellegrini, N. Tolou, M. Schenk, *et al.*, Bistable vibration energy harvesters: a review, *Journal of Intelligent Material Systems and Structures* **24**, 1303–1312 (2013).

[7] H.M. Ouakad, M.I. Younis, On using the dynamic snap-through motion of MEMS initially curved microbeams for filtering applications, *Journal of Sound and Vibration* **333**, 555–568 (2014).

[8] M. Bagheri, M. Poot, L. Mo, *et al.*, Dynamic manipulation of nanomechanical resonators in the high-amplitude regime and non-volatile mechanical memory operation, *Nature Nanotechnology* **6**, 726–732 (2011).

[9] M.T.A. Saif, On a tunable bistable MEMS – theory and experiment, *Journal of Microelectromechanical Systems* **9**, 157–170 (2000).

[10] S. Krylov, B.R. Ilic, D. Schreiber, *et al.*, The pull-in behavior of electrostatically actuated bistable microstructures, *Journal of Micromechanics and Microengineering* **18**, 055026 (2008).

[11] R.L. Harne, K.W. Wang, Robust sensing methodology for detecting change with bistable circuitry dynamics tailoring, *Applied Physics Letters* **102**, 203506 (2013).

[12] F. Moss, P.V.E. McClintock, eds, *Noise in Nonlinear Dynamical Systems, Volume 3: Experiments and Simulations* (Cambridge University Press, Cambridge, 1989).

[13] S.P. Timoshenko, *Theory of Elastic Stability* (McGraw-Hill, New York, 1936).

[14] I. Kovacic, M.J. Brennan, eds, *The Duffing Equation: Nonlinear Oscillators and their Behaviour* 1st edn (John Wiley & Sons, Ltd, Chichester, 2011).

[15] R. Seydel, *Practical Bifurcation and Stability Analysis* (Springer, New York, 2010).

[16] Y.A. Kuznetsov, *Elements of Applied Bifurcation Theory* (Springer, New York, 1998).

[17] M.I. Younis, *MEMS Linear and Nonlinear Statics and Dynamics* (Springer, New York, 2011).

[18] R. Lifshitz, M.C. Cross. *Nonlinear Dynamics of Nanomechanical and Micromechanical Resonators*. In H.G. Schuster, ed. *Reviews of Nonlinear Dynamics and Complexity: Volume 1* (Wiley-VCH Verlag GmbH, Weinheim 2008) pp. 1–52.

[19] F.C. Moon, *Chaotic and Fractal Dynamics : an Introduction for Applied Scientists and Engineers* (John Wiley & Sons, Inc., New York, 1992).

[20] S.H. Strogatz, *Nonlinear Dynamics and Chaos: with Applications to Physics, Biology, Chemistry, and Engineering* (Addison-Wesley, Reading, Massachusetts, 1994).

[21] P. Glendinning, *Stability, Instability and Chaos: an Introduction to the Theory of Nonlinear Differential Equations* (Cambridge University Press, Cambridge, 1994).

[22] J. Gleick, *Chaos* (Viking, New York, 1987).

[23] E. Lorenz, *The Essence of Chaos* (CRC Press, Florence, Kentucky, 1995).

[24] L. Gammaitoni, P. Hänggi, P. Jung, *et al.*, Stochasic resonance, *Reviews of Modern Physics* **70**(1), 223–287 (1998).

[25] T. Wellens, V. Shatokhin, A. Buchleitner, Stochastic resonance, *Reports on Progress in Physics* **67**, 45–105 (2004).

[26] S. Piggott, *The Earliest Wheeled Transport: From the Atlantic Coast to the Caspian Sea* (Cornell University Press, Ithaca, NY, 1983).

[27] H. Frahm. Means for damping the rolling motion of ships. U.S. Patent 970,368. U.S. Patent Office; 1910.

[28] D.P. Taylor, 1996 Fluid dampers for applications of seismic energy dissipation and seismic isolation. In *Proceedings of the Eleventh World Conference on Earthquake Engineering*; Acapulco, Mexico.

[29] B.B. Holmes. Shock absorber. U.S. Patent 1,751,479. United States Patent Office; 1930.
[30] D.J. Wagg, L.N. Virgin, *Exploiting Nonlinear Behavior in Structural Dynamics* (Springer, New York, 2012).
[31] A.F. Vakakis, O.V. Gendelman, L.A. Bergman, *et al.*, *Nonlinear Targeted Energy Transfer in Mechanical and Structural Systems* (Springer, Dordrecht, 2008).
[32] R.S. Lakes, Extreme damping in composite materials with a negative stiffness phase, *Physical Review Letters* **86**, 2897–2900 (2001).
[33] M.E. Pontecorvo, S. Barbarino, F.S. Gandhi, *et al.*, 2013 A load-bearing structural element with energy dissipation capability under harmonic excitation. In *Proceedings of the ASME 2013 Conference on Smart Materials, Adaptive Structures and Intelligent Systems*; Snowbird, Utah, USA. SMASIS2013-3060.
[34] D.R. Johnson, M. Thota, F. Semperlotti, *et al.*, On achieving high and adaptable damping via a bistable oscillator, *Smart Materials and Structures* **22**, 115027 (2013).
[35] G.J. Murray, F. Gandhi, 2011 The use of damping to mitigate violent snap-through of bistable systems. In *Proceedings of the ASME 2011 Conference on Smart Materials, Adaptive Structures and Intelligent Systems SMASIS2011*; Scottsdale, Arizona. SMASIS2011-4997.
[36] R. Wiebe, L.N. Virgin, I. Stanciulescu, *et al.*, Characterizing dynamic transitions associated with snap-through: a discrete system, *Journal of Computational and Nonlinear Dynamics* **8**, 011010 (2013).
[37] J. Winterflood, T. Barber, D.G. Blair, Using Euler buckling springs for vibration isolation, *Classical and Quantum Gravity* **19**, 1639–1645 (2002).
[38] L.N. Virgin, R.B. Davis, Vibration isolation using buckled struts, *Journal of Sound and Vibration* **260**, 965–973 (2003).
[39] D.T.R. Pasala, A.A. Sarlis, S. Nagarajaiah, *et al.*, Adaptive negative stiffness: new structural modification approach for seismic protection, *Journal of Structural Engineering* **139**, 1112–1123 (2013).
[40] A. Sarlis, D. Pasala, M. Constantinou, *et al.*, Negative stiffness device for seismic protection of structures, *Journal of Structural Engineering* **139**, 1124–1133 (2013).
[41] O.V. Gendelman, C.H. Lamarque, Dynamics of linear oscillator coupled to strongly nonlinear attachment with multiple states of equilibrium, *Chaos, Solitons and Fractals* **24**, 501–509 (2005).
[42] D.R. Johnson, R.L. Harne, K.W. Wang, A disturbance cancellation perspective on vibration control using a bistable snap-through attachment, *Journal of Vibration and Acoustics* **136**, 031006 (2014).
[43] L.I. Manevitch, G. Sigalov, F. Romeo, *et al.*, Dynamics of a linear oscillator coupled to a bistable light attachment: analytical study, *Journal of Applied Mechanics* **81**, 041011 (2014).
[44] S. Priya, D.J. Inman, *Energy Harvesting Technologies* (Springer, New York, 2009).
[45] A. Erturk, D.J. Inman, *Piezoelectric Energy Harvesting* (John Wiley & Sons, Ltd, Chichester, 2011).
[46] N. Elvin, A. Erturk, *Advances in Energy Harvesting Methods* (Springer, New York, 2013).

[47] J. Cruz, *Ocean Wave Energy: Current Status and Future Perspectives* (Springer, Berlin, 2008).
[48] J. Scruggs, P. Jacob, Harvesting ocean wave energy, *Science* **323**, 1176–1178 (2009).
[49] L. Zuo, X. Tang, Large-scale vibration energy harvesting, *Journal of Intelligent Material Systems and Structures* **24**, 1405–1430 (2013).
[50] L. Tang, Y. Yang, C.K. Soh, Toward broadband vibration-based energy harvesting, *Journal of Intelligent Material Systems and Structures* **21**, 1867–1897 (2010).
[51] Z. Li, L. Zuo, J. Kuang, G. Luhrs, Energy-harvesting shock absorber with a mechanical motion rectifier, *Smart Materials and Structures* **22**, 025008 (2013).
[52] P.L. Green, E. Papatheou, N.D. Sims, Energy harvesting from human motion and bridge vibrations: an evaluation of current nonlinear energy harvesting solutions, *Journal of Intelligent Material Systems and Structures* **24**, 1494–1505 (2013).
[53] M.F. Daqaq, R. Masana, A. Erturk, D.D. Quinn, On the role of nonlinearities in vibratory energy harvesting: a critical review and discussion, *Applied Mechanics Reviews* **66**, 040801 (2014).
[54] C.R. McInnes, D.G. Gorman, M.P. Cartmell, Enhanced vibrational energy harvesting using nonlinear stochastic resonance, *Journal of Sound and Vibration* **318**, 655–662 (2008).
[55] S. Zhao, A. Erturk, On the stochastic excitation of monostable and bistable electroelastic power generators: relative advantages and tradeoffs in a physical system, *Applied Physics Letters* **102**, 103902 (2013).
[56] R. Masana, M.F. Daqaq, Response of Duffing-type harvesters to band-limited noise, *Journal of Sound and Vibrations* **332**, 67556767 (2013).
[57] J.W. Gardner, V.K. Varadan, O.O. Awadelkarim, *Microsensors, MEMS, and Smart Devices* (John Wiley & Sons, Ltd, Chichester, 2001).
[58] F.K. Chang, *Structural Health Monitoring* (Technomic, Lancaster, Pennsylvania, 1999).
[59] F. Ansari, *Sensing Issues in Civil Structural Health Monitoring* (Springer, Dordrecht, The Netherlands, 2005).
[60] T. Thundat, E.A. Wachter, S.L. Sharp, R.J. Warmack, Detection of mercury vapor using resonating microcantilevers, *Applied Physics Letters* **66**, 1695–1697 (1995).
[61] W. Zhang, K.L. Turner, 2004 A mass sensor based on parametric resonance. In *Proceedings of Solid-State Sensor, Actuator and Microsystems Workshop*; Hilton Head Island, South Carolina. 49–52.
[62] M.I. Younis, F. Alsaleem, Exploration of new concepts for mass detection in electrostatically-actuated structures based on nonlinear phenomena, *Journal of Computational and Nonlinear Dynamics* **4**, 021010 (2009).
[63] V. Kumar, *et al.*, Bifurcation-based mass sensing using piezoelectrically-actuated microcantilevers, *Applied Physics Letters* **98**, 153510 (2011).
[64] A.B. Zorin, Quantum-limited electrometer based on single cooper pair tunneling, *Physical Review Letters* **76**, 4408–4411 (1996).
[65] R. Vijay, M.H. Devoret, I. Siddiqi, Invited review article: the Josephson bifurcation amplifier, *Review of Scientific Instruments* **80**, 111101 (2009).

[66] J.F. Rhoads, S.W. Shaw, K.L. Turner, Nonlinear dynamics and its applications in micro- and nanoresonators, *Journal of Dynamic Systems, Measurement, and Control* **132**, 034001 (2010).
[67] W. Zhang, K.L. Turner, Application of parametric resonance amplification in a single-crystal silicon micro-oscillator based mass sensor, *Sensors and Actuators A* **122**, 23–30 (2005).
[68] K.L. Turner, C. Burgner, Z. Yie, *et al.*, Nonlinear dynamics of MEMS systems, *AIP Conference Proceedings* **1339**, 111 (2011).
[69] J. Qiu, J.H. Lang, A.H. Slocum, A curved-beam bistable mechanism, *Journal of Microelectromechanical Systems* **13**, 137–146 (2004).
[70] H.M. Ouakad, M.I. Younis, The dynamic behavior of MEMS arch resonators actuated electrically, *International Journal of Non-Linear Mechanics* **45**, 704–713 (2010).
[71] J. Kim, R.L. Harne, K.W. Wang, Enhancing damage identification robustness to noise and damping using integrated bistable and adaptive piezoelectric circuitry, **137**, 011005 (2015).
[72] M.I. Younis, H.M. Ouakad, F.M. Alsaleem, R. Miles, W. Cui, Nonlinear dynamics of MEMS arches under harmonic electrostatic actuation, *Journal of Microelectromechanical Systems* **19**, 647–656 (2010).
[73] S. Krylov, B.R. Ilic, S. Lulinsky, Bistability of curved microbeams actuated by fringing electrostatic fields, *Nonlinear Dynamics* **66**, 403–426 (2011).
[74] F.D. Bannon, J.R. Clark, C.T.C. Nguyen, High-Q HF micromechanical filters, *IEEE Journal of Solid-State Circuits* **35**, 512–526 (2000).
[75] J.F. Rhoads, S.W. Shaw, K.L. Turner, Nonlinear response of resonant microbeam systems with purely-parametric electrostatic actuation, *Journal of Micromechanics and Microengineering* **16**, 890–899 (2006).
[76] B.K. Hammad, A.H. Nayfeh, E.M. Abdel-Rahman, On the use of the subharmonic resonance as a method for filtration, *Journal of Computational and Nonlinear Dynamics* **6**, 041007 (2011).

2

Mathematical Modeling and Analysis of Bistable Structural Dynamics

This chapter reviews the analytical assessment methods for linear and nonlinear dynamic systems that are used throughout the remainder of the book. Following a survey of methods to derive governing equations of motion and means for equation system normalization, the state space representation is introduced and shown to be a basis for characterizing system stability. Then, time series trajectories and phase portraits are shown to reveal essential dependences on common system and excitation parameters. Several methods are then detailed for analytically examining the equation governing the motion of the conventional bistable oscillator, considering steady-state dynamics near- and far-from-equilibrium, including the dynamic snap-through phenomenon. Techniques to explicitly investigate the onset of dynamic bifurcations are described; these techniques are foundational tools for determining critical operating conditions beyond which the potential to realize particular dynamic regimes will drastically change. Then, an overview is given regarding the extension of the analytical methods to multi-degree-of-freedom and electromechanical bistable structures.

2.1 A Linear Oscillator

2.1.1 Free Response

One of the most fundamental vibratory systems is the single degree-of-freedom (DOF) harmonic oscillator, Figure 2.1a. By virtue of design [1], modal analysis [2], or reduced-order modeling [3], a near-endless variety of dynamic systems may be examined according to the principal deflections of equivalent lumped-parameter constituents around one or more stable equilibria, which are encapsulated by the representation of the seemingly simple schematic in Figure 2.1a. Throughout this book, the general form of restoring force for such an oscillatory system is given by a smooth polynomial $F(y)=k_1y+k_2y^2+k_3y^3+\cdots$ where the deformation of the spring from an undeformed length is y. The constants k_i may be positive, negative, or zero. The values are often identified during experimentation or derived by employing a series expansion of the exact force expression written in terms of physically meaningful system parameters, for instance a beam length, thickness, and so forth.

Harnessing Bistable Structural Dynamics: For Vibration Control, Energy Harvesting and Sensing, First Edition. Ryan L. Harne and K.W. Wang.

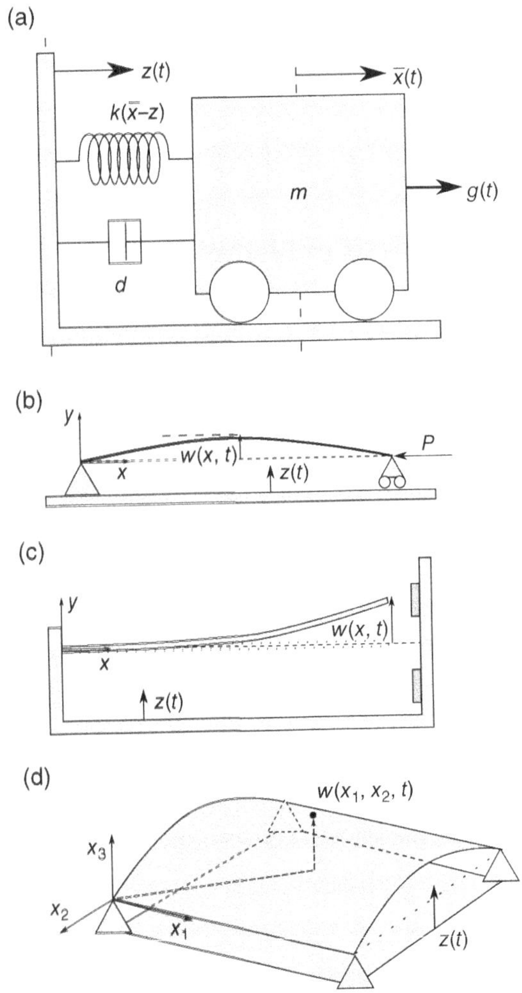

Figure 2.1 General schematics of several base-excited systems. (a) An oscillator with arbitrary spring restoring force $F(\bar{x} - z)$. (b) The fundamental mode response of a simply-supported beam with axial load. (c) The fundamental mode response of a ferromagnetic beam tip in the presence of magnetic fields. (d) A bistable composite plate mid-point response.

As an example, when the base excitation is $z = 0$, the restoring force for the one-dimensional, *absolute* motion X of the mass-spring system shown in Figure 1.1 is $F(X) = 2kX\left[1 - l_o/\sqrt{X^2 + d^2}\right]$ where k is the spring stiffness constant. If the base excitation is non-zero $z \neq 0$, the prior relation for the restoring force *remains the same* following the substitution $x \rightarrow y = X - z$ where y is the *relative* motion of the mass with respect to the base motion, which is likewise the extent that the spring is relatively deformed. Considering again $z = 0$, the restoring force smoothly varies for smooth changes in X, but the force is not in the polynomial form. Thus, a Taylor series expansion is applied to convert the

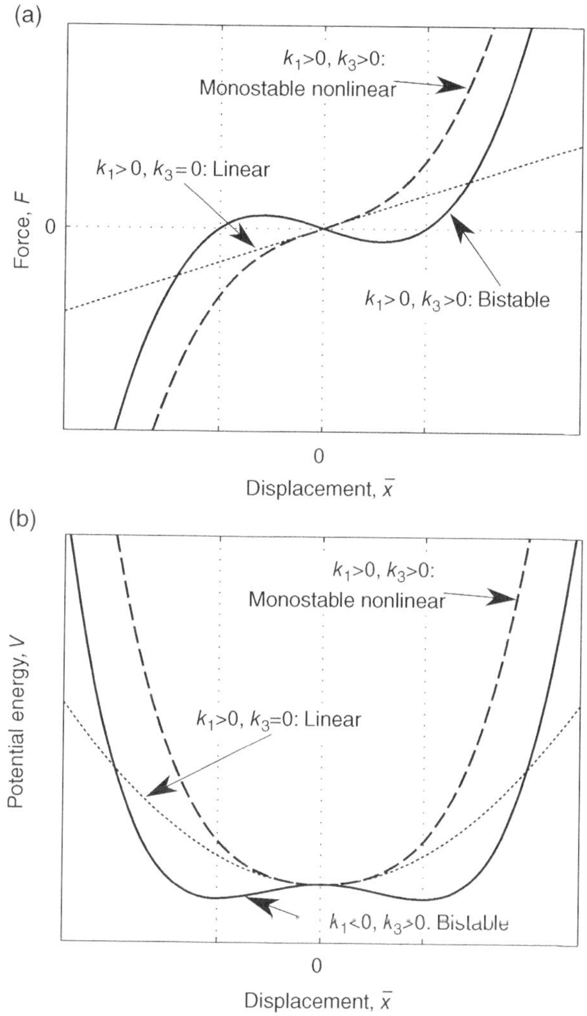

Figure 2.2 (a) Spring restoring force profiles for linear, monostable hardening Duffing, and bistable oscillators, and (b) corresponding potential energies.

force into the polynomial $F(X) = k_1 X + k_3 X^3$ while terms of higher order in X are eliminated from the expansion based on the assumption that the displacements from x=0 are finite but small enough such that the higher order contributions to the restoring force are negligible. By the expansion, the coefficients are found to be $k_1 = 2k(1 - l_o/d)$ and $k_3 = kl_o/d^3$. Finally, although it is not necessary to leverage the theoretical techniques as elaborated upon below, the restoring force polynomials considered in this book will be *odd* which means that $F(X) = -F(-X)$.

When the oscillator has spring stiffness constant $k_1 = k > 0$ and $k_{2,3,...} = 0$, the linear force-displacement profile is presented in Figure 2.2a, with the corresponding parabolic potential energy function in Figure 2.2b. Application of

Newton's second law to predict the *free response* of the system, in the absence of damping, $d=0$, ground motions, $z(t)=0$, and force $g(t)$, yields the governing equation for the absolute motion $\bar{x}(t)$ of the inertial mass m:

$$m\ddot{\bar{x}}+k\bar{x}=0 \tag{2.1}$$

where the overdot operator is d/dt. The use of Hamilton's principle and Lagrange's equations is an alternative strategy to determine the governing equations of dynamic systems based on energy methods of analysis. Hamilton's principle states that the traversed path of a body over any given time interval is that which minimizes a time integral of the difference between kinetic and potential energies, subject to relevant constraints [4]. Using variational calculus, application of Hamilton's principle leads to Lagrange's equations

$$\frac{\partial}{\partial t}\left[\frac{\partial L}{\partial \dot{\bar{x}}}\right]-\frac{\partial L}{\partial \bar{x}}=0 \tag{2.2}$$

where the Lagrangian L is defined as $L=T-V$, which is the difference between the kinetic and potential energies. For the undamped and unforced harmonic oscillator, the Lagrangian is

$$L=T-V=\frac{1}{2}m\dot{\bar{x}}^2-\frac{1}{2}k\bar{x}^2 \tag{2.3}$$

Consequently, substitution of Eq. (2.3) into Eq. (2.2) leads to

$$m\ddot{\bar{x}}+k\bar{x}=0 \tag{2.4}$$

There are numerous connections among Hamiltonian mechanics, Lagrangian mechanics, and Newton's laws, and the interested reader is recommended to consider additional texts providing focused treatment of these foundations [5,6].

A common goal when investigating system dynamics is to predict the response due to known initial conditions $(\bar{x}_0,\dot{\bar{x}}_0)=(\bar{x}(t=0),\dot{\bar{x}}(t=0))$ in the absence of external force $(g(t)=0)$ or ground motions $(z(t)=0)$, which is termed the *free response*. For this case, an assumed solution of the displacement response $\bar{x}(t)=a\sin\omega_n t+b\cos\omega_n t$ is substituted into Eq. (2.4), where the selection of this response form is based on the experience that the free vibration of a mass-spring oscillator is sinusoidal. Then, application of the initial conditions leads to the exact expression for the undamped linear oscillator response which is given by

$$\bar{x}(t)=\frac{\dot{\bar{x}}_0}{\omega_n}\sin\omega_n t+\bar{x}_0\cos\omega_n t \tag{2.5}$$

where it is also found that $\omega_n=\sqrt{k/m}$, which is referred to as the *natural frequency*.

Almost all real-world systems are dissipative and as a result damping is included in the model. For the oscillator having viscous damping constant d, the damping ratio is defined by $\zeta = d/2\sqrt{km}$. The damping ratio is the ratio of the damping constant d with respect to a critical quantity expressed as $d_c = 2\sqrt{km} = 2m\omega_n$. Systems with $\zeta > 1$ are termed *overdamped*; those with $\zeta = 1$ are *critically damped*; and those where $\zeta < 1$ are *underdamped*. Viscous damping effects are proportional to velocity, so a damping force $d\dot{x}$ is included in the force balance Eq. (2.1):

$$m\ddot{\bar{x}} + d\dot{\bar{x}} + k\bar{x} = 0 \tag{2.6}$$

Then, a similar strategy as before is applied, whereas for damped systems a different assumed solution is utilized: $\bar{x}(t) = ae^{\lambda t}$. With this assumed solution based on a *complex exponential* form, a and λ are complex in the most general case. Then, the substitution yields

$$\left[m\lambda^2 + d\lambda + k\right]ae^{\lambda t} = 0 \tag{2.7}$$

which is alternatively expressed by

$$\left[\lambda^2 + 2\zeta\omega_n\lambda + \omega_n^2\right]ae^{\lambda t} = 0 \tag{2.8}$$

Because $a = 0$ indicates the trivial, static solution, the dynamics of the system are therefore determined from the *characteristic equation*:

$$\lambda^2 + 2\zeta\omega_n\lambda + \omega_n^2 = 0 \tag{2.9}$$

The form of the roots of Eq. (2.9) indicates the type of motions that the system will exhibit. The roots of the quadratic characteristic equation polynomial in Eq. (2.9) are computed by $\lambda_{1,2} = -\zeta\omega_n \pm \omega_n\sqrt{\zeta^2 - 1}$. Since two roots are found, the total response of the system is the sum or superposition of the solutions representing the individual roots

$$\bar{x}(t) = a_1 e^{\lambda_1 t} + a_2 e^{\lambda_2 t} \tag{2.10a}$$

The quadratic polynomial Eq. (2.9) will have two real roots, two repeated roots, or a complex-conjugate pair of roots. When both roots are real, the system is overdamped, as indicated by the fact that Eq. (2.10a) is then the sum of two exponential functions having negative, real arguments as time increases. When the roots are repeated, the system is critically damped, and the response is expressed by $\bar{x}(t) = (a_1 + a_2 t)\,e^{-\omega_n t}$. When the roots are a complex conjugate pair, the system is underdamped, which is the most common structural damping characteristic encountered throughout the remainder of this book. For underdamped oscillations, application of Euler's identity $e^{i\theta} = \cos\theta + i\sin\theta$, where $i = \sqrt{-1}$ is the imaginary number, leads to the *real* response given by

$$\bar{x}(t) = Ae^{-\zeta\omega_n t}\sin\left[\omega_d t + \phi\right] \tag{2.10b}$$

where $\omega_d = \omega_n\sqrt{1-\zeta^2}$ is the *damped natural frequency* and A and ϕ are determined by application of the initial conditions $(\bar{x}_0, \dot{\bar{x}}_0)$:

$$A = \sqrt{\bar{x}_0^2 + \left[\frac{\dot{\bar{x}}_0 + \zeta\omega_n\bar{x}_0}{\omega_d}\right]^2} \tag{2.11}$$

$$\tan\phi = \frac{\bar{x}_0\omega_d}{\dot{\bar{x}}_0 + \zeta\omega_n\bar{x}_0} \tag{2.12}$$

While the result of Eq. (2.10a) is the most general and correct form of solution, only the *real* part of the response is the actual, or physically-observable response. Such a finding is confirmed by application of the initial conditions in order to identify the terms $a_{1,2}$ and subsequent use of Euler's identity [7], for example as found from Eq. (2.10a) to (2.10b).

There are a variety of graphical means by which to evaluate system dynamics. Most commonly, time series and phase portraits assist in understanding the trajectories that the inertial mass may follow for given initial conditions. Figure 2.3a,b show example responses in terms of displacement over time and velocity and displacement parametrically plotted (the phase portrait), respectively. The representative underdamped oscillator has ζ=0.05 and $\omega_n = 2\pi\cdot 5$ rad s^{-1}, which is alternatively expressed as $\omega_n/2\pi = f_n$=5 Hz or cycles/s. As plainly observed, as a consequence of the dissipative damping force, the responses decay to a final displacement value of $\bar{x}$=0 m as time $t \to \infty$, and the frequency of vibration is the damped natural frequency of the oscillator, which is slightly less than 5 Hz.

When applying energy and variational methods for damped vibratory systems, a dissipation function is included in Lagrange's equation. For example, the dissipation function for viscous damping is $D = (1/2)d\dot{\bar{x}}^2$

$$\frac{\partial}{\partial t}\left[\frac{\partial L}{\partial \dot{\bar{x}}}\right] + \frac{\partial D}{\partial \dot{\bar{x}}} - \frac{\partial L}{\partial \bar{x}} = 0 \tag{2.13}$$

Further details on these foundational developments may be found in other vibrations texts [2,8–12].

2.1.2 Base-excited Response

When the damped oscillator is excited by constant-amplitude base acceleration of frequency ω, such that the motion $z(t)$ shown in Figure 2.1a is expressed by $\ddot{z}(t) = -P_0 \sin\omega t$, the governing equation becomes

$$m\ddot{\bar{x}} + d\dot{x} + kx = 0 \tag{2.14a}$$

$$m(\ddot{x} + \ddot{z}) + d\dot{x} + kx = 0 \tag{2.14b}$$

$$m\ddot{x} + d\dot{x} + kx = mP_0 \sin\omega t \tag{2.14c}$$

thus introducing the relative coordinate $x = \bar{x} - z$. The Eq. (2.14) is alternatively expressed as

$$\ddot{x} + 2\zeta\omega_n\dot{x} + \omega_n^2 x = p_0 \sin\omega t \tag{2.15}$$

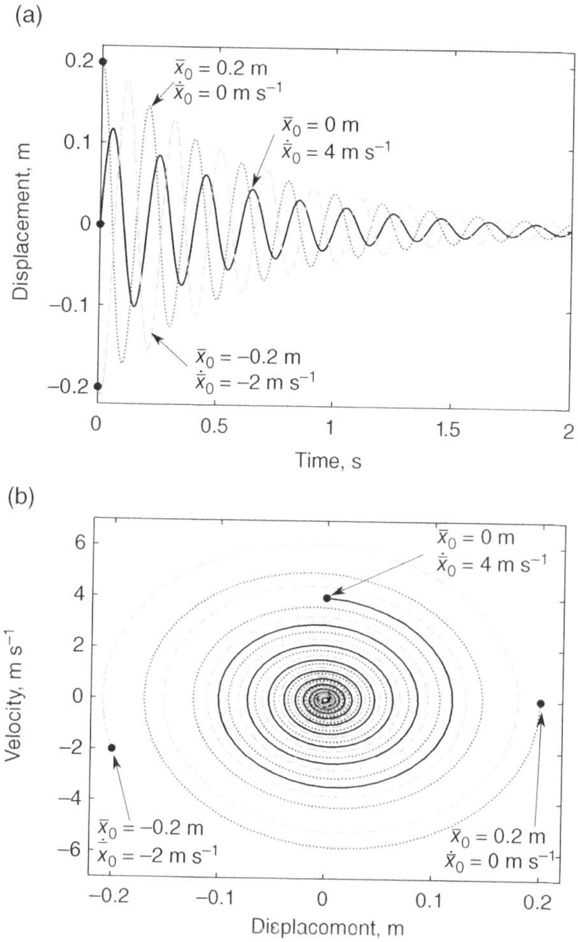

Figure 2.3 Example free response (a) time series and (b) phase portraits, of a linear oscillator for various initial conditions and $\zeta = 0.05$ and $\omega_n = 2\pi \cdot 5$ rad s^{-1}.

where $p_0 = P_0$. Note that if the excitation is constant-amplitude base *displacement*, $z(t) = P_0 \sin \omega t$, then Eq. (2.15) is modified such that $p_0 = \omega^2 P_0$. It is also worth noting that, in the absence of base motion $z = 0$ but the presence of a finite excitation force $g(t) \neq 0$, we may express the response of the oscillator in terms of the absolute coordinate $\bar{x}$ and apply Newton's second law to obtain the governing equation $m\ddot{\bar{x}} + d\dot{\bar{x}} + k\bar{x} = g(t)$. Comparing this to (2.14c), there is a one-to-one analogy between *base-excited response* considering the *relative coordinate of motion* and *force-excited response* considering the *absolute coordinate of motion.*

Returning to the case of the constant-amplitude base acceleration as the input, the total response of the linear oscillator to initial conditions and excitations is found to be the sum of the free, Eq. (2.10b), and forced responses:

$$x(t) = Ae^{-\zeta\omega_n t}\sin\left[\omega_d t + \phi\right] + X\sin\left[\omega t - \psi\right] \tag{2.16}$$

where

$$X = \frac{p_0}{\sqrt{\left[\omega_n^2 - \omega^2\right]^2 + \left[2\zeta\omega_n\omega\right]^2}} \tag{2.17}$$

$$\tan\psi = \frac{2\zeta\omega_n\omega}{\omega_n^2 - \omega^2} \tag{2.18}$$

Often, it is desired to explore the steady-state dynamics of systems, in other words the motions which remain due to the excitations long after initial transient behaviors have decayed due to dissipative influences. By assuming a solution form of $x(t) = Xe^{\lambda t}$ and substituting it into Eq. (2.15), the steady-state dynamics may be solved to yield a frequency response function (frf) of the oscillator response to the input excitation level where the FRF magnitude is

$$\left|\frac{\omega_n^2 X}{p_0}\right| = \frac{1}{\sqrt{\left[1-\alpha^2\right]^2 + \left[2\zeta\alpha\right]^2}} \tag{2.19}$$

where $\lambda = i\omega$ and $\alpha = \omega/\omega_n$ determines the nearness of the excitation frequency ω to resonance. *Resonance* is the condition such that the excitation frequency is identical to the oscillator natural frequency, $\omega = \omega_n$. Note that the amplitude of the underdamped linear oscillator frf is independent of excitation level p_0, so that the FRF magnitude at resonance is governed only by the system damping.

2.2 Stability

For analytical and computational convenience, it is useful to express the governing equations in a first order form, commonly referred to as state space. Therefore, the *states* of Eq. (2.15) are selected to be $\boldsymbol{x} = [x_1, x_2]^{\mathrm{T}} = [x, \dot{x}]^{\mathrm{T}}$, where the bold denotes a vector quantity and T denotes the transpose operator. Then, Eq. (2.15) is alternatively expressed as

$$\dot{\boldsymbol{x}} = \begin{bmatrix} \dot{x}_1 \\ \dot{x}_2 \end{bmatrix} = \begin{bmatrix} 0 & 1 \\ -\omega_n^2 & -2\zeta\omega_n \end{bmatrix} \begin{bmatrix} x_1 \\ x_2 \end{bmatrix} + \begin{bmatrix} 0 \\ p_0 \sin\omega t \end{bmatrix} \tag{2.20}$$

More generally, this representation is given by $\dot{\boldsymbol{x}} = \boldsymbol{f}(\boldsymbol{x},t)$ or decomposed into state-only and excitation-only components such that

$$\dot{\boldsymbol{x}} = \boldsymbol{A}(\boldsymbol{x}) + \boldsymbol{F}(t) \tag{2.21}$$

Equilibria, or fixed points, $\boldsymbol{x}^* = \left[x_1^*, x_2^*\right]^{\mathrm{T}}$, of a dynamic system are determined by

$$\boldsymbol{f}(\boldsymbol{x}^*, t) = \boldsymbol{0} \tag{2.22}$$

For the unforced, $p_0 = 0$, system (2.20), the equilibrium points satisfy

$$\begin{bmatrix} 0 & 1 \\ -\omega_n^2 & -2\zeta\omega_n \end{bmatrix} \begin{bmatrix} x_2^* \\ x_2^* \end{bmatrix} = \begin{bmatrix} 0 \\ 0 \end{bmatrix} \tag{2.23}$$

Solving Eq. (2.23) indicates that at equilibrium the linear oscillator has zero velocity $x_2^* = 0$, and tends to zero displacement $x_1^* = 0$.

The stability of a fixed point is determined by evaluating the local response around an equilibrium point, expressed as $\boldsymbol{y} = \boldsymbol{x} - \boldsymbol{x}^*$. The local response $\boldsymbol{y}$ is substituted into the unforced Eq. (2.21), in other words when $\boldsymbol{F}(t) = \boldsymbol{0}$. If required, a Taylor series expansion about the fixed point $\boldsymbol{x}^*$ is applied to linearize the resulting equation. This method is also called a coordinate transformation. Then, a solution of form $\boldsymbol{y}(t) = \boldsymbol{a}e^{\lambda t}$ is assumed. For the linear oscillator, the prior steps are trivial because linearization of the system is not required and the fixed point is identically zero for both states. Nevertheless, the result is an equation system of the form

$$\lambda \boldsymbol{a}e^{\lambda t} = \mathrm{A}\boldsymbol{a}e^{\lambda t} \rightarrow \left[\mathrm{A} - \lambda \mathrm{I}\right]\boldsymbol{a}e^{\lambda t} = 0 \tag{2.24}$$

The capitalized, non-bold script denotes a matrix, including I which is the identity matrix. Eq. (2.24) is alternatively expressed as an *eigenvalue problem*. according to $\mathrm{A}\boldsymbol{a} = \lambda\boldsymbol{a}$.

Because $\boldsymbol{a} = 0$ indicates only a trivial solution, Eq. (2.25) must be satisfied:

$$\det\left[\mathrm{A} - \lambda \mathrm{I}\right] = 0 \tag{2.25}$$

where det[] is the determinant, and using $A_{ij} = a_{ij}$ is given by

$$\det\begin{bmatrix} a_{11} - \lambda & a_{12} \\ a_{21} & a_{22} - \lambda \end{bmatrix} = 0 \tag{2.26}$$

Eq. (2.25) leads to the characteristic equation

$$\lambda^2 - (a_{11} + a_{22})\lambda + (a_{11}a_{22} - a_{21}a_{12}) = 0 \tag{2.27}$$

The quantity $\mathrm{tr}[\mathrm{A}] = (a_{11} + a_{22})$ is called the trace of A, while $\det[\mathrm{A}] = (a_{11}a_{22} - a_{12}a_{21})$ is the determinant of A. As a result, the characteristic equation may also be written as

$$\lambda^2 - \mathrm{tr}\left[\mathrm{A}\right]\lambda + \det\left[\mathrm{A}\right] = 0 \tag{2.28}$$

where the eigenvalues (solutions) of (2.28) are

$$\lambda_{1,2} = \frac{1}{2}\left[\mathrm{tr}\left[\mathrm{A}\right] \pm \sqrt{\mathrm{tr}\left[\mathrm{A}\right]^2 - 4\det\left[\mathrm{A}\right]}\right] \tag{2.29}$$

The discriminant of the eigenvalues is $\Delta = \text{tr}[A]^2 - 4\det[A]$, such that $\lambda_{1,2} = \frac{1}{2}\left[\text{tr}[A] \pm \sqrt{\Delta}\right]$. The sign of Δ determines whether the eigenvalues $\lambda_{1,2}$ are real, $\Delta>0$, or complex, $\Delta<0$. Moreover, the signs of the eigenvalues are critical in defining the stability of the system. Because the assumed solution is $\boldsymbol{y}(t) = \boldsymbol{a}e^{\lambda t}$, the presence of an eigenvalue having a positive, real component indicates that the system will be unstable because the response will grow exponentially in time. This is the approach to determine stability of a system response or configuration used throughout this book.

For the underdamped linear oscillator, the eigenvalues may be expressed following Section 2.1.1 as $\lambda_{1,2} = -\zeta\omega_n \pm i\omega_d$. Then, substituting this into the assumed solution $\boldsymbol{x}(t) = \boldsymbol{a}e^{\lambda t}$ – noting that for the linear oscillator $\boldsymbol{x}(t) = \boldsymbol{y}(t)$ because the fixed point coordinates are both zero for $F(x) = kx$ – and applying Euler's identity leads to

$$\boldsymbol{x}(t) = \boldsymbol{a}e^{-\zeta\omega_n t} \sin\omega_d t \tag{2.30}$$

As a result, if the linear oscillator damping constant is negative, such that $\zeta = d/2\sqrt{km} < 0$, the response of the system grows exponentially in time. This unbounded response indicates the linear oscillator will be unstable as a consequence of negative damping, $d<0$.

2.3 A Monostable Nonlinear Oscillator

Consider the restoring force for an oscillator expressed using $F(x) = k_1 x + k_3 x^3$ where both $k_1, k_3 > 0$, such as in Figure 2.2a. This refers to a system known as a monostable, nonlinear, *hardening* Duffing oscillator [13]. The case of $k_1>0$ and $k_3<0$ is the *softening* Duffing oscillator. Either way, the nonlinear governing equation of the Duffing oscillator mass motion is

$$m\ddot{x} + d\dot{x} + k_1 x + k_3 x^3 = 0 \tag{2.31}$$

The fixed points of the Duffing oscillator are found by considering the form $\dot{\boldsymbol{x}} = \boldsymbol{f}(\boldsymbol{x}, t)$ and solving $\boldsymbol{f}(\boldsymbol{x}^*, t) = 0$, where as before $\boldsymbol{x} = [x_1, x_2]^T = [x, \dot{x}]^T$.

$$\begin{bmatrix} x_2^* \\ -\frac{k_1}{m}x_1^* - \frac{k_3}{m}x_1^{3*} - \frac{d}{m}x_2^* \end{bmatrix} = \begin{bmatrix} 0 \\ 0 \end{bmatrix} \tag{2.32}$$

It is found that only one fixed point exists such that the velocity is zero $x_2^* = 0$, while the displacement is also zero $x_1^* = 0$. Thus, just like the linear oscillator, the Duffing oscillator is monostable. In contrast to the linear system, determination of stability for the nonlinear oscillator requires a linearized equation formulation. Expressing the state vector of a fixed point as $\boldsymbol{x}^*$, and defining a small perturbation about equilibrium $\boldsymbol{y} = \boldsymbol{x} - \boldsymbol{x}^*$, a Taylor series expansion of the autonomous (time-independent) governing equation leads to

$$\dot{\boldsymbol{y}} = \boldsymbol{f}(\boldsymbol{x}^*) + \mathbf{D}_x\boldsymbol{f}(\boldsymbol{x}^*)\boldsymbol{y} + O\left(\|\boldsymbol{y}\|^2\right) \tag{2.33}$$

Here, $\mathbf{D}_x\boldsymbol{f}$ is the Jacobian of $\boldsymbol{f}$, and $O(||\boldsymbol{y}||^2)$ are the higher order terms of the Taylor series that negligibly contribute to the total response. The components of the Jacobian are determined by the partial derivatives

$$\mathbf{D}_x\boldsymbol{f}(\boldsymbol{x}^*) = \begin{bmatrix} \frac{\partial f_1}{\partial x_1} & \frac{\partial f_1}{\partial x_2} & \cdots & \frac{\partial f_1}{\partial x_n} \\ \frac{\partial f_2}{\partial x_1} & \frac{\partial f_2}{\partial x_2} & \cdots & \frac{\partial f_2}{\partial x_n} \\ \vdots & \vdots & \ddots & \vdots \\ \frac{\partial f_n}{\partial x_1} & \frac{\partial f_n}{\partial x_2} & \cdots & \frac{\partial f_n}{\partial x_n} \end{bmatrix}_{\mathbf{x}=\mathbf{x}^*} \tag{2.34}$$

where n is the number of states. By definition, $\boldsymbol{f}(\boldsymbol{x}^*) = 0$, so that the linearized response neglecting higher-order terms is

$$\dot{\boldsymbol{y}} = \mathbf{D}_x\boldsymbol{f}(\boldsymbol{x}^*)\boldsymbol{y} \tag{2.35}$$

Then, the stability of the fixed point $\boldsymbol{x}^*$ is determined by applying the principles presented in Section 2.2 to equation system (2.35). An assumed solution $\boldsymbol{y}(t) = \boldsymbol{a}e^{\lambda t}$ is therefore substituted into Eq. (2.35) and the real and imaginary values of the eigenvalues of the characteristic equation are found so as to classify system stability [14].

For the monostable Duffing system, the fixed point is $\boldsymbol{x}^* = [x_1^*, x_2^*]^{\mathrm{T}} = [0, 0]^{\mathrm{T}}$, as described above. The Jacobian evaluated at the equilibrium is

$$\mathbf{D}_x\boldsymbol{f}(\boldsymbol{x}^*) = \begin{bmatrix} 0 & 1 \\ -\frac{k_1}{m} - 3\frac{k_3}{m}x_1^2 & -\frac{d}{m} \end{bmatrix}_{x=x^*} = \begin{bmatrix} 0 & 1 \\ -\frac{k_1}{m} & -\frac{d}{m} \end{bmatrix} \tag{2.36}$$

It is evident, then, that the linearized response of the hardening Duffing oscillator governed by Eq. (2.35) is identical to that of a linear oscillator, and the stability of the fixed point is likewise in correspondence.

Since a closed form solution for the free response Eq. (2.31) is not possible, the governing equation must be integrated to exactly predict the dynamics. A common strategy is to utilize a time-stepping Runge–Kutta algorithm to numerically integrate the nonlinear governing equations. Several orders of this algorithm are available in a number of computational software packages such as MATLAB, Mathcad, or Mathematica. Following this direct simulation approach, Figure 2.4a shows that the period of oscillations of the free response of a hardening Duffing oscillator is reduced over time as the system settles towards equilibrium, indicated by growing representative free response periods T_1 and T_2. Ultimately,

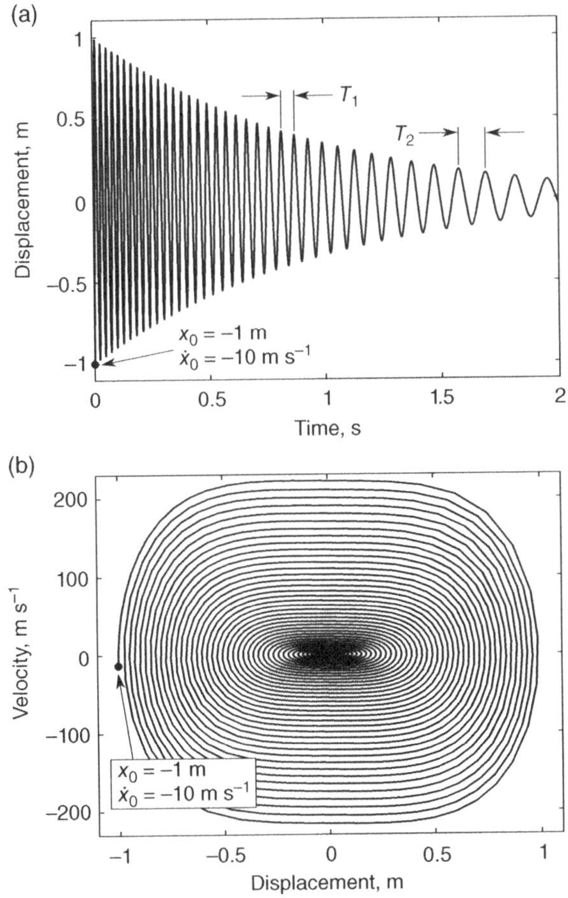

Figure 2.4 Monostable nonlinear oscillator free response (a) time series and (b) phase portrait, for $m = 1$ kg, $d = \pi$ N s m^{-1}, $k_1 = (2\pi 5)^2$ N m^{-1}, $k_3 = 100(2\pi 5)^2$ N m^{-1}.

for very small displacements the natural frequency ω_n of the hardening Duffing oscillator tends to the same frequency as that of the linear oscillator. This was found above following linearization of the nonlinear system about equilibrium.

2.4 A Bistable Oscillator

2.4.1 Free Response and Stability

The symmetric, bistable oscillator force–displacement profile is shown in Figure 2.2a where negative linear $k_1<0$ and positive nonlinear $k_3>0$ terms contribute to the stiffness of the system. Often, the governing equation is expressed using positive stiffness coefficients such that

$$m\ddot{x} + d\dot{x} - k_1 x + k_3 x^3 = 0 \tag{2.37}$$

where the more common convention of $k_1>0$ and $k_3>0$ is used. As will be demonstrated throughout this book, the principal dynamics of numerous bistable systems are adequately characterized according to the governing Eq. (2.37). These bistable systems include, but are by no means limited to, the fundamental transverse buckling mode response of beams, Figure 2.1b; the primary dynamics of a ferromagnetic beam tip acted upon by two attractive magnetic fields, Figure 2.1c; and the principal out-of-plane vibrations of asymmetrically-laminated composite plates, Figure 2.1d.

The fixed points $\boldsymbol{x}^* = [x_1^*, x_2^*]^{\mathrm{T}} = [x^*, \dot{x}^*]^{\mathrm{T}}$ of the bistable system are determined from

$$\begin{bmatrix} x_2^* \\ \frac{k_1}{m}x_1^* - \frac{k_3}{m}x_1^{3*} - \frac{d}{m}x_2^* \end{bmatrix} = \begin{bmatrix} 0 \\ 0 \end{bmatrix} \tag{2.38}$$

Solving Eq. (2.38) reveals three fixed points: $\boldsymbol{x}_1^* = [0, 0]^{\mathrm{T}}$, $\boldsymbol{x}_{2,3}^* = \left[\pm\sqrt{k_1/k_3}, 0\right]^{\mathrm{T}}$. The stability of the configurations must be evaluated following linearization of the governing equation. The transformed variable denoting small displacements around one of the equilibria, $\mathbf{y} = \mathbf{x} - \mathbf{x}^*$, is again employed. The Jacobian is then

$$\mathbf{D}_{\boldsymbol{x}}\boldsymbol{f}(\boldsymbol{x}^*) = \begin{bmatrix} 0 & 1 \\ \frac{k_1}{m} - 3\frac{k_3}{m}x_1^2 & -\frac{d}{m} \end{bmatrix}_{\boldsymbol{x}=\boldsymbol{x}^*} \tag{2.39}$$

Some parameter values are now inserted for clarity of the following developments: $m=1$ kg, $k_1=1$ N m^{-1}, and $d=\gamma$ N s m^{-1}. Then, for $\boldsymbol{x}_1^* = [0, 0]^{\mathrm{T}}$, the linearized system is

$$\dot{\boldsymbol{y}} = \mathbf{D}_{\boldsymbol{x}}\boldsymbol{f}(\boldsymbol{x}_1^*)\boldsymbol{y} = \begin{bmatrix} 0 & 1 \\ 1 & -\gamma \end{bmatrix}\boldsymbol{y} \tag{2.40}$$

while for $\boldsymbol{x}_{2,3}^* = \left[\pm\sqrt{k_1/k_3}, 0\right]^{\mathrm{T}} = \left[\pm\sqrt{1/k_3}, 0\right]^{\mathrm{T}}$, the linearized systems are both found to be

$$\dot{\boldsymbol{y}} = \mathbf{D}_{\boldsymbol{x}}\boldsymbol{f}\left(\boldsymbol{x}_{2,3}^*\right)\boldsymbol{y} = \begin{bmatrix} 0 & 1 \\ \frac{k_1}{m} - 3\frac{k_3}{m}\frac{k_1}{k_3} & -\frac{d}{m} \end{bmatrix}\boldsymbol{y} = \begin{bmatrix} 0 & 1 \\ -2 & -\gamma \end{bmatrix}\boldsymbol{y} \tag{2.41}$$

Recall that the eigenvalues of the characteristic equation of the linearized governing equations (for the underdamped system, $0<\gamma<1$) are determined using $\lambda_{1,2} = \frac{1}{2}\left[\mathrm{tr}[\mathrm{A}] \pm \sqrt{\Delta}\right]$. Here $\mathrm{A} = \mathbf{D}_{\boldsymbol{x}}\boldsymbol{f}(\boldsymbol{x}_1^*)$ and $\Delta = \mathrm{tr}[\mathrm{A}]^2 - 4\det[\mathrm{A}]$. Then, considering Eq. (2.40), it is found that $\mathrm{tr}\left[\mathrm{A}_{x_1^*}\right] = -\gamma$, while $\det\left[\mathrm{A}_{x_1^*}\right] = -1$. Consequently, the eigenvalues are $\lambda_{1,2} = \frac{1}{2}\left[-\gamma \pm \sqrt{\gamma^2 + 4}\right]$. For such underdamped systems, the eigenvalues thus contain a positive real component, which indicates the local equilibrium $\boldsymbol{x}_1^* = [0, 0]^{\mathrm{T}}$ is unstable when subjected to small perturbation.

Repeating this procedure for the fixed points $\boldsymbol{x}_{2,3}^*$, the eigenvalues are $\lambda_{1,2} = \frac{1}{2}\left[-\gamma \pm \sqrt{\gamma^2 - 8}\right]$. Therefore, the eigenvalues always contain negative real components, indicating *stability* of the *two fixed points* $\boldsymbol{x}_{2,3}^*$: hence, *bistability*.

In summary, the above theoretical developments show that the bistable system is characterized as having two stable configurations separated by an unstable equilibrium configuration. Small perturbations to the system when it is initially in the unstable position cause the mass of the system to move towards one of the surrounding, stable configurations.

An alternative conception of stability pertaining to the statics and mechanics of structures may be adopted by considering the potential energy function of an oscillatory system. This conception is particularly enlightening when considering bistable systems. The potential energy $V(x)$ of a system is the integral of the work done by conservative forces moving the particle along a trajectory 0 to x

$$V(x) = \int_0^x F(x)\mathrm{d}x \tag{2.42}$$

For the bistable oscillator, the conservative forces are given by $F(x) = -k_1 x + k_3 x^3$, and thus the potential energy is

$$V(x) = -\frac{1}{2}k_1 x^2 + \frac{1}{4}k_3 x^4 \tag{2.43}$$

The force–displacement and potential energy–displacement profiles are shown in Figure 2.2a,b, respectively. The potential energy profile of the bistable system is historically referred to as the "double-well" potential. From the force–displacement profile, Figure 2.2a, it is clear that three displacement configurations lead to zero force reaction from the spring. One central static equilibrium position is characterized as exhibiting negative stiffness, that is, a negative slope in the force–displacement profile, $\partial F(x)/\partial x < 0$. In contrast, the surrounding static equilibria exhibit positive stiffnesses. The central equilibrium appears as a "hilltop" of potential energy whereas the surrounding equilibria are "valleys" of the potential energy profile.

Considering Lagrange's equations (2.2), static equilibrium occurs where $\partial V/\partial x = 0$. Lagrange observed that for an equilibrium to be locally stable it must be at a "complete relative minimum of the total potential energy" [1,15]. Therefore, stability of an equilibrium requires that $\partial^2 V/\partial x^2 > 0$. Using the perspective shown by Figure 2.2b, the convexity of the positive and negative static equilibria "valleys" demonstrates their stability, whereas the concavity of the central configuration indicates it is an unstable position. A common thought experiment to illustrate the energy concept is that of a ball carefully positioned on a hilltop adjacent to two valleys: a small push from the top of the hill will destabilize its equilibrium and cause it to fall into one of the valleys. Further details and other perspectives on stability may be found in a number of other texts [4,14,16–19].

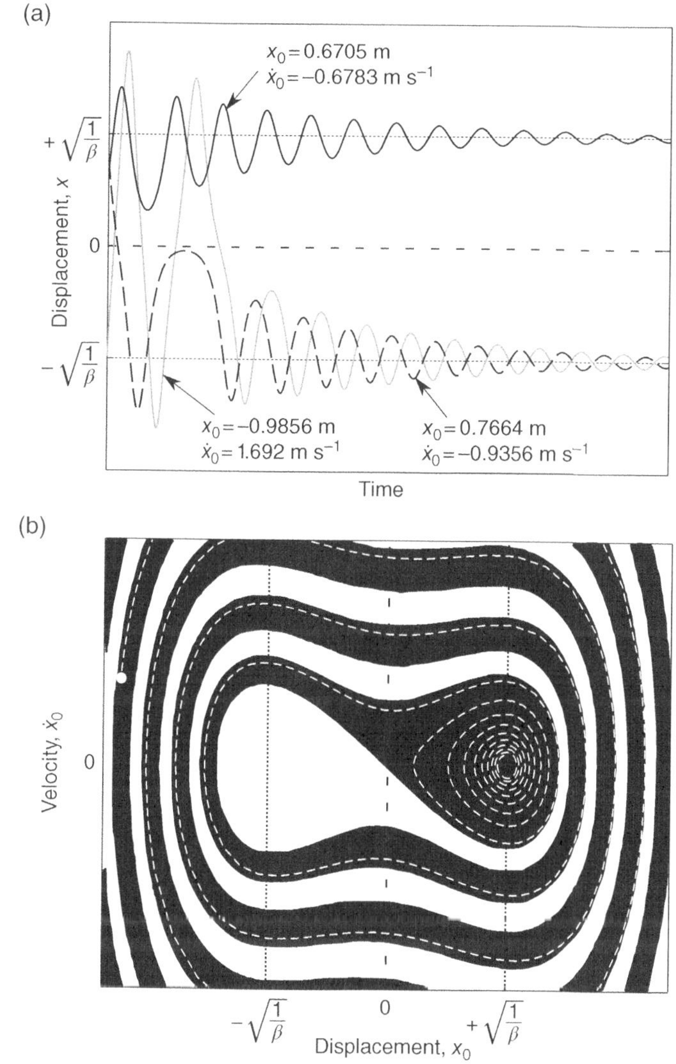

Figure 2.5 (a) Example free response displacement time series of bistable system. (b) Basin of attraction map with single example case of phase portrait free response overlaid as light dashed curve. $\beta = 1$, $\gamma = 0.1$.

The general free response of a damped bistable system does not have a closed form solution, and thus numerical approaches are often used to determine the dynamics resulting from a set of initial conditions. Several directly simulated free responses of a bistable system are shown in Figure 2.5a, where the parameters $m=1$ kg, $k_1=1$ N m^{-1}, $k_3=1$ N m^{-3}, and $d=\gamma$ N s m^{-1} are provided. These parameters lead to a governing equation of the form $\ddot{x}+\gamma\dot{x}-x+\beta x^3=0$. Figure 2.5a reveals the non-uniqueness of the final static configuration of the bistable system for the case in which no "cross-well" dynamics occur (dark solid curve), whereas single (dashed curve) or multiple (light solid curve) cross-well events may occur prior to the mass settling gradually towards one of the equilibria positions surrounding the central unstable configuration. For the parameters selected, the equilibria are located at $x_{2,3}^{*} = \pm\sqrt{1/\beta}$.

It is clear that the non-uniqueness of the final configuration of a bistable system following free vibration requires evaluation of many initial conditions to form more broad conclusions as to general trends amongst the influence of different starting states. These investigations lead to the study of "basins of attraction" in which case the outcomes of a large number of direct simulations, each employing unique initial conditions, are plotted with a shading notation that represents which stable orientation the bistable system settled towards. A representative basin of attraction map, for the system parameters as given above, is shown in Figure 2.5b. The horizontal and vertical axes denote the starting initial displacement x_0 and velocity $\dot{x}_0$, respectively. The dark shading indicates the mass settled into the stable position $x_2^{*} = +\sqrt{1/\beta}$ whereas unshaded regions denote a final position at $x_3^{*} = -\sqrt{1/\beta}$. It is worthwhile to note the relationship between phase portraits of the free response and a basin of attraction map. Figure 2.5b shows an example of a free response phase portrait for an initial condition given by the light circle. The light dashed curve follows precisely within the bounds of the space enclosed by the dark shading and ultimately settles into the equilibrium $x_2^{*} = +\sqrt{1/\beta}$. Consider, then that at any time during the representative time series we sample the instantaneous displacement and velocity of the mass. If we restart our simulation using this displacement and velocity as initial conditions, then the mass would continue along the same trajectory and likewise settle into the same final configuration. Thus, there is a close link between phase portraits and basin of attraction maps to evaluate the free response.

2.4.2 Base-excited Response

The dynamics of a sinusoidally-excited bistable system vary in numerous ways as compared to the free response. Intrawell oscillations (single-well dynamics); interwell vibrations (snap-through, cross-well dynamics); various amplitudes of intra- and interwell response; multiharmonic, aperiodic and chaotic oscillations: a great variety of dynamics may be exhibited by a bistable system undergoing single-frequency harmonic excitation.

Here, we begin the analytical formulation from a more generic representation and proceed through a complete non-dimensionalization procedure which is used throughout the remainder of this book to more rigorously assess the dynamics of excited bistable systems. The governing equation for the relative inertial mass motion $\hat{x}(t)$ of the bistable structure when excited by base acceleration of frequency Ω (rad s^{-1}) and amplitude P_0 (m s^{-2}), $\ddot{z}(t) = -P_0 \cos\Omega t$, is given by

$$m\ddot{\hat{x}} + d\dot{\hat{x}} - k_1\hat{x} + k_3\hat{x}^3 = mP_0 \cos\Omega t \tag{2.44}$$

For the purpose of obtaining greater insight from the analysis, it is beneficial to non-dimensionalize the governing equation. In this way, by using variables and parameters which have no specific time or length scale, more general conclusions may be derived from later investigations. Thus, the following non-dimensional length and time variables are defined in Eq. (2.45), respectively,

$$x = \hat{x}/\hat{x}_c \,;\, \tau = t\omega_n \tag{2.45}$$

Here, $\hat{x}_c$ is a characteristic length of the system. Additional dimensionless parameters are specified:

$$\gamma = d/m\omega_n \,;\, \beta = k_3\hat{x}_c^2/m\omega_n^2 \,;\, \omega = \Omega/\omega_n \,;\, \omega_n = \sqrt{k_1/m} \,;\, p = P_0/\hat{x}_c\omega_n^2 \tag{2.46}$$

Substitution of Eqs. (2.45) and (2.46) into Eq. (2.44) leads to the non-dimensional governing equation for an excited bistable system.

$$x'' + \gamma x' - x + \beta x^3 = p\cos\omega\tau \tag{2.47}$$

The apostrophe operator indicates differentiation with respect to non-dimensional time τ, $\partial/\partial\tau$. Exact analytical solutions to Eq. (2.47) cannot be determined for all combinations of parameters (γ,β,p,ω), and numerical techniques are required to "exactly" predict the precise response. The examples in Figure 1.4 through Figure 1.12 highlighted the large range and various forms of the dynamics of a bistable system that may be exhibited due to single-frequency harmonic excitation. Below, we explore features of harmonically-excited bistable systems that become evident through a more encompassing study of the forced responses.

Like basin of attraction maps for the free response of a bistable system, comparable plots may be realized for sinusoidal excitation. As before, a large range of initial conditions is selected and the time response of the system is directly simulated by numerically integrating Eq. (2.47) using a parameter set (γ,β,p,ω). Unlike maps computed for the free response where the shading indicates a final system equilibrium configuration, the various regions of shading on maps for sinusoidally-excited systems correspond to the unique *steady-state* behaviors that are realized once transients decay. Therefore, there may be more than two shaded regions for a base-excited (or, generally, forced) bistable system, which individually indicate a unique steady-state oscillation, be it snap-through motions, intrawell oscillations, multiharmonic vibrations, aperiodic oscillation, and so forth.

An example parameter set is considered for which only snap-through dynamics and low amplitude intrawell oscillations are uncovered, (γ,β,p,ω)=(0.1, 1, 0.2, 0.8). Figure 2.6 presents a basin of attraction map for a sinusoidally-excited bistable system where the dark shading indicates the steady-state response becomes a low amplitude intrawell oscillation after the decay of the transient vibrations, whereas unshaded regions denote the initial conditions led to interwell, snap-through vibrations. In Figure 2.6a, the basin map computed over a large range of initial conditions is presented showing banded regions of initial states which lead to one of the distinct steady-state oscillation forms that are possible for this parameter set. Figure 2.6b presents a close-up view of the boxed region in Figure 2.6a, revealing that the banded nature of the plot continues to finer scale. Another zoom in Figure 2.6c of the boxed region in Figure 2.6b further exemplifies the multi-scale,

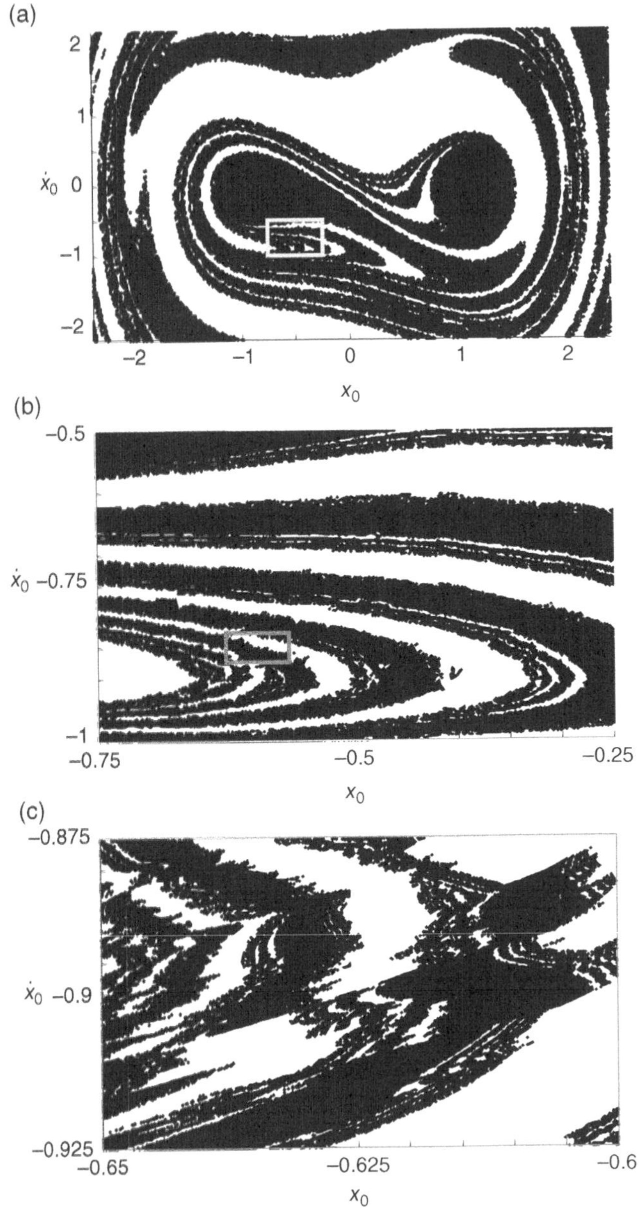

Figure 2.6 Basin of attraction maps for an excited bistable system where $(\gamma, \beta, p, \omega) = (0.1, 1, 0.2, 0.8)$. Plot (b) is a close-up view of (a) indicated by the bounded region in (a). Plot (c) is a close-up view of (b) indicated by the bounded region in (b).

banded character of the basin map structure. Specifically, it is seen that the structure of the basins is *fractal* which indicates a scalable repetition of form, often of highly complex geometry; readers with greater interest in the fractal characteristics are encouraged to consider [20]. In total, the findings give strong evidence of the extreme sensitivity to initial conditions in forced oscillations of bistable systems: even as the initial condition precision is refined within a banded region of the basin of attraction map, still further banded areas are identified.

When the operating parameters $(\gamma, \beta, p, \omega)$ change, the basins likewise alter in terms of the relative percentage of initial conditions leading to the unique steady-state oscillations as well as regarding the number of unique responses which may exist. As described in Section 1.2, single-periodic, multiharmonic, and chaotic vibrations are possible as a consequence of single-frequency excitation of a bistable system, such that the observed outcome is critically dependent on initial conditions and the parameters involved. Therefore, taking advantage of steady-state bistable dynamics for a practical application is reliant on effective design strategies which minimize the possibility that multiple (or specifically, undesired) dynamics coexist for the same operating condition.

There is an additional graphical means to characterize steady-state dynamics that assists in determining the periodicity and structure of a given response. A Poincaré map is generated by sampling the resulting time series at integer increments of the excitation period, $2\pi n/\omega$ where $n=1, 2, \ldots$. At these times, a data point is plotted on the phase portrait plane $(x, \dot{x})$ in terms of the response displacement and velocity evaluated at that instant, Figure 2.7a. When the bistable oscillator undergoes single-periodic dynamics, like ordinary intrawell or interwell vibrations, the Poincaré map is merely a single point which does not change in location $(x, \dot{x})$. For a multiharmonic response of order $1/n$, there will be n points plotted. In contrast, for order n, there will only be one point since multiple response periods occur over a given excitation period although the sampling is conducted only according to the excitation cycle. Interestingly, a unique result is found when generating a Poincaré map when a bistable system undergoes chaotic vibration. Although the system oscillates chaotically, the Poincaré map demonstrates remarkable organization, as shown for one example in Figure 2.7b. The structure of the chaotic Poincaré map is referred to as a strange attractor. This indicates that, unlike an orbital attractor denoting a complete trace of a phase portrait, the strange attractor exists in a sub-sampled state space and is the result of the fractal nature of the system [20]. Although critical assessment of strange attractors is not common in the research fields considered in this book, it is valuable to highlight this unique characteristic of bistable dynamics which is historically a topic of great interest [17,21,22].

2.5 Analytical Methods for Steady-state Dynamics

Apart from undertaking numerous, computationally-expensive *numerical simulations* or costly *experiments* to determine the dynamics of a bistable system resulting from a given set of system and excitation parameters, more insightful and significantly more cost- and time-efficient findings may be obtained by making appropriate assumptions about the anticipated device response and applying an *analytical* treatment to the governing Eq. (2.47) to obtain an approximate solution. From an engineering perspective, the models formulated from an analytical strategy provide substantially valuable tools in the early design-stage

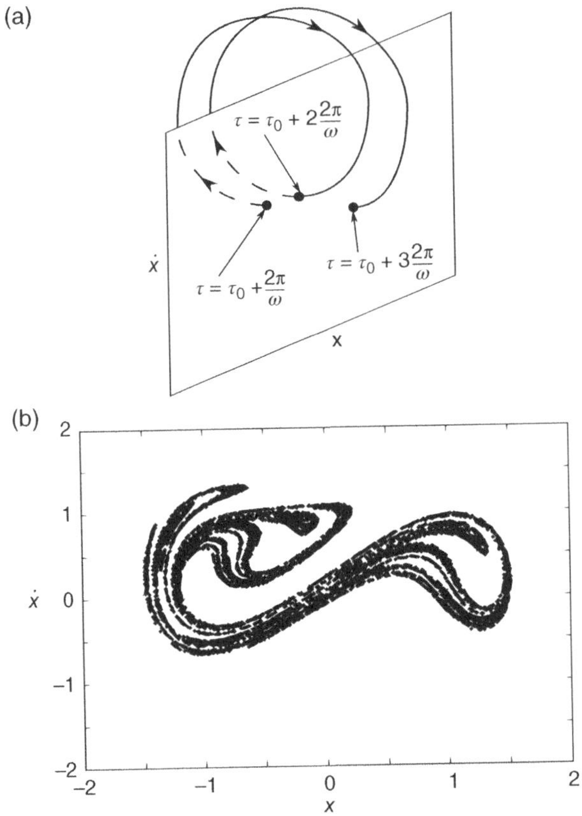

Figure 2.7 (a) Illustration of the point sampling strategy to generate a Poincaré map. (b) Poincaré map for excited bistable system with parameters. $(\gamma, \beta, p, \omega) = (0.1, 1, 0.37, 1.29)$.

evaluation of parameters which optimize performance of a bistable system in operational environments associated with excitation conditions characterized *a priori*. There are effectively two classes of methods available to approximately solve for the steady-state response of an underdamped bistable system governed by Eq. (2.47): perturbation methods and methods that are suitable for predictions far-from-equilibria. The selection of the appropriate method is made by applying an informed *a priori* knowledge regarding the anticipated type of steady-state response. The following sections introduce two approaches that may be applied to approximately determine the small or large amplitude, steady-state dynamics of a forced, bistable system. Readers interested in still further approximate analytical solution strategies are directed to a variety of detailed texts [23–28].

2.5.1 Small Oscillations

The basic approach of perturbation methods is to approximate the system response x using the first few terms of an expansion according to the order of a small parameter ε. This is often a power series expansion such that $x = x_0 + \varepsilon x_1 + \varepsilon^2 x_2 + \cdots$, where the lowest order term in the expansion x_0 is the response of the counterpart linear system. Substitution of this assumed solution into the governing equation then yields a system of equations, each in terms of an order of ε, which are solved sequentially up to the highest order of interest [27]. For the bistable system governing Eq. (2.47), direct application of a perturbation method requires selecting which terms are assumed to be small, and thus of order ε^1. Since the non-dimensional damping, nonlinearity, and excitation level are typically considered the small contributors to the weakly nonlinear dynamics, one such modification to Eq. (2.47) is given by

$$x'' + \varepsilon\gamma x' - x + \varepsilon\beta x^3 = \varepsilon p\cos\omega\tau \tag{2.48}$$

It is apparent that when ε=0, Eq. (2.48) *does not* yield the governing equation of a linear oscillator: $x''+x=0$. Consequently, a coordinate transformation is necessary to estimate the local, intrawell dynamics of bistable systems when employing perturbation methods. Moreover, this indicates perturbation methods are not ordinarily suitable to predict the interwell dynamics of bistable structures since such behaviors involve responses that are inherently far from equilibrium.

The coordinate $y = x \mp |x^*|$ is defined where $|x^*| = \sqrt{1/\beta}$ is the amplitude of the stable equilibrium away from the central unstable configuration. Substitution of y into Eq. (2.47) leads to governing Eq. (2.49) which is sufficient for application of one of many perturbation methods to determine approximate analytical solutions for the small, local oscillations around that equilibrium x^*.

$$y'' + \gamma y' + 2y \pm 3\sqrt{\beta}y^2 + \beta y^3 = p\cos\omega\tau \tag{2.49}$$

For the following, the specific coordinate transformation $y = x - x^*$ is used, thus leading to the governing equation which is scaled by small parameter ε:

$$y'' + \varepsilon^2\gamma y' + 2y + 3\varepsilon\sqrt{\beta}y^2 + \varepsilon^2\beta y^3 = \varepsilon^2 p\cos\omega\tau \tag{2.50}$$

In this case, two orders of ϵ are required due to the combination of quadratic and cubic nonlinearities. Thus an assumed solution requires at least a three term expansion: terms of order ε^0, ε^1, and ε^2.

The perturbation strategy which is illustrated here is the *method of multiple scales* [24], although alternative approaches are available [23,25]. By the method of multiple scales, the response expansion is

$$y(\tau) = y_0(T_0, T_1, T_2) + \varepsilon y_1(T_0, T_1, T_2) + \varepsilon^2 y_2(T_0, T_1, T_2) + \cdots \tag{2.51}$$

where terms $T_0 = \tau$, $T_1 = \varepsilon\tau$, and $T_2 = \varepsilon^2\tau$ denote progressively *slower* non-dimensional times. Thus, using an illustrative metaphor [23], the critical linear response may be envisioned to occur on the fastest time (e.g., second hand on a clock), while the next most important correction to the solution occurs on the next slowest time (minute hand on the clock), and the smallest correction exists on the slowest time scale (hour hand on the clock). Note that by this notation, the operator $\partial/\partial\tau$ requires a chain rule of differentiation such that

$$\frac{\partial y_i}{\partial\tau} = \frac{\partial y_i}{\partial T_0} + \varepsilon\frac{\partial y_i}{\partial T_1} + \varepsilon^2\frac{\partial y_i}{\partial T_2}; \frac{\partial^2 y_i}{\partial\tau^2} = \frac{\partial^2 y_i}{\partial T_0^2} + 2\varepsilon\frac{\partial^2 y_i}{\partial T_0\partial T_1} + \varepsilon^2\left(\frac{\partial^2 y_i}{\partial T_1^2} + 2\frac{\partial^2 y_i}{\partial T_0\partial T_2}\right) + \cdots \tag{2.52}$$

Then, substitution of Eq. (2.51) into (2.50), and eliminating terms of order greater than ε^2 leads to

$$\left[\frac{\partial^2 y_0}{\partial T_0^2}+2y_0\right]+\varepsilon\left[\frac{\partial^2 y_1}{\partial T_0^2}+2y_1+3\sqrt{\beta}y_0^2+2\frac{\partial^2 y_0}{\partial T_0\partial T_1}\right]$$
$$+\varepsilon^2\left[\frac{\partial^2 y_2}{\partial T_0^2}+2y_2+2\frac{\partial^2 y_1}{\partial T_0\partial T_1}+6\sqrt{\beta}y_0y_1+\gamma\frac{\partial y_0}{\partial T_0}+2\frac{\partial^2 y_0}{\partial T_0\partial T_2}\right.$$
$$\left.+\frac{\partial^2 y_0}{\partial T_1^2}+\beta y_0^3-p\cos\omega\tau\right]=0 \tag{2.53}$$

Because of the independence of terms having unique order ε in Eq. (2.53), they must individually be equal to zero. Thus, we equate each coefficient of unique order in ε to zero which yields a system of ordinary differential equations (2.54–2.56) that are solved sequentially, beginning with order ε^0.

$$\frac{\partial^2 y_0}{\partial T_0^2}+2y_0=0 \tag{2.54}$$

$$\frac{\partial^2 y_1}{\partial T_0^2}+2y_1+3\sqrt{\beta}y_0^2+2\frac{\partial^2 y_0}{\partial T_0\partial T_1}=0 \tag{2.55}$$

$$\frac{\partial^2 y_2}{\partial T_0^2}+2y_2+2\frac{\partial^2 y_1}{\partial T_0\partial T_1}+6\sqrt{\beta}y_0y_1+\gamma\frac{\partial y_0}{\partial T_0}+2\frac{\partial^2 y_0}{\partial T_0\partial T_2}$$
$$+\frac{\partial^2 y_0}{\partial T_1^2}+\beta y_0^3-p\cos\omega\tau=0 \tag{2.56}$$

A detuning parameter is introduced, σ, which indicates a shift away from a resonance of the system to be explored. In the following, focus is given to the primary resonance, $\omega=\sqrt{2}+\varepsilon^2\sigma$, although application of this approach to other resonances of order $1/n$ or order n, where $n=1, 2,\ldots$, is just as straightforward to undertake.

Then, the solution to (2.54) is found to be

$$y_0=a_1(T_1,T_2)\sin\sqrt{2}\tau+a_2(T_1,T_2)\cos\sqrt{2}\tau \tag{2.57}$$

where the coefficients $a_{1,2}$ are to determined in subsequent steps. Equation (2.57) is then substituted into (2.55).

$$\frac{\partial^2 y_1}{\partial T_0^2}+2y_1+\frac{3}{2}\sqrt{\beta}\left(a_2^2-a_1^2\right)\cos 2\sqrt{2}\tau+3\sqrt{\beta}a_1a_2\sin 2\sqrt{2}\tau$$
$$+2\sqrt{2}\frac{\partial a_1}{\partial T_1}\cos\sqrt{2}\tau-2\sqrt{2}\frac{\partial a_2}{\partial T_1}\sin\sqrt{2}\tau+\frac{3}{2}\sqrt{\beta}\left(a_1^2+a_2^2\right)=0 \tag{2.58}$$

The terms proportional to $\sin\sqrt{2}\tau$ and $\cos\sqrt{2}\tau$ lead to particular solutions proportional to $y_1\propto\tau\sin\sqrt{2}\tau$ which is unbounded in time. The method of multiple scales then states that such *secular* terms must be eliminated from the solution. To do so, the additional constraints are found to be

$$\frac{\partial a_1}{\partial T_1} = \frac{\partial a_2}{\partial T_1} = 0 \tag{2.59}$$

indicating that the coefficients $a_{1,2}$ are functions only of T_2. Then, the resulting solution to (2.58) leads to

$$y_1 = \frac{1}{4}\sqrt{\beta}\left[\left(a_2^2 - a_1^2\right)\cos 2\sqrt{2}\tau + 2a_1 a_2 \sin 2\sqrt{2}\tau - 3\left(a_1^2 + a_2^2\right)\right] \tag{2.60}$$

The procedure repeats by substituting Eqs (2.57) and (2.60) into (2.56) and eliminating secular terms. At this step, it can be shown that solution to the equation following such substitution is not necessary and only the elimination of the secular terms is required [23,24]. For brevity sake, following several steps of simplifications [13,23], it is found that the steady-state amplitude defined by $A = \sqrt{a_1^2 + a_2^2}$ and phase lag with respect to the excitation ϕ are determined from

$$9\beta^2 A^6 + 12\sqrt{2}\beta\sigma A^4 + 2[\gamma^2 + 4\sigma^2]A^2 = p^2 \tag{2.61}$$

$$\tan\phi = \frac{2\sqrt{2}\sigma + 3\beta A^2}{\sqrt{2}\gamma} \tag{2.62}$$

such that the total solution is given by

$$y = y_0 + \varepsilon y_1 + \cdots = A\sin[\omega\tau - \phi] - \frac{1}{4}\varepsilon\sqrt{\beta}A^2\{3 + \cos[2\omega\tau - 2\phi]\} + \cdots \tag{2.63}$$

Equation (2.61) is a cubic polynomial in terms of A^2 whose roots indicate mathematically feasible solutions, although stability of the resulting amplitudes must be checked. Stability determination will be revisited in the following section. Once the amplitude is determined, the phase ϕ may be computed. Thus, one may then calculate an approximated time series response of the local, intrawell oscillations of the bistable system as given by Eq. (2.63), or plot the steady-state amplitude and phase of the motions. One should recall that the predictions are determined around the primary resonance $\omega = \sqrt{2}$, thereby limiting the validity of the results to the single-periodic responses around this frequency, in addition to the constraint that the system not be far from equilibrium (the stable configuration).

Figure 2.8a,b presents a sample of displacement amplitudes A and phases ϕ, respectively, as functions of the detuning parameter σ and for various levels of excitation p. Here, the remaining parameters are $\gamma = 0.1$ and $\beta = 1$. The bistable system exhibits softening-type of nonlinear responses when undergoing intrawell oscillations around one of the equilibrium positions, thus the peak amplitude is achieved at smaller and smaller values of frequency as excitation amplitude is increased. A comparison of the representative steady-state phase portraits computed via the multiple scales method solution, Eq. (2.63), and those from direct simulations are shown in Figure 2.8c. The predictions by multiple

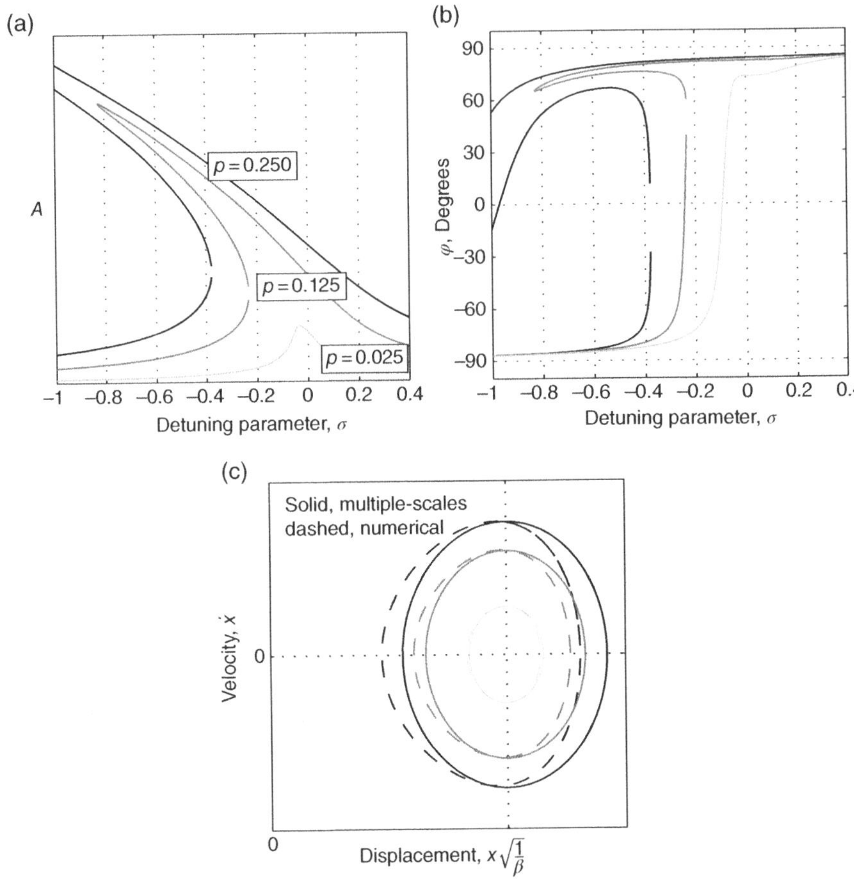

Figure 2.8 (a) Intrawell oscillation displacement amplitude, and (b) phase. (c) Comparison of directly simulated results (dashed curves) against those predicted by the method of multiple scales (solid curves) for $\omega = \sqrt{2}$. In all results $\gamma = 0.1$, $\beta = 1$, and $\varepsilon = 0.1$.

scales (solid curves) progressively become less accurate than the numerically integrated responses (dashed curves) as the excitation level increases. As will be shown throughout this book, approximate analytical treatments of a wide variety of bistable systems tend to more accurately predict *amplitudes* and *phases*, rather than to effectively reconstruct complete time series of the results. Consequently, for a more comprehensive assessment of the steady-state dynamics, it is useful and informative to present analytical results alongside numerical findings.

2.5.2 Large Oscillations

As described in Chapter 1, the large amplitude, interwell, or snap-through motions are a principal bistable dynamic form that is favorably exploited throughout the research fields of interest in this book. Analytical approximations of snap-through dynamics thereby require that the theoretical method is not limited to the assumption that the response represents a small deviation or perturbation around an otherwise linear vibration. The analytical tools available to solve an equation like (2.47) for the excited bistable oscillator must therefore adopt additional guiding assumptions to maintain validity for large ranges of motion, relative to a characteristic length of the system.

One common, primary guiding assumption to leverage such analytical tools is that the system dynamics are periodic and steady-state. As such, based on the observations regarding the response trajectories described in Section 1.2, the useful information obtained is realistically limited to amplitudes and phases rather than accurately reconstructing time series. While a number of techniques exist to predict the full range of periodic responses of strongly nonlinear systems [27], the strategy demonstrated here, and used to great advantage in the analysis of a variety of bistable structures across research disciplines, is the *method of harmonic balance.*

Using this approach and based upon the order of nonlinearity observed in the governing equation, the response of the system is expanded in a Fourier series

$$x(\tau)=c_0(\tau)+\sum_{n=1}^{N} a_n(\tau)\sin n\omega\tau + b_n(\tau)\cos n\omega\tau \tag{2.64}$$

In general, one may employ several terms of the expansion (2.64) but the tractability of the resulting equations is rapidly diminished. Fortunately, it is often the case that a first order expansion yields predictions of adequate accuracy with respect to experimentally observed fundamental dynamic amplitudes, such that the first approximation of the response may be given by

$$x(\tau)=c(\tau)+a(\tau)\sin\omega\tau + b(\tau)\cos\omega\tau \tag{2.65}$$

Equation (2.65) is then substituted into (2.47). The coefficients are assumed to vary slowly in non dimensional time such that c', a', $b' \neq 0$, but c'', a'', $b'' \approx 0$. Since only the fundamental harmonic terms are assumed to result from the single frequency excitation at ω, via (2.65), the higher order harmonics are neglected in the substitution. Then, the coefficients of the constant, sine, and cosine terms are collected together. Since the resulting equation is composed of the harmonic terms whose sum is zero, this indicates the individual coefficient expressions must individually be equal to zero in order to avoid the trivial solution. Thus, one yields three modulation equations according to the coefficient expressions:

$$-\gamma c' = c\left[-1+\beta\left(c^2+\frac{3}{2}r^2\right)\right] \tag{2.66}$$

$$-\gamma a' + 2\omega b' = a\left[-1-\omega^2+\beta\left(3c^2+\frac{3}{4}r^2\right)\right]-\gamma\omega b \tag{2.67}$$

$$-2\omega a' - \gamma b' = \gamma\omega a + b\left[-1-\omega^2+\beta\left(3c^2+\frac{3}{4}r^2\right)\right]-p \tag{2.68}$$

where $r^2 = a^2 + b^2$. The term *modulation* refers to the fact that the slow modulations of variables c, a, and b are governed by these equations (2.66–2.68). Note that Eqs (2.66–2.68) may be expressed using $\boldsymbol{q} = [c, a, b]^{\mathrm{T}}$ such that

$$\mathbf{P}\boldsymbol{q}' = \boldsymbol{G}(\boldsymbol{q}) \tag{2.69}$$

For steady-state oscillations, $\boldsymbol{q}' = 0$, solutions to Eq. (2.66) are

$$c^2 = \begin{cases} \dfrac{1}{\beta} - \dfrac{3}{2} r^2 \\ 0 \end{cases} \tag{2.70}$$

which indicates the response either has a displacement bias ($c^2 \neq 0$ for intrawell oscillations) or has a zero mean (c^2=0 for interwell vibrations). Defining the term

$$\Lambda = -1 - \omega^2 + \beta\left(3c^2 + \frac{3}{4} r^2\right) \tag{2.71}$$

then Eqs. (2.67) and (2.68) are squared and added together to yield

$$\left[\Lambda^2 + \gamma^2 \omega^2\right] r^2 = p^2 \tag{2.72}$$

Equation (2.72) is a cubic polynomial in terms of r^2 which is solved by substitution of one of the selections of c^2 given in (2.70). Then, the remaining coefficients are determined by

$$a = \frac{\gamma\omega}{\Lambda^2 + \gamma^2 \omega^2} p \tag{2.73}$$

$$b = \frac{\Lambda}{\Lambda^2 + \gamma^2 \omega^2} p \tag{2.74}$$

The roots of polynomial (2.72) are easily computed because third order polynomials have closed form expressions for their roots [29]. Some roots r^2 may be complex conjugate pairs or real and negative, which are not meaningful ($r \in \mathbb{R}$ such that $r^2 \in \mathbb{R}$). Of the remaining real roots, the solutions must be checked for stability. The approach used here is to linearize Eq. (2.69) around one of the steady-state solutions $\boldsymbol{q}^*$ and assess the eigenvalues of the Jacobian

$$\mathrm{J} = \left.\frac{\mathrm{d}\left(\mathrm{P}^{-1}\boldsymbol{G}\right)}{\mathrm{d}\boldsymbol{q}}\right|_{\mathbf{q}=\mathbf{q}^*} \tag{2.75}$$

As discussed in Section 2.2, eigenvalues having negative real components are stable solutions whereas positive real components indicate unstable results.

Figure 2.9 plots the predicted results of the normalized displacement amplitude r following the application of the method of harmonic balance to the bistable oscillator equation (2.47). The system parameters considered are included in each respective sub-figure. For Figure 2.9a–c, the nonlinearity strength, β, increases by two orders of magnitude and the functional dependence of the response amplitude r on excitation frequency ω is shown. In Figure 2.9d, the dependence of response amplitude r on the excitation level p is plotted. Results predicted by the analysis are shown as solid or dashed curves whereas results obtained by numerically integrating the governing Eq. (2.47) are plotted as unfilled circles for verification purposes. The numerical results represent the unique steady-state responses observed from the outcomes of 20 simulations conducted at each excitation frequency or level, where the initial conditions are randomized for each run.

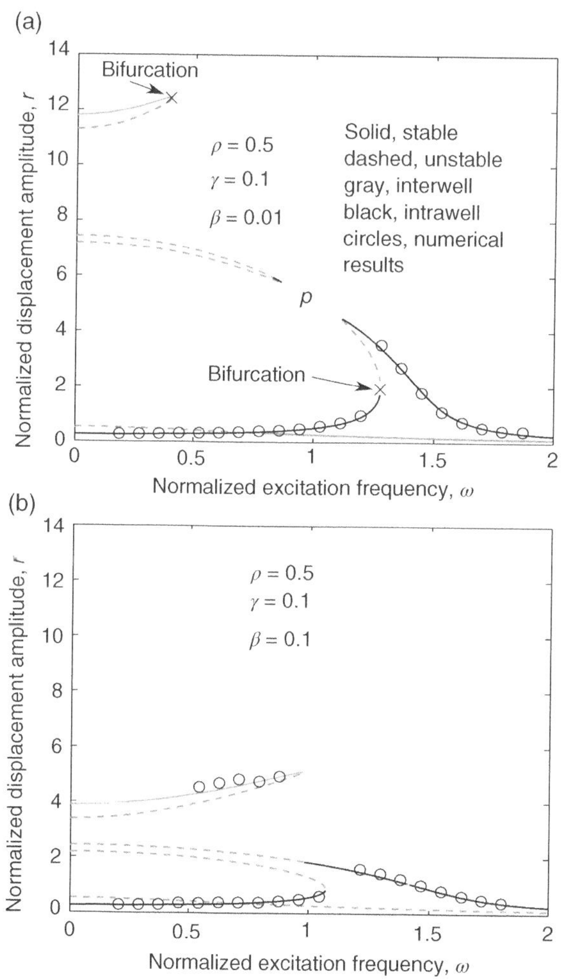

Figure 2.9 Analytical and numerical predictions of steady-state intra- and interwell bistable system dynamics as nonlinearity strength β increases from (a) 0.01 to (b) 0.1 to (c) 1, as a function of normalized excitation frequency ω. (d) Predictions of dependence on excitation level p. Numerical results plotted as unfilled circles. Notable bifurcations denoted by crosses and labels.

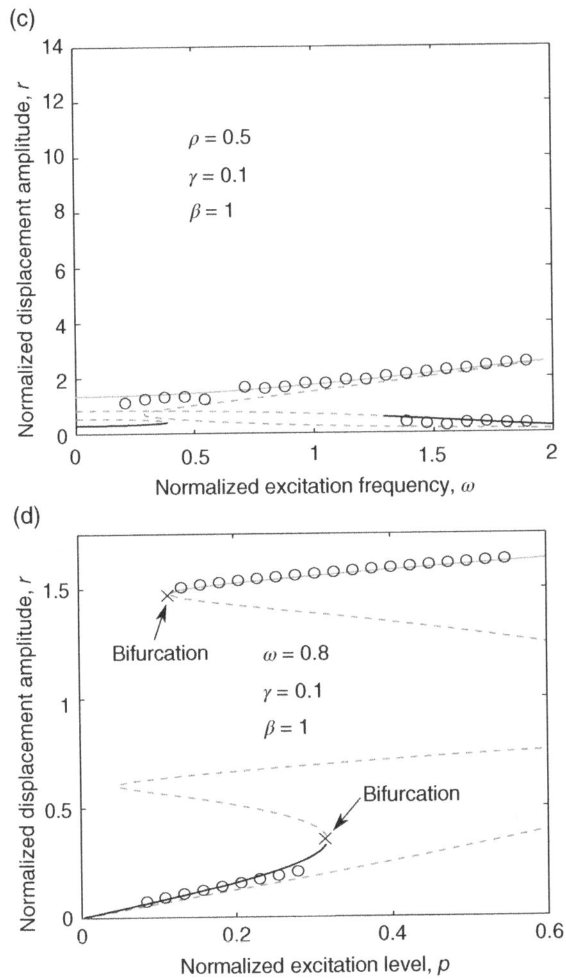

Figure 2.9 (Continued)

In contrast to perturbation techniques, the method of harmonic balance is capable of predicting the steady-state dynamics of a bistable system regardless of whether they are intra- or interwell behaviors. In other words, the harmonic balance method yields the *entire range of dynamic response* via its solution output. Based upon the specific system and excitation parameters, the accuracy of the analytical result varies as compared to the data determined from direct simulation of the governing equation (2.47). For example, interwell responses are predicted for each case of nonlinearity strength β in Figure 2.9a–c, but only across some ranges of frequency are the energetic cross-well behaviors identified from the simulated results. Recalling the assumptions employed in the formulation of the assumed solution (2.65) – namely, a single term Fourier expansion proportional to the driving frequency ω – it is known that higher-order Fourier expansions enable more accurate prediction, particularly with respect to lower frequency bounds on the interwell responses [30].

Nevertheless, this analytical solution strategy provides a convenient pathway to predict the intra- *and* interwell vibrations of bistable systems from the same formulation, which contrasts with perturbation techniques that are generally limited to estimating features of the small, local oscillations. Therefore, the method of harmonic balance serves as a core analytical tool in the investigation of bistable system dynamics considered throughout this book.

2.6 Bifurcations of Bistable Systems

The response predictions from the method of harmonic balance indicate that intra- or interwell responses occur over certain ranges of excitation conditions according to the system parameter set under consideration. For certain parameters and excitations, two predicted response profiles – typically one stable result and one unstable solution – are observed to meet for a given excitation/parameter set. This is observed in Figure 2.9a,d as denoted by the cross marks for two bifurcations that are of the class termed *saddle-nodes* [31–34].

For these bifurcations, due to a small shift in excitation condition, whether by frequency or level, a large change in response amplitude may occur. For example, consider Figure 2.9a where the low amplitude intrawell state will destabilize if the excitation frequency exceeds $\omega \sim 1.25$. For frequencies infinitesimally greater than this critical point, the analytical results only predict the occurrence of the high amplitude intrawell dynamic. Or consider Figure 2.9d, where a bistable system initialized in the high amplitude interwell state will suddenly drop back down in response amplitude to the low amplitude intrawell state if the excitation level reduces below $p \sim 0.12$.

Bifurcations of nonlinear systems are important to investigate and carefully understand because they represent an extreme sensitivity in response amplitude in consequence to infinitesimally small changes in system parameters or excitations. In some contexts, as will be detailed in later parts of this book, bifurcations may be advantageously exploited for parameter tracking purposes, or, conversely, they may represent structural catastrophe if the system is known to be unable to withstand excessive deflection that could be induced by a high amplitude dynamic response.

As a result, analytical investigation of bifurcations is of great utility for researchers and engineers. Whether by perturbation methods that enable the investigation of "jumps" from low amplitude to high amplitude intrawell oscillations (and vice versa in amplitude) or by the method of harmonic balance which may consider all ranges of response types and their transitions, closed-form expressions may be derived for the critical system parameters and excitations that induce bifurcations. The results from such bifurcation analyses are subject to the same guiding assumptions as required in the employment of the analytical tools, and therefore have the same limitations of fidelity and accuracy as those results obtained from the formulations explored in Section 2.5.

To exemplify such a bifurcation analysis [35], consider the use of the harmonic balance method which finally leads to the response polynomial Eq. (2.72) expressed uniquely when assuming intra- or interwell response forms. As a result, we take advantage of the critical case for which the roots of a cubic polynomial transition from a real and complex conjugate pair to three real values. These are the saddle-node form of bifurcations denoted by the cross marks in Figure 2.9a,d. The critical conditions determined from each polynomial will be distinct with respect to the assumption that the initial dynamic was of intra- or interwell form such that the transition thereafter leads to the alternate response type.

Considering a general cubic polynomial equation $\alpha_1 s^3 + \alpha_2 s^2 + \alpha_3 s + \alpha_4 = 0$, the conditions for which two roots s coalesce into one is expressed by [29]

$$\left[\left(2\alpha_2^3 - 9\alpha_1\alpha_2\alpha_3 + 27\alpha_1^2\alpha_4\right)^2 - 4\left(\alpha_2^2 - 3\alpha_1\alpha_3\right)^3\right]^{1/2} = 0 \tag{2.76}$$

Applying the cubic polynomial satisfying interwell vibrations to Eq. (2.76) leads to the following critical conditions, here defined in terms of the square of the excitation level:

$$p^2 = \frac{8}{81}\frac{1}{\beta}\left\{\left(1+\omega^2\right)\left[\left(1+\omega^2\right)^2 + (3\gamma\omega)^2\right] - \left[\left(1+\omega^2\right)^2 - 3(\gamma\omega)^2\right]^{3/2}\right\} \tag{2.77}$$

The excitation level expressed by Eq. (2.77) indicates the minimum level required such that, using given system parameters and sinusoidal excitation frequency, the energetic interwell vibrations are sustained, as in the top-most cross mark in Figure 2.9d for the interwell response.

By this same strategy, we likewise determine the minimum excitation level required to transition from the low amplitude state up to the interwell dynamic, as in the lowest-amplitude cross mark in Figure 2.9d. For this case, the double root expression itself is employed

$$s_d = \left(\alpha_2\alpha_3 - 9\alpha_1\alpha_4\right) / \left[2\left(3\alpha_1\alpha_3 - \alpha_2^2\right)\right] \tag{2.78}$$

Once the intrawell response polynomial coefficients, determined in Eq. (2.76), are substituted into (2.78), it is found that the denominator is always negative for most of the cases of system damping $\gamma \leq 0.75$ considered. Therefore, if the numerator becomes positive, then the roots r^2 would be imaginary. Thus, the second bifurcation is simply determined through solution to $\alpha_2\alpha_3 - 9\alpha_1\alpha_4 = 0$ which yields

$$p^2 = \frac{8}{135}\frac{1}{\beta}\left(2-\omega^2\right)\left[\left(2-\omega^2\right)^2 + (\gamma\omega)^2\right] \tag{2.79}$$

The expressions in Eqs. (2.77) and (2.79) may of course be written in terms of changing damping γ, nonlinearity strength β, or excitation frequency ω, in addition to excitation level p, thus providing a comprehensive means to assess the sensitivity of the critical conditions with respect to changes in each parameter. This example bifurcation analysis is a conventional and representative pathway along which researchers may gain insight into the influence of shifting parameters – perhaps those over which one may have the least control – upon the successful utilization of the preferred bistable dynamic form.

2.7 Multiple Degrees-of-Freedom Systems

When dynamic systems contain multiple, coupled bodies, or alternatively for continuous elastic structures across a wide range of excitation frequencies, distinct *modes* of response may be evident. For lumped-parameter systems, such as the example of the bistable oscillator bearing attached to the surrounding frame and bearing oscillator shown in Figure 1.3b, the unique modal dynamics indicate specific phase and amplitude relationships that occur depending on the excitation frequency. For continuous elastic structures, the modal dynamics refer to specific displacement and phase variations along the structural dimensions.

The modal dynamics of linear multi-DOF systems have been historically studied [8]. Their investigation is considerably assisted through the implementation of modal analysis [2,11] which enables the decomposition of a fully-coupled system of N equations into N decoupled ordinary differential equations (ODEs) by using a strategic coordinate transformation. Thereafter, such ODEs are solved sequentially using elementary procedures. Then, the solutions are transformed back into the original system response coordinates. Similar modal analysis approaches or approximation methods, such as the assumed modes and Galerkin's methods, are often used to discretize a system of partial differential equations for continuous elastic structures into a series of ODEs [36,37].

Upon derivation, it is discovered that the same convenient outcomes do not occur when applying traditional linear system analysis methods in the investigation of the nonlinear response of multi-DOF structures. Fortunately, there are a variety of analytical tools with which to explore modal behaviors of multi-DOF nonlinear systems, or the multi-mode response of nonlinear continuous structures (which will hereafter not be described uniquely with respect to the more general framework of multi-DOF dynamics). These include the concept of nonlinear normal modes [38,39] and normal forms [40]. Moreover, the methods of multiple-scales and harmonic balance are likewise applicable to higher DOF systems. Yet, due to the additional phenomena exhibited by multi-DOF nonlinear systems such as internal resonances, combination resonances, energy localization, and the introduction of new bifurcation types due to the coupled bodies, all such analytical tools have limitations in thoroughly characterizing the full range of observed and relevant dynamics.

Whether by modal formulations or higher-order implementations of alternative strategies described earlier in this book, the analytical study of multi-DOF nonlinear systems is often a challenging trade-off between the complexity of the resulting formulation and the accuracy and fidelity of the predictions. From this perspective, the conclusions drawn from guided numerical studies and careful experimentation may sometimes be more insightful and lead to more timely design-based intuition than results from analytical methods. This is particularly the case when the geometry or composition of the system does not permit the derivation of a compact set of equations that effectively govern the primary dynamic responses of relevance.

As a result, in this work the use of analytical tools in the study of multi-DOF nonlinear systems containing bistable structures is limited to acquiring initial insight complemented with additional numerical and experimental investigations. Due to the success of the approach, higher-order implementations of the method of harmonic balance for multi-DOF structures will be presented in later sections of this book, for instance Sections 3.5 and 4.6. For certain combinations of the strongly nonlinear bistable systems with other dynamic bodies, this strategy is found to provide a viable trade-off between practicality of mathematical derivation and the utility of the insight acquired. For further analytical tools, the interested reader is encouraged to consider a range of other works that explore the intricate dynamics of multi-DOF nonlinear systems in greater detail [1,3,18,24,38–41].

2.8 An Electromechanical Bistable System

Be it for vibration control, energy harvesting, or sensing purposes, the multiphysics of electromechanical system dynamics is critical across many research disciplines. In terms of analytical investigation, the inclusion of an electrical DOF, through the integration of electromechanical transduction and attachment of the output electrical product (i.e., current or voltage) to an external circuitry, introduces a comparable increase in difficulty in the derivation and solution of the theoretical formulation as compared to coupling an additional dynamic mechanical DOF to the primary structure. Indeed, there is perfect analogy between the dynamics of multi-DOF mechanical systems and electromechanical systems [42] once one arrives at the basic set of governing ODEs.

As pertains to the bistable systems considered in this work, the electromechanical transductions of interest are applied in parallel with the restoring and damping elements of the mechanical structures, like the implementation illustrated in Figure 2.10. In this example, the transduction is assumed to be a piezoelectric inclusion such that velocity $\dot{x}$ of the moving mass m induces a flow of current proportional to a coupling coefficient Ψ. The inherent piezoelectric capacitance is C and the load in the circuit is modeled as a generic resistance R, across which is measured the electrical generalized coordinate, the voltage $v(t)$. When the bistable system has a symmetric restoring spring force around the unstable equilibrium position, the coupled electromechanical governing equations may be derived through the combination of Newton's and Kirchhoff's laws.

$$m\ddot{x} + d\dot{x} - k_1 x + k_3 x^3 = mP_0 \cos \Omega t - \Psi v \tag{2.80}$$

$$C\dot{v} + \frac{1}{R} v = \Psi \dot{x} \tag{2.81}$$

Investigating equation systems comparable to (2.80) and (2.81) is a primary aim of many vibration energy harvesting studies. Only minor alterations to the

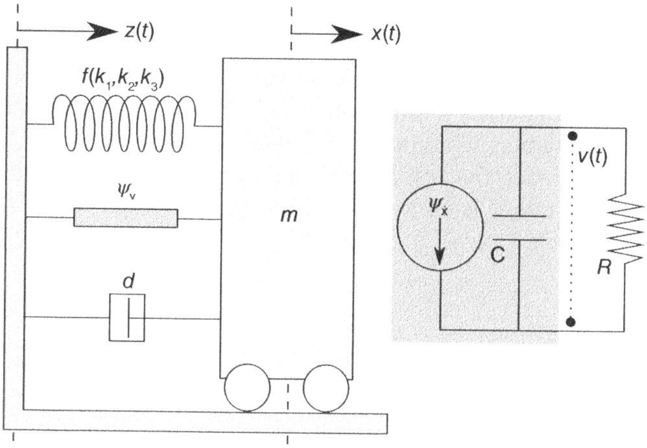

Figure 2.10 Schematic of electromechanical bistable system having piezoelectric transduction in parallel with restoring and damping forces. The corresponding circuit schematic, where the resulting transduced voltage is loaded by a resistance R, is shown on the right.

governing equations are required if considering electromagnetic transduction, as in a magnetic mass moving through the axis of a coil, and indeed the non-dimensional analysis of the systems is identical, see for instance [43–45] and Section 4.3.

The implementation of analytical techniques to predict the response of electromechanical bistable systems generally described by the coupled Eqs. (2.80) and (2.81) follows similar tracks as described in Section 2.5. Since the coupled electrical equation contains only first-order derivatives, it is found that implementation of the method of harmonic balance to this equation system leads only to coefficient equations which may be conveniently substituted back into the mechanical equation which finally yields the response polynomial [44,46]. Thus, while introducing additional steps to the overall model formulation, the additional challenge to obtain a solution is minimal. For clarity, this procedure will be explicitly conducted in Chapter 4 as part of the exploration of a variety of vibration energy harvesting systems. Should the external electric circuitry be of higher order, for example *RLC* type circuits, the difficulty in the analytical formulation is directly comparable to investigating a 2-DOF dynamic system where one of the sub-systems is bistable.

2.9 Summary

To investigate the bistable structures pertaining to the important vibration control, energy harvesting, and sensing contexts in this book, a well-balanced combination of analytical, numerical, and experimental efforts is preferred. This chapter introduced the fundamental components of the primary analytical tools leveraged in the individual case studies and the rationales for supporting such theoretical predictions with valuable simulation and experimental work. Considering the strong nonlinearities that promote far-from-equilibrium dynamics in bistable structures, neither analytical, numerical, nor experimental examinations may suitably stand alone if one seeks comprehensive understanding of the complex system dynamics. The approaches must be combined to form a complementary set of evidence that promotes conclusive discovery of the underlying behaviors. This strategy is carried out throughout the case studies hereafter presented in this book.

References

[1] L.N. Virgin, *Vibration of Axially Loaded Structures* (Cambridge University Press, Cambridge, 2007).

[2] D.J. Inman, *Engineering Vibration* (Prentice Hall, Saddle River, New Jersey, 2001).

[3] G. Kerschen, editor, *Nonlinear Dynamics, Volume 1, Proceedings of the 33rd IMAC, A Conference and Exposition on Structural Dynamics* (Springer, New York, 2015).

[4] L.N. Virgin, *Introduction to Experimental Nonlinear Dynamics* (Cambridge University Press, Cambridge, 2000).

[5] L. Meirovitch, *Methods of Analytical Dynamics* (McGraw-Hill, New York, 1970).

[6] J.N. Reddy, *Energy and Variational Methods in Applied Mechanics* (John Wiley & Sons, Inc., New York, 1984).

[7] L.E. Kinsler, A.R. Frey, A.B. Coppens, J.V. Sanders, *Fundamentals of Acoustics* (John Wiley and Sons, Inc., New York, 2000).

[8] J.P. Den Hartog, *Mechanical Vibrations* (McGraw-Hill, New York, 1947).

[9] L. Meirovitch, *Principles and Techniques of Vibrations* (Prentice Hall, Upper Saddle River, New Jersey, 1996).

[10] K. Ogata, *System Dynamics* (Prentice Hall, Englewood Cliffs, New Jersey, 1992).

[11] S.S. Rao, *Mechanical Vibrations* 4th edn (Pearson Prentice Hall, Upper Saddle River, New Jersey, 2004).

[12] R.W. Clough, J. Penzien, *Dynamics of Structures* (John Wiley and Sons, Inc., New York, 1975).

[13] I. Kovacic, M.J. Brennan, editors, *The Duffing Equation: Nonlinear Oscillators and their Behaviour* 1st edn (John Wiley & Sons, Ltd, Chichester, 2011).

[14] A.H. Nayfeh, B. Balachandran, *Applied Nonlinear Dynamics: Analytical, Computational, and Experimental Methods* (Wiley-VCH Verlag GmbH, Weinheim, 1995).

[15] H.L. Langhaar, *Energy Methods in Applied Mechanics* (John Wiley and Sons, Inc., New York, 1962).

[16] J. Guckenheimer, P.J. Holmes, *Nonlinear Oscillations, Dynamical Systems, and Bifurcations of Vector Fields* (Springer, New York, 1983).

[17] S.H. Strogatz, *Nonlinear Dynamics and Chaos: with Applications to Physics, Biology, Chemistry, and Engineering* (Addison-Wesley, Reading, Massachusetts, 1994).

[18] D. Wagg, S. Neild, *Nonlinear Vibration with Control: For Flexible and Adaptive Structures* (Springer, Dordrecht, 2010).
[19] P. Glendinning, *Stability, Instability and Chaos: an Introduction to the Theory of Nonlinear Differential Equations* (Cambridge University Press, Cambridge, 1994).
[20] F.C. Moon, *Chaotic and Fractal Dynamics : an Introduction for Applied Scientists and Engineers* (John Wiley & Sons, Inc., New York, 1992).
[21] J. Gleick, *Chaos* (Viking, New York, 1987).
[22] E. Lorenz, *The Essence of Chaos* (CRC Press, Florence, Kentucky, 1995).
[23] A.H. Nayfeh, *Introduction to Perturbation Techniques* (John Wiley & Sons, Inc., New York, 1981).
[24] A.H. Nayfeh, D.T. Mook, *Nonlinear Oscillations* (Wiley-VCH Verlag GmbH, Weinheim, 1995).
[25] A.H. Nayfeh, *Perturbation Methods* (Wiley-VCH Verlag GmbH, Weinheim, 1973).
[26] R.H. Rand, D. Armbruster, *Perturbation Methods, Bifurcation Theory, and Computer Algebra* (Springer-Verlag, New York, 1987).
[27] R.E. Mickens, *Truly Nonlinear Oscillations: Harmonic Balance, Parameter Expansions, Iteration, and Averaging Methods* (World Scientific, Hackensack, New Jersey, 2010).
[28] A.W. Bush, *Perturbation Methods for Engineers and Scientists* (CRC Press, Boca Raton, 1992).
[29] R.S. Burington, *Handbook of Mathematical Tables and Formulas* (McGraw-Hill Inc., New York, 1973).
[30] W. Szemplińska-Stupnicka, J. Rudowski, Steady states in the twin-well potential oscillator: computer simulations and approximate analytical studies, *Chaos* **3**, 375–385 (1993).
[31] Y.A. Kuznetsov, *Elements of Applied Bifurcation Theory* (Springer, New York, 1998).
[32] S. Wiggins, *Introduction to Applied Nonlinear Dynamical Systems and Chaos* (Springer, New York, 2003).
[33] E. Benoît, *Dynamics Bifurcations* (Springer-Verlag, Berlin, 1991).
[34] M. Fečkan, *Bifurcation and Chaos in Discontinuous and Continuous Systems* (Springer-Verlag, Berlin, 2011).
[35] R.L. Harne, M. Thota, K.W. Wang, Concise and high-fidelity predictive criteria for maximizing performance and robustness of bistable energy harvesters, *Applied Physics Letters* **102**, 053903 (2013).
[36] L. Meirovitch, *Analytical Methods in Vibration* (Macmillan, New York, 1967).
[37] D.J. Inman, *Vibration with Control* (John Wiley & Sons, Ltd, Chichester, 2006).
[38] S.W. Shaw, C. Pierre, Normal modes for non-linear vibratory systems, *Journal of Sound and Vibration* **164**, 85–124 (1993).
[39] A.F. Vakakis, L.I. Manevitch, Y.V. Mikhlin, *et al.*, *Normal Modes and Localization in Nonlinear Systems* (John Wiley & Sons, Inc., New York, 1996).
[40] A.H. Nayfeh, *Method of Normal Forms* (John Wiley & Sons, New York, 1993).
[41] G. Kerschen, ed., *Nonlinear Dynamics, Volume 2, Proceedings of the 33rd IMAC, A Conference and Exposition on Structural Dynamics* (Springer, New York, 2015).

[42] K.W. Wang, J. Tang, *Adaptive Structural Systems with Piezoelectric Transducer Circuitry* (Springer, New York, 2008).
[43] R.L. Harne, K.W. Wang, Prospects for nonlinear energy harvesting systems designed near the elastic stability limit when driven by colored noise, *ASME Journal of Vibration and Acoustics* **136**, 021009 (2014).
[44] R.L. Harne, K.W. Wang, On the fundamental and superharmonic effects in bistable energy harvesting, *Journal of Intelligent Material Systems and Structures* **25**, 937–950 (2014).
[45] Q. He, M.F. Daqaq, Influence of potential function asymmetries on the performance of nonlinear energy harvesters under white noise, *Journal of Sound and Vibration* **333**, 3479–3489 (2014).
[46] B.P. Mann, D.A.W. Barton, B.A.M. Owens, Uncertainty in performance for linear and nonlinear energy harvesting strategies, *Journal of Intelligent Material Systems and Structures* **23**, 1451–1460 (2012).

3

Vibration Control

This chapter explores the recent advances in the use of bistable structural dynamics for controlling vibrations. The investigations have sought to achieve exceptional damping, isolation, or absorption of the vibrations by selectively exploiting the dynamics of bistable members that are strategically integrated into the engineered structural or material systems. The first part of this chapter highlights the particular dynamical features of bistable structures that the respective vibration control applications have leveraged to advance their aims. Then, a series of sections present detailed case studies which show how the dynamics may be harnessed to effect high and adaptable damping, to prevent the transmission of harmonic excitations, and to absorb the forced vibrations.

3.1 Topic Review

Structural vibrations concern most people on an everyday basis: they may result in human discomfort, system performance degradation, and even catastrophic failure, to name but a few of the possible consequences. The sources and transmissions of the vibrations vary considerably. For example, road–tire–suspension interactions may jolt vehicle passengers, compressors of air conditioning units vibrate machine components and may propagate unwanted sound into the household through duct works, and wind-induced vibration of bridges or tall buildings could induce a disastrous chain effect. To prevent these negative effects, scientists and engineers have devoted enormous attention over the years to developing materials, systems, and strategies that dampen, isolate, or absorb structural vibrations.

Regardless of the specific goal, the vibration control solutions that have been investigated and commercialized may be roughly categorized as *active*, *passive*, or *passive-adaptive* methods. Active approaches typically comprise systems that include controllers, sensors and actuators, and necessary external power sources to provide actions (e.g., forces or moments) upon the vibrating host architecture so as to maintain favorable dynamic behavior [1,2]. Passive solutions attenuate or dissipate oscillations via the design or inherent composition of the system or material itself [3–5]. The advantages of the passive approach are that the

Harnessing Bistable Structural Dynamics: For Vibration Control, Energy Harvesting and Sensing, First Edition. Ryan L. Harne and K.W. Wang.

treatments are usually relatively simple to design and implement and inexpensive to manufacture and put into practice, and the passive solutions will not destabilize the host structure which they are designed to control. However, passive designs have fixed capabilities, and so the performance may not be optimal when the structural characteristics or operating condition changes. On the other hand, while active systems with feedback/feedforward actions are generally more effective in adapting to variations and uncertainties, they have the disadvantages of being costly, complex, high maintenance, requiring relatively high power inputs, and have the potential to become unstable. Comparing the two methods above, one can see the merit of a self-contained *passive-adaptive* approach, sometimes referred to as *semi-active*. In such vibration control solutions, passive parameters or characteristics can be adapted in real-time to improve or optimize performance features, whether the adaptations are triggered by environmental or disturbance changes or based on feedback/feedforward controls. Although the adaptation may also be effected using actuators, the complexity and power requirements are typically much less and the actuations are less likely to destabilize the system as compared to purely active treatments. For example, the adaptive solutions may be as simple as deliberately exploiting the temperature dependence of material damping properties [6], externally changing a flow orifice area to tune the energy dissipation characteristics in a structure–fluid integrated system [7], or adjusting automotive suspension properties using electric-field-dependent magneto-rheological devices [8]. In this book, the major focus is on passive and passive-adaptive approaches.

In certain vibration control applications, it has been recognized that the use of linear material and structural features may pose limitations to the effectiveness of the resulting solution. Consequently, researchers have actively investigated the exploitation of nonlinear material properties and structural features that might improve vibration damping and attenuation by leaps and bounds [9–12]. Deliberately utilizing the dynamics of bistable systems has become a particularly engaging idea in the recent research on leveraging nonlinearities for the purposes of vibration control. It has motivated investigators worldwide who represent scientific disciplines that span materials science, mechanics, physics, and dynamical systems, with a notable inspiration from biology found in the mix. The capacity to switch between two stable states is the essential feature of bistable systems that has become a foundation for such interest. Unique, energetic properties are harnessed by switching from one stable state to the other across the negative stiffness region, and the selective activation of low amplitude oscillations or snap-through behaviors enables an adaptability of dynamical properties.

By utilizing the built-in energizing feature of bistability – in other words, negative stiffness – researchers have uncovered a host of new opportunities to cultivate towards realizing advanced passive and passive-adaptive vibration control solutions. The following sections review the recent research efforts that have taken advantage of the characteristics of bistable structures or material systems to empower exceptional vibration damping, isolation, or absorption properties, which lay a groundwork for safer, more versatile, and higher performance engineering systems.

3.1.1 Damping

Using the dynamics of bistable structures to realize large and adaptive damping characteristics has followed two tracks of implementation: *materials* and *structures*. The first approach employs composite materials with embedded negative stiffness inclusions in a host material matrix which has positive stiffness. In contrast to the lumped, one-dimensional examples of bistability described earlier, a material characterized as having negative stiffness may transition from one stable volume to another volume while the temperature remains constant. How aggregating materials with negative and positive stiffnesses is capable of achieving high damping, in addition to maintaining material stability, can be concisely explained using a fundamental understanding of mechanics. Using one-dimensional counterparts as an example, consider stiffnesses (springs) arranged together with respect to a displacement constraint. By combining a negative stiffness with a positive stiffness of equal magnitude, the net material has vanishing composite stiffness. If the material includes inherent damping, an enormous growth in apparent energy dissipation is effected due to the vanishing magnitude of the dynamic bulk modulus [13].

The intriguing phenomenon is analogous to one outcome of structural members loaded at the elastic stability limit. This is the critical loading condition such that the magnitude of the linear stiffness of the system asymptotically approaches zero due to a balance of positive and negative stiffness effects [14], and the damping ratio grows explosively [15]. But the analogy is not exact because the damping ratio of structures loaded at the elastic stability limit is a function of inertia, and mass does not contribute to the theoretical derivations which demonstrate that extremely large damping may be achieved using negative stiffness inclusions within a positive stiffness matrix [16]. "Stabilizing" the negative stiffness phase of the composite material therefore allows one to exploit the harmonically excited snap-through behaviors of the inclusions as they greatly deform the surrounding material matrix, which leads to large energy dissipation characteristics. For readers unsatisfied by these simplified descriptions, Drugan [17] presents a rigorous proof to verify the potential for stable, two-phase matter containing a negative stiffness phase.

Lakes and colleagues have been at the forefront of the efforts to theoretically and experimentally explore the opportunities afforded via harnessing bistable, negative stiffness inclusions for material systems [18–22]. One of the experimental realizations of extreme damping composite materials that has been investigated is created by embedding vanadium dioxide (VO_2) ferroelastic inclusions in a surrounding tin (Sn) matrix. The VO_2 inclusions undergo a phase transition at 67 °C, which entails a sudden volumetric change by the passage of temperature through the critical value. Hence, the inclusions effect a negative stiffness in the vicinity of the transition temperature. The phenomenological outcome of the negative stiffness-induced volumetric change can be conceived as a three-dimensional snap-through behavior. The tests conducted to quantify the elastic and damping properties of the composite use a torsional excitation upon one end of a material sample at frequencies sufficiently below resonance while measurements of the sample's torsional response are acquired from the opposite end as the excitation [22]. When the temperature of the experimental set-up is gradually cooled from a value greater than the transition temperature, the inclusions undergo a phase transition near the critical temperature. Thus, in the time period which transpires during the phase transition, the VO_2 inclusions effect a negative stiffness which is of a magnitude that is capable of being stabilized by the positive stiffness of the host Sn matrix [19]. By straightforward relationships, the instantaneous modulus and loss tangent of the material are quantified from the measurements [22].

Using terms employed earlier in regards to bistable structural systems, the harmonic torsional excitations upon the cylindrical material sample induce repeated snap-through behaviors of the negative stiffness inclusions at the critical phase transition temperature. It is found that the material energy dissipation characteristics significantly increase due to the local, energetic volumetric deformations induced by the inclusions and balance between negative and positive stiffnesses. Due to the reliance on the phase transition in order to effect the unique damping properties, one challenge to extend this concept to applications for practical composite materials is to broaden the temperature range across which the phase transition features occur [22]. On the other hand, the temperature tuning represents an adaptation mechanism that could be exploited to selectively trigger the high damping. In total, these investigations provided an early and thorough elucidation of the opportunities that can be harnessed by exploiting snap-through dynamics in material systems for unusually large energy dissipation characteristics.

Other research efforts have been undertaken to explore such possibilities for future composite materials. Klatt and Haberman [23] have developed a multiscale material model that spans from microscale negative and positive stiffness constituents up to a macroscale homogenization procedure to quantify the effective properties which are invested in the composite. The method uses a finite element analysis at all scales, and in fact employs mechanical bistable and linear spring analogues in the microscale framework. Nevertheless, the resulting model composition can be employed to predict the multiscale characteristics of materials composed of bistable and monostable constituents, which can be compared to the continuum domain investigations undertaken by Lakes and colleagues. A recent study by Matthews *et al.* [24] has taken advantage of the proposed multiscale model to investigate design strategies based upon Bayesian network classifiers that identify material compositions which would lead to the homogenized mechanical properties of interest.

Due to the non-intuitive nature of effecting extreme material damping via instability phenomena, there have been several rigorous efforts by Kochmann and Drugan [25] and Kochmann [26] devoted to the investigation of the overall material stability and properties of two-phase matter consisting of negative stiffness inclusions. Such studies are justifiably motivated because the previous proofs adopted certain assumptions in their undertaking, including assumed geometric relations between the negative and positive stiffness material phases, like negative stiffness spheres encapsulated by a positive stiffness material. Thus, the continued efforts have shown that the achievement of extreme damping is indeed a generalized result without regard to the geometric shapes of the constituents; nevertheless, it is found that certain volumetric ratios are critical [27,28]. Although the vanishing dynamic bulk modulus is the cause of the sudden, large apparent damping at the point of phase transition, Kochmann and colleagues [27] have shown that the frequency at which the greatest damping occurs is considerably reduced. On the whole, the developments by the several research teams have thoroughly demonstrated the opportunity of realizing unusually high damping in composite materials via harnessing the dynamics of bistable, negative stiffness inclusions.

The other approach to exploit bistable system dynamics as a means for large damping properties has focused on discrete and structural bistable members. Interestingly, some of the studies undertaken in this spirit have gathered inspiration from the phenomena exhibited by phase transition materials that have a history of being modeled as chains of bistable links [29,30]. In these established investigations, the bistable springs are *bilinear* and thus have discontinuities in reaction force more closely representative of phase change properties than the continuous, smooth polynomial relationship which describes the traditional bistable spring restoring force. Figure 3.1 illustrates the differences between

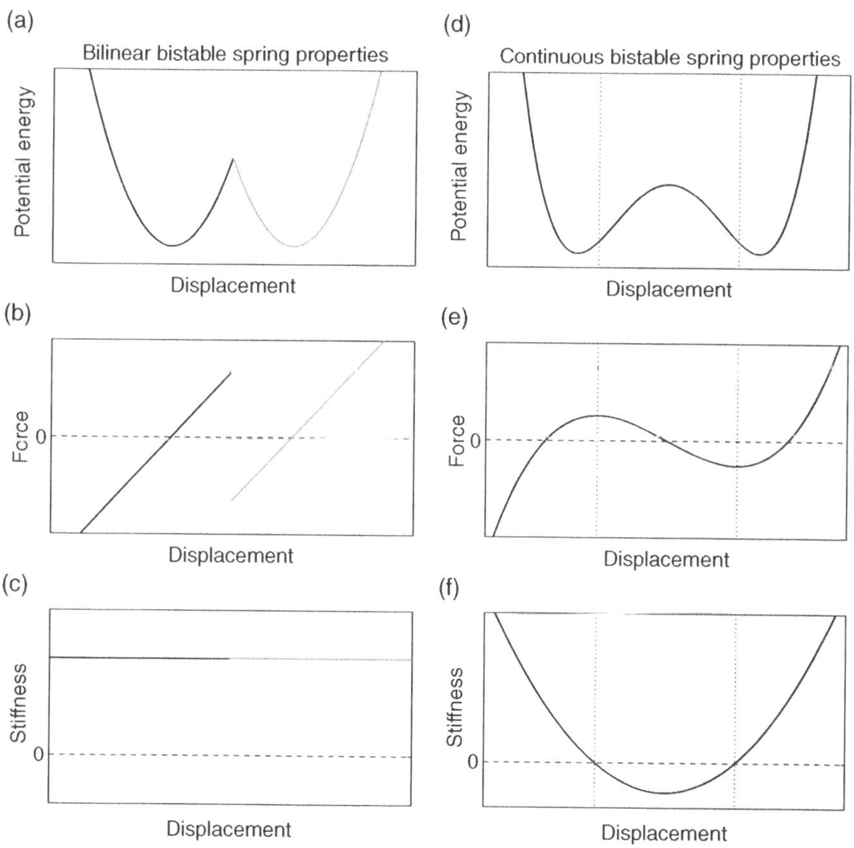

Figure 3.1 (a–c) Bilinear bistable spring properties where the force–displacement relationship is described by a discontinuous pair of linear functions. (d–f) Bistable spring properties where the force–displacement relationship is defined by a smooth cubic polynomial.

bilinear and continuous bistable spring properties. Although both spring types exhibit the characteristic double-well potential energy profile, only the continuous bistable spring force will effect a negative stiffness feature for certain ranges of displacement. This distinction is most prominent due to harmonic excitation of a bistable spring–mass system because the softening stiffness features of the continuous bistable spring will allow the system to snap-through for lower excitation levels than for the bistable, bilinear spring.

Following the studies on phase transition materials, researchers have identified opportunities for a unique energy dissipation strategy resulting from the large hysteretic properties of periodically-excited bistable, bilinear link chains that were identified by Puglisi and Truskinovsky [31]. Thus, the recent efforts have proposed using successively activated snap-through dynamics of mechanical bistable link chains, which are acted upon by displacement or force control at an end, to dissipate large energy via damper elements arranged in parallel with the bistable spring links. The efforts have considered impulsive and harmonic excitations and find that the bistable link chains are able to absorb and dissipate much greater energy than chains which lack such "waiting links". The findings are detailed in works by Cherkaev and Zhornitskaya [32], Cherkaev *et al.* [33], and Slepyan *et al.* [34]. Additionally, intriguing wave propagation features are observed in the chains due to impulsive excitations [33] which indicate that the dissipation features include local oscillations activated by snap-through dynamics as well as cascade-type phenomena that diffuse the energy along the chain.

Cohen and Givli [35] have examined important structural characteristics involved in using bistable link chains as shock absorbing mechanisms and have observed how tailoring the damping ratio influences the maximum extension and internal tension the chain is subjected to due to impulse events. These are meaningful design-based insights to reflect upon for those who wish to effectively employ such structures as energy dissipation devices because many applications have peak stroke limitations and also fatigue or failure concerns that must be considered. From these investigations, researchers have identified novel means to exploit series configurations of bistable structural members that enable large dissipation properties for a variety of excitation types.

Snap-through may also be exploited by integrating individual bistable devices into a vibrating host system to effect large damping characteristics. In a straightforward implementation of the concept, Dong and Lakes [36,37] have proposed multi-column modules that serve as structural support members; alongside stiff columns, the modules include weaker constituents (for a given axial compression) that are held near the threshold of buckling. Thus, as the modules are compressed both statically and harmonically, the weaker constituents undergo dynamic snap-through, dissipating a large energy in the process. The damping features are quantified using the hysteretic area enclosed in the experimentally measured load–displacement profiles. It is found that when the weaker columns additionally undergo an end tilt buckling – where the flat ends of the columns lose flush contact with the frame surface and thereafter touch the frame only by the column edges – the damping properties of the module system are enhanced [37]. As a consequence of the high strength columns which are in parallel with the weaker members, the modules concurrently provide a high static stiffness.

Thus, by an effective design and knowledge of the loading conditions, both high stiffness and damping may be realized.

Gandhi and colleagues [38–41] have investigated several design strategies to exploit discrete, bistable compressed spring-bar mechanisms – the von Mises truss – to provide simultaneous high stiffness and damping capacity. The studies include a variety of numerical and experimental works that investigate the energy dissipation characteristics of bistable device concepts in response to impulsive and harmonic excitations, and explore how integrating the devices into greater systems may improve the damping capacities. The importance of an appropriate level of damping was detailed for its influence in tailoring the hysteretic force–displacement profiles measured under harmonic excitations [38,42]. Because the beneficial energetic and dissipative responses characteristic of snap-through are able to be harnessed only when the bistable components are acted upon by forces, as compared to displacement control, it was stressed that an ideal design must employ bistable constituents in a force-controlled manner to maximize damping properties [40].

3.1.2 Isolation

In addition to exploiting negative stiffness to induce energetic motions, there is a marked engagement of research that aims to exploit this unique spring force characteristic for different purposes. Across many applications, there is extensive interest to isolate systems from the excitations which may lead to their vibration. The achievement of this goal depends upon the energy transmission path in the application: for example, isolating a floor from the vibrations caused by reciprocating machinery, or isolating a sensitive microscope from the motions of the enclosing building. In the simplest realization, a suspension system represents the restoring force members (springs), often arranged together with dissipative elements, which support a mass and attempt to eliminate vibrations across the transmission path. According to basic vibration theory, best isolation is achieved by tuning the suspension system (with respect to suspended mass) so as to have a vanishingly small natural frequency; under this condition, one may take advantage of the vibration attenuating "roll-off" feature of a mass–spring system excited far above resonance [43]. On the other hand, a suspension must maintain sufficient static stiffness to support high loads. Because the suspended weight will statically deflect the isolator, the suspension must be able to prevent severe displacement biases that could result in the system "bottoming out". Bistable members have been found to open up a breadth of opportunities to successfully achieve these mechanical properties for dramatic improvements in vibration isolation performance [44].

In one implementation, buckled, bistable components are used on their own as the suspending members. As described in Section 3.1.1, the process of loading a structure near to the elastic stability limit reduces the magnitude of the linear stiffness to a vanishing value; consequently, the natural frequency of the system approaches zero. While this result is favorable to minimize the resonance frequency of an isolation system, it does not provide for means to support high static load since undesirably large suspension displacements might occur. However, loading the system *beyond* the elastic stability limit to make the structure bistable can alleviate this concern. Structural modeling shows that while the system achieves very low natural frequency due to its small magnitude of linear stiffness when slightly post-buckled, the maximum static displacement that is traveled after carrying the supported mass will also reduce since the nonlinear restoring forces in the post-buckled configuration are greater than before buckling for the same magnitude of compressive force from the buckling critical load [45].

Therefore, researchers have investigated vibration isolation systems constructed using buckled, bistable components as the supporting spring components and have explored the small and large amplitude dynamics of the suspension systems. To demonstrate the effective support of large loads, Winterflood, Blair, and colleagues have investigated the use of multiple buckled beams in parallel [46–48]. The resulting experimental platforms are found to provide significant vibration isolation performance when excited at frequencies above their exceptionally low resonances. Contemporary studies by Virgin, Plaut, and collaborators have examined different bistable isolator designs and modifications thereof – for example, a Winkler-type foundation connecting two buckled beams – to assess how the isolation performance of such conceptions varies as well as how the structures could become vulnerable to adverse dynamic behaviors [49–51]. Because the static stiffness of a buckled strut isolator changes with the applied "dead" load, the dynamic properties of oscillations about the statically-deflected position will likewise be influenced. Thus, the vibration transmissibility of such isolators excited by single- and two-frequency ground motions was characterized to determine the several influences of the dead load upon the performance of the suspension platforms [50].

An early exploitation of bistability for vibration isolation was developed and commercialized by D.L. Platus of Minus K® Technology Inc. [52]. As first reported in 1991 [53], the isolator employed bistable, negative stiffness springs in parallel with linear, positive stiffness springs. While the linear springs carry the static load, the cancellation of the magnitude of the linear stiffness leads to extremely low natural frequencies according to the supported mass, such that the isolator resonant frequencies can be around 1 Hz. Hence, the systems can provide dramatic vibration isolation performance whether from seismic activity, structural motions, or vibrations excited by nearby systems. Numerous designs have been investigated using this isolation system architecture to achieve exceptional reduction in vertical as well as horizontal motions [54]. The effectiveness of exploiting bistable components for vibration isolation in this way is attested by more than two decades of commercial success of Minus K® in providing high performance isolation systems to a large number of university and government laboratories [52].

Adopting the ideas of this early effort, researchers have investigated numerous isolator designs which exploit bistable spring mechanisms in parallel with positive stiffness components, including Carrella *et al.* [55], Shaw *et al.* [56], and Zhou and Liu [57]. One of the intriguing factors of such designs is the variety of nonlinear dynamical behaviors which may be excited. These include large amplitude, single-periodic motions, multiharmonic behaviors, or aperiodic responses, all of which are responses which would deteriorate suspension performance. These dynamical behaviors have been examined from several perspectives by Carrella *et al.* [58,59], Yang *et al.* [60], and Huang *et al.* [61].

The phenomena may become exacerbated when using bistable spring components that are significantly buckled; thus, when employing continuous structural components, like buckled beams or bistable composite plates, higher-order post-buckled modal behaviors may be triggered if the systems are not suitably employed [62,63]. As a result, a general conclusion is that the bistable, negative stiffness members of the vibration isolation system must undergo only low excitation levels with respect to the sensitivity of activating snap-through behaviors. It is clear that the bistable dynamics that are deliberately exploited to effect large damping in structural and material systems must on the whole be carefully avoided by intelligent design and deployment when using bistable members for high performance vibration isolation.

There are means available to inhibit the onset of adverse dynamical phenomena in isolation systems that integrate negative and positive stiffnesses. An increase in the level of damping in parallel with the spring components can suppress the nonlinear behaviors, although one consequence is that the isolation performance will be degraded. Seepersad, Haberman, Wilson, and collaborators have investigated the construction of bistable and linear spring isolation platforms using three-dimensional (3D) printing techniques established in the rapid prototyping field [64,65]. The polymer devices that result from this fabrication process are significantly damped as compared to the alloy or composite materials employed in the systems described previously. Consequently, the potential to activate the steady-state bifurcations leading to the unwanted behaviors is reduced for the isolation systems composed of 3D printed materials, since such phenomena are dependent upon the energy dissipation levels [66].

In addition to increased damping levels in suspension system design, more complex architectures are capable of sufficiently alleviating concern of undesired nonlinear dynamics. For example, Sarlis *et al.* [67] and Pasala *et al.* [68] have developed large-scale seismic protection platforms that integrate pre-compressed springs for the negative stiffness members alongside other damper and spring components. Measurements of the isolator reaction force as a function of displacement are particularly unique when compared to a traditional bistable spring restoring force profile; this is due to several force discontinuities which occur as a result of components in the system that become active based upon the magnitude of the sway motion. The conclusions drawn from the experimentally measured and numerically predicted performance of these negative stiffness seismic protection platforms are comparable to the efforts conducted using bistable beams and plates: the performance enhancement realized by the combination of negative and positive stiffness springs is generally more pronounced when the excitation levels are small with respect to the onset of large amplitude motions in the bistable spring members [67,68].

3.1.3 Absorption

While the use of dynamics induced via the negative stiffness of bistable members differs between those interests that pursue high damping or exceptional vibration isolation, the conceptual implementations are comparable since the focus is primarily upon the spring and damper elements – that is, the (visco)*elastic* components. On the other hand, the study and application of vibration absorption strategies requires consideration of the dynamic moving mass since the investigations are inherently multi-body problems. The exploitation of bistability for these vibration control aims therefore adopts different dynamic perspectives.

Frahm's historic anti-roll tank [69] and mechanical [70] vibration absorber patents set the stage for over a century's worth of investigations to develop dynamic attachments which may completely eliminate the harmonically-excited vibrations of structures to which they are attached. For those who read the patents and marvel at the figures, the works indicate an exceptional understanding of the mechanics and dynamics involved. Yet, no equations are used; only a richly described intuition supports the claims. Den Hartog and Ormondroyd [71] are the often-cited originators of the theoretical underpinnings of the vibration absorber: the vibrations of the excited, host system are suppressed due to a cancellation of forces between the driving motions and those induced by the dynamic vibration absorber.

Thus, the reactive attenuation of vibrations in this manner necessitates that one considers the dynamically coupled system. At a minimum, the design and study of vibration absorbers is a two DOF problem. As has been well-established, the effective employment of nonlinear vibration absorbers is more intricate than merely harnessing or avoiding the unique nonlinear phenomena exhibited by the absorber when examined on its own. The nonlinear dynamics of a two DOF system are rich with features not evident with a single nonlinear oscillator [72]. A significant collection of academic and applied research has explored the implementation of nonlinear vibration absorbers [10,73–75]. In contrast, the use of bistable absorber concepts is a comparably recent idea that has been stimulated by the possibility to exploit the broadband and energetic snap-through behaviors for an enhancement of the vibration absorption characteristics.

As an early precursor to the more recent investigations, Buzhinskii [76] examined a bistable membrane and mass system that was attached to a vibrating host structure to ideally suppress the motions of the host. A model of this coupled system was formulated that reduced the spring force of the buckled membrane to a force-activated switch dependent upon the displacement of the attachment mass from one stable position to the inverted configuration. Both the predicted and measured responses indicated that the bistable absorber could significantly attenuate the resonant vibrations of the host structure for a large range of excitation amplitudes. On the other hand, the results showed that large amplitude single-periodic or multiharmonic dynamics could be triggered at a certain threshold of excitation level. It was also found that a design of the bistable membrane which reduced the stroke of the attachment mass from one stable position to the next would activate adverse large amplitude dynamics for a smaller level of excitation. Unfortunately, the work did not present experimental or numerical results of the time-dependent responses of the system to identify conclusively what dynamics of the bistable membrane absorber were the favorable mechanisms for vibration attenuation [76]; recall from Section 1.2.1 that there are four different, steady-state, single-periodic responses that a bistable oscillator may exhibit due to single-frequency excitations. Yet, all in all, this early study promoted an idea that later researchers have begun to explore in earnest.

Several teams have investigated the dynamics of a host linear structure with a bistable attachment intended for reactive attenuation of the host's vibrations, whether excited externally or due to initial impulses. Among them, the use of nonlinear normal mode analysis has been a recurring tool to probe the dynamical regimes which the two DOF system may exhibit. Using this analytical method, Gendelman and Lamarque [77] identified zones of response which were presented as functions of the attachment nonlinearity and the linear coupling between the two bodies. These zones included a parameter region of bistable attachment design that could provide smooth damping features due to a gradual transition from snap-through dynamics to small amplitude intrawell oscillations. Another region was one characterized by an efficient "pumping" of the energy due to snap-through effects that were activated continually until the energy of the host structure was sufficiently decayed; after this period, the snap-through behaviors of the attachment suddenly transitioned to the intrawell responses. Lastly, a design parameter region was shown to induce transient chaos during which period the bistable attachment intermittently transitioned from intra- to interwell behaviors; this was the least efficient of the dynamic regimes [77]. Another investigation by Gourdon and Lamarque [78] concludes differently, however. By using Melnikov's method and numerical simulations, the researchers concluded that transient chaos provides "very good efficiency" in the pumping of energy from the host structure to the bistable attachment.

Avramov and Mikhlin have employed nonlinear normal mode analysis and show that there are, indeed, effective operating conditions and design parameters for which a bistable snap-through attachment provides appreciable attenuation of the host structure [79,80]. On the other hand, the continued efforts recognized that for a large range of realistic design parameters, a significant proportion of the potential dynamics which are suitable for absorbing harmonically-excited vibrations are multiharmonic in nature, in other words they are not periodic behaviors occurring at the same frequency as the excitation [81]. Moreover, in general the dynamics may be slowly modulating, whereby the kinetic energies repeatedly transfer between the host and bistable systems without reaching a meaningful "steady state". From these results one may ask what is the benefit of a vibration absorber if the host structure continually modulates between a low and high energy dynamic state? One clear conclusion from the research was that more investigation was required to decisively determine the benefit of the multiharmonic dynamics as a vibration control tool [81].

Using several analytical strategies, including nonlinear normal modes, Manevitch and colleagues have concurred that multiharmonic behaviors of the two DOF system may exist that effect a strong exchange of energy from the host structure to the bistable attachment [82]. The study examined the critical threshold of impulsive excitation level which activates the distinct dynamical operating regimes resulting in either weak or efficient transfer of the vibrational energies. A series of numerical studies by Romeo *et al.* [83] verified the analytical findings. The numerical results also identified several dynamical behaviors of the impulsively excited two DOF system which depend upon the level of the sudden input: low input energy levels activate intermittent snap-through dynamics and hence energy damping, moderate energy levels induce transient modulation of the energies between the two bodies, and high impulsive energy levels may activate either single-periodic or multiharmonic snap-through motions where the prior was stated to be efficient [83].

In summary, a variety of conclusions have been reached as to the effectiveness of implementing a bistable vibration absorber, particularly as relate to the benefit of activating (or avoiding) the snap-through dynamics. At the same time, one must recognize that only the earliest work included experimental efforts, albeit briefly reported [76]. Due to the focus on and employment of only a few analytical tools with numerical simulation examples as supporting evidence, the more recent efforts do not yet appear to provide a decisive and fully coherent elucidation of the practical strategies one should employ towards the implementation of a bistable structure as a vibration absorption device.

3.1.4 Summary

From the reviews above, it can be concluded that researchers have shown that exploiting the dynamics of bistable structures could open up new possibilities for the effective damping, isolation, and absorption of vibrations. The strategic utilization of negative stiffness is a common theme amongst the research areas, whether it is leveraged to activate large amplitude snap-through motions or as a means to suspend structures or isolate vibrations. To more specifically elucidate the rationales for employing such dynamic behaviors in vibration control applications, the following sections present recent research case studies on the investigation and harnessing of the dynamics of bistable structures to dampen (Section 3.2), isolate (Sections 3.3, 3.4, and 3.5), or reactively attenuate vibrations (Section 3.6).

3.2 High and Adaptable Damping Using Bistable Snap-through Dynamics[1]

Structural and material systems can be considerably enhanced by a passive adaptability of damping properties, such that they can switch from very low to very high levels in response to certain operating conditions. In addition to the rotorcraft example described in Section 1.3.1, another strong candidate where these properties are favorable is a microphone. The audible, dynamic acoustic pressures that many people are exposed to on a daily basis span about six orders of magnitude [84], indicating that a vast range of transducer motions are activated through fluid–structure interactions. A high-quality microphone must accurately quantify this enormous dynamic range. Thus, for a microphone, low damping is advantageous at whisper sounds levels to provide sensitive transduction, while greater damping is required to prevent excessive and damaging

1 Material from this section has been reprinted by permission of IOP Publishing, from D.R. Johnson, M. Thota, F. Semperlotti and K.W. Wang, On achieving high and adaptable damping via a bistable oscillator. Smart Materials and Structures 22:115027, 2013. doi: 10.1088/0964-1726/22/11/115027.

vibrations of the capacitive plate when sound levels are high. Many researchers have identified the large dissipation characteristics of structures and materials containing bistable, negative stiffness constituents [19,26,40]. The unusually high losses are recognized to be more readily induced as a result of greater levels of excitation. This excitation level influence upon the damping properties closely matches the adaptable damping profile that could benefit many structural and material systems.

On the other hand, it is well-known that bistable oscillators may undergo a rich assortment of dynamic phenomena that are not necessarily periodic with the same frequency as the harmonic excitation source [85,86]. Thus, to effectively employ an adaptable damping feature in a structure by the integration of bistable constituents, one must closely investigate the nature of the snap-through motions. The following study explores these dynamic characteristics by considering a conventional, lumped-parameter bistable device, which is both a simple mechanical construction as well as representative of many other bistable systems due to the common dynamics they share (see Section 1.2).

3.2.1 Model Formulation of the Bistable Device

Figure 3.2 presents a schematic of the bistable device. A frame oscillates with single frequency harmonic motion according to $z(t) = Z\cos\omega t$. Connected to the frame is a rotating pivot to which is attached a spring that has linear stiffness constant k and natural length l_0. The opposite end of the spring is attached to a sliding mass, m, by another pivot which is always a vertical distance d from the top pivot. A motion coordinate for the sliding mass is defined as $x(t)$, where $x = 0$ when the two pivots are vertically aligned. Thus, the coordinate x describes the relative motions between the frame and the mass.

The viscous dissipation force by the damper is assumed to be linearly proportional to the relative velocity between the frame and mass, indicating that $F_c = c\dot{x}$.

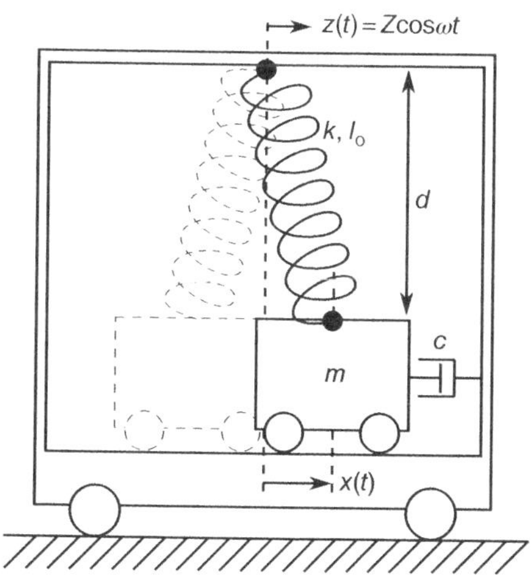

Figure 3.2 Schematic of the mass–spring-damper bistable oscillator excited by a ground motion $z(t)$. The pivoting points are shown as black dots.

Assuming that the spring only undergoes deformations along its length, the spring force which restores the motion of the mass in the x axis may be expressed as

$$F_k = kx\left(1-\frac{l_o}{\sqrt{x^2+d^2}}\right) \tag{3.1}$$

A Taylor series expansion of the force may be taken about $x=0$ to express the spring force by

$$F_k = k_1x + k_3x^3 + \cdots \tag{3.2}$$

Here, $k_1 = -k[(l_o/d)-1]$ and $k_3 = kl_o/2d^3$. By this reduction from Eq. (3.1) to (3.2), it is assumed that $x/d \ll 1$. This is not a restrictive limitation because it is known that the representation of the spring force in Eq. (3.2), retaining only the first and third-orders of the power series, is sufficiently accurate with respect to the exact dynamics [87,88]. It is seen from Eq. (3.2) that when the natural length of the spring is equal to the vertical distance between pivots, $l_o = d$, the linear term in the spring force expansion vanishes. Under this condition, the system is said to be *essentially nonlinear* because the spring force profile given by Eq. (3.1) is no longer able to be linearized. However, the following investigations consider the case of $l_o > d$ which is a geometric constraint that induces bistability. By this condition, the spring is relaxed only when the mass is positioned laterally with respect to the vertical alignment of the pivots, such as the case illustrated in Figure 3.2. Using Eq. (3.1), the static equilibria of the mass, x^*, are be found to be

$$x^* = \begin{cases} 0 \\ \pm\sqrt{l_o^2 - d^2} \end{cases} \tag{3.3}$$

An application of Newton's laws of motion provides the governing equation for the motion of the sliding mass.

$$m(\ddot{x}+\ddot{z}) = -F_c - F_k \tag{3.4}$$

Where the overdot indicates differentiation with respect to time, *d*/*dt*. By substitution of the previous definitions, and using the exact expression for the spring force given in Eq. (3.1), the governing equation is derived to be

$$m\ddot{x} + c\dot{x} + kx\left(1-\frac{l_o}{\sqrt{x^2+d^2}}\right) = mZ\omega^2\cos\omega t \tag{3.5}$$

In Eq. (3.5), the exact representation of the restoring spring force, Eq. (3.1), is employed because the following model investigations of this study derive insight using numerical/simulation methods. In contrast, certain analytical solution strategies for the governing equations of motion (multiple-scales, harmonic balance, etc.) require a power series representation of the spring force, Eq. (3.2), in order to successfully predict the dynamics via useful closed-form expressions. The consequence of employing the simplified spring force expression may be a reduction of fidelity of the analytically-predicted dynamic behaviors with respect to the exact results determined from simulations. Although this study focuses on simulation strategies to solve Eq. (3.5), later sections of this book will directly compare the results of analytical and numerical solution strategies to evaluate the errors generated by adopting the analytical tools.

Statically-stable equilibria satisfy the inequality $dF_k(x^*)/dx > 0$. For the device such that $l_o > d$, the calculations indicate that the equilibrium at $x^* = 0$ is unstable whereas the other two positions at $x^* = \pm\sqrt{l_o^2 - d^2}$ are stable. Thus, the stable equilibria are equidistant from the unstable configuration: a symmetric bistable device.

3.2.2 A Metric for Energy Dissipation Capacity

There are numerous measures employed to quantify the energy damping properties of materials and structures, including loss factor, logarithmic decrement, and loss tangent. However, the implementation of these metrics requires that the dynamics are periodic. These constraints may not hold for nonlinear systems that undergo steady-state (or aperiodic) vibrations with dynamics exhibiting frequencies of oscillation other than those of the harmonic excitation. Yet, all periodically-excited dissipative systems deplete a finite energy during a cycle of excitation. Thus, to employ a metric that is meaningful for the present bistable device which is anticipated to exhibit strongly nonlinear vibrations that may or may not be coincident with the excitation frequency, the following study characterizes energy damping capacity by computing the average power dissipated by the damper over a prescribed period of time. The instantaneous power dissipated by the bistable device shown in Figure 3.2 is then computed by the product of the viscous force and relative velocity of the mass with respect to the frame:

$$P_i(t) = c\dot{x}^2 \tag{3.6}$$

The instantaneous power is averaged over many cycles of excitation, lasting a time t^*, to provide an indication of long-term energy dissipation capacity. Thus, the average power dissipated is computed via

$$P_a = \frac{1}{t^*}\int_o^{t^*} P_i(t)\,d(t) \tag{3.7}$$

3.2.3 Numerical Analysis of the Base-excited Response

Using a fourth-order Runge–Kutta algorithm, Eq. (3.5) is numerically integrated to investigate the bistable device response when excited under different conditions. The interest is to explore how the average power dissipation varies due to the different types of dynamics that may be exhibited. The system parameters used in the following examples are $m = 0.27$ kg, $c = 2.79$ N s m^{-1}, $k = 30$ kN m^{-1}, $l_o = 11.440$ mm, $d = 11.346$ mm. In the examination below, the initial conditions are held constant such that the bistable device begins the simulations at rest, $\dot{x}(t=0)=0$ m s^{-1}, and at one of the stable equilibria, $x(t=0)=\sqrt{l_o^2-d^2}$ m. The excitation frequency is $\omega/2\pi=12$ Hz while several amplitudes of excitation are considered, $Z=[1\quad 6\quad 7\quad 13\quad 15]$ mm. A collection of the results after long time numerical integration routines is provided in Figure 3.3, presented in terms of the time dependent response characteristics.

In the left column of Figure 3.3 the time series of the bistable device relative displacement x is shown for a time period that occurred once the transient behaviors related to initial conditions had decayed. Thick dotted lines indicate the stable equilibria positions. The right column of Figure 3.3 shows the corresponding phase portraits of this time span; here, the circle points indicate the stable equilibria configurations in phase space. From top to bottom rows, the excitation amplitude Z is increased according to the increments described above.

Although the excitation frequency remains constant, the variation in excitation amplitude is extremely influential in determining the resulting qualitative dynamic behavior of the bistable device. The intrawell oscillations revealed in Figure 3.3a are particularly distinct with respect to the other dynamics in that the intrawell motions exhibit a large bias in displacement as the mass vibrates around one of the stable equilibria. For an excitation amplitude of $Z=6$ mm, a multiharmonic, interwell motion is evident in Figure 3.3b. This is a third-order superharmonic because the excitation occurs at 12 Hz although the bistable device undergoes an unmistakable response at 4 Hz. Aperiodic motions dominate the dynamics for a slight increase in excitation amplitude to $Z=7$ mm, Figure 3.3c. Phase portraits of such aperiodic or chaotic dynamics reveal apparently nonrepetitive trajectories. This result is clear in Figure 3.3c2 as evidenced by the fact that the inertial mass traverses across a large area of the phase space without consistently retracing its step from one cycle to the next.

Considering Figure 3.3d, the excitation level is now $Z=13$ mm but the dynamic response is of reduced amplitude. Specifically, the response is a low amplitude interwell dynamic. In some contexts, this is referred to as excitation-induced stability (EIS) [89–91] because certain excitation conditions lead to the stabilization of an otherwise unstable dynamic state. Yet, increasing the excitation amplitude beyond this value for the same initial conditions leads to the energetic snap-through motions, Figure 3.3e, which clearly exhibit very large velocity trajectories. Following the damping performance metric of Eq. (3.7), the large variation in velocities of the distinct dynamic types observed in Figure 3.3 may be anticipated to contribute to a considerable change in average power dissipation.

3.2.4 Energy Dissipation Features of the Dynamic Types

Figure 3.4 plots the average power dissipated by the bistable device damper as determined using Eq. (3.7). The results in Figure 3.4a–e correspond to the counterpart time series shown in Figure 3.3a–e, although the cumulative integration time t^* in Figure 3.4 is shown for the full span computed here rather than the truncated length shown in Figure 3.3. To provide an indication of how quickly the results of the integration converge, Figure 3.4 includes long- and short-dash lines that denote $\pm 5\%$ and $\pm 2\%$ of the final value of average power dissipated, respectively.

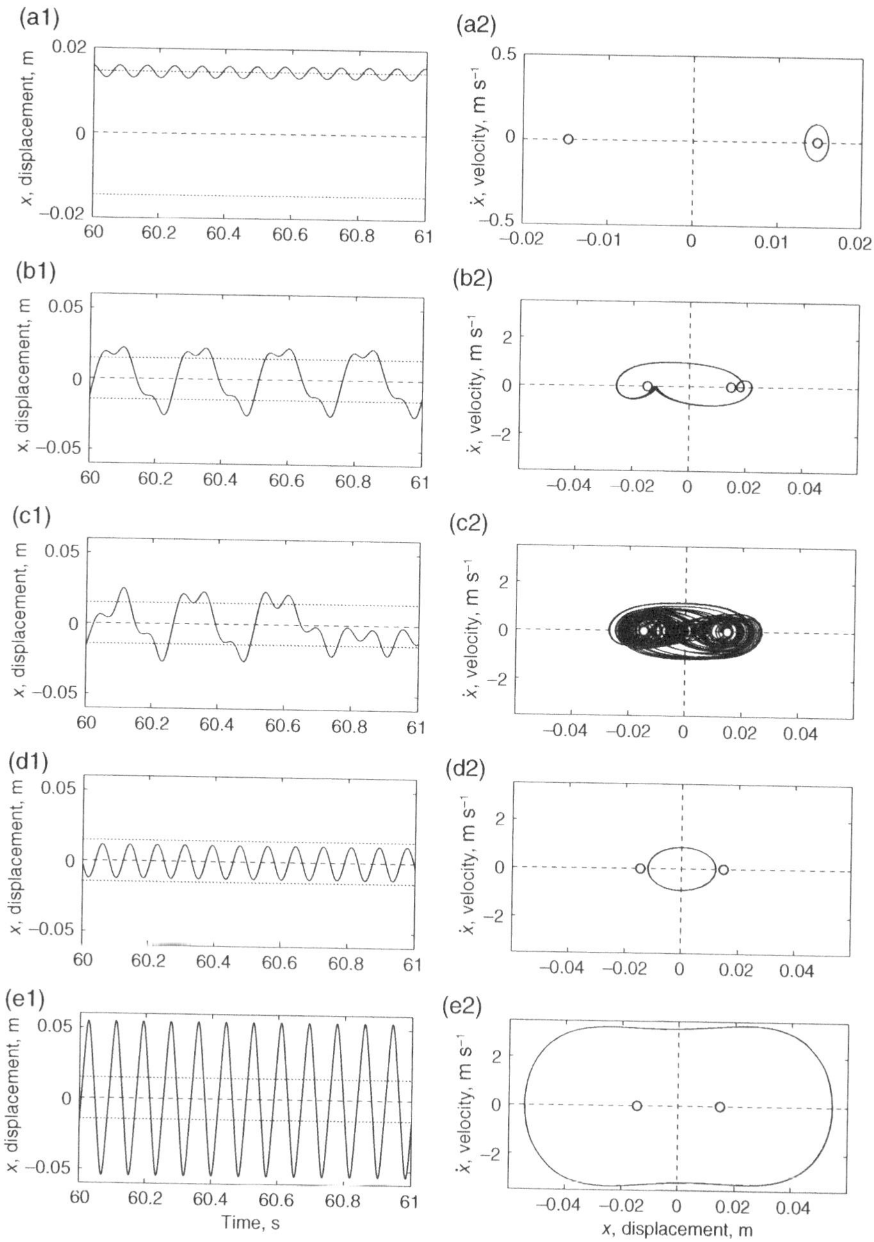

Figure 3.3 (a1–e1) Plots of the displacements of the bistable device as functions of time. Thick dotted lines indicate the stable equilibria positions. (a2–e2) Phase portraits of displacement and velocity. Circle points indicate the stable equilibria. All results computed for excitation frequency $\omega / 2\pi = 12$ Hz. Excitation amplitude Z is (a) 1 mm, (b) 6 mm, (c) 7 mm, (d) 13 mm, and (e) 15 mm.

For the periodic motions, the average power dissipation values converge rapidly. In contrast, the integration of the instantaneous power computed for the aperiodic motions takes much longer to converge to ±2% of the final value, Figure 3.4c. In terms of the magnitude of the power dissipated, the interwell dynamics which

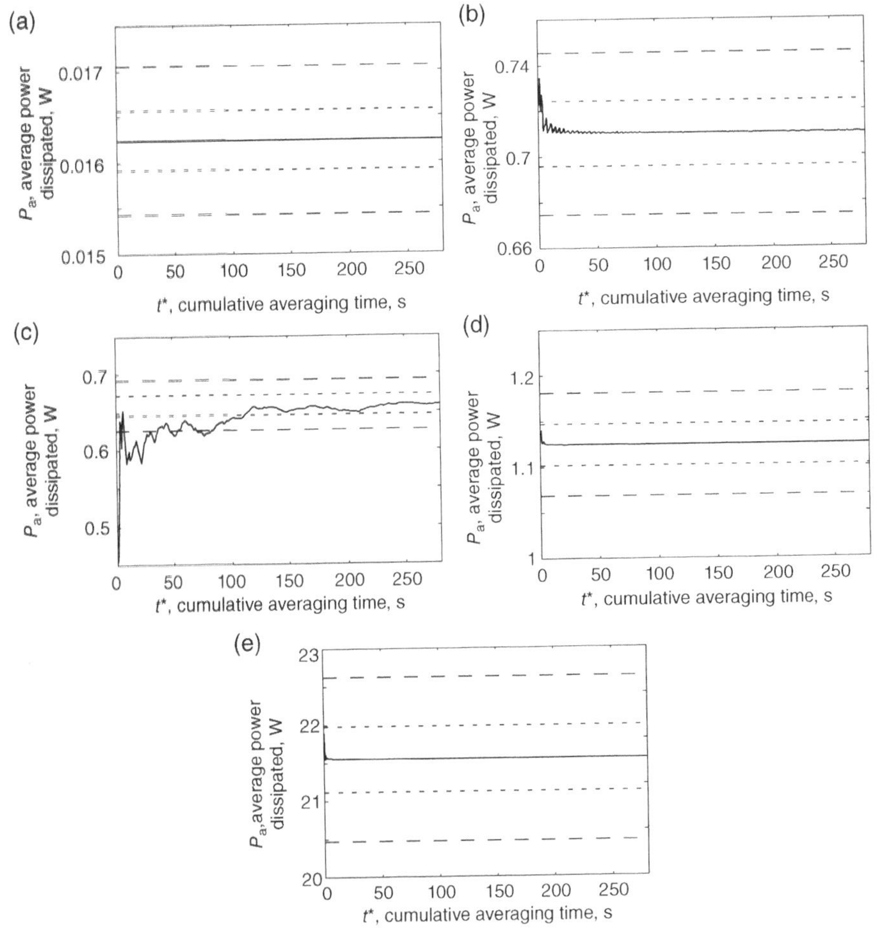

Figure 3.4 Average power dissipated by the bistable device damper. The plots (a–e) here correspond to the counterpart results (a–e) in Figure 3.3. Long-dash lines are ±5% of the final value, short-dash lines are ±2% of the final value.

occur primarily at a single frequency – in other words, both the EIS responses, Figure 3.4d; and the snap-through vibrations, Figure 3.4e – dissipate much more power than the interwell dynamics which are not synchronous with the excitation frequency. Interestingly, the multiharmonic motions and the aperiodic dynamics, Figure 3.4b,c, respectively, lead to similar magnitudes of average dissipated power although they are very different dynamic behaviors, as is exemplified comparing the phase portraits in Figure 3.3b,c. Between the intrawell and snap-through responses, three orders of magnitude of change in average power dissipation is computed, as shown in Figure 3.4a and e. This result affirms the prior intuition that the dynamic types, activated uniquely with respect to the excitation amplitude Z, lead to dramatic adaptation of energy dissipation properties.

The aperiodic motions shown in Figure 3.3c require a long period of integration time to converge towards a final value of average power dissipated,

Figure 3.4c. Of course, a meaningful metric of damping capacity requires that an average value is determined. At the least, the measure indicates an expected long-term performance for the bistable device as a structure used for vibration damping, although the instantaneous character of the system may be ever-changing. To assess the consistency of convergence in the average dissipated power computation Eq. (3.7) for aperiodic motions, Figure 3.5 plots results of additional simulations of the bistable device response which were computed for three sets of system and excitation parameters that induce aperiodic motions. The top row of Figure 3.5 provides a confirmation of the aperiodicity evident by way of the phase portraits and the bottom row presents the results of calculating average power dissipated. The parameters employed are provided in the Figure 3.5 caption. Although the responses each require a long period of integration time to converge to a final value, a time comparable to the case shown in Figure 3.4c, the results of Figure 3.5 indicate that the aperiodic motions do provide an average power dissipation when considering the long-term characteristics. This finding is confirmed by other focused examinations of chaos; indeed, the long-term behavior of a bistable system undergoing chaotic vibrations has remarkable structure and repeatability in phase space, due to its unique strange attractor [92]. Thus, although it requires a longer period of integration to reach a convergent value for aperiodic dynamics, the average power dissipation of the bistable device can be employed as a broadly meaningful measure of the magnitude of adaptability in vibration damping performance.

3.2.5 Influences Due to Change in Frequency and Initial Conditions

As was briefly shown in Section 1.2, the dynamics of harmonically-excited bistable systems are known to be sensitive to changes in the excitation conditions – frequency as well as amplitude – and steady-state dynamics may be influenced by the configuration of the system at the initial time of excitation. To investigate these features as they relate to maintaining an effective adaptability of energy dissipation properties, the model equation of the bistable device is explored further and the results of average power dissipation are presented in Figure 3.6. The data shown in Figure 3.6a are computed from excitations at $\omega / 2\pi = 6$ Hz, while data in Figure 3.6b,c are derived using frequencies 9 and 12 Hz, respectively. The averaging time of $t^* = 180$ s is taken for all cases. The stars are individual results found by using 20 randomly selected initial conditions. The normal distributions of the initial displacements have a standard deviation of 150% of the magnitude of the stable equilibrium distance, while the distribution of initial velocities has a standard deviation of 1 m s^{-1}. The circles indicate the mean of the 20 cases computed for each excitation amplitude, Z, while the various gradients of background shading indicate the dynamic types which are exhibited for a given excitation amplitude.

As shown in Figure 3.6a, at 6 Hz the average power dissipation suddenly transitions from a low to high level due to the change in excitation amplitude around $Z \sim 0.008$ m. A narrow range of excitation amplitudes in this vicinity reveals possible coexistence of the dynamic types: either snap-through or aperiodic motions are observed. The range of excitation amplitudes which lead to coexistence of the

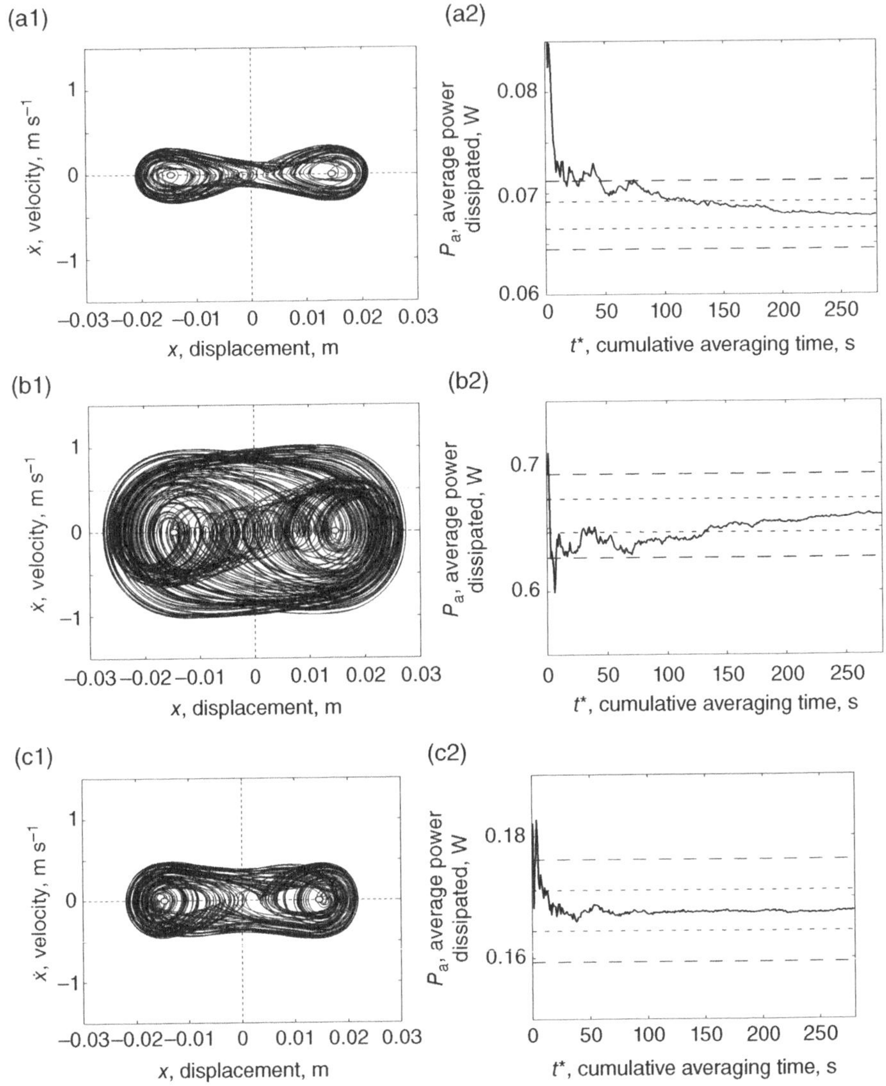

Figure 3.5 (a1–c1) Phase portraits and (a2–c2) computations of average power dissipated for bistable device parameters that induce aperiodic motions. Long-dash lines are ±5% of the final value, short-dash lines are ±2% of the final value. The results show that average power dissipation of aperiodic dynamics does converge given a long enough window of the integration used in Eq. (3.7). (a) $m = 0.4$ kg, $k = 1.4$ kN m^{-1}, $\omega / 2\pi = 5$ Hz, $Z = 9.3$ mm. (b) $m = 0.4$ kg, $k = 1.4$ kN m^{-1}, $\omega / 2\pi = 6$ Hz, $Z = 12$ mm. (c) $m = 0.6$ kg, $k = 2.1$ kN m^{-1}, $\omega / 2\pi = 4$ Hz, $Z = 7$ mm.

dynamic types is increased as the excitation frequency is changed to 9 and 12 Hz, Figure 3.6b,c respectively. For these latter cases, EIS responses may be induced which lead to significantly lower power dissipation than the snap-through motions. This is shown by the individual star data points in the dark gray shaded areas in

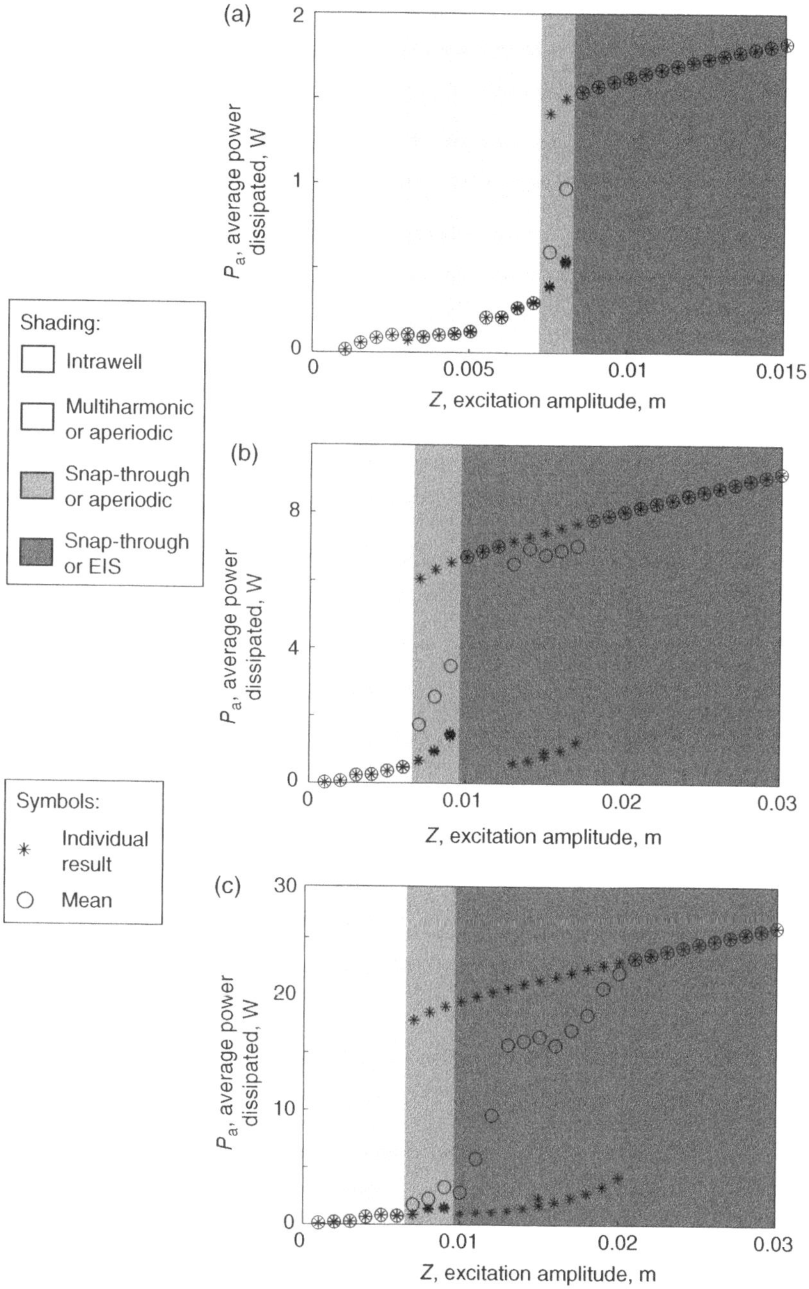

Figure 3.6 Average power dissipated as a function of excitation amplitude at a frequency $\omega/2\pi$ of (a) 6, (b) 9, and (c) 12 Hz. Results for 20 different initial conditions are shown as individual stars whereas the circles indicate the mean power dissipation for a given excitation amplitude. The shadings indicate (white) intrawell oscillations, (light gray) multiharmonic or aperiodic motions, (medium gray) snap-through or aperiodic behaviors, and (dark gray) snap-through or EIS dynamics. To distinguish aperiodic or EIS from the snap-through dynamics, the aperiodic or EIS motions yield much smaller values of average power dissipation.

Figure 3.6b,c where low and high levels of energy dissipation are found to coexist. By considering the change of the mean (circles) as compared to the individual results (stars), it is clear that at 12 Hz, Figure 3.6c, the likelihood of activating a low energy dissipating EIS dynamic is reduced as the excitation amplitude is increased. In one sense, the mean values indicate a long-term trend in power dissipation performance because the bistable device may be exposed to perturbations in the working environment which excite it away from steady-state behavior; this could include interferences such as impulsive jolts to the structure, ambient vibrations, and so on. The mean values shown in Figure 3.6 also provide a statistical measure based upon how close in value the means are to the distinct low or high power dissipation levels at a given excitation amplitude, which are characterized by the EIS or snap-through motions, respectively. Clearly, at lower frequencies – where the mean values jump suddenly from low to high energy dissipation levels, such as in Figure 3.6a - this adaptability of energy damping can be exploited with greater confidence of the outcome. On the other hand, the magnitude of change in energy dissipation properties is more pronounced at higher frequencies, such as in Figure 3.6c. Thus, as a mechanism for large and adaptable damping, the requirements of the application may drive greater attention to one feature or the other and thus tailor design and implementation accordingly.

3.2.6 Experimental Studies

The findings of the previous sections indicate the adaptability of energy dissipation characteristics of the bistable device is influenced by excitation amplitude and frequency and by the initial configuration of the system at the start of harmonic excitation. To verify these results and derive practical conclusions as to the attainment of high and adaptable damping properties, experimental studies are conducted. The bistable device examined in the experiments is shown in Figure 3.7. The structure is fabricated using the same lumped constituents that were used in the model schematic of Figure 3.2. A frame is excited by a base motion (a controlled electrodynamic shaker platform) and at the top of the frame is a rotational pivot. In between this pivot and a second is attached a linear coil spring of stiffness $k = 5.88$ kN m^{-1}. Connected to the second pivot is a sliding bearing mass, $m = 0.36$ kg, which moves relative to the frame. The distance between the rotational pivots when the spring is relaxed is approximately 102 mm; this length includes both the natural length of the spring and stand-off distances associated with the mounting hardware. Because this distance is greater than the vertical span between pivots, $d = 83.7$ mm, the two stable equilibria of the device displacement from the vertically-aligned pivots are $x^* = \pm 29.1$ mm. From the decay of free vibrations, the damping constant of the system is approximated to be $c = 1.6$ N s m^{-1}.

The device is excited by a wide variety of shaker platform amplitudes and frequencies and some results are shown in Figure 3.8. The left column presents time series of the distinct dynamics found at an excitation frequency of $\omega / 2\pi = 6$ Hz while the right column shows the corresponding phase portraits. From top to bottom, the excitation amplitude increases according to $Z = [1.0 \quad 6.8 \quad 8.1 \quad 11.6]$ mm. The qualitative trends of the measured responses are in good agreement with the

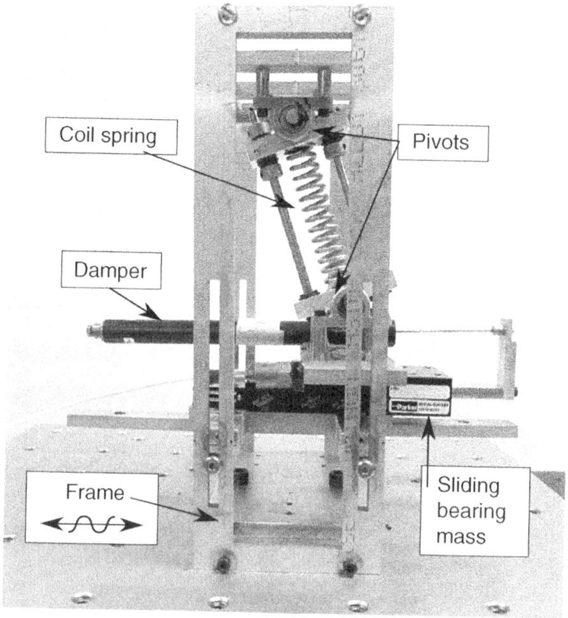

Figure 3.7 Bistable device composed using a frame excited by shaker platform motions in the direction indicated by the double-arrow and a sliding bearing mass which moves with respect to the frame according to the pivot point attachments of the coil spring. A dashpot damper dissipates the relative motion between the bearing mass and frame.

results obtained by numerical integration, as shown in Figure 3.3. For the small excitation amplitude, an intrawell dynamic is induced, characterized by the low amplitude oscillations around one of the stable equilibria seen in Figure 3.8a. When the amplitude is increased to 6.8 mm, Figure 3.8b, a multiharmonic response is activated having predominant frequency of $\omega/4$; whereas the excitation is at 6 Hz, three predominant waveforms are shown to occur over the time of 2 s, and hence these induced motions are at 1.5 Hz. Aperiodic dynamics are triggered when the excitation amplitude is around 8.1 mm, indicated by the seemingly random trajectories of the phase portrait, Figure 3.8c2. Finally, the energetic snap-through response is activated for the highest level shown here, Figure 3.8d. Between the intrawell and the snap-through dynamics, there is clearly a dramatic increase in the relative velocity of the mass of the bistable device. As prior numerical findings would suggest, there is likely a substantial increase in the energy dissipation properties. Prior to consideration of the power dissipation features, it is noted that the EIS dynamics were not observed, in contrast to the numerical study. One explanation for the difference is that the model does not fully account for all of the intricate features of the experimental bistable device, such as the small sliding bearing friction. Considering the discrepancy from a practical perspective, the absence of EIS can be perceived as an advantage in the context of the current study. By minimizing the number of potential coexisting responses which might occur, the consistency of activating a low or high

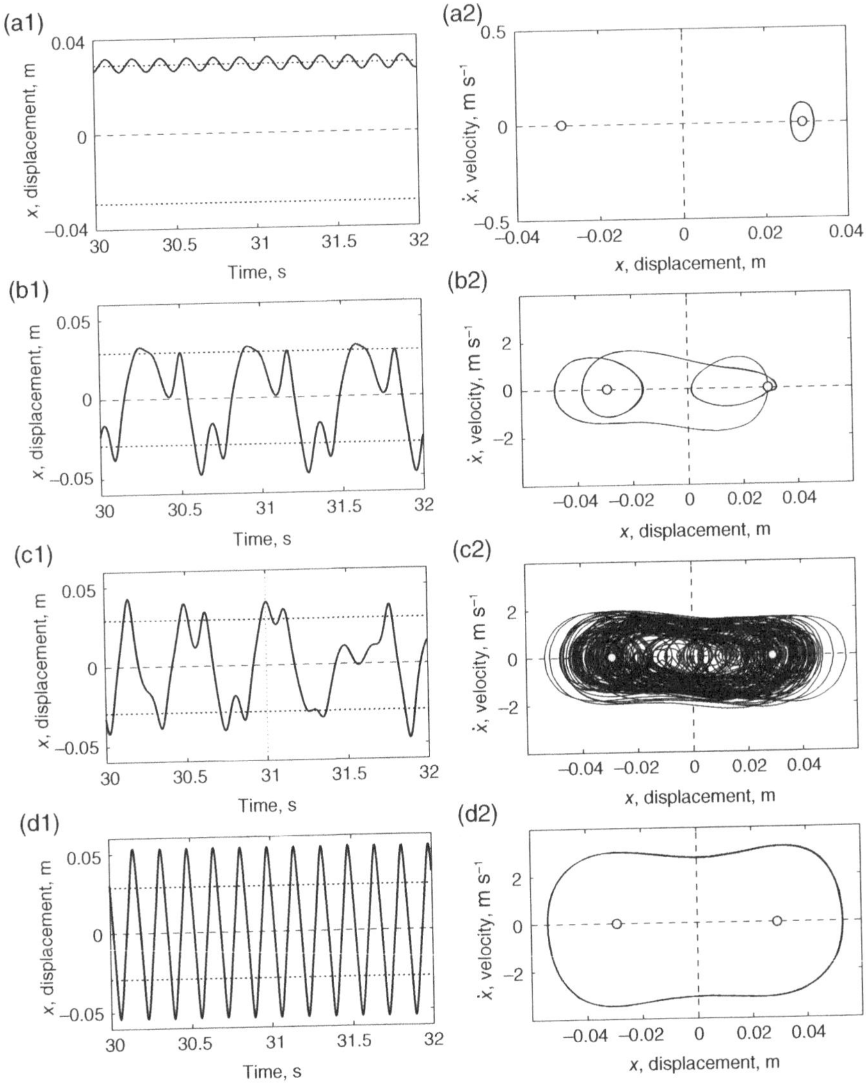

Figure 3.8 (a1–d1) Plots of the measured displacements of the bistable device as functions of time. Thick dotted lines indicate the stable equilibria positions. (a2–d2) Phase portraits of displacement and velocity. Circle points indicate the stable equilibria. All results measured for excitation frequency $\omega / 2\pi = 6$ Hz. Excitation amplitude Z is (a) 1.0, (b) 6.8, (c) 8.1, (d) 11.6 mm.

level of damping based upon one of the preferred dynamic regimes is enhanced. Thus, the viability of utilizing the bistable device as an adaptable damping mechanism is improved by minimizing the possibility for coexisting dynamics.

Comparable to the numerically computed data in Figure 3.6a, Figure 3.9 presents results of average power dissipation as derived from the experimental measurements, for a large range of excitation amplitudes when the excitation frequency is 6 Hz. To activate all of the potential coexisting dynamics that could be identified for this frequency, the excitation amplitude was incrementally increased from a

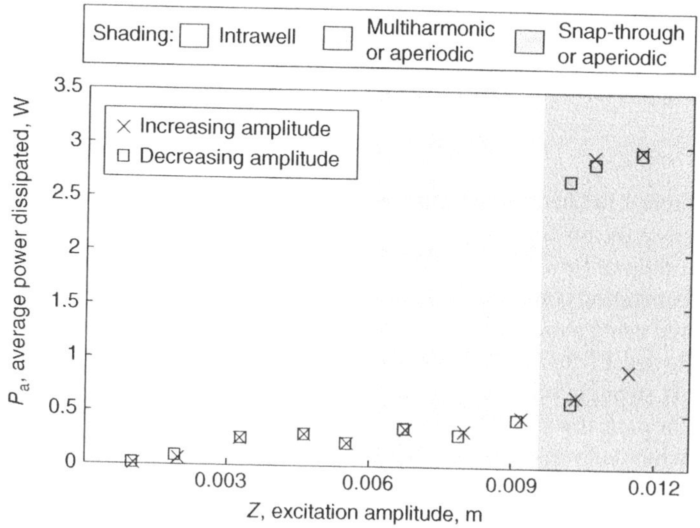

Figure 3.9 Experimentally determined average power dissipation as the excitation amplitude of shaker motions is varied at 6 Hz. The cross (square) points indicate the average power measures for an increasing (decreasing) amplitude of excitation. The shadings indicate: white, intrawell oscillations; medium gray, multiharmonic or aperiodic motions; and dark gray, snap-through or aperiodic behaviors.

minimum and decreased incrementally from a maximum level, with sufficient time in between changes to allow the system to reach a steady-state (if it existed). The increasing excitation amplitude results are shown as crosses in Figure 3.9 while the results for decreasing level are indicated by squares.

In Figure 3.9, the magnitude of power dissipation provided by the bistable device is seen to vary according to excitation amplitude similarly to the trends found for the numerically predicted case presented in Figure 3.6a. The range of excitation amplitudes leading to coexistence of a low power dissipation and high power dissipation dynamic is fairly narrow. Figure 3.9 shows that the responses that dissipate less energy are aperiodic or multiharmonic motions and not an EIS behavior. For excitation amplitudes greater than approximately 11.7 mm, the snap-through dynamics were the only dynamics that were activated but the motions were so energetic that the mass was found to be potentially capable of sliding off the track it traveled upon. As a result, no sufficiently long data series for excitation amplitudes greater than this value could be acquired for the latter determination of average power dissipation.

In terms of a power dissipation strategy based upon selectively triggering an energetic and high power dissipation snap-through dynamic, one must employ the bistable device to minimize the potential for coexisting responses. The results shown in Figure 3.9 are one such example where only a small variation in the amplitude of excitation is required to confidently transition from a low to high power dissipation dynamic regime. Thus, when compared to the examples provided by the numerical results in Figure 3.6b,c where coexisting dynamics are evident across a large range of excitation amplitudes, it is found from the modeling as well as experimental studies that using a bistable device at lower frequencies is a better approach to ensure that damping properties are effectively tailored to their target levels.

3.2.7 Summary

The appreciable interest in harnessing snap-through dynamics of bistable materials and structures as a means for high and passive-adaptive damping requires an effective understanding of how to trigger the desired dynamics according to the system design and operating conditions. From examining the simulated and measured results acquired using a mechanical bistable device as the token architecture, it is shown that the sensitivity of activating dynamics which induce large energy dissipation depends strongly on excitation amplitude and frequency, and on the initial conditions. In part, the sensitivity becomes manifest in terms of coexisting dynamic regimes which may include intrawell oscillations, multiharmonic or aperiodic motions, excitation-induced stability responses, or the energetic snap-through vibrations. In spite of the numerous behaviors, only the snap-through dynamics provide large dissipation of power. Thus, in general, one may conclude that the adaptability of damping properties can be classified as whether or not snap-through is activated: when it is, a large power dissipation is enabled, and vice versa. It is shown that lower frequencies of excitation are a preferred operating condition to minimize the potential for coexisting dynamic forms. These findings provide practical design-based insight into the deployment of bistable systems as a means for high and adaptable damping in a passive architecture.

3.3 Isolating Structures Under Large Amplitude Excitations Through Activation of Low Amplitude Interwell Dynamics: Criteria for Excitation-induced Stability[2]

In the context of vibration isolation, a suspension system must provide significant reduction in the transmitted vibrations for a wide range of amplitudes and broad spectrum of harmonic exciting motions. Due to the interest in bistable members as constituents in isolation systems, the numerous single-periodic, steady-state dynamics that could be activated for a given single frequency of excitation may pose concern unless the preferred responses are the ones that become manifest. As described in Section 3.1.2, many isolation systems which incorporate bistable members are designed very close to the limit of elastic stability, and in some cases are loaded beyond the threshold. Of course, a post-buckled vibration isolator design may exhibit energetic snap-through motions if the excitation levels increase beyond the magnitudes that were originally

2 Material from this section has been reprinted by permission of ASME, from Z.Wu, R.L. Harne and K.W. Wang, Excitation-induced stability in a bistable duffing oscillator: analysis and experiments. Journal of Computational and Nonlinear Dynamics 10:011016, 2015. doi: 10.1115/1.4026974. Copyright © 2015 by ASME.

anticipated. Under such larger amplitudes of excitation, effective understanding in the design and implementation of the isolator becomes critical.

As introduced in Section 3.2.3, the dynamic behavior referred to as excitation-induced stability (EIS) may be activated instead of snap-through responses when the magnitude of the excitation increases substantially. This is because both responses are *interwell* motions, although EIS is the low amplitude version of the cross-well behaviors. Explorations regarding EIS in bistable systems originated with a study by Holmes [93] which demonstrated that if the forcing level exceeds a threshold value, the unstable saddle fixed point on the Poincaré map can become stabilized through a pitchfork bifurcation. From this finding, EIS is a steady-state dynamic in bistable systems that coexists under the same harmonic excitations as a snap-through response. Blair *et al.* [94] also documented the phenomenon for a forced bistable oscillator which was later demonstrated experimentally by Kim *et al.* [91,95] using analog bistable circuit studies and bifurcation analyses.

To date, there are no explicitly reported criteria that elucidate the roles of system and excitation parameters that may lead to the activation of EIS behaviors in a bistable system. Due to the recent interest in exploiting bistable structural members in isolation systems, there is great merit for an investigation that uncovers the factors which govern an effective activation of low amplitude interwell EIS oscillations. In other words, should the excitation levels acting upon a vibration isolation system with bistable constituents exceed the range for which the system was preferentially designed, what are the design means available to ensure that low amplitude EIS oscillations are activated instead of the adverse snap-through dynamics? The answers to such a question are of critical importance for preserving the integrity of isolation systems that exploit bistable members. Therefore, the following sections present a comprehensive investigation to derive and verify criteria that delineate the system and excitation parameters necessary to activate EIS responses.

3.3.1 Governing Equation Formulation of the Bistable Oscillator

A generalized bistable oscillator is the analytical model structure of interest for the ease with which the governing equations of many bistable systems may be reduced to such a general form. The non-dimensional governing equation of the harmonically excited bistable oscillator may be written as

$$\ddot{x}+\gamma\dot{x}-x+x^3=p\cos\omega\tau \tag{3.8}$$

where x is the normalized relative displacement between an inertial frame and the oscillator mass as measured from the central unstable equilibrium; γ is the damping factor; p is the excitation level; and the dot operator represents the derivative with respect to non-dimensional time τ. Several previous studies [96–99] have shown that the most fundamental, periodic vibration of a bistable oscillator can be captured by a single term Fourier expansion, in other words a one-term harmonic balance. The displacement response is then expressed using $x(\tau)=c(\tau)+r(\tau)\cos[\omega\tau-\phi(\tau)]$. The offset term $c(\tau)$ is selected as nonzero or zero based on whether the dynamics of interest are intrawell or interwell, respectively. Therefore, the symmetry of the response about the unstable equilibrium serves as an important criterion to distinguish between the dynamical responses. Substituting the expression for $x(\tau)$ into Eq. (3.8), assuming slow-varying coefficients, and eliminating higher-order terms, the steady-state vibration amplitude r satisfies [97,100]

$$\frac{225}{16}r^6 - \frac{15}{2}\left(2-\omega^2\right)r^4 + \left[\left(2-\omega^2\right)^2 + (\gamma\omega)^2\right]r^2 = p^2 \tag{3.9}$$

or

$$\frac{9}{16}r^6 - \frac{3}{2}\left(1+\omega^2\right)r^4 + \left[\left(1+\omega^2\right)^2 + (\gamma\omega)^2\right]r^2 = p^2 \tag{3.10}$$

where Eqs. (3.9) and (3.10) govern the vibration amplitude of intrawell and interwell dynamics, respectively.

In the context of vibration isolation, the amplitude of the structural motions or the magnitude of the transmitted force are desired to be minimized at the steady-state operating point. These measures are both dependent upon the amplitude of the system vibration, which in this study is the variable r. Consequently, in the following, it is understood that favorable vibration isolation for the bistable system is achieved when r is small; thus, for high excitation levels, small r corresponds to the EIS dynamic regime rather than the adverse snap-through motions that lead to a much greater magnitude of vibration.

3.3.2 Stability of the Analytically Predicted Interwell Dynamics

Since EIS dynamics represent oscillations crossing the unstable equilibrium position twice per excitation period [91,95], such phenomena are classified as a form of interwell response. To determine the stability of the interwell response, a perturbation $y(\tau)$ is applied to the steady-state response $x_{ss} = r\cos[\omega\tau - \phi]$ where r is the steady-state response amplitude determined from Eq. (3.10). Substituting the expression $x(\tau) = x_{ss}(\tau) + y(\tau)$ into Eq. (3.8) and neglecting higher order terms of $y(\tau)$, the resulting governing equation for the perturbation $y(\tau)$ becomes

$$\ddot{y} + \gamma\dot{y} + \left[\left(\frac{3}{2}r^2 - 1\right) + \frac{3}{2}r^2\cos[2\omega\tau - 2\phi]\right]y = 0 \tag{3.11}$$

The stability of the interwell response can therefore be determined from the stability of the perturbation $y(\tau)$. In other words, stable perturbation $y(\tau)$ corresponds to a stable steady-state response r and vice versa. To better analyze the stability of $y(\tau)$, Eq. (3.11) is first expressed as

$$y'' + \frac{\gamma}{2\omega}y' + \left(\frac{\alpha}{4\omega^2} + \frac{\beta}{4\omega^2}\cos T\right)y = 0 \tag{3.12}$$

with $T = 2\omega\tau - 2\phi$, $\alpha = 3r^2/2 - 1$, $\beta = 3r^2/2$, and the operator' represents a derivative with respect to time T. Applying a change of variable for the perturbation such that $\varphi = y\exp\left[\frac{1}{2}\int_0^T \frac{\gamma}{2\omega}\mathrm{d}s\right]$ [101], Eq. (3.12) is thus transformed into a standard Mathieu's form

$$\varphi'' + \left[\delta + \varepsilon \cos T\right]\varphi = 0 \tag{3.13}$$

where the following terms are defined as

$$\delta = \frac{\alpha}{4\omega^2} - \frac{\gamma^2}{16\omega^2}$$

$$\varepsilon = \frac{\beta}{4\omega^2}$$

By this transformation, the stability of the interwell dynamics of the bistable system can be examined through a corresponding study of the stable boundaries of the Mathieu's equation. A few of the boundaries are shown schematically in Figure 3.10 where the hatched areas correspond to stable domains of Eq. (3.13) which are specifically explored in this work. Applying the Poincaré–Lindstedt method [72,102], the three lowest-order transition curves labeled B_1, B_2, and B_3 may be approximated as:

$$B_1 : \delta = -\frac{1}{2}\varepsilon^2 + O\left(\varepsilon^4\right) \tag{3.14a}$$

$$B_2 : \delta = \frac{1}{4} - \frac{1}{2}\varepsilon - \frac{1}{8}\varepsilon^2 + O\left(\varepsilon^3\right) \tag{3.14b}$$

$$B_3 : \delta = \frac{1}{4} + \frac{1}{2}\varepsilon - \frac{1}{8}\varepsilon^2 + O\left(\varepsilon^3\right) \tag{3.14c}$$

Therefore, from Figure 3.10, the perturbation φ is stable if δ is between B_1 and B_2 or above B_3. These conditions are respectively by the inequalities of Eq. (3.15).

$$-\frac{1}{2}\varepsilon^2 < \delta < \frac{1}{4} - \frac{1}{2}\varepsilon - \frac{1}{8}\varepsilon^2; \delta > \frac{1}{4} + \frac{1}{2}\varepsilon - \frac{1}{8}\varepsilon^2 \tag{3.15}$$

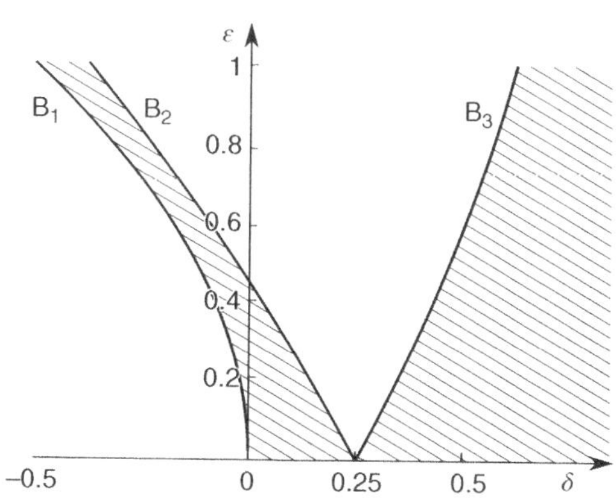

Figure 3.10 Schematic of stable and unstable domains of Mathieu's equation, where the stable dynamics occur in the hatched regions. B_1, B_2, and B_3 are the first three transition curves on the $\delta - \varepsilon$ plane.

Neglecting terms ε of order 2 or higher in Eq. (3.15) and substituting expressions for δ and ε, the steady-state response r is stable if

$$S_1 < r < S_2;\ r > S_3 \tag{3.16}$$

where

$$S_1 = \sqrt{\left(4+\gamma^2\right)/6} \tag{3.17a}$$

$$S_2 = \sqrt{\left(4+\gamma^2+4\omega^2\right)/9} \tag{3.17b}$$

$$S_3 = \sqrt{\left(4+\gamma^2+4\omega^2\right)/3} \tag{3.17c}$$

The three boundaries in Eq. (3.17), expressed in terms of system parameters p, γ, and excitation frequency ω are criteria to determine the stability of the interwell responses for a bistable oscillator. It will be found that vibration bounded by the curves S_1 and S_2 are the EIS oscillations. The intersections of the steady-state response amplitude curve $r(\gamma, p, \omega)$ – which are determined from the roots of Eq. (3.10) – and the three transition curves govern the frequencies at which interwell dynamics change stability. For instance, the upper frequency bound for EIS, at which responses transition between interwell and intrawell dynamics is determined by substituting $r = S_1$ into Eq. (3.10), and the resulting expression is

$$\omega_{\mathrm{Ub}} = \sqrt{\frac{-16-16\gamma^2-3\gamma^4+2\sqrt{2}\sqrt{192p^2+64\gamma^2+48p^2\gamma^2+48\gamma^4+12\gamma^6+\gamma^8}}{8\left(4+\gamma^2\right)}} \tag{3.18}$$

which can be simplified into $\omega_{\mathrm{Ub}} = \sqrt{(\sqrt{6}p-1)/2}$ assuming the damping is small, $\gamma << 1$.

By the same token, the lower frequency bound of EIS ω_{Lb}, which separates the snap-through and EIS dynamics, and is determined by combining S_2 and Eq. (3.10), is found to satisfy

$$\frac{1}{9}\left(4+\gamma^2+4\omega_{\mathrm{Lb}}^2\right)\left[\gamma^2\omega_{\mathrm{Lb}}^2+\left(-\frac{2}{3}-2\omega_{\mathrm{Lb}}^2/3+\gamma^2/12\right)^2\right]=p^2 \tag{3.19}$$

which is a cubic characteristic polynomial of variable ω_{Lb}^2. Once again assuming small damping, the expression is simplified as $\omega_{\mathrm{Lb}} = \sqrt{-1+(3/2)^{4/3}p^{2/3}}$.

Likewise, the frequency at which the large amplitude interwell, snap-through response becomes unstable is determined by replacing r with S_3 in Eq. (3.10) and the resulting expression is

$$\omega_{\mathrm{inter}} = \sqrt{\frac{-\gamma^2+\sqrt{3p^2\gamma^2+\gamma^4}}{2\gamma^2}} \tag{3.20}$$

Lastly, the minimum forcing level p_{cr} required to maintain EIS is determined by equating S_1 and S_2. This is identical to equating the lower frequency ω_{Lb} and upper frequency ω_{Ub} bounds defined in Eq. (3.18) and (3.19). Substituting the resulting frequency ω into Eq. (3.10) yields an expression for the critical forcing level

$$p_{cr} = \frac{\sqrt{32+24\gamma^2+8\gamma^4+\gamma^6}}{4\sqrt{3}} \tag{3.21}$$

Similarly, Eq. (3.21) can be further simplified into $p_{cr} = \sqrt{2/3}$ for small damping γ.

3.3.3 Validation of the Stability Criteria Using Numerical Simulations

In this section, the accuracy of the stability boundaries developed in the prior section is validated using numerical tools. First, the steady-state response amplitudes of the bistable system are computed by solution of Eq. (3.9) and (3.10). The stability of these dynamics is then determined based upon the sign of the eigenvalues of the Jacobian matrix of the system linearized around a given stationary point [72,98,100]. Figure 3.11a shows the predicted response amplitudes with system parameters $p=5$ and $\gamma=0.1$. Dotted curves are the stability transition criteria, Eq. (3.17). The solid (dashed) lines correspond to interwell (intrawell) responses which are computed using Eq. (3.9) and (3.10). Finally, circles (crosses) are interwell (intrawell) vibration amplitudes determined from direct numerical integration of Eq. (3.8) using a fourth-order Runge–Kutta algorithm.

To exemplify the characteristic responses of the two interwell dynamic forms, points A and B are considered in Figure 3.11a at $\omega=2.3$ where snap-through (point A) and EIS (point B) responses coexist. Figure 3.11b, c, and d depict the corresponding phase portraits, time series trajectories of displacement, and frequency spectra of the responses, respectively. Black and gray curves represent snap-through and EIS oscillations, respectively. Filled circles in Figure 3.11b and dashed lines in Figure 3.11c represent the stable equilibrium positions for reference. As indicated in Figure 3.11b,c, both responses are symmetric about the central unstable equilibrium, which is the defining characteristic of interwell response as compared to intrawell. However, apart from this, the snap-through interwell vibrations and the EIS behaviors are vastly different from each other. First, it is clear that snap-through motions exhibit much greater amplitudes of vibration. Next, Figure 3.11c demonstrates that the responses are almost 180° out-of-phase with each other, implying a phase transition between the two different interwell responses [89,98]. As shown through the phase portrait and time-series, the EIS response is very nearly sinusoidal in contrast to snap-through dynamics which excite other prominent spectral components, including a harmonic that is 3 times the excitation frequency, 3ω, as shown in Figure 3.11d.

In Figure 3.11a, the criteria expressed by Eq. (3.17) predict steady-state snap-through motions satisfying $r > S_3$ to be stable, which coincides with stability limits determined via Jacobian analysis. Both of these results are also in good agreement with the numerically predicted responses. The criteria inequality $S_1 < r < S_2$ predicts stable EIS to be manifest from $\omega \approx 2$ to $\omega \approx 2.37$. Figure 3.11a shows that these criteria are also in good agreement with numerical results and the Jacobian analysis, further validating the accuracy of the approximated stability criteria.

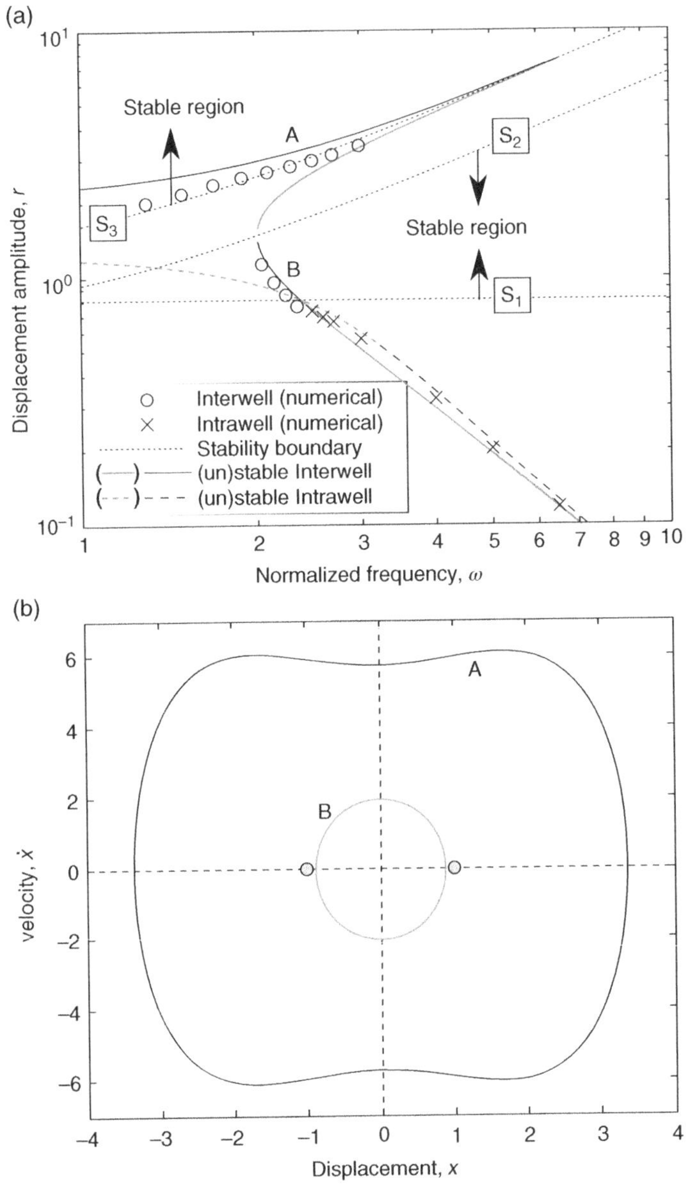

Figure 3.11 (a) Steady-state response amplitude. Black solid (dashed) lines are stable interwell (intrawell) responses whose stability is determined via Jacobian analysis. Dotted lines are the first three approximated transition curves, S_1, S_2, and S_3. Circles (crosses) correspond to interwell (intrawell) response amplitudes computed via direct numerical integration. Gray lines are analytically predicted unstable responses. Criteria predict stable interwell response regions between S_1 and S_2 or above S_3 (as indicated by arrows). Two interwell responses A and B coexist at $\omega = 2.3$ and the corresponding (b) phase portraits, (c) time series and (d) frequency spectra are presented, with black and gray representing snap-through and EIS dynamics, respectively. Filled circles or dashed lines indicate the positions of the stable equilibria in (b) and (c), respectively.

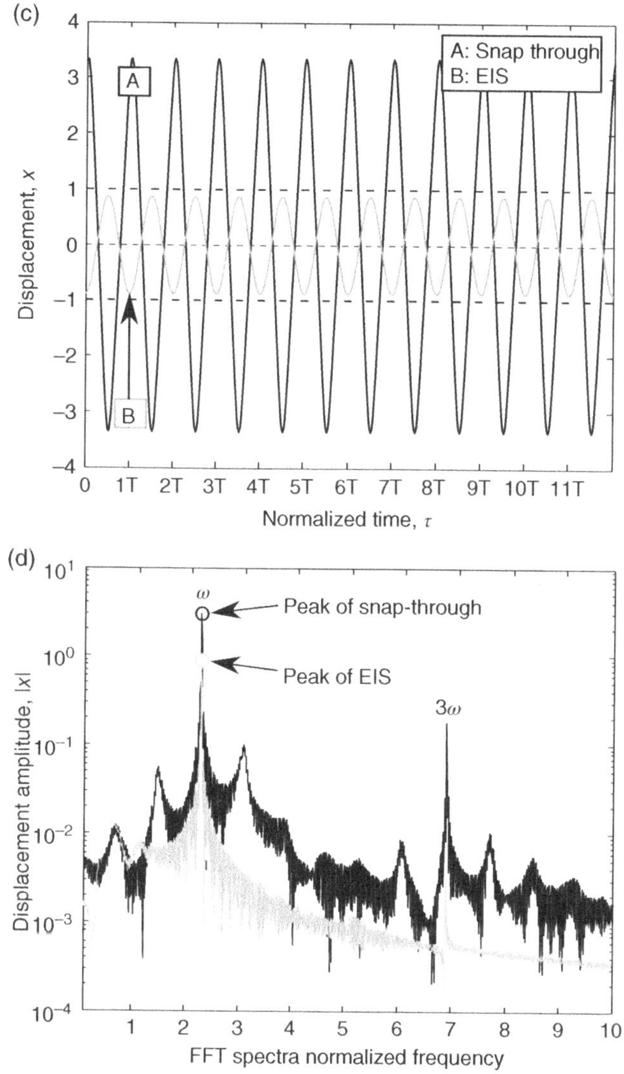

Figure 3.11 (Continued)

To investigate the influence of damping, responses were determined using system characteristics $p = 5$ and damping γ ranging from 0.05 to 0.5. Figure 3.12a presents the EIS response amplitude as a function of damping and input frequency, where these values are predicted analytically from Eq. (3.10) and a corresponding Jacobian analysis for determination of stability. Increasing lightness of the contour represents increasing response amplitude. The approximated upper and lower frequency bounds ω_{Ub} and ω_{Lb}, determined using simplified forms of Eqs. (3.18) and (3.20), respectively, are presented as dashed lines in Figure 3.12a.

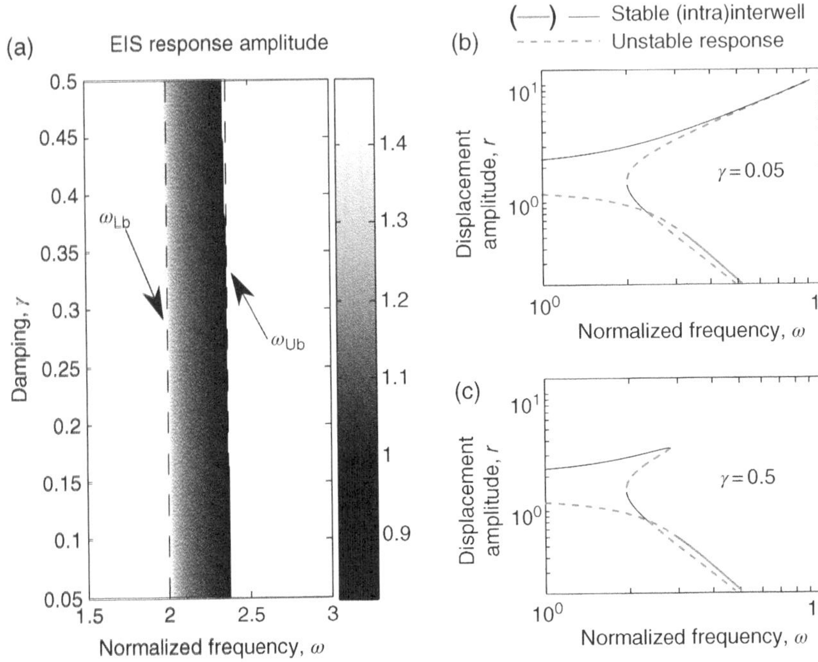

Figure 3.12 (a) EIS response amplitude as a function of damping and frequency with system parameter $p = 5$. Increasing lightness of the contour represents increasing response amplitude. Dashed lines correspond to estimated frequency boundaries using simplified form of Eqs (3.18) and (3.19). (b) and (c) are two analytically predicted, representative responses with damping $\gamma = 0.05$ and $\gamma = 0.5$, respectively. Solid black (gray) lines indicate interwell (intrawell) responses and dashed gray lines are unstable responses.

As damping γ changes from 0.05 to 0.5, the analysis predicts that EIS will persist although the stable bandwidth of frequencies across which it exists is seen to shift slightly downwards in frequency. However, as damping changes, the predicted steady-state response amplitude does not alter to a significant degree at a given frequency of excitation, which indicates that damping has little influence on the stability and vibration amplitude of the EIS dynamic. Nonetheless, a drastic degradation on the bandwidth of stable, large amplitude snap-through response is observed as damping γ increases. This is depicted in Figure 3.12b and c which show representative response amplitudes, determined from Eqs (3.9) and (3.10); here, solid black (gray) lines represent interwell (intrawell) responses and dashed gray lines denote unstable responses. For damping $\gamma = 0.05$, as seen in Figure 3.12b, the analysis predicts the snap-through dynamics are destabilized for excitation frequencies $\omega > 9.27$, whereas the behaviors destabilize for frequencies $\omega > 2.86$ when damping is $\gamma = 0.5$, Figure 3.12c. Overall, in Figure 3.12a, the criteria boundary definitions (thick dashed lines) and analytically-determined boundaries (defined by the limits of the contour) are in good agreement for smaller values of damping γ. Collectively, the results of Figure 3.12 indicate that the EIS oscillations, favorable for vibration isolation performance, are insensitive to changes in damping whereas the energetic snap-through dynamics, detrimental to vibration isolation performance, may be destabilized due to a higher degree of damping;

For the system parameters under study, EIS always coexists with the regular interwell response, as depicted in Figure 3.12b and c. To determine which outcome is more likely on changing the initial conditions, the basins of attraction are investigated at an excitation frequency $\omega = 2.3$ and level $p = 5$. Figure 3.13a1, b1, and c1 show numerically determined basins for damping $\gamma = [0.1 \;\; 0.25 \;\; 0.5]$. Three distinct responses are observed. In the basin plots, the unshaded areas correspond to steady-state snap-through dynamics, gray shaded regions indicate EIS, and black shading denotes a period-3 harmonic interwell response. As damping increases from $\gamma = 0.1$ to $\gamma = 0.25$, a greater number of initial conditions leads to EIS. For the highest damping considered $\gamma = 0.5$, period-3 harmonic interwell response is induced, appearing to primarily coalesce in a fractal manner amongst the initial conditions that lead to EIS, clearly seen in Figure 3.13c1. The three dynamic responses are illustrated via steady-state time-series and phase portraits – middle and bottom rows of Figure 3.13, respectively – where the damping is $\gamma = 0.5$. Figure 3.13 exemplifies the initial condition and damping dependence of either obtaining or avoiding EIS phenomena. From this evaluation, it is clear that vibration isolation systems exploiting bistable members are best served by incorporating a high degree of damping. On the other hand, a strict control of initial conditions may be required, as indicated by the fractal spread of period-3 dynamics amongst the initial conditions which lead to EIS oscillations, Figure 3.13c1.

3.3.4 Experimental Validation of the Stability Criteria

The influence of the excitation level $0.1 \leq p \leq 15$ is evaluated while damping remains $\gamma = 0.1$. Figure 3.14 presents the EIS response amplitude as a function of excitation level and frequency. The contour in Figure 3.14a shows that, for a given excitation level, the amplitude of the EIS response is reduced with increasing frequency of excitation. The frequency bounds ω_{Ub} and ω_{Lb} determined from the criteria of Eqs (3.18) and (3.19), respectively, match very well with the limits of the EIS response which are predicted analytically using Eq. (3.10). When the excitation level p is less than the critical value p_{cr}, determined from Eq. (3.21), no stable EIS region exists. This is verified by the corresponding response predictions from the harmonic balance method, with one exemplary case provided in Figure 3.14b. As p increases above the threshold value, Figure 3.14a shows that the frequency range of EIS responses correspondingly increases. The harmonic balance response predictions in Figure 3.14c show that the EIS branch indeed materializes for an excitation level above the threshold value p_{cr}. These results help verify the accuracy of the criteria in defining the existence and stability of EIS oscillations. As they pertain to the activation of these low amplitude responses for bistable vibration isolation systems with respect to change in excitation level, the criteria provide clear guidelines as to when the EIS oscillations may be activated instead of the potentially damaging snap-through dynamics.

To validate the criteria governing the inducement of EIS response, a ferromagnetic bistable cantilever beam system is fabricated and a series of experiments

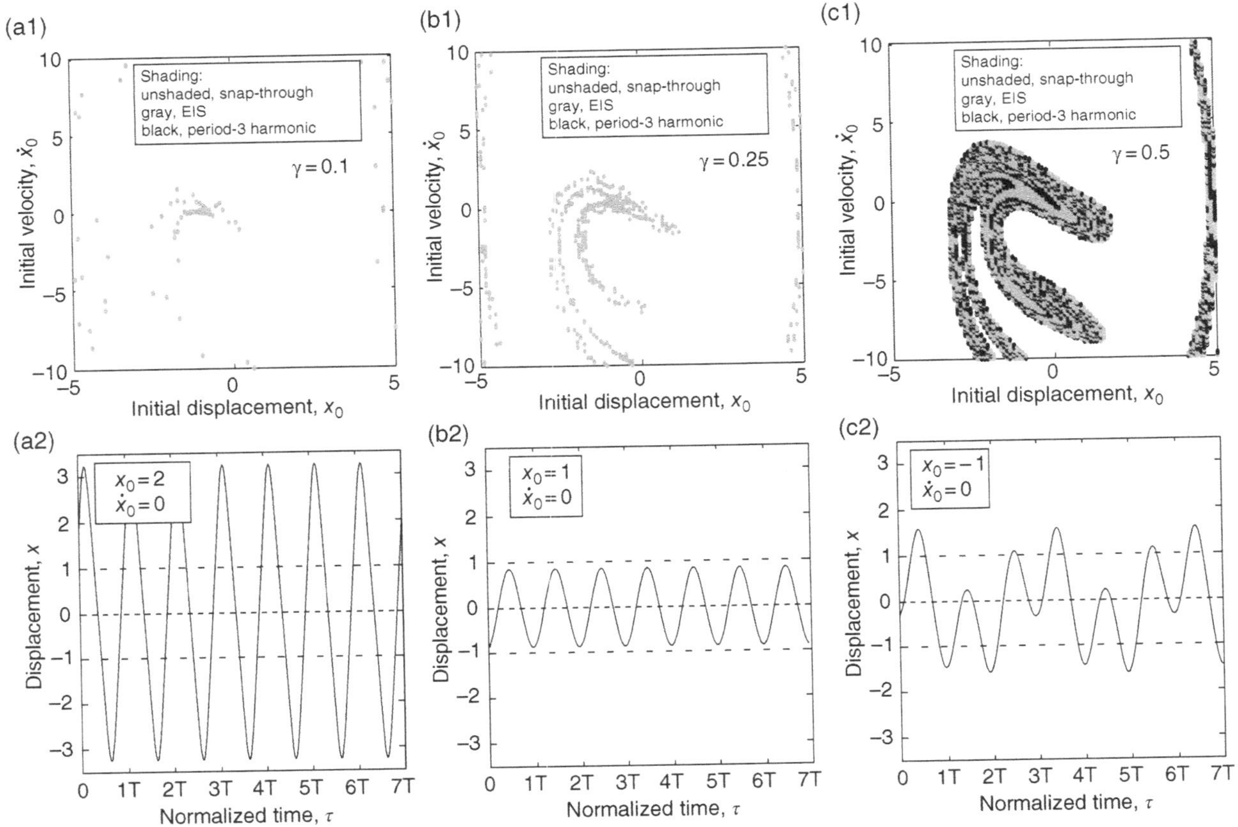

Figure 3.13 (a) Basin of attraction map for system parameters $p = 5$, $\omega = 2.3$, where unshaded, gray, and black shaded areas indicate snap-through, EIS, and period-3 harmonic interwell responses, respectively. (b) Steady-state time domain responses where $T = 2\pi / \omega$ is the excitation period, and (c) corresponding phase portraits for different initial conditions with damping $\gamma = 0.5$. In (b) and (c) thick dashed lines and filled circles, respectively, represent the two statically-stable equilibria positions.

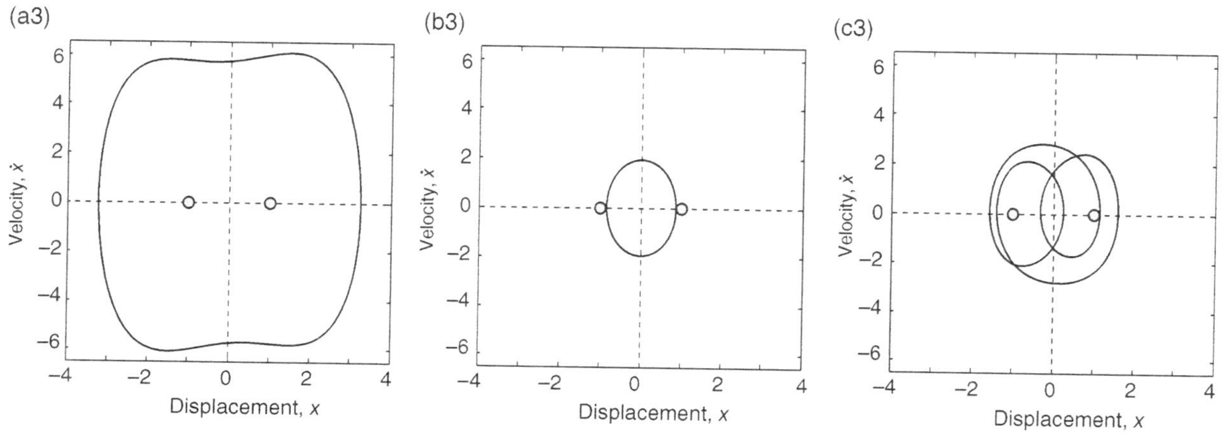

Figure 3.13 (Continued)

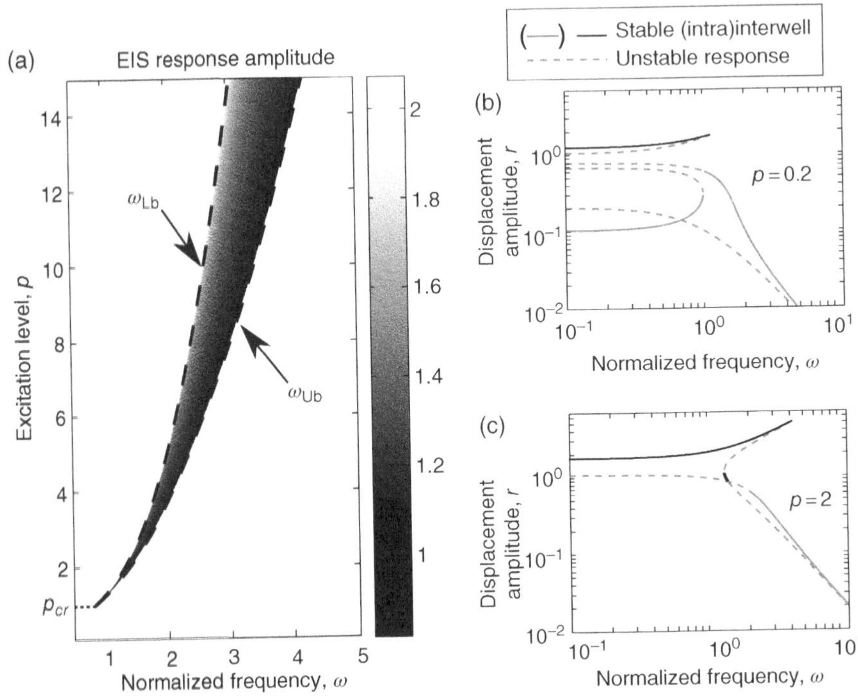

Figure 3.14 (a) EIS response amplitude as a function of excitation level and frequency with system parameter $\gamma = 0.1$. Increasing lightness of the contour represents increasing response amplitude. Dashed lines correspond to estimated frequency boundaries using simplified form of Eqs (3.18) and (3.19). (b) and (c) are two analytically predicted, representative responses with excitation level $p = 0.2$ and $p = 2$, respectively. Solid black (gray) lines indicate interwell (intrawell) responses and dashed gray lines are unstable responses.

performed. A schematic and photograph of the experimental set-up are provided in Figure 3.15. A supporting structural frame, from which the beam is cantilevered, is attached to an electrodynamic shaker table activated in the direction indicated in Figure 3.15. The motion of the beam tip x is the response coordinate of interest. Bistability of the beam free tip deflection is generated using attractive magnetic fields [103]. A 0.05 cm thick and 1.27 cm wide spring steel beam is clamped, and has a cantilevered length of 14.0 cm. Two 0.64 cm thick magnets of diameter 1.27 cm are placed $d = 2.54$ cm apart from center to center to realize a double-well potential. At rest, the normal distance between the tip of the beam and the magnet is approximately $h = 0.8$ cm. To better attain EIS phenomena, as suggested in Section 3.3.3, the beam damping is increased; to this end, a dissipative tape is applied close to the clamped end of the beam and extends along the length, as shown in Figure 3.15. It is assumed that only fundamental buckling mode dynamics are activated in the experiments and this assumption is borne out from examination of the data. The fundamental resonance frequency of the buckled beam around each magnet is approximately 19.5 Hz and the damping loss factor is identified to be $\gamma \approx 0.16$. Acceleration of the harmonic shaker excitation and displacement of the cantilever beam tip are recorded.

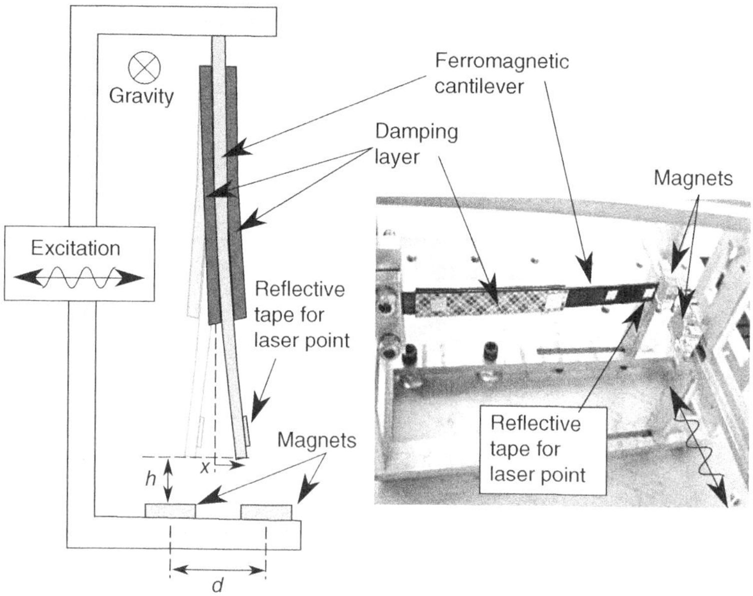

Figure 3.15 Schematic and photograph of the experimental system.

The shaker supplies a backward-swept sinusoid ($-0.05\,\mathrm{Hz\,s^{-1}}$) such that beam response is measured in the bandwidth of 12 to 24 Hz. The upper frequency bound ω_{Ub} is determined based on whether the oscillation is around one of the stable equilibria or the unstable equilibrium. For small oscillations, this characteristic change determines the transition between intrawell and the EIS response. The lower frequency bound ω_{Lb} is identified depending on the response amplitude: a sudden increase in amplitude is the transition from EIS to regular high orbit interwell during the backward frequency sweep. Figure 3.16a plots the experimentally recorded phase portraits for intrawell, EIS, and snap-through dynamics. Snap-through and EIS responses oscillate with near symmetric phase trajectories around the unstable equilibrium $(x, \dot{x}) = (0, 0)$, while the intrawell response orbits one of the stable equilibria. The displacement frequency response function (frf) magnitude of the EIS response is presented as the contour in Figure 3.16b as the shaker input frequency f and level $\ddot{u}$ are changed incrementally. The frf is determined as the ratio of the relative cantilever beam tip displacement and the shaker acceleration. The contour plot represents increasing displacement frf magnitude by increasing lightness of shading. The tests are conducted for an average base acceleration ranging from 2 to $3.74\,\mathrm{m\,s^{-2}}$. The frequency range of EIS responses is presented in Figure 3.16b by the distance spanned by the two boundary points for a given excitation level. The bandwidth is seen to increase as the excitation level increases; this result corresponds directly to the analytical predictions as shown in Figure 3.14. Experimental measurements also corroborate the analytical finding that EIS response amplitudes for a given excitation level decrease as excitation frequency increases. Overall, the experimental results are in very good qualitative agreement with the

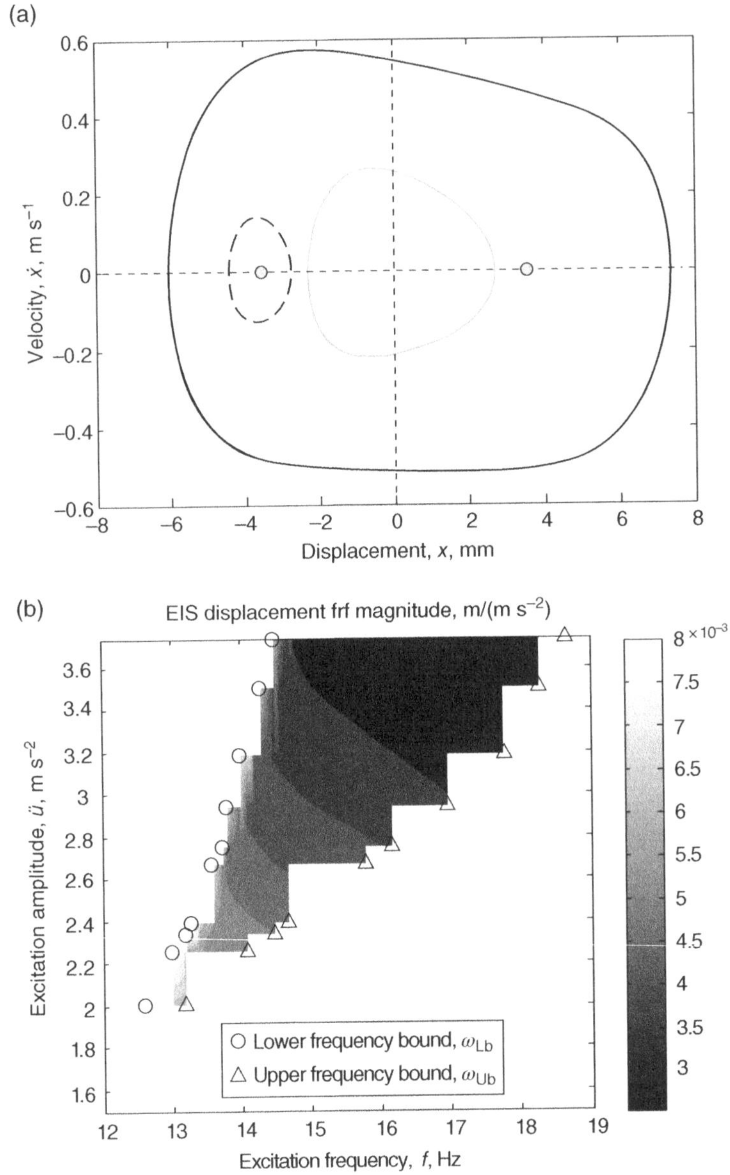

Figure 3.16 (a) Experimentally determined phase portraits with dashed black, solid gray, and black curves representing intrawell, EIS, and snap-through dynamics, respectively. Filled circles represent two stable equilibrium positions of the bistable oscillator. (b) Experimentally determined displacement frf of the bistable oscillator as a function of shaker input frequency and amplitude. Circles and triangles are experimentally determined frequency boundaries of stable EIS region.

analytical studies and demonstrate the prominent role of excitations in the realization and characteristics of activating EIS for a bistable system. By designing a bistable vibration isolation system in accordance with the criteria derived here, one may ensure that the low amplitude interwell EIS dynamics are triggered in the isolation platform should the excitation levels increase beyond those intended for ordinary operation of the system.

3.3.5 Summary

Robustness in the design of a vibration isolation system utilizing bistable members requires a complete assessment of the potential dynamic behaviors. Because the exploitation of bistable constituents introduces the concern that snap-through behaviors may be activated – which is detrimental to the integrity of the suspension system as well as unacceptable for isolation performance – the effective design of bistable isolators should consider means such that low amplitude EIS behaviors may be induced instead of snap-through. To elucidate this critical operating regime, criteria are derived which determine the critical forcing level to induce EIS dynamics and the frequency range across which EIS occurs. Examining these criteria shows that the existence and frequency range of EIS responses are primarily influenced by the level of excitation; for levels greater than the critical threshold the frequency range of EIS steadily increases. It is also found that the degree of damping in the bistable structure is not influential in governing the existence and frequency range of EIS. On the other hand, greater levels of damping increase the likelihood of obtaining EIS response when multiple dynamics coexist for the same excitation conditions. These findings are corroborated through direct simulations as well as a series of experiments, and clearly demonstrate the accuracy of the derived criteria. Through consideration of the criteria, one may more effectively harness the low amplitude EIS dynamics of a bistable vibration suspension system.

3.4 Exploiting Excitation-induced Stability for Dual-stage Vibration Isolation[3]

Following Platus' conceptual and commercial development of vibration isolation systems which incorporate negative stiffness members in parallel with positive linear stiffness springs [53,54], there was an effective explosion of interest in developing high-performance isolators inspired by this mechanical design [53,58–60,104–106]. Recently, Lu *et al.* [106] observed that the negative stiffness mechanism could provide an even greater advantage when configured as the upper stage of a dual-stage isolator; it was shown that the resulting dual-stage isolator is more effective than the single-stage design at higher frequencies. In all of the previous studies, the negative stiffness components are employed in parallel with linear springs of sufficient stiffness to ensure that the systems ultimately provide a *small, positive linear stiffness* for the anticipated set of working conditions (e.g., supported load, excitation frequency, etc.). Such configuration provides for the low dynamic and high static stiffnesses of

3 Material from this section has been reprinted by permission of IOP Publishing, from K. Yang, R.L. Harne, K.W. Wang and H. Huang, Investigation of a bistable dual-stage vibration isolator under harmonic excitation. Smart Materials and Structures 23:045033, 2014. doi: 10.1088/0964-1726/23/4/045033.

interest – in other words, to minimize the isolator resonance and provide ample nonlinear static stiffness for supporting large "dead" loads – while avoiding designs that exhibit bistability since the potential for jump-type phenomena is typically undesirable in relation to minimizing displacement or force transmission.

On the other hand, one may hypothesize that the contribution of dynamic forces from the negative stiffness member and the positive stiffnesses may be exploited in such a way as to *cancel* the *dynamic force* transmitted through the system. Indeed, the oscillations associated with the EIS phenomenon partially realize the idea of force counterbalance [89]. To date, no investigations of vibration isolation systems that deliberately exploit EIS have been reported. From one perspective, the design guidelines to activate EIS that were developed in Section 3.3 are sufficient means around which to develop a single-stage vibration isolator that harnesses the low amplitude dynamics of EIS. On the other hand, the recent study described above [106] found that nonlinear vibration isolators employing negative stiffness members (albeit with a net positive linear stiffness) were greatly enhanced by the dual-stage design. Bringing the performance advantages together from the insights of the prior research, this study investigates the development of dual-stage vibration isolation systems which retain *bistable* characteristics to exploit a *dynamic force counterbalance* based upon EIS. By integrating the beneficial dynamic features of dual-stage designs with EIS, the aim is to provide further leaps forward in performance and robustness via intelligent elucidation and deployment of the architectures.

The following sections investigate the opportunities for the proposed isolation system through a collection of analytical, numerical, and experimental examinations. To quantify the performance of the proposed platform and compare results to other architectures, the traditional metric of force transmissibility is employed.

3.4.1 Governing Equation Formulation of a Bistable Dual-stage Vibration Isolator

Figure 3.17 depicts the modeled geometry of the bistable dual-stage isolator explored here. The system comprises a bistable first stage and a linear second stage, where the stages are numbered according to their proximity to the harmonic exciting forces. The bistable stage consists of a mass m_1, a damper d_1, and a bistable spring mechanism which has a restoring potential energy expressed by $U(X_t) = -\frac{1}{2}k_1 X_t^2 + \frac{1}{4}k_3 X_t^4$, where X_t is the stroke motion of the bistable stage. Thus, the bistable stage has two stable equilibria located at $X_t = \pm\sqrt{k_1/k_3}$. The two stable equilibria are symmetric and adjacent to the unstable equilibrium at $X_t = 0$. The use of such a generalized formulation of potential energy allows the findings of this study to apply to a variety of common physical realizations of bistable spring mechanisms including axial pre-load, electrostatic effects, and magnetic interactions, as exemplified in Chapter 1. The mass, stiffness, and damping of the linear second stage are m_2, k_2, and d_2, respectively. The stroke motion of the linear stage is Y_t. The bistable stage mass m_1 is subjected to

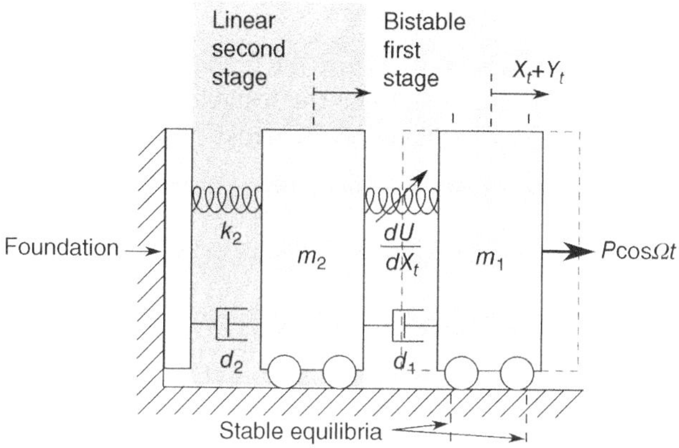

Figure 3.17 Schematic of the bistable dual-stage isolator and the harmonic exciting forces.

harmonic excitation force denoted by $P\cos\Omega t$, in which P, Ω, and t represent the amplitude, frequency, and time, respectively. The governing equations are therefore:

$$m_1\left(\ddot{X}_t+\ddot{Y}_t\right)+d_1\dot{X}_t-k_1X_t+k_3X_t^3=P\cos\Omega t \tag{3.22}$$

$$m_2\ddot{Y}_t+d_2\dot{Y}_t+k_2Y_t-d_1\dot{X}_t+k_1X_t-k_3X_t^3=0 \tag{3.23}$$

The transmitted force imposed on the foundation is P_t:

$$P_t=-\left[k_2Y_t+d_2\dot{Y}_t\right] \tag{3.24}$$

The following normalized parameters are defined:

$$\omega_1=\sqrt{k_1/m_1};\ \omega_2=\sqrt{k_2/m_2};\ f=\omega_1/\omega_2$$

$$\gamma_1=d_1/m_1\omega_1;\ \gamma_2=d_2/m_2\omega_2;\ \mu=m_1/m_2$$

$$x=X_t\sqrt{k_3/k_1};\ y=Y_t\sqrt{k_3/k_1}$$

$$p_0=P\sqrt{k_3}/k_2\sqrt{k_1};\ p_t=P_t\sqrt{k_3}/k_2\sqrt{k_1}$$

$$\omega=\Omega/\omega_2;\ \tau=\omega_2t$$

Substitution of the normalized parameters into Eqs (3.22–24) leads to a new equation system defined with respect to the normalized time τ:

$$x''+(1+\mu)f\gamma_1x'-(1+\mu)f^2x+(1+\mu)f^2x^3-\gamma_2y'-y-(p_0/\mu)\cos\omega\tau=0 \tag{3.25}$$

$$y''+\gamma_2y'+y-\mu f\gamma_1x'+\mu f^2x-\mu f^2x^3=0 \tag{3.26}$$

$$p_t=-(y+\gamma_2y') \tag{3.27}$$

where the operator' denotes the derivative with respect to τ.

3.4.2 Analytical Solution of the Governing Equations

A first order harmonic balance is employed to capture the fundamental dynamics of the system defined by Eqs (3.25–3.27). Since the bistable stage has two dynamic responses, inter- and intrawell oscillations, the stroke motions x and y can be expanded as

$$x = c_1(\tau) + a_1(\tau)\sin\omega\tau + b_1(\tau)\cos\omega\tau \tag{3.28}$$

$$y = a_2(\tau)\sin\omega\tau + b_2\cos\omega\tau \tag{3.29}$$

where $c_1(\tau) = 0$ and $c_1(\tau) \neq 0$ represent the inter- and intrawell response, respectively.

Eqs (3.28) and (3.29) are substituted into Eqs (3.25) and (3.26). Then one assumes that the leading coefficients slowly vary in time τ, and that the higher-order terms provide negligible contribution to the resulting responses. Next, the leading coefficients are grouped together according to the associated harmonic and corresponding equation. From this process, a series of modulation equations are derived, expressed in terms of the response coefficients.

$$-\gamma_1 c_1' = f\Lambda_c c_1 \tag{3.30}$$

$$-(1+\mu)f\gamma_1 a_1' + 2\omega b_1' + \gamma_2 a_2' = \Sigma a_1 - \sigma b_1 - a_2 + \omega\gamma_2 b_2 \tag{3.31}$$

$$-2\omega a_1' - (1+\mu)f\gamma_1 b_1' + \gamma_2 b_2' = \sigma a_1 + \Sigma b_1 - \omega\gamma_2 a_2 - b_2 - p_0/\mu \tag{3.32}$$

$$\mu f\gamma_1 a_1' - \gamma_2 a_2' + 2\omega b_2' = -\mu f^2\Lambda a_1 + \mu f\omega\gamma_1 b_1 + (1-\omega^2)a_2 - \omega\gamma_2 b_2 \tag{3.33}$$

$$\mu f\gamma_1 b_1' - 2\omega a_2' - \gamma_2 b_2' = -\mu f\omega\gamma_1 a_1 - \mu f^2\Lambda b_1 + \omega\gamma_2 a_2 + (1-\omega^2)b_2 \tag{3.34}$$

where the terms are defined as

$$\Lambda_c = -1 + c_1^2 + \frac{3}{2}r_1^2;\ \Lambda = -1 + 3c_1^2 + \frac{3}{4}r_1^2$$

$$\Sigma = (1+\mu)f^2\Lambda - \omega^2;\ \sigma = (1+\mu)f\omega\gamma_1$$

$$r_1 = \sqrt{a_1^2 + b_1^2};\ r_2 = \sqrt{a_2^2 + b_2^2}$$

where r_1 and r_2 represent the oscillation amplitudes of the motions x and y, respectively.

The steady-state response of the system is now determined. After reduction of Eqs (3.31–3.34), a polynomial is obtained; the roots of the polynomial are the responses of the bistable stage amplitude r_1. The polynomial is found to be

$$r_1^2\left(\alpha_1\Lambda^2 + \alpha_2\Lambda + \alpha_3\right) = (p_0/\mu)^2\left[\left(1-\omega^2\right)^2 + (\omega\gamma_2)^2\right] \tag{3.35}$$

where,

$$\alpha_1 = \left[f^2 - (1+\mu)f^2\omega^2\right]^2 + \left(f^2\omega\gamma_2\right)^2 \tag{3.36}$$

$$\alpha_2 = -2f^2\omega^2\left\{\left(1-\omega^2\right)\left[1-\left(1+\mu\right)\omega^2\right]+\left(\omega\gamma_2\right)^2\right\} \tag{3.37}$$

$$\begin{aligned}\alpha_3 = \omega^2 & \left\{\left(1-\omega^2\right)^2\omega^2+\left(f\gamma_1\right)^2\left[1-\left(1+\mu\right)\omega^2\right]^2\right. \\ & \left.+\omega^2\left[\left(f\gamma_1\gamma_2\right)^2+\left(\omega\gamma_2\right)^2+2\mu f\gamma_1\gamma_2\omega^2\right]\right\}\end{aligned} \tag{3.38}$$

Note that Λ in Eq. (3.35) contains the coefficient c_1^2 which is determined by solving the steady-state response of Eq. (3.30), where $c_1^2 = 0$ or $c_1^2 = 1 - \frac{3}{2}r_1^2$. Hence, the amplitude r_1 of either interwell (i.e., $c_1^2 = 0$) or intrawell (i.e., $c_1^2 = 1 - \frac{3}{2}r_1^2$) response is derived. According to Eq. (3.27), the force transmissibility is

$$|\mathrm{TR}| = \left|\frac{P_t}{P}\right| = \left|\frac{p_t}{p_0}\right| = \sqrt{1+\left(\omega\gamma_2\right)^2}\left|\frac{r_2}{p_0}\right| \tag{3.39}$$

3.4.3 Examining the Stability of Analytical Predictions

It is necessary to identify the stable and unstable solutions of the polynomial Eq. (3.35). Equations (3.30—3.34) may be rewritten in the form,

$$\mathbf{z}' = \mathbf{G}^{-1}\mathbf{F}\left(\mathbf{z}\right) \tag{3.40}$$

where,

$$\mathbf{z} = \begin{bmatrix} c_1 \\ a_1 \\ b_1 \\ a_2 \\ b_2 \end{bmatrix}; \mathbf{G} = \begin{bmatrix} -\gamma_1 & 0 & 0 & 0 & 0 \\ 0 & -\left(1+\mu\right)f\gamma_1 & 2\omega & \gamma_2 & 0 \\ 0 & -2\omega & -\left(1+\mu\right)f\gamma_1 & 0 & \gamma_2 \\ 0 & \mu f\gamma_1 & 0 & -\gamma_2 & 2\omega \\ 0 & 0 & \mu f\gamma_1 & -2\omega & -\gamma_2 \end{bmatrix}$$

and $\mathbf{F}(\mathbf{z})$ is the right hand side of Eq (3.30–3.34). The Jacobian matrix of Eq. (3.40) is determined from

$$\mathbf{J} = \mathbf{G}^{-1}\left.\frac{\partial\mathbf{F}}{\partial\mathbf{z}}\right|_{\mathbf{z}=\mathbf{z}_s} \tag{3.41}$$

where $\mathbf{z}_s$ denotes the vector of steady-state solutions. The stable response solutions are ascertained if all eigenvalues of the matrix $\mathbf{J}$ have negative real parts.

3.4.4 Comparison of Isolator Performance with a Counterpart Linear Design

When the first stage of the isolator is replaced by a linear oscillator, the system becomes a linear dual-stage isolator. To provide meaningful comparison between the isolator designs, it is intuitive that the isolators must provide an identical static loading capacity. Equations (3.42) and (3.43), respectively, represent the static loading of the bistable and linear dual-stage isolators.

$$-k_1 X + k_3 X^3 = P \tag{3.42}$$

$$k_L X = P \tag{3.43}$$

Here, P is the static load; k_L is stiffness of the equivalent linear first stage, and thus the natural frequency of this stage is $\omega_L = \sqrt{k_L / m_1}$; X is the deformation induced by the static load. Using the same approach that normalizes Eqs (3.22) and (3.23), and defining $f_L = \omega_L / \omega_2$ as the frequency tuning ratio of the equivalent linear dual-stage isolator, where ω_2 is the natural frequency of the linear second stage, the parametric relationship between the isolators is

$$f_L^6 + f^2 f_L^4 - f^2 \left(p_0 / \mu\right)^2 = 0 \tag{3.44}$$

Figure 3.18a plots the force transmissibility of the bistable dual-stage isolator, the equivalent linear dual-stage isolator, and a linear single-stage isolator having the same system parameters as the second stage of the linear dual-stage isolator. Figure 3.18b presents the stroke motion amplitudes of the first stages of the dual-stage isolators. For the bistable system, only the stable predictions from the analytical solution strategy are plotted. Parameters used to predict the responses are: mass ratio of the first stage mass to the second stage mass is $\mu = 1.0$; frequency tuning ratio of the first stage to the second stage of the bistable dual-stage isolator is $f = 0.1$; the normalized excitation force amplitude is $p_0 = 1.0$; based on Eq. (3.44), the frequency tuning ratio of the equivalent linear dual-stage isolator is $f_L = 0.46$; loss factors of the first and second stages are $\gamma_1 = 0.10$, and $\gamma_2 = 0.05$, respectively. To validate the analytical predictions of the bistable dual-stage isolator, Figure 3.18 also presents the single-periodic responses determined through numerical integration of governing Eqs (3.25–3.27) using a fourth-order Runge–Kutta algorithm.

Using these system parameters, as shown in Figure 3.18 when $\omega \leq 1.08$ the bistable first stage exhibits the energetic, snap-through interwell response, where the mass oscillates symmetrically around the unstable equilibrium $x = 0$. For the bistable dual-stage isolator, there are two interwell response branches: the snap-through dynamics where the first stage has the large oscillation amplitude, and the much smaller amplitude motions of excitation-induced stability (EIS), which occurs across the bandwidth $0.55 \leq \omega \leq 1.08$ in this example. When the excitation frequency $\omega > 1.08$, the bistable stage vibrates around either equilibrium (i.e., intrawell response) for the present set of system parameters. In Figure 3.18(a), for the bistable dual-stage isolator, between the frequencies of the two peaks ($0.55 \leq \omega \leq 1$), there exists a *valley* response in the EIS frequency band where the vibration is effectively isolated and the maximum attenuation can be nearly 30 dB. In contrast, the linear dual-stage isolator provides much less isolation in the bandwidth between its respective peaks, Figure 3.18a. This valley response indicates that dynamic stiffness of the bistable dual-stage isolator is significantly less than that for the linear dual-stage isolator. A detailed interpretation of this unique feature will be presented in Section 3.4.5. After the second peak, both

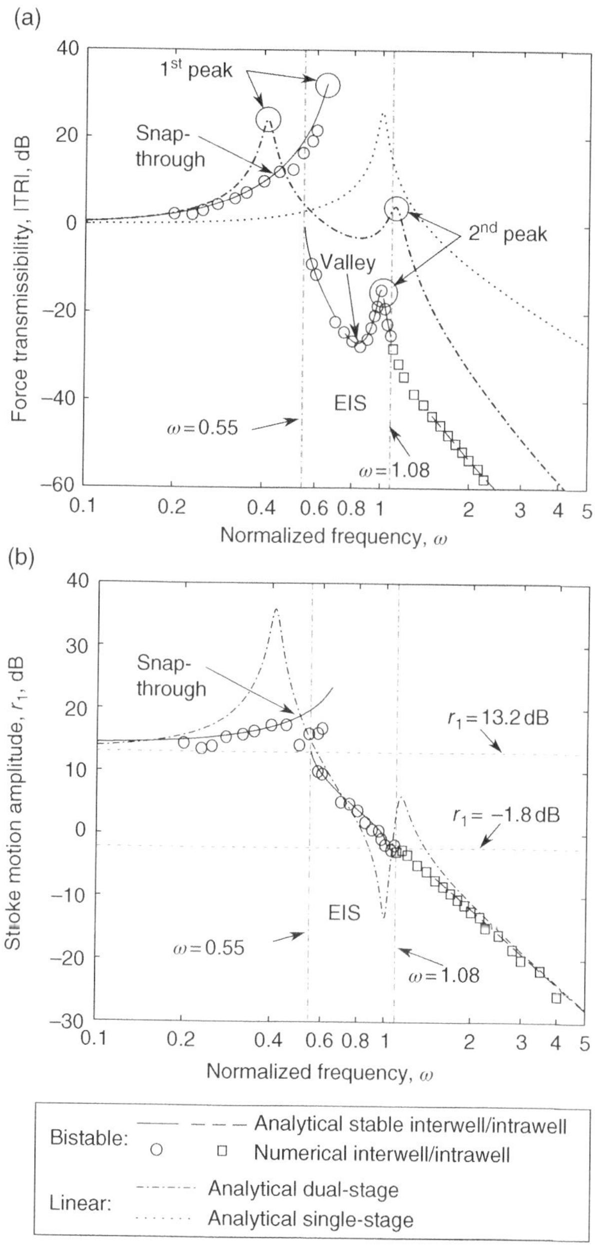

Figure 3.18 Force transmissibility comparison of a bistable dual-stage isolator with an equivalent linear dual-stage isolator and a linear single-stage isolator: (a) force transmissibility; (b) stroke motion amplitudes of the first stage.

dual-stage isolators outperform the single stage isolator in the high frequency band, because the vibration attenuation rates with respect to ω of the bistable and linear dual-stage isolators are twice that of the single stage isolator, as shown in Figure 3.18a. The clear difference between the amplitudes of the dual-stage isolators' transmissibilities at high frequencies indicates that the bistable dual-stage isolator provides a more effective isolation performance when the static loading capacities of the dual-stage systems are identical. Moreover, compared with the linear dual-stage isolator, Figure 3.18b shows that the bistable dual-stage isolator can achieve much smaller stroke motion, which reflects the smaller deflection of the bistable oscillator's nonlinear spring. Hence, the bistable dual-stage isolator is more suitable for applications having compact space constraints. These factors encourage further investigation of the source of these advantages for the bistable dual-stage isolator and if they may be suitably harnessed for greater degrees of performance.

For the numerical data points in Figure 3.18 presented as circles and squares for interwell and intrawell response, respectively, the amplitude at each frequency is the fundamental component of a single-periodic response computed by taking the fast-Fourier transform (FFT) of the last 35% of the long-time response. Across frequencies, the numerical results are in good agreement with the analytical predictions, validating the model composition and assumptions. Around the valley response in $0.6 \le \omega \le 1.0$ and the intrawell response in $1.1 \le \omega \le 1.6$, numerical simulations uncovered potential multiharmonic responses. By repeated simulations of the model, these behaviors are observed to be strongly dependent on the initial conditions. The influence of the initial conditions is carefully investigated and discussed in Section 3.4.7.

3.4.5 Explanation of the Valley Response

The valley response exists within the frequency bandwidth corresponding to EIS and is a unique phenomenon due to bistability. Because the restoring force of the bistable stage spring includes contributions from negative and positive stiffness members – negative linear and positive cubic potential forces – the dynamic force supplied by one component may be counterbalanced by the other. Therefore, the linear stage of the isolator, which is subjected to the dynamic force of the bistable stage, will also possess a valley response and consequently transmit a minimum of vibration to the foundation. A quantitative explanation of the phenomenon is provided by rewriting Eq. (3.26) as

$$y'' + \gamma_2 y' + y = \mu f F_{\text{bistable}} \tag{3.45}$$

$$F_{\text{bistable}} = \gamma_1 x' + f\left(-x + x^3\right) \tag{3.46}$$

Thus, the vibration of the linear stage depends on the dynamic force F_{bistable} which is determined by the stroke motion of the bistable stage. By substituting the interwell solution into Eq. (3.46) and neglecting higher order terms, the fundamental amplitude of the dynamic force is determined to be:

$$\left|F_{\text{bistable}}\right| = r_1 \sqrt{\left(\omega \gamma_1\right)^2 + f^2\left[(3/4) r_1^2 - 1\right]^2}$$

If $r_1 \to \sqrt{4/3}$ and $\gamma_1 \to 0$, the dynamic force will approach zero, and thus excitation transmitted to the linear stage, and hence the foundation, will

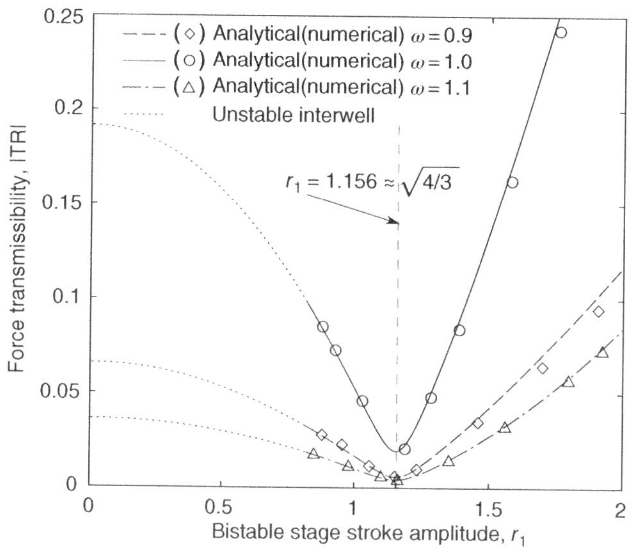

Figure 3.19 Force transmissibility of the valley response as a function of the bistable stage stroke amplitude.

correspondingly approach zero. This is the valley response characteristic. Based on this interpretation, for a lightly damped bistable stage, the valley point (minimum of the valley response) will be obtained for any combination of excitation force p_0 and frequency ω so long as the bistable stage stroke amplitude $r_1 = \sqrt{4/3}$. To verify this conclusion, Figure 3.19 plots the force transmissibility of the valley response as a function of bistable stage stroke amplitude r_1 at three excitation frequencies, $\omega = [0.9 \quad 1.0 \quad 1.1]$. The loss factor $\gamma_1 = 0.01$ is selected for the bistable stage for the computations. Across each curve, excitation forces p_0 are varied to implement different stroke amplitude r_1, whereas the excitation frequency ω is fixed. Therefore, for each curve in Figure 3.19, r_1 in the horizontal axis is varied only with respect to the excitation force p_0 according to Eq. (3.35). In this figure, results computed from numerical simulations are denoted by circles for verification of the analytical findings. Figure 3.19 shows that the valley point locations at the three frequencies are identical, and the corresponding stroke amplitudes are $1.156 \approx \sqrt{4/3}$. Numerical results are in good agreement with the analytical predictions. This verifies that the first order harmonic balance method is accurate to obtain the steady-state responses of the bistable dual-stage isolator. The results also validate the consistent occurrence of the valley point minimum when the amplitude of the bistable stage stroke tends to $r_1 = \sqrt{4/3}$ due to specific combinations of the excitation force p_0 and frequency ω, which serves as a useful design guideline for vibration isolation improvement.

3.4.6 Investigating the Design Parameter Influences

To develop insights into an effective design of the bistable dual-stage isolator, three parameters which may be feasibly tailored in such an isolator platform are

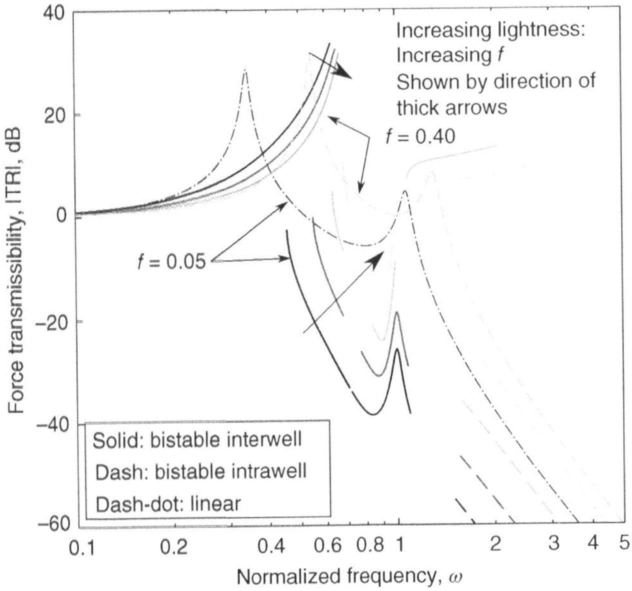

Figure 3.20 Force transmissibility comparison of the bistable and linear dual-stage isolators corresponding to variation in frequency tuning ratio where $f = [0.05 \quad 0.10 \quad 0.20 \quad 0.40]$.

selected for parametric study: the frequency tuning ratio of the first stage to the second stage f, the mass ratio of the first stage mass to the second stage mass μ, and the loss factor of the first stage γ_1. The initial parameters for the investigations are $\mu = 1.0$, $f = 0.1$, $\gamma_1 = 0.05$, $\gamma_2 = 0.05$, $p_0 = 1.0$. In the following studies, when one parameter changes, the remaining parameters are kept constant at the initial values described above.

Figure 3.20 shows the responses corresponding to variation in frequency tuning ratio across the range $f = [0.05 \quad 0.10 \quad 0.20 \quad 0.40]$. The corresponding linear dual-stage isolator responses with frequency tuning ratios calculated by Eq. (3.44) are shown for the cases of frequency tuning ratios of the bistable dual-stage isolator $f = 0.05$ and $f = 0.40$. As f increases, the isolation performances of the bistable and linear systems are both found to degrade. Unlike the linear counterpart, the location of the second peak of the bistable dual-stage isolator is unaffected by a change of the tuning ratio. For some higher values of the tuning ratio, $f \geq 0.20$, another branch of interwell response may emerge and potentially coexist across portions of the intrawell frequency band. For larger values of tuning ratio f, the bandwidth of the stable responses within the valley response diminishes. This indicates that it becomes more difficult to retain single-periodic vibration in the valley response bandwidth so that beneficial isolation performance is less ensured. Therefore, for better isolation performance, the frequency tuning ratio needs to be small.

The influence of the mass ratio μ is shown in Figure 3.21, where $\mu = [0.1 \quad 0.5 \quad 1.0 \quad 2.0]$. The first peak of the bistable dual-stage isolator shifts to lower frequency as μ increases, which is the same trend exhibited by the linear counterpart's first peak. By increasing the mass ratio for the bistable dual-stage

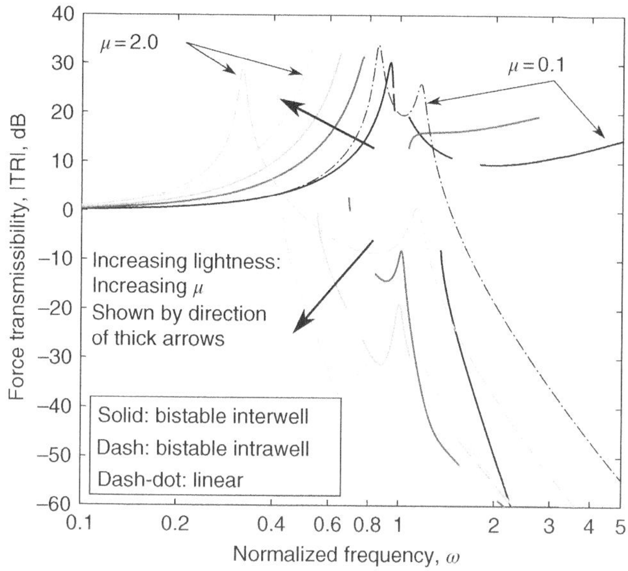

Figure 3.21 Force transmissibility comparison of the bistable and linear dual-stage isolators corresponding to variation in mass ratio where $\mu = [0.1 \quad 0.5 \quad 1.0 \quad 2.0]$.

isolator, the beneficial valley response is stabilized, and the valley point frequency location moves to lower frequency. It can be also seen that a large mass ratio may destabilize another interwell response which overlaps the beneficial response for vibration isolation in the high frequency band where $\omega > 1.0$. Therefore, a large mass ratio improves the reliability of the bistable dual-stage isolator.

Figure 3.22 shows how changing the first stage loss factor γ_1 influences the response, where $\gamma_1 = [0.01 \quad 0.05 \quad 0.10 \quad 0.20]$. When γ_1 increases, both resonant peaks of the linear dual-stage isolator are attenuated and the isolation performance at high frequencies degrades. Although the bistable dual-stage isolator has the same trend as the linear counterpart in the intrawell response regime at higher frequencies, the variation in the frequency band of the interwell response is far different. The large amplitude snap-through dynamics are seen to be insensitive to changing loss factor because the response bandwidths and amplitudes for the several cases of damping are nearly coincident, whereas the valley responses may become greater with increase in γ_1. Around frequencies slightly greater than the second response peak for the bistable dual-stage isolator, an increase in loss factor γ_1 extends the bandwidth of existence of the intrawell response region towards the second peak, which suggests that it is more difficult to maintain an interwell response. The response of the valley point is lowest for smaller values of γ_1 which agrees with the interpretation in Section 3.4.5 regarding significant valley point attenuation magnitude if $\gamma_1 \to 0$. However, a too small bistable stage loss factor γ_1 (e.g., $\gamma_1 = 0.01$) may activate a high-energy branch of interwell response for frequencies $\omega > 1.0$. The stable frequency bandwidth for the valley response is also narrower for decreased loss factor. Practically speaking, due to the strong nonlinearity, a very small loss factor γ_1 might lead to chaotic or

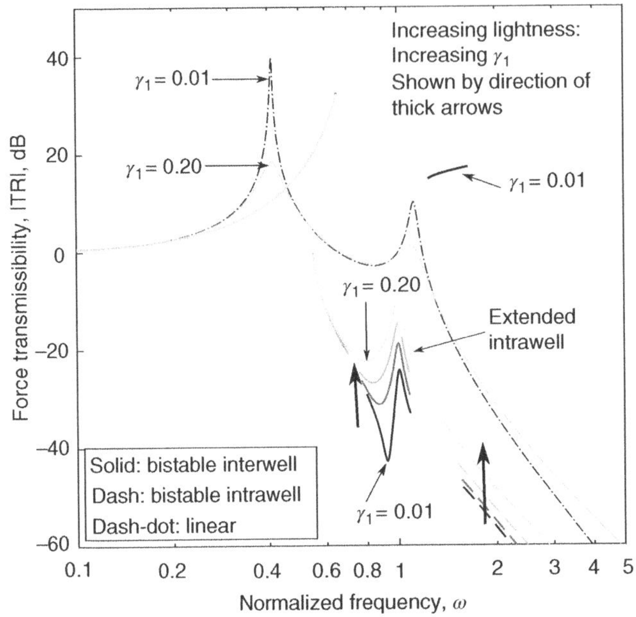

Figure 3.22 Force transmissibility comparison of the bistable and linear dual-stage isolators corresponding to variation in the loss factor of the first stage where $\gamma_1 = [0.01 \quad 0.05 \quad 0.10 \quad 0.20]$.

quasi-periodic vibrations which are not conducive to isolation performance. Due to these observations, the influence of loss factor upon the likelihood of steady-state response will be further studied in the following section.

3.4.7 Influence of Initial Conditions

The initial conditions may affect solution outcomes for the bistable dual-stage isolator, which is a common factor for nonlinear systems [66]. Therefore, in this section, a large number of random initial conditions are chosen for simulations to uncover the most repeatable nonlinear behaviors and determine the likelihood of the beneficial responses for vibration isolation. The parameters for numerical integration are identical to those in Figure 3.18. Taking $\omega = 0.8$ which is in the frequency band of the valley response as an example, the numerical responses of the bistable first stage are computed when the bistable dual-stage isolator has the initial conditions $(x_0, x_0', y_0, y_0') = (-1, 0, 0, 0)$ in Figure 3.23a and $(x_0, x_0', y_0, y_0') = (0, 0, 0, 0)$ in Figure 3.23b.

In the top row of Figure 3.23 are time series plots of the transmitted force p_t where $T = 2\pi / \omega$ in the horizontal axis represents the normalized period of excitation; the middle row of plots show the phase portraits of bistable stage motion x and velocity x'; while the bottom row of the figure presents the FFT spectra of the response x. When the initial conditions are $(-1, 0, 0, 0)$, Figure 3.23a, the oscillation is a single-periodic valley response. When the initial conditions are changed to $(0, 0, 0, 0)$, Figure 3.23b, order-3 harmonic behaviors emerge at frequencies $n\omega/3$ where $n = 1, 3, 5, \ldots$, and become a dominant contribution to the spectrum. As shown in the horizontal axes of Figure 3.23a2 and b2, the overall

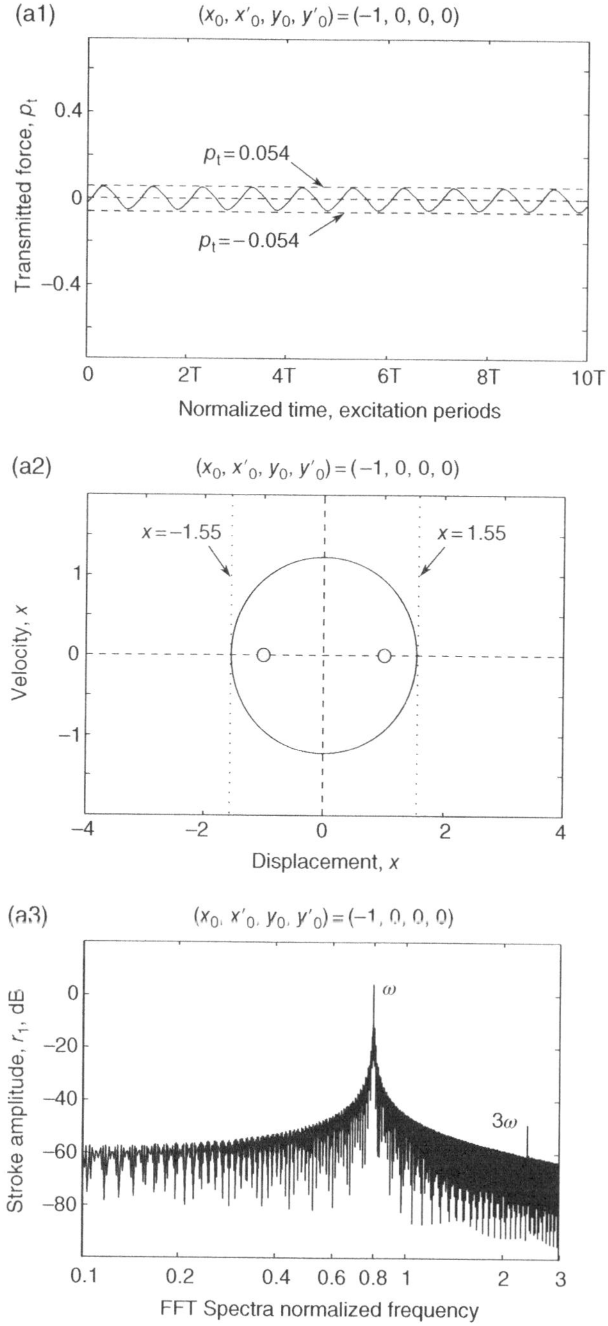

Figure 3.23 Numerically predicted responses of the bistable dual-stage isolator at $\omega = 0.8$ subjected to two different initial conditions (x_0, x'_0, y_0, y'_0): the top row, (−1, 0, 0, 0); the bottom row: (0, 0, 0, 0). From left to right: (a) and (d), time series plots of the transmitted force p_t; (b) and (e), phase plots of the bistable stage motion x and velocity x'; (c) and (f), FFT plots of the bistable stage motion x.

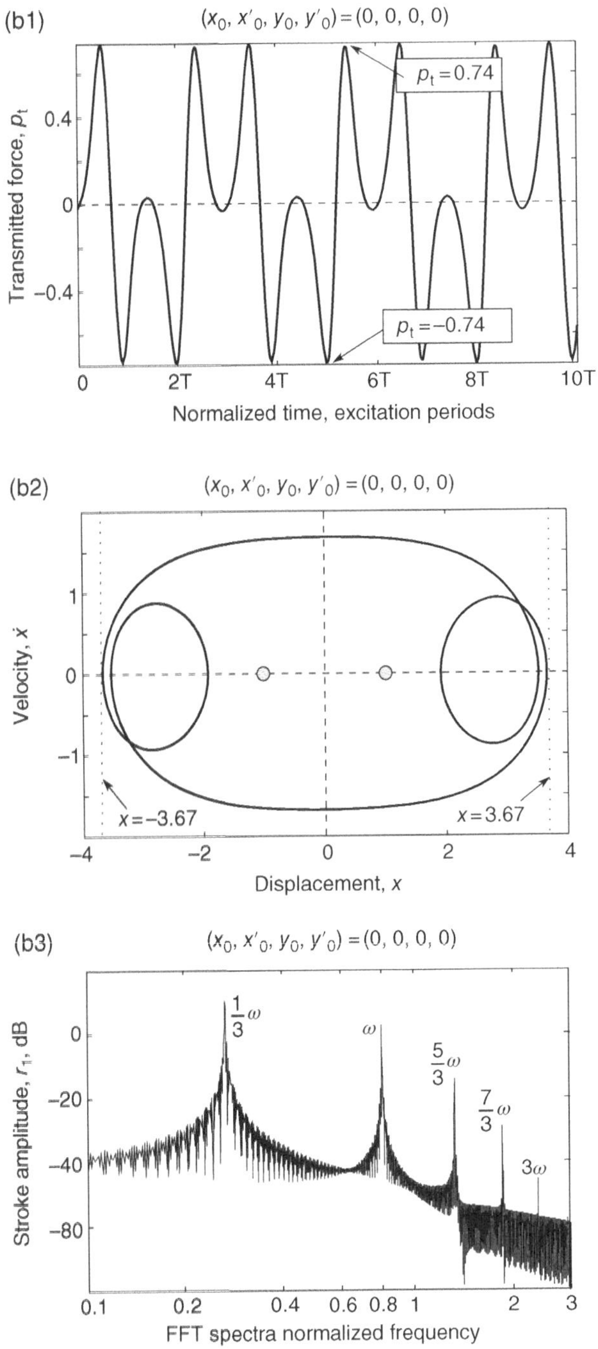

Figure 3.23 (Continued)

amplitude of the bistable stage response x for initial conditions (0, 0, 0, 0) is nearly twice that for (−1, 0, 0, 0), which leads to a larger dynamic force imposed on the linear stage so that the isolation performance is significantly degraded, that is, the maximum amplitude of the transmitted force $|p_t| = 0.054$ in Figure 3.23a1 while $|p_t| = 0.74$ in Figure 3.23b1. Figure 3.23 clearly demonstrates the large variation in response amplitude induced by either the single-periodic valley response or the multiharmonic behaviors. This suggests that the initial conditions should be closely investigated for design guidelines by which favorable isolation performance is ensured from the desired single-periodic valley response.

To know the initial condition range which leads to the single-periodic valley response and best vibration isolation, a basin of attraction map at $\omega = 0.8$ is constructed using a 300 by 300 grid of (x_0, y_0) displacement initial conditions. The map is determined by simulations where initial displacements of the stages are varied but all initial velocities are zero. This represents the isolator having various static loads and being excited from an initial state of rest. Figure 3.24 plots the basin map, where the black-shaded areas represent the initial conditions that lead to the single-periodic valley response beneficial for vibration isolation and unshaded areas indicate the responses tending to multiharmonic behaviors. Two unbroken regions of initial conditions stand out, which represent a large range of configurations for which small deviations within these regions lead to the same final steady-state response. Since the relative areas of these two regions leading to either single-periodic valley or multiharmonic response are similar, the task of the designer is to ensure the initial state of the isolator remains in the region leading to the beneficial single-periodic valley response (the black-shaded region).

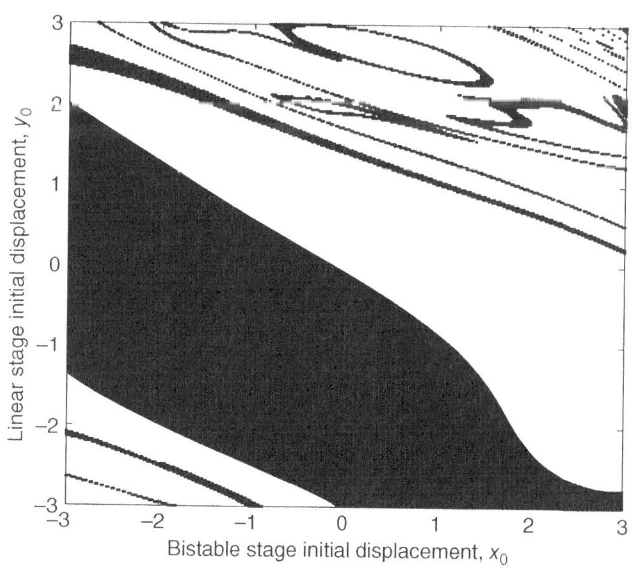

Figure 3.24 Basin of attraction map at $\omega = 0.8$. The initial condition ranges of (x_0, y_0) are both from −3 to 3. Single-periodic valley response, black-shaded area. Multiharmonic response with $\omega/3$ component, unshaded area.

It is seen that many such initial conditions in the unbroken black-shaded region have negative values, indicating the bistable stage mass is closer to the linear stage mass from the start of excitation, as envisioned using the schematic of Figure 3.17. This finding may be advantageous in application because it represents that the static load upon the bistable dual-stage isolator is in the direction of the foundation for the starting initial conditions that lead to the favorable single-periodic valley response. Such a static loading situation may be a common result, for example due to gravitational body forces applied because of a mass supported by the bistable stage.

As described in Section 3.4.4, the numerical results in Figure 3.18 uncovered the potential for multiharmonic responses in the frequency bandwidths $0.6 \le \omega \le 1.0$ and $1.1 \le \omega \le 1.6$. The activation of such multiharmonic responses is dependent on the initial conditions and their inducement will degrade isolation performance. For effective vibration isolation, the bistable dual-stage isolator must avoid these multiharmonic responses and exhibit the beneficial dynamics, namely the single-periodic valley and intrawell responses, depending upon the frequency of excitation. Therefore, it is insightful to study the likelihood of the single-periodic responses in the frequency band where multiharmonic responses may occur. In the following numerical examinations, the initial conditions of the time integration of governing Eqs (3.25 and 3.26) are $(x_0, x_0', y_0, y_0') = (-1 + \tilde{R}, 0, \tilde{R}, 0)$, where $\tilde{R}$ is a random variable satisfying a normal distribution with zero mean and unit standard deviation. The initial conditions indicate that, at the beginning of the application of the excitation the system is as rest, whereas the displacement of either stage is randomly distributed around its negative equilibrium position, representative of various static loads applied to the isolator.

Figure 3.25 presents the percentage likelihood that the bistable dual-stage isolator exhibits single-periodic valley and intrawell responses for several cases of the bistable stage loss factor γ_1. These two forms of single-periodic response – valley behaviors and intrawell oscillations – represent the favorable dynamics for isolation performance as demonstrated above, for example, in Figure 3.18. In Figure 3.25, the discrete excitation frequencies from $\omega = 0.6$ to $\omega = 1.6$ are chosen, and 200 long-time responses under the random initial conditions are simulated at each frequency. The results in $0.6 \le \omega \le 1.0$ represent the percentage likelihood of single-periodic valley responses, whereas the data in $1.1 \le \omega \le 1.6$ represent percentage likelihood of single-periodic intrawell responses. The alternative response form in both of these bandwidths (i.e., the remaining percentage) represents multiharmonic response outcomes which are adverse to vibration isolation performance. When $\gamma_1 = 0.05$, it is difficult to realize the single-periodic valley response at $\omega = 0.7$ and single-periodic intrawell response at $\omega = 1.1$. Overall, the likelihood of obtaining single-periodic valley and intrawell response is increased for greater γ_1, and is exceptionally likely to occur across the whole band of frequencies when the loss factor is $\gamma_1 = 0.20$. In fact, this represents a loss factor that could be easily realized in practice. The findings in Figure 3.25 indicate that greater damping of the bistable stage is a useful design strategy to stabilize the single-periodic dynamics and avoid multiharmonic response so that effective isolation performance may be ensured.

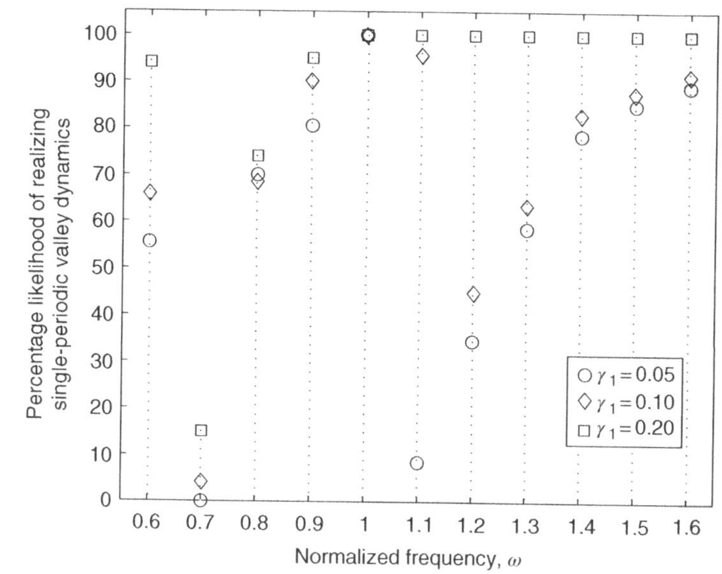

Figure 3.25 Percentage likelihood of the single-periodic valley responses in $0.6 \leq \omega \leq 1.0$ and single-periodic intrawell responses in $1.1 \leq \omega \leq 1.6$ for three cases of the bistable stage loss factor γ_1.

3.4.8 Prototype Investigations: Numerical and Experimental Validation

In this section, a prototype of the bistable dual-stage isolator configured by cantilevered beams and mutually repulsive magnets is developed to verify the prior analytical predictions by numerical simulation and experiment.

The prototype of the bistable dual-stage isolator is shown in Figure 3.26a. The bistable stage consists of two identical magnets and a cantilevered beam providing an effective, one-dimensional stiffness k_0. Damping tape is applied to the beam for the bistable stage damping d_1. The frame containing the bistable stage is connected to a second cantilevered beam which represents the linear stage stiffness k_2. Therefore, the frame and a portion of the second cantilevered beam represent the mass of the linear stage, and the displacement of the frame and second cantilever beam tip represent the stroke motion of the linear stage. X_t and Y_t are the stroke motions of the bistable and linear stages, respectively. One magnet is fixed at the tip of the bistable stage's beam, whereas the other is fixed on the frame, and they are oriented so as to repulse each other. By decreasing the distance Δ between the magnets, the repulsive magnetic force buckles the beam and induces two stable equilibria which are symmetric about the unstable equilibrium $X_t = 0$. The restoring force of the bistable stage is [107]:

$$F = k_0 X_t - 3HX_t\left(X_t^2 + \Delta^2\right)^{-\frac{5}{2}} \tag{3.47}$$

$$H = \frac{\eta_0 M^2}{2\pi} \tag{3.48}$$

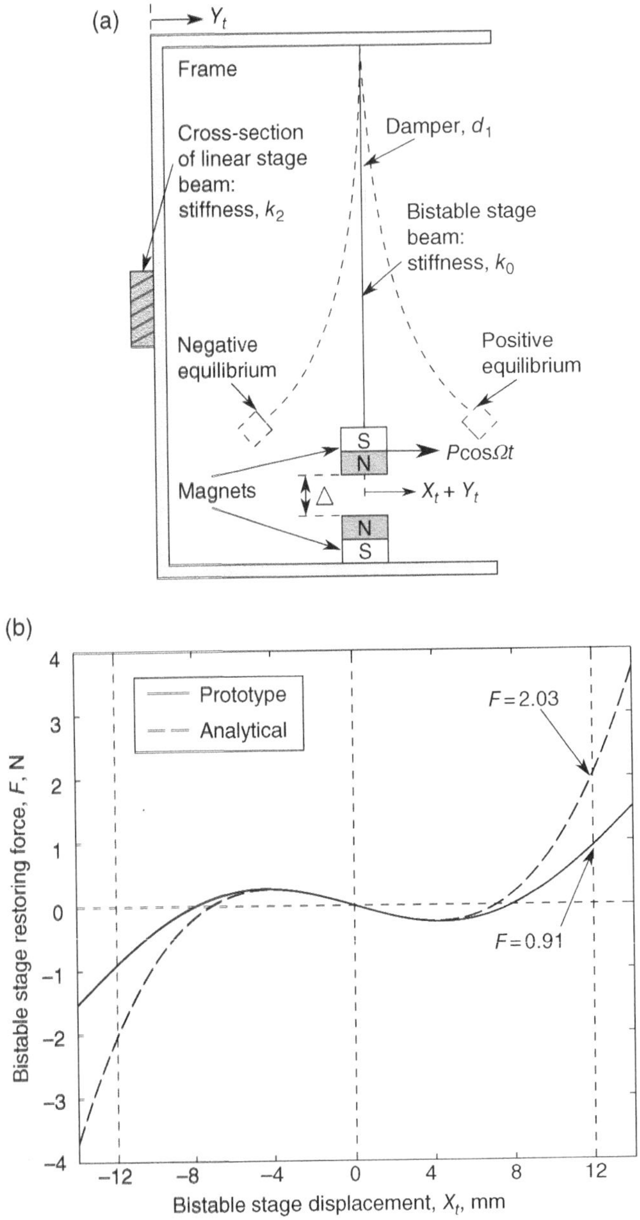

Figure 3.26 Schematic of a prototype of the bistable dual-stage isolator configured by cantilevered beams and mutually repulsive magnets (a) and bistable stage restoring spring forces F of the prototype and analytical model with respect to the bistable stage's stroke X_t.

where η_0 and M are free permeability and magnetic pole strength, respectively. The restoring force F may be expanded by a Taylor series such that

$$F = -k_1 X_t + k_3 X_t^3 - \frac{105H}{8\Delta^9} X_t^5 + \cdots \tag{3.49}$$

where,

$$k_1 = \left(\frac{3H}{\Delta^5} - k_0\right) \tag{3.50}$$

$$k_3 = \frac{15H}{2\Delta^7} \tag{3.51}$$

and

$$\Delta < \left(\frac{3H}{k_0}\right)^{1/5}$$

Thus, it is evident that there exists a deviation between the restoring force employed in the analytical model of Section 3.4.2 and the exact force expression of the specific prototype design considered in this section, which is given by Eq. (3.47). The deviation is reflected by the higher order terms in the Taylor series expansion which are neglected by the analytical model: $-(105H/8\Delta^9)X_t^5 + \cdots$. Thus, as regards the prototype configuration, the analytical model employs a restoring spring force expression of $F = -k_1X_t + k_3X_t^3$ for the bistable stage where k_1 and k_3 are given by Eqs (3.50) and (3.51). Figure 3.26b compares the two force expressions – Eq. (3.47) and the prior analytical model expression – to demonstrate the deviation. The parameters are $\eta_0 = 4\pi \times 10^{-7}$ N A^{-2} and $M = 1.74$ A m^2 while the distance between the magnets is $\Delta = 0.022$ m. The stiffness of the bistable stage cantilevered beam is $k_0 = 260$ N m^{-1}; therefore, by the analytical model expression, the stable equilibria of the bistable stage are approximately located at $X_t = \pm\sqrt{k_1/k_3} = \pm 7$ mm. Figure 3.26b shows that the deviation of the bistable stage restoring forces between the prototype and analytical model becomes non-trivial if the stroke X_t is considerably larger than the stable equilibria value. The result indicates that the analytical model predictions of the prototype response, as studied in this section, will be accurate so long as the bistable stage exhibits EIS or intrawell dynamics in which case stroke motion remains small.

To evaluate the validity of the earlier analytical findings that employed the reduced-order restoring spring force expression for the bistable stage against the exact expression in Eq. (3.47) for the prototype studied in this section, numerical simulations of the governing equations of the bistable dual-stage isolator are conducted using the full restoring force expression in Eq. (3.47) for the bistable stage. In the simulations, the bistable stage mass is $m_1 = 0.23$ kg. The linear stage mass is $m_2 = 0.23$ kg and stiffness is $k_2 = 9270$ N m^{-1}. As observed through the results shown in Figure 3.22, a large bistable stage loss factor γ_1 is conducive to stabilization of the single-periodic dynamics; thus, to ensure that the single-periodic EIS and intrawell responses are obtained, γ_1 is chosen to be 0.4 in the simulation. Finally, damping of the linear stage is $\gamma_2 = 0.05$, while the excitation amplitude is $P = 30$ N.

Figure 3.27 presents the force transmissibility of the transmitted force P_t to the excitation force P. The numerical simulation results of the prototype are in very good agreement with the analytical predictions across the bandwidth. Due to the stabilization effect of the large loss factor γ_1, the simulations predict that the

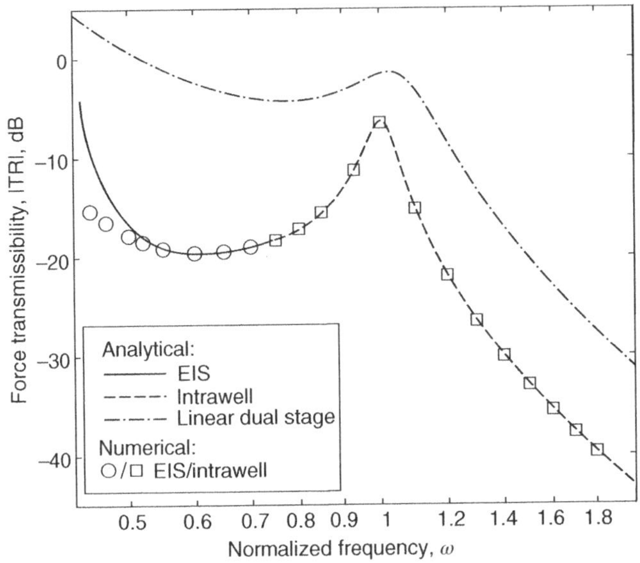

Figure 3.27 Force transmissibility of the force P_t to the excitation force P as a function of normalized excitation frequency ω.

bistable dual-stage isolator exhibits only the single-periodic EIS or intrawell responses. Figure 3.27 indicates that the bistable dual-stage isolator significantly outperforms the linear counterpart in the frequency band under consideration. The good agreement between numerical and analytical results validates the conclusions drawn from the earlier analytical investigations using the fundamental harmonic balance approach and simplified restoring force expression.

In addition to the numerical verification, experimental studies are conducted for a practical validation of the results. The experimental set-up is constructed to demonstrate the valley response attenuation capability of the bistable dual-stage isolator prototype shown in Figure 3.26a. A detailed schematic and photograph of the set-up are provided in Figure 3.28a and b, respectively. The bistable stage consists of a thin, cantilevered aluminum beam (length: 99.0 mm, width: 13.0 mm, thickness: 1.3 mm), two repulsive magnets, and a brass mass (0.225 kg). One of the magnets is embedded into the brass mass which is fastened to the tip of the beam, whereas the other magnet is fixed to a frame. The two magnets are oriented so as to repulse each other. The stable equilibria of the bistable stage are ±5 mm from the centerline.

In this experiment, the brass mass of the bistable stage is connected to the shaker armature with a ball joint to minimize torsional influences during shaker excitation. In fact, during experimentation the actual rotation of the mass is very small due to the large radius of curvature for the thin beam and the short range of motion considered here (i.e., the schematic of Figure 3.28a greatly exaggerates the mass translation and rotation for the sake of illustrating the overall dynamic behavior). The frame and a portion of the thick cantilevered beam represent the mass of the linear stage; thus, the thick cantilevered beam tip displacement represents the stroke motion of the linear stage. The shaker excitation constitutes displacement-controlled input (or velocity- or acceleration-controlled) as

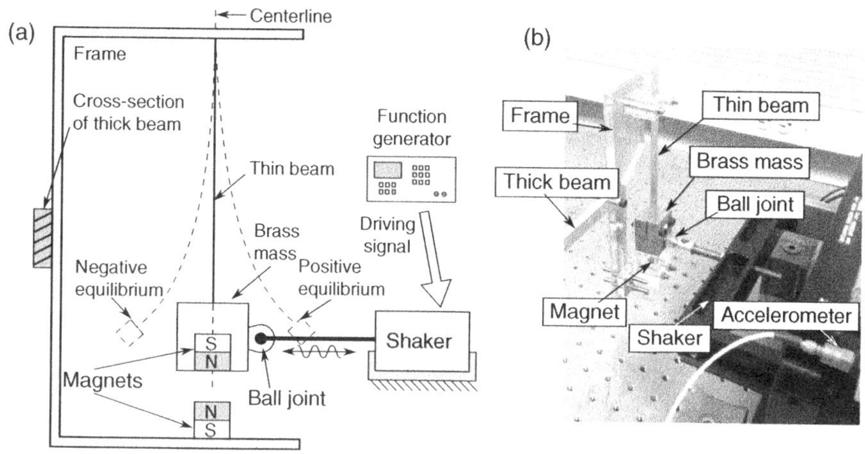

Figure 3.28 Experimental set-up: (a) schematic and (b) photograph, to validate the valley response of a bistable dual-stage isolator prototype.

compared to a constant periodic force. Therefore, the bistable stage mass connected to the shaker armature is driven to exhibit single-periodic vibration. By varying the control signal of the shaker, the bistable stage will exhibit harmonic vibrations of varied stroke amplitudes. This allows direct verification of the valley phenomenon as earlier studied through Figure 3.19. By Eq. (3.39), the transmitted vibration force is proportional to the vibration amplitude of the linear stage at a specified frequency. Thus, the vibration isolation can be expressed by dividing the root mean square (RMS) of the linear stage's steady-state stroke by the RMS of the shaker's steady-state stroke. Therefore, the RMS transmissibility is:

$$|\mathrm{TR}|_{\mathrm{RMS}} = \frac{\sqrt{\dfrac{1}{T}\displaystyle\int_0^T |Y_t|^2\,\mathrm{d}t}}{\sqrt{\dfrac{1}{T}\displaystyle\int_0^T |A_t|^2\,\mathrm{d}t}} \tag{3.52}$$

where Y_t and A_t are steady-state strokes of the linear stage and the shaker, and T is the sampling time used for the integration.

The parameters of the bistable stage are obtained by system identification. Then, the length of the thick beam is adjusted so that the frequency tuning ratio $f \approx 0.1$, which was also used in Figure 3.19 to interpret the valley response. The ratio of the bistable first stage mass to the linear second stage mass is $\mu \approx 1.0$. The loss factors of the bistable stage and the linear stage are $\gamma_1 \approx 0.025$ and $\gamma_2 \approx 0.05$, respectively. A function generator and amplifier drive the shaker at frequencies 4, 5, and 6 Hz, where the shaker is able to produce sufficient sinusoidal stroke. An accelerometer and laser interferometer measure the shaker acceleration and the vibration displacement of the thick beam tip (i.e., response of the linear stage), respectively.

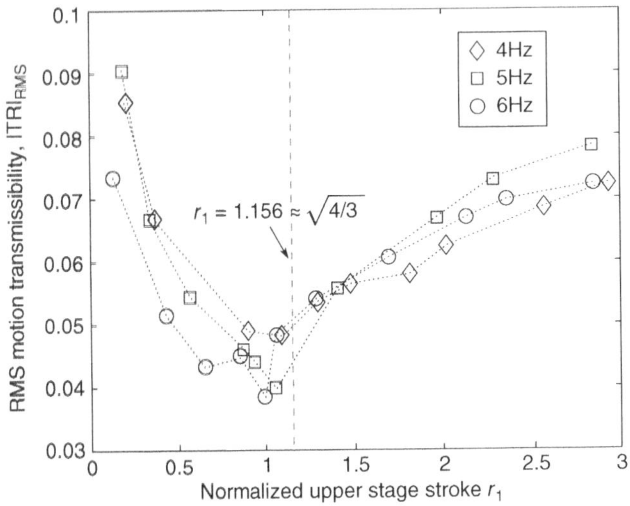

Figure 3.29 RMS transmissibility as a function of normalized bistable stage stroke.

Figure 3.29 presents the force transmissibility as a function of the bistable stage stroke, the latter being normalized by the magnitude of the stable equilibrium distance. The experimental results show that the minimum of the valley response is consistently obtained so long as the key bistable stage stroke amplitude r_1 is achieved, verifying the analytical observations in Section 3.4.5. The normalized bistable stage stroke corresponding to the minimum at each frequency is found to be between 1.0 and 1.1, which is also in good agreement with the analytically predicted value $\sqrt{4/3} \approx 1.156$. There is deviation between the trend of the slopes of the curves predicted analytically and measured experimentally; this is attributed to the experimental configuration for which the repulsive magnetic dipole and bistable stage beam do not exactly realize negative linear and positive cubic potential forces and the use of the ball joint containing small damping material that dissipates higher frequency vibration. Apart from this minor discrepancy in trends, the experimental results are in very good agreement with the analyses regarding attainment of the valley response and its specific minimum as defined by the bistable stage stroke.

3.4.9 Summary

By integrating the advantages of dual-stage nonlinear vibration isolator design with a configuration that maintains bistability for exploitation of EIS phenomena, this study found that the force counterbalance enabled by EIS can greatly improve isolation performance. Based on analyses of the fundamental periodic vibration, it is found that the bistable dual-stage isolator may outperform an equivalent linear dual-stage isolator over a significant frequency bandwidth. It is shown that a valley response, the unique phenomenon of the bistable dual-stage isolator due to counterbalance of its potential forces, improves isolation performance. The means to realize and maintain the valley response minimum for favorable isolation performance is elucidated through the model formulation, and the strategy is verified through simulations and experiments. The initial condition dependence to realize the beneficial single-periodic valley behavior is examined and it is found that increasing the bistable stage damping is conducive to stabilization of the single-periodic responses for effective vibration isolation. Numerical and experimental investigations of a prototype both quantitatively and qualitatively validate the analytical predictions, indicating that the insights obtained from the analytical studies can be used to guide the favorable implementation of a bistable dual-stage vibration isolation system.

3.5 Dynamic Stabilization of a Vibration Suspension Platform Attached to an Excited Host Structure[4]

A vibration suspension system is inevitably attached or grounded to a moving structure or environment, such as the configuration shown in Figure 3.30. Depending on the relative inertial difference between the suspension platform and the host structure, the dynamics of the suspension may be strongly coupled to the host. Indeed, a fundamental assessment of the two DOF vibration problem indicates that only when the inertia of the host structure is many orders of magnitude greater than the suspended mass will the coupling between bodies become negligible [3]. Coupled suspension-to-host structure dynamics are exacerbated when the inertias of the two bodies are close in magnitude. This scenario occurs

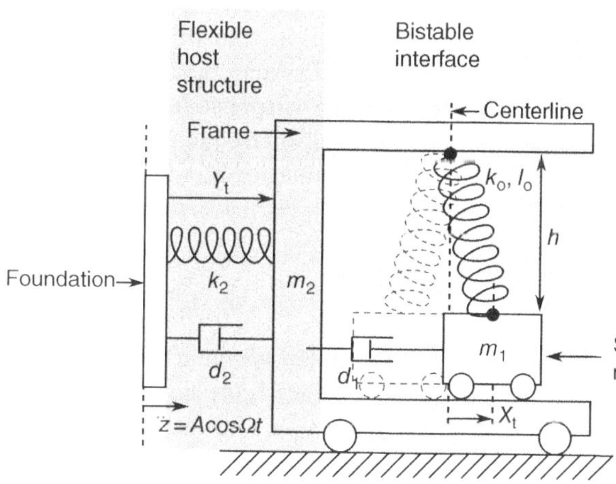

Figure 3.30 Schematic of the two DOF coupled system where a bistable suspension is attached to a flexible host structure.

4 Material from this section has been reproduced with permission from K.Yang, R.L. Harne, K.W. Wang, and H. Huang, Dynamic stabilization of a bistable suspension system attached to a flexible host structure for operational safety enhancement. Journal of Sound and Vibration 333 pp. 6651-6661, copyright 2014 Elsevier Ltd. Reproduced with permission. All rights reserved. doi: 10.1016/j.jsv.2014.07.033

in a variety of engineering contexts, such as for vibration isolation systems employed on off-shore platforms or buoys, in which case the suspended mass is dynamically coupled to the moving platform or buoy mass that is excited by the local wave activity. This scenario is also encountered for automotive suspension systems, in which case the sprung vehicle mass is coupled to an unsprung mass that is acted upon by the dynamic variations in terrain.

As such, the combination of the suspension, its load, and the host structure dynamics together can be approximately represented by a two DOF system. This two DOF system introduces two resonant features, and the second resonance occurs adjacent to the roll-off frequency band that is useful for vibration attenuation [108]. In practice, due to design uncertainties and changing excitation conditions, the suspension may not always be operated as preferred in the roll-off frequency band. When the excitation frequency shifts downward and approaches the resonance bandwidth, the vibration transmission may be significantly amplified by the resonance feature. It is well-known that increasing the damping in a linear suspension platform is ineffective to attenuate or avoid such adverse resonance, but instead significantly deteriorates isolation performance in the roll-off bandwidth [108].

The prior studies of Sections 3.3 and 3.4 found that EIS dynamics in a bistable isolation system can enable an innovative means to reduce the dynamic response of the isolation architecture and hence attenuate the amplitude of transmitted force or displacement through the system. Motivated by these insights, the aim of this study is to investigate the potential for a bistable suspension system when its dynamics are strongly coupled to a flexible host structure. From this perspective, the understanding sought in this study are the design strategies for such a suspension architecture that enable a large reduction in transmitted vibration as well as the effective means to ensure robustness of system performance. The following sections thus investigate how the bistable suspension is influenced around the critical resonance feature adjacent to the roll-off frequency bandwidth, and employs a thorough approach to analytically, numerically, and experimentally verify the design insights.

3.5.1 Model Formulation of the Bistable Suspension Coupled to a Flexible Structure

Figure 3.30 presents a schematic of the two DOF coupled system where a bistable suspension is attached to a single DOF host system. Each body is modeled using lumped inertial, damping, and stiffness elements to represent the principal and often dominant vibration characteristics of the coupled system. The bistable suspension interface suspends a mass m_1, which may represent a sensitive instrument to isolate from vibrations. The bistable interface comprises a damper with damping constant d_1 and a spring with stiffness k_0 having undeformed length l_0. One end of the spring suspends the mass m_1, while the opposing end connects to a frame mounted to the flexible host structure. With respect to the motion of the suspended mass, the height of the frame is h, where $h < l_0$. Hence the spring of stiffness k_0 is compressed in the upright position and induces two stable equilibria, positioned symmetrically about a central, unstable configuration denoted by the location of the centerline in Figure 3.30.

The host structure is modeled as a single DOF linear oscillator, and its mass (which includes the supporting frame) is m_2. The stiffness and damping of the host structure are k_2 and d_2, respectively. When the two DOF nonlinear coupled system is subjected to harmonic base acceleration excitation $\ddot{z} = A\cos\Omega t$ at the foundation, the governing equations are derived as

$$m_1\left(\ddot{X}_t + \ddot{Y}_t + A\cos\Omega t\right) + d_1\dot{X}_t + F_\mathrm{p} = 0 \tag{3.53}$$

$$m_2\left(\ddot{Y}_t + A\cos\Omega t\right) + d_2\dot{Y}_t + k_2Y_t - d_1\dot{X}_t - F_\mathrm{p} = 0 \tag{3.54}$$

$$F_\mathrm{p} = k_0X_t\left(1 - \frac{l_0}{\sqrt{X_t^2 + h^2}}\right) \tag{3.55}$$

where A, Ω, and t are the excitation amplitude, excitation frequency, and time, respectively. X_t and Y_t are the relative oscillations of the suspended mass and host structure mass, respectively, and F_p is the restoring spring force of the bistable interface. From Eq. (3.55), it is seen that the bistable interface induces two stable configurations for the suspended mass relative to the frame: $X_t^* = \pm\sqrt{l_0^2 - h^2}$. The governing equations may be simplified by utilizing a Taylor series expansion of F_p with respect to X_t around X_t^* and reserving terms up to the third order. Eq. (3.53) and (3.54) are then simplified to be

$$m_1\left(\ddot{X}_t + \ddot{Y}_t + A\cos\Omega t\right) + d_1\dot{X}_t - k_1X_t + k_3X_t^3 = 0 \tag{3.56}$$

$$m_2\left(\ddot{Y}_t + A\cos\Omega t\right) + d_2\dot{Y}_t + k_2Y_t - d_1\dot{X}_t + k_1X_t - k_3X_t^3 = 0 \tag{3.57}$$

where the simplified bistable spring has a negative linear stiffness of magnitude $k_1 = k_0[(l_0 / h) - 1]$ and positive cubic stiffness of $k_3 = k_0l_0 / 2h^3$. When suspending the mass m_1, the simplified bistable suspension has a linear resonant frequency $\omega_1 = \sqrt{k_1 / m_1}$, loss factor $\gamma_1 = d_1 / m_1\omega_1$, and produces two symmetric equilibria

$$X_\mathrm{eq} = \pm\sqrt{k_1 / k_3} = \pm\sqrt{2}h\sqrt{1 - (h / l_0)} \tag{3.58}$$

that are symmetrically adjacent to an unstable equilibrium at $X_t = 0$. In the absence of the bistable interface and supported mass, the host structure has a natural frequency $\omega_2 = \sqrt{k_2 / m_2}$, and its loss factor is $\gamma_2 = d_2 / m_2\omega_2$. Additional non-dimensional parameters are defined, including frequency tuning ratio between the bistable suspension and the host structure $f = \omega_1 / \omega_2$; mass ratio between the suspended mass and the host structure mass $\mu = m_1 / m_2$; normalized excitation frequency $\omega = \Omega / \omega_2$; and normalized time $\tau = \omega_2 t$. Using these variables, a non-dimensional form of Eqs (3.56) and (3.57) yields

$$x'' + y'' + p_0\cos\omega\tau + \gamma_1 fx' - f^2x + f^2x^3 = 0 \tag{3.59}$$

$$y'' + p_0\cos\omega\tau + \gamma_2 y' + y - \mu\gamma_1 fx' + \mu f^2x - \mu f^2x^3 = 0 \tag{3.60}$$

where $p_0 = A/|X_{eq}|\omega_2^2$; $x = X_t/|X_{eq}|$; $y = Y_t/|X_{eq}|$; and operator ( )' denotes the derivative with respect to normalized time τ.

Transmissibility from the base excitation to the absolute acceleration of the suspended mass m_1 is the index to evaluate the vibration transmission attenuation of the two DOF nonlinear coupled system. Assuming single frequency sinusoidal excitation, the transmissibility is defined by the frequency response function (frf) of the absolute acceleration of the suspended mass to the exciting acceleration:

$$|T| = \frac{|\ddot{X}_t + \ddot{Y}_t + A\cos\Omega t|}{|A\cos\Omega t|} = \frac{|x'' + y'' + p_0\cos\omega\tau|}{p_0} \tag{3.61}$$

3.5.2 Analytical Solution of the Governing Equations

Systems that exhibit bistability are strongly nonlinear and require appropriate analytical tools that do not rely on assumption of small perturbations about a linear response. When it is anticipated that the system undergoes a periodic response due to sinusoidal excitation, an effective method to analytically predict steady-state dynamic behaviors of a strongly nonlinear system is the method of harmonic balance. In this study, a fundamental Fourier series expansion is employed to predict the steady-state response of the suspended mass and host structure mass relative oscillations, x and y, respectively.

$$x = c_1(\tau) + a_1(\tau)\sin\omega\tau + b_1(\tau)\cos\omega\tau \tag{3.62}$$

$$y = a_2(\tau)\sin\omega\tau + b_2\cos\omega\tau \tag{3.63}$$

In Eq. (3.62), $c_1(\tau) = 0$ indicates that the suspended mass oscillates symmetrically about the unstable equilibrium induced by the bistable interface at $x = 0$, that is, an interwell response. In contrast, $c_1(\tau) \neq 0$ represents an intrawell response where the suspended mass has non-zero displacement away from the central configuration and undergoes vibrations about this stable equilibrium. The moduli $r_1 = \sqrt{a_1^2 + b_1^2}$ and $r_2 = \sqrt{a_2^2 + b_2^2}$ are the oscillation amplitudes of x and y, respectively.

Substituting Eqs (3.62) and (3.63) into Eqs (3.59) and (3.60), five modulation equations with respect to the coefficients $\mathbf{q} = [c_1 \quad a_1 \quad b_1 \quad a_2 \quad b_2]^{\mathrm{T}}$ are obtained. Then, by assuming steady-state response and reducing the set of equations through incremental substitution, a polynomial is obtained whose roots are the amplitude r_1 of the suspended mass connected to the bistable interface:

$$r_1^2\left(\alpha_1\Lambda^2 + \alpha_2\Lambda + \alpha_3\right) = \left(1 + \omega^2\gamma_2^2\right)p_0^2 \tag{3.64}$$

where,

$$\Lambda = -1 + 3c_1^2 + \frac{3}{4}r_1^2 \tag{3.65a}$$

$$\alpha_1 = \left[f^2 - (1+\mu)f^2\omega^2\right]^2 + \left(f^2\omega\gamma_2\right)^2 \tag{3.65b}$$

$$\alpha_2 = -2f^2\omega^2\left\{\left(1-\omega^2\right)\left[1-\left(1+\mu\right)\omega^2\right]+\left(\omega\gamma_2\right)^2\right\} \tag{3.65c}$$

$$\begin{aligned}\alpha_3 = \omega^2\Big\{&\left(1-\omega^2\right)^2\omega^2+\left(f\gamma_1\right)^2\left[1-\left(1+\mu\right)\omega^2\right]^2\\&+\omega^2\left[\left(f\gamma_1\gamma_2\right)^2+\left(\omega\gamma_2\right)^2+2\mu f\gamma_1\gamma_2\omega^2\right]\Big\}\end{aligned} \tag{3.65d}$$

Similar to the solution strategy steps in Section 3.4.2, one of the modulation equations with respect to coefficient set $\mathbf{q}$ has two distinct solutions for c_1^2. The solutions satisfy $c_1^2 = 0$ for interwell oscillations and $c_1^2 = 1-(3/2)r_1^2$ for intrawell oscillations. By substituting one such solution for c_1^2 into Eqs (3.64) and (3.65), the roots of the polynomial Eq. (3.64) are computed, and the coefficient set $\mathbf{q}$ is thereafter back-calculated. Using the same method as employed in Section 3.4.3, the negative real components of the eigenvalues of the Jacobian matrix determined from a linearization of the modulation equations around a solution set of coefficients provide an indication of the stability of the response coefficient set. In this study, presentations of the analytically predicted results only include the stable and, hence physically meaningful, solutions.

Finally, once the stable solution coefficients are determined from this procedure, the analytical results of the transmissibility $|T|$ are expressed using

$$|T| = \frac{\sqrt{\left(a_1+a_2\right)^2\omega^4+\left(b_1\omega^2+b_2\omega^2-p_0\right)^2}}{p_0} \tag{3.66}$$

3.5.3 Description of the Linear Suspension for Comparison

In this section, a counterpart linear suspension is defined for performance comparison when attached to the single DOF host structure. For meaningful comparison, it is required that the bistable and linear suspensions provide an identical isolation performance in the roll-off frequency band. Because vibration from the foundation may be highly attenuated in the roll-off frequency band, the suspension oscillation may be small. When a nonlinear system undergoes small oscillations, the nonlinearity can be neglected so that its response is nearly identical to that of a linearized counterpart. Therefore, the counterpart linear suspension is defined from a linearization of the bistable suspension characteristics. When the bistable suspension exhibits small oscillations, the suspended mass m_1 vibrates around either stable equilibrium: the intrawell response. The response motion may then be expressed by $X_t = \delta x \pm \sqrt{l_0^2 - h^2}$, where $\delta x << l_0$. Hence, the restoring spring force of the bistable interface becomes

$$F_p = k_0\left(\delta x \pm \sqrt{l_0^2-h^2}\right)\left[1-\frac{l_0}{\sqrt{\delta x^2 \pm 2\delta x\sqrt{l_0^2-h^2}+l_0^2}}\right] \tag{3.67}$$

Taking a Taylor expansion of Eq. (3.67) with respect to the small quantity $\delta x/l_0$, and neglecting the higher order terms, the linearization of Eq. (3.67) is

$$F_p \approx k_0\left(1-\frac{h^2}{l_0^2}\right)\delta x = Lk_1\delta x \tag{3.68}$$

where,

$$L=\left(1-\frac{h^2}{l_0^2}\right)/\left(\frac{l_0}{h}-1\right) \tag{3.69}$$

Eq. (3.68) defines a linearized restoring force of the bistable suspension. Thus, the linear suspension counterpart is realized with an interface composed of damping d_1 and stiffness Lk_1. In this way, the linear suspension counterpart serves as the benchmark for meaningful comparison against the bistable design. The additional non-dimensional parameters of interest for the corresponding two DOF linear coupled system are the frequency tuning ratio $f_l=\sqrt{L}f$, and loss factor of the linear counterpart $\gamma_{l1}=\gamma_1/\sqrt{L}$.

3.5.4 Analytical and Numerical Assessment of Key Suspension Dynamics

The governing equations of the two DOF coupled systems are first evaluated by comparing a series of numerical and analytical results, to provide model validation and to investigate the difference in vibration transmission achieved by the bistable and linear suspensions. For the following numerical simulations, the original length of the spring in the bistable interface is $l_0=0.08$ m, and the ratio between the frame height and the original spring length is $h/l_0=0.99$. Therefore, according to the expressions in Section 3.5.1, which express the bistable interface restoring forces before and after Taylor series expansion, the stable equilibria of the original and simplified bistable interface are ±0.0113 m and ±0.0112 m, respectively. The mass ratio and frequency tuning ratio from the bistable suspension to the host structure are $\mu=0.34$ and $f=0.28$, respectively. The bistable suspension loss factor is $\gamma_1=0.15$, while the host structure has a small loss factor $\gamma_2=0.03$. For the counterpart linear suspension, the mass ratio is $\mu=0.34$, whereas the frequency tuning ratio and loss factor are $f_l=\sqrt{L}f=0.39$ and $\gamma_{l1}=\gamma_1/\sqrt{L}=0.107$, respectively. The normalized amplitude of the excitation is $p_0=0.06$ in the following simulations.

Figure 3.31 presents analytically predicted transmissibility of the two DOF coupled systems across a wide frequency band $0.8\le\omega\le 4$ which encompasses the roll-off frequency band and the adjacent second resonance induced by the linear coupled system. The responses of the linear coupled system are shown as the dash-dot curve, whereas responses corresponding to employment of the bistable suspension are solid or dashed curves, representing inter- and intrawell dynamics, respectively. Apart from the results of the linear counterpart with the equivalent loss factor, the figure also shows the responses when applying the linear suspensions with two other loss factors $\gamma_{l1}=0.3$ and $\gamma_{l1}=0.6$, to evaluate whether increased linear suspension damping may improve the operational effectiveness near the second resonance. In the roll-off frequency band, the nonlinear and the linear suspensions realize the same isolation performance of 75 dB/decade transmissibility attenuation rate, which provides favorable isolation performance when the suspensions are operated in this bandwidth. However, it is seen that the

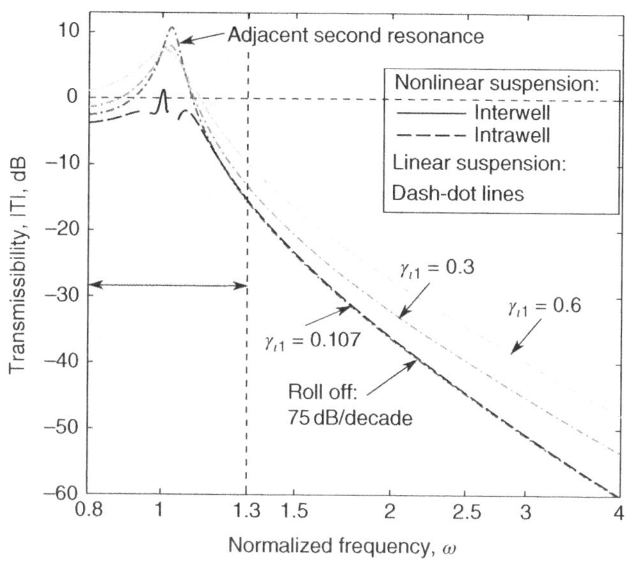

Figure 3.31 Analytically predicted transmissibility $|T|$ of the two DOF coupled systems with the bistable and linear suspensions in bandwidth $0.8 \leq \omega \leq 4$.

second resonance induced by the two DOF coupled system with the linear suspension significantly amplifies the vibration transmission across a fairly wide bandwidth around $\omega = 1$ adjacent to the roll-off frequency band. In practice, the suspension cannot always be ensured to be operated in the favorable isolation bandwidth at higher frequencies due to the inevitable deviation of excitation conditions away from the roll-off bandwidth. The amplified vibration transmission resulting from such deviations may be detrimental to the operational safety and performance integrity of the suspension. Therefore, this resonance should be reduced or avoided altogether. As is evident from prior studies [108], Figure 3.31 confirms that increasing linear suspension damping is ineffective in reducing or avoiding the detrimental resonance, and has the undesirable consequence of increasing the transmitted vibration levels in the roll-off bandwidth. In contrast to the suspension performance realized with the linear designs, when the bistable suspension is applied the dramatic resonance feature is substantially suppressed in the resonance bandwidth denoted by the shaded background in Figure 3.31. In fact, the nonlinear coupled system almost does not exhibit a second resonance around excitation frequency $\omega = 1$, indicated by the fact that the transmissibility of the two DOF system with the bistable, nonlinear suspension (black solid and dashed curves) is almost always less than 0 dB. This enhances operational safety and simultaneously retains the favorable isolation performance in the roll-off frequency band characterized by the high rate of attenuation per decade increase in excitation frequency.

To study the suspension performance improvement of the bistable suspension more closely, Figure 3.32 focuses on the results of Figure 3.31 specifically in the shaded bandwidth $0.8 \leq \omega \leq 1.3$ which encompasses the conventional resonance bandwidth of the two DOF coupled system with linear suspensions. Figure 3.32a,b

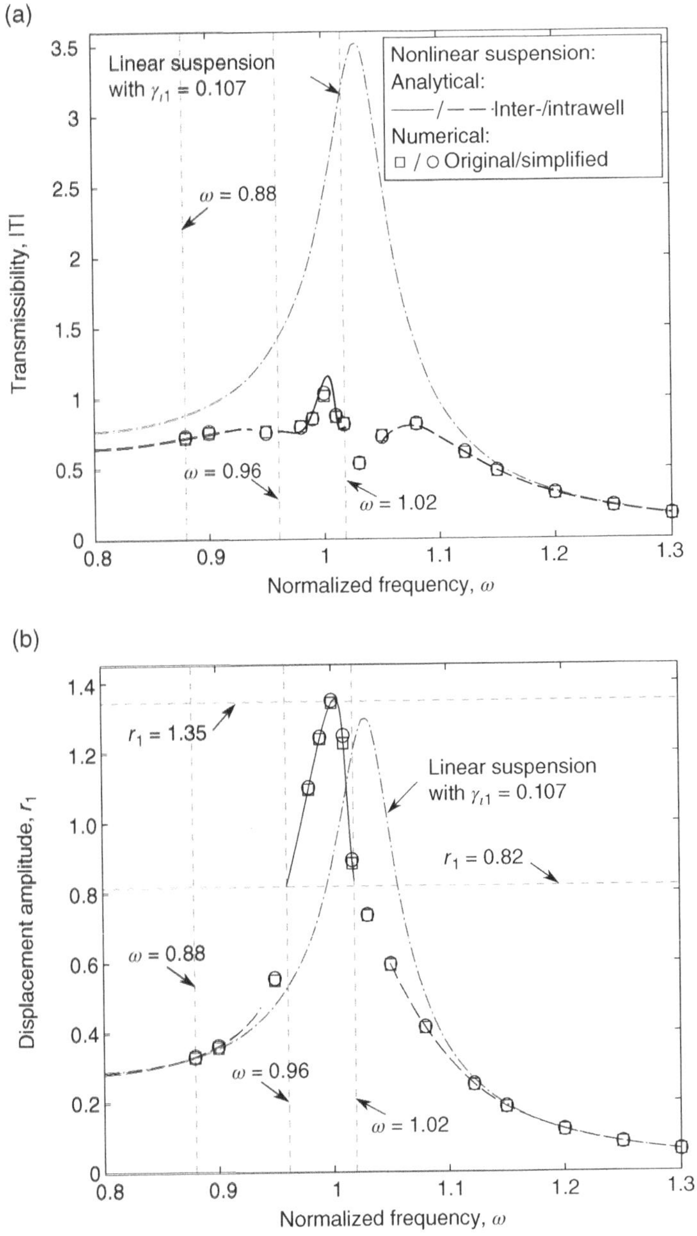

Figure 3.32 (a) Transmissibility $|T|$ and (b) normalized suspension relative amplitudes r_1 of the two DOF coupled systems with the bistable and linear suspensions.

present the transmissibility $|T|$ and normalized suspended mass relative amplitude r_1 of the two DOF coupled systems, respectively, having either the bistable or linear suspensions. In these results, the linear counterpart employs the equivalent loss factor $\gamma_{l1} = 0.107$. To thoroughly verify the analytical results, numerical responses of the coupled system with the bistable suspension are obtained by numerically integrating the original governing Eqs (3.53) and (3.54) as well as Eqs (3.56) and (3.57) following Taylor series expansion. A fourth order Runge–Kutta algorithm is employed in MATLAB to acquire the numerical results and the response amplitudes required in the figures are determined from FFT spectra of the displacements after transient behaviors are decayed in the simulations.

In the bandwidth $0.96 \le \omega < 1.02$, the analytically predicted responses of the coupled system with the bistable suspension indicate that the suspended mass undergoes interwell oscillation where it vibrates symmetrically about the central unstable equilibrium. For frequencies outside this bandwidth, the responses trend to intrawell response, in which case the suspended mass oscillates around one of the stable equilibria. In the interwell bandwidth $0.96 \le \omega < 1.02$, Figure 3.32b shows that the relative oscillation amplitude of the bistable suspension is $0.82 \le r_1 \le 1.35$, which is close to the value of the normalized stable equilibrium amplitude of the bistable interface $r_1^* = 1$. As a result, this interwell response can be identified to be EIS, as compared to energetic snap-through dynamics. As shown in Figure 3.32a, the two DOF coupled system with the linear suspension undergoes a resonance that greatly amplifies the vibration transmitted to the suspended mass in the vicinity of $\omega = 1$ which is adjacent to the roll-off frequency band. In contrast, in the EIS frequency band $0.96 \le \omega < 1.02$, the maximum transmissibility of the coupled system with the bistable suspension is much smaller than the linear counterpart. Additionally, Figure 3.32b finds that the maximum relative amplitude of the bistable suspension oscillation is similar to that of the linear counterpart, indicating that the improved performance enabled by the bistable suspension does not come at the cost of requiring greater dynamic stroke range.

Figure 3.32a,b also present results of numerically integrating the original governing Eqs (3.53) and (3.54) and the simplified Eqs (3.56) and (3.57) by squares and circles, respectively. Both numerical results are in good agreement with the analytical findings, validating the model composition and demonstrating that the simplification of the governing equations does not detrimentally reduce fidelity of the model formulation under these dynamic conditions. In the bandwidth $0.8 \le \omega < 0.88$, numerical simulations predict multiharmonic responses may occur depending on initial conditions. The maximum amplitude of the multiharmonic response at each excitation frequency is greater than the analytically predicted single-periodic response, and thus the multiharmonic response is not desired. Note that the normalized excitation frequency in Figure 3.32a,b is relative to the natural frequency of the host structure $\omega_2 = \sqrt{k_2 / m_2}$. Thus, in practice, by designing appropriate parameters the undesirable multiharmonic responses might be outside the operating bandwidth, so that the system exhibits only the single-periodic response. Since the present study is focused on the single-periodic response that is conducive to effective and robust performance of the suspension platform, the multiharmonic results are omitted in the figures.

3.5.5 Excitation Condition Influences

Unlike the two DOF coupled system with the linear suspension, the transmissibility of the coupled system with the bistable suspension is dependent on excitation level. To evaluate this sensitivity as normalized excitation level varies from $0.001 \le p_0 \le 0.1$, Figure 3.33a presents a contour map of the analytically predicted single-periodic transmissibility of the coupled system with the bistable suspension across the bandwidth $0.8 \le \omega \le 1.3$. Three response regions are shown, each defined within the bounds of the dashed curves, which represent the stable inter- and intrawell responses. In the absence of stable single-periodic dynamics predicted by the analysis, the areas are left unshaded. In the map, darker shading indicates increased transmissibility. When $p_0 \le 0.035$, the system exhibits intrawell response across the full bandwidth. When $p_0 > 0.035$, the interwell response emerges in the vicinity of $\omega = 1$, and the interwell bandwidth is broadened by increasing the excitation level p_0. As shown in the contour map, the white solid line presents the transmissibility maxima for each level of excitation, and it is seen that the transmissibility maxima are all in the vicinity of $\omega = 1$. The transmissibility maxima indicate the maximum vibration transmission amplification under respective excitation levels in the bandwidth; in other words, this is the *worst*-case performance result for the bistable suspension. Therefore, the *smaller* the value obtained, the *better* the suspension performance. Following the white solid line of maximum transmissibility, it is found that moderate excitation levels, such as p near 0.05, suppress the maxima.

For clarity, Figure 3.33b plots the transmissibility maxima that is shown as the white solid line in Figure 3.33a. In Figure 3.33b, the transmissibility maxima of the two DOF coupled system with the linear suspension are also provided. As anticipated, the transmissibility of the two DOF linear coupled system does not vary as the excitation level is changed. However, the transmissibility when using the bistable suspension exhibits a minimum across the range of excitation levels considered. It is seen that the transmitted vibration with the bistable suspension is minimized around an excitation level of $p_0 \approx 0.05$. This feature indicates that the bistable suspension could provide a maximum of vibration isolation performance when the coupled system is subjected to moderate excitation levels. Numerical results in Figure 3.33b are in very good agreement with the analytical findings, which validates the viability of the appropriate utilization of bistability as a means to enhance the operational safety of the suspension via the reduction of transmissibility maxima.

The trend of minimized transmissibility provided by the bistable suspension may be explained by considering the components of the normalized restoring spring force of the bistable interface, as shown in Eqs (3.59) and (3.60). According to Eq. (3.59), Eq. (3.61) can be rewritten as follows:

$$|T| = \frac{|\gamma_1 f x' + f_p|}{p_0} \tag{3.70}$$

where,

$$f_p = f^2\left(-x + x^3\right) \tag{3.71}$$

As shown in Eqs (3.70) and (3.71), the transmissibility $|T|$ is affected by the normalized damping force $\gamma_1 f x'$ and potential force f_p. When the bistable suspension

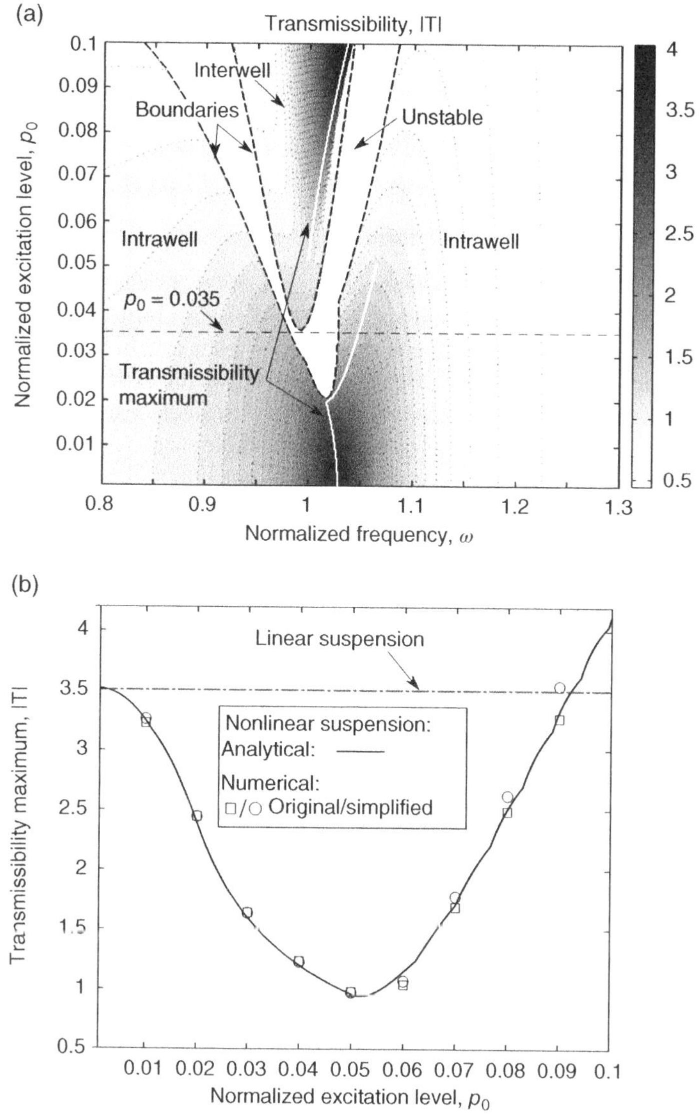

Figure 3.33 (a) Transmissibility contour map of the two DOF coupled system with the bistable suspension and (b) the transmissibility maximum $|T|$ corresponding to variation of normalized excitation level p_0.

exhibits EIS under a specific excitation level, the normalized relative oscillation amplitude $r_1 = |x|$ is in the vicinity of the normalized equilibrium value $r_1^* = |x_{eq}| = 1$, and thus the positive nonlinear force may counterbalance the negative linear force $-f^2 x$ so that the transmissibility is minimized, as shown in Figure 3.33b. The result is a minimization of the amplified vibration levels in the resonance bandwidth.

Note that the normalized excitation level is defined here using $p_0 = A/|X_{eq}|\omega_2^2$, where the equilibrium of the simplified bistable interface X_{eq} is related to the undeformed spring length l_0 and frame height h. Therefore, if the excitation level A and natural frequency of the host structure ω_2 are known, the best suspension performance may be achieved by properly selecting the spring length l_0 and frame height h.

3.5.6 Experimental Suspension System Platform

A prototype of the bistable suspension is fabricated, and the two DOF coupled system with the bistable suspension attached to a single DOF flexible host structure is constructed to replicate the overall system architecture investigated in the prior sections (see schematic model in Figure 3.30). Figure 3.34 shows the experimental set-up. The two DOF nonlinear coupled system is attached to an electrodynamic shaker which provides harmonic base acceleration. The bistable suspension interface consists of a spring compressed in its upright vertical position and an adjustable damper. Between two rotational bearings, the spring connects the surrounding frame to the suspended mass, which is the top bearing and rotational bearing attached to it. The spring is initially compressed when the top bearing is aligned with the centerline. In this way the geometric constraints induce two stable equilibria of the suspended mass displacement that are symmetric about the centerline. The frame is mounted on the bottom bearing and is connected to the shaker by a spring which represents the single DOF host structure stiffness.

The combined mass of the top bearing and rotational bearing connected to it represents the suspended mass $m_1 = 0.347$ kg. The total mass of the frame, bottom bearing, and other hardware rigidly attached to the frame represents the host structure mass $m_2 = 1.029$ kg. The linear natural frequency of the bistable suspension with respect to the suspended mass is $\omega_1 \approx 10.56$ rad s^{-1} (1.68 Hz) and the stable equilibria are located at ± 0.012 m with respect to the centerline, while the natural frequency of the host structure is $\omega_2 \approx 37.70$ rad s^{-1} (6 Hz). The shaker is controlled to prescribe a slowly backward swept sinusoidal base acceleration from 4.8 to 7.8 Hz at a rate 0.025 Hz s^{-1}. This frequency bandwidth encompasses the resonance of the two DOF coupled system with the linear suspension, which is adjacent to the roll-off frequency band. Accelerometers are mounted on the shaker and the suspended mass, which provide a suitable set of data for the latter determination of vibration transmission from the shaker to the suspended mass m_1.

3.5.7 Experimental and Analytical Comparisons of Isolation Performance

A series of experimental investigations are performed to demonstrate the important suspension characteristics uncovered in Sections 3.5.4 and 3.5.5 and to further validate the analytical model through qualitative comparison with measured data. The parametric features which are investigated are strongly relevant factors in the employment of vibration suspension systems, whether linear or nonlinear, and include the level of excitation, the damping of the bistable suspension, and the magnitude of the suspended mass.

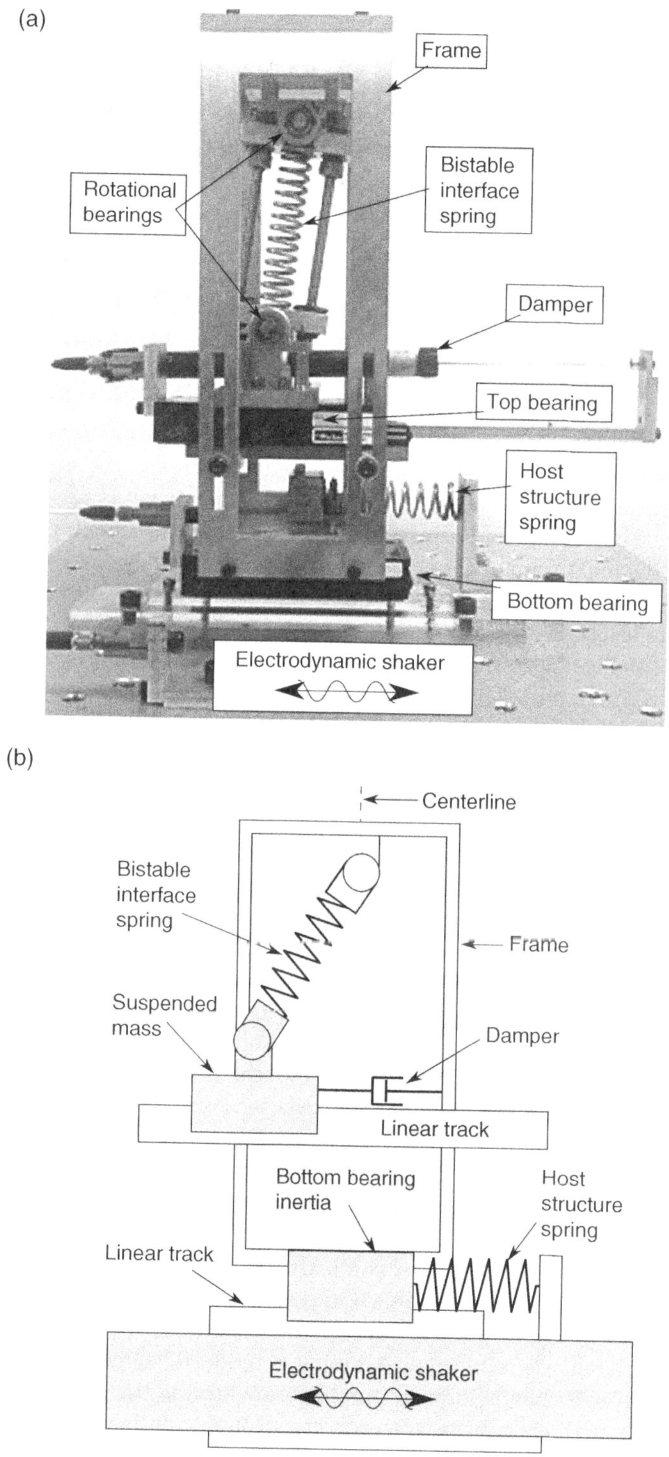

Figure 3.34 (a) Photograph and (b) schematic of experimental set-up.

Three excitation levels are chosen, and the root mean square (RMS) values of the excitations during the entire swept frequency band are $[A^{(1)} \quad A^{(2)} \quad A^{(3)}] = [1.02 \quad 1.66 \quad 2.37]$ m s^{-2}. For the analytical predictions, the steady-state excitation amplitudes are $[A^{(1)} \quad A^{(2)} \quad A^{(3)}] = [0.61 \quad 1.02 \quad 1.43]$ m s^{-2} (i.e. the normalized amplitudes are $[0.036 \quad 0.06 \quad 0.084]$). In the following experiments, the mass ratio and frequency tuning ratio are $\mu \approx 0.34$ and $f \approx 0.28$, respectively. The other parameters which are used in analyses are the same as those used to generate Figure 3.31.

Figure 3.35a,b present experimental and analytical transmissibilities $|T|$, respectively, as the excitation level is varied. The horizontal axis shows the excitation frequency normalized by the natural frequency of the host structure. It is seen experimentally that when the smallest excitation level $A^{(1)}$ is used, the bistable suspension only exhibits an intrawell response (unfilled data points) across the entire bandwidth. By increasing excitation level to $A^{(2)}$, the interwell response (filled data points) appears throughout $0.95 \leq \omega \leq 1.08$, and the maximum of $|T|$ is decreased in consequence. When the excitation level is further increased, the interwell frequency bandwidth is broadened to encompass $0.86 \leq \omega \leq 1.13$, and the maximum of $|T|$ is increased. Similar to the discovery by the analytical model as shown in Figure 3.33a,b, the experimental results also reveal that moderate levels of excitation can lead to the minimization of transmissibility in the resonance bandwidth which enhances the overall operational safety of the suspension system. There is a strong agreement between experimental and analytical results in Figure 3.35a,b across the several salient trends, thus providing clear support for the main conclusions regarding excitation level dependence of the bistable suspension performance when attached to the host structure.

The influence of the bistable suspension damping is then considered. The adjustable damper is used with three settings denoted here as $d_1^{(1)}$, $d_1^{(2)}$, and $d_1^{(3)}$. Due to the frequencies involved and test apparatus available, there is difficulty in precisely identifying the damper constants employed in the experimentation, although the general trend is that $d_1^{(1)} < d_1^{(2)} < d_1^{(3)}$. The RMS of excitation level in the experiment is $A = 1.66$ m s^{-2}, which is the moderate excitation as used for the prior experiments. For the comparable analytical results, damping constants are $[d_1^{(1)} \quad d_1^{(2)} \quad d_1^{(3)}] = [0.55 \quad 0.92 \quad 1.28]$ N s m^{-1} (i.e. the bistable suspension loss factors are: [0.15, 0.25, 0.35]), and the moderate excitation level is $A = 1.02$ m s^{-2}. The mass and frequency tuning ratios remain the same as in the prior experiment set.

In the experiments, the bistable suspension exhibits interwell response in the range $0.95 \leq \omega \leq 1.08$, as shown in Figure 3.36a, and the frequency bandwidth across which the interwell dynamics occur is mostly insensitive to variation in the bistable suspension damping. This finding is in very good agreement with the analytical results shown in Figure 3.36b. Both experimental and analytical findings indicate transmissibility is reduced as damping is reduced, suggesting this design change is one way to avoid vibration amplification around the resonance peak bandwidth. Because the bistable suspension is a strongly nonlinear system, too small damping may increase the activation likelihood of undesirable multi-harmonic or chaotic responses. From this perspective, achieving robustness in

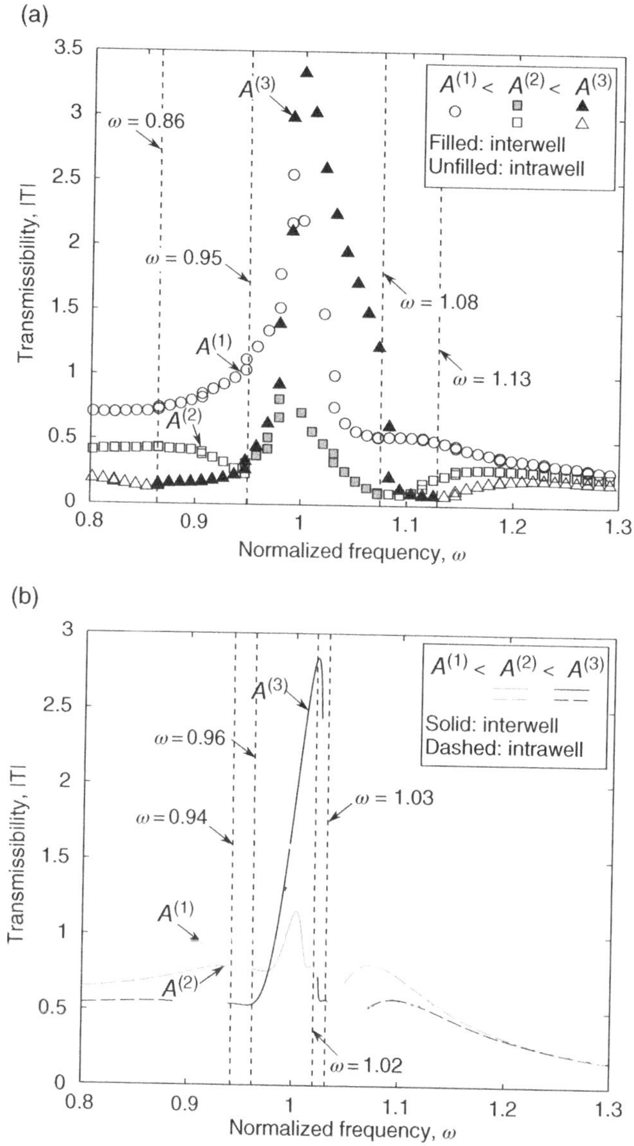

Figure 3.35 (a) Experimental and (b) analytical transmissibility $|T|$ corresponding to excitation level variation.

suspension performance suggests selection of moderate damping: appropriately small bistable interface damping to adequately enhance the operational safety of the system, but not too small to activate undesired responses.

A final critical factor to evaluate is the performance variation induced by changes in the suspended mass, which is a common change of working condition for a vibration suspension system. While the mass of the host structure remains $m_2 = 1.029$ kg, the suspended mass is experimentally varied by

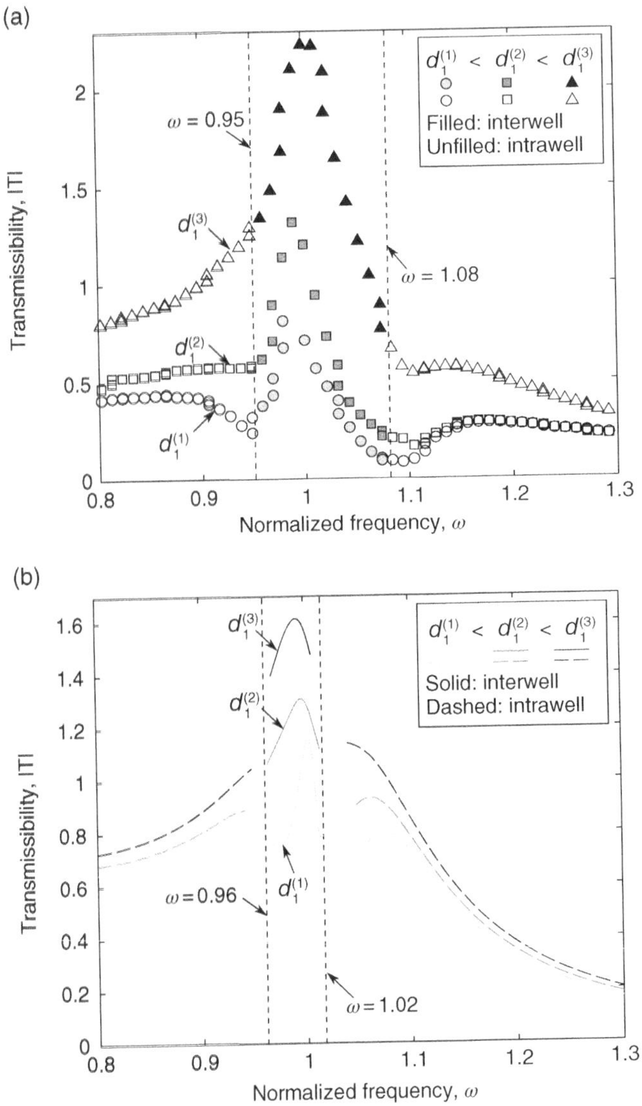

Figure 3.36 (a) Experimental and (b) analytical transmissibility $|T|$ corresponding to variation of bistable suspension damping.

connecting blocks of brass to the top bearing. In this way, three suspended masses are realized: $[m_1^{(1)} \quad m_1^{(2)} \quad m_1^{(3)}] = [0.347 \quad 0.6 \quad 0.847]$ kg. The RMS of the excitation levels for the experimental and analytical results are the same as those employed in the prior section. For the analytical results, six values of suspended mass are chosen: $[m_1^{(1)} \quad m_1^{(2)} \quad m_1^{(3)}] = [0.347 \quad 0.6 \quad 0.847]$ kg and $[m_1^{(4)} \quad m_1^{(5)} \quad m_1^{(6)}] = [2.0 \quad 4.0 \quad 6.0]$ kg, where the final three values are greater than were able to be evaluated experimentally. The bistable suspension damping values in the experiment and analysis are chosen to be the smallest cases used in the prior section, respectively.

Figure 3.37a,b present experimental and analytical transmissibilities, respectively, as the suspended mass is varied. There is good qualitative agreement between analytical and measured data and both results indicate that the frequency bandwidths leading to interwell response do not vary corresponding to change in the suspended mass. By closely considering the lighter suspended masses $m_1^{(1)}$, $m_1^{(2)}$, and $m_1^{(3)}$, the experimental and analytical results show that the maximum $|T|$ is decreased as the suspended mass is increased. Analytical results in Figure 3.37b show that when the suspended mass becomes greater than the

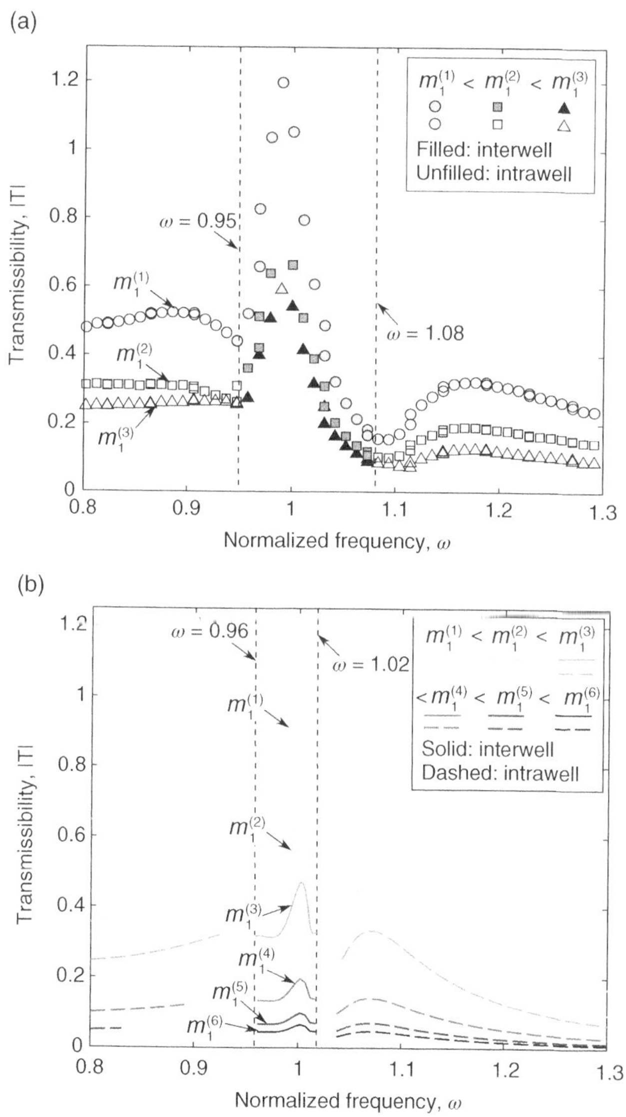

Figure 3.37 (a) Experimental and (b) analytical transmissibility $|T|$ corresponding to suspended mass variation.

frame mass – $m_1^{(4)}$, $m_1^{(5)}$, and $m_1^{(6)}$ – the transmissibility continues to reduce. While the trend shows that large suspended mass is beneficial to achieve small vibration transmission in the resonance bandwidth, too heavy a suspended load may lead to large deflection of the bistable interface when the suspension is subjected to high body force, for example inertial force due to aerospace vehicle launch acceleration or the introduction of a great load to an automobile truck bed. In practice, supporting too large a mass may not satisfy static deflection constraints of the bistable spring. Therefore, the trade-off between bistable spring deflection constraints and suspension performance should be considered to appropriately support the suspended mass and ensure the operational safety of the suspension system.

3.5.8 Summary

In this study, it is found that the dynamic stabilization phenomenon of bistable suspension systems is a powerful means to drastically reduce the transmitted vibration when the dynamics of the host system and suspended mass are strongly coupled. Such performance enhancements are shown to be due to the dynamic force counterbalance feature that is distinctively activated in a bistable suspension due to the phenomena known as excitation-induced stability. The result is a significant improvement in suspension performance when compared to an equivalent linear suspension. Experimental and numerical results strongly concur with the analytical predictions and provide thorough evidence of the enhancements in operational safety which may be effected by harnessing the EIS dynamics of a bistable suspension when it is strongly coupled to the motions of the vibrating, host structure.

3.6 Snap-through Dynamics for Vibration Absorption[5]

One approach to vibration control is to apply a force to a primary structure which opposes the excitation, effectively canceling the external disturbance. A familiar and passive example of this approach is the utilization of a linear, tuned mass absorber [71]. The recent interest in exploiting the dynamics of bistable structures has contributed to new ideas formulated around employing snap-through dynamics as an enhanced means for reactive attenuation of forced vibrations. Following the early exploration by Buzhinskii [76], several researchers have analytically examined the nonlinear normal mode characteristics of bistable devices attached to host, linear structures for the purposes of vibration absorption [77,80,81].

5 Material from this section has been reprinted by permission of ASME, from D.R. Johnson, R.L. Harne and K.W. Wang, A disturbance cancellation perspective on vibration control using a bistable snap-through attachment. Journal of Vibration and Acoustics 136:031006, 2014. doi: 10.1115/1.4026673. Copyright © 2014 by ASME.

Overall, the conclusions from these explorations suggest that there are readily accessible operating ranges in which the bistable attachment provides sufficient absorption of the input energy. On the other hand, the studies to date on steady-state behaviors have given more attention to nonlinear modal dynamics, which may be challenging to realize for such a strongly nonlinear system that is reduced in practice due to the reliance upon precise initial conditions to effect such oscillations [109]. Moreover, the lack of experimental studies from the prior investigations indicate that important insights on the nonlinear dynamics of the coupled system as realized in practice have yet to be uncovered.

As a result, to shed new light on the problem, the investigation presented below evaluates the effectiveness of a bistable attachment and snap-through dynamics from the perspective that an effective attachment *must necessarily* provide an opposing or destructive dynamic force, that is, an induced force which is substantially in opposition to the applied excitation [3]. In fact, the bistable oscillator has some potentially challenging characteristics when it comes to achieving opposing dynamic force, especially for the energetic snap-through response. For instance, the displacement–force profile of a bistable system is strongly nonlinear and includes a region of negative stiffness, so a conventional intuition as to the relation between displacements and forces based upon linear systems is not applicable. Moreover, as detailed in Section 1.2, a bistable oscillator may permit coexisting steady-state dynamics, where more than one response dynamic is possible at a particular forcing frequency and amplitude. Although the previous studies on the utilization of snap-through dynamics for vibration absorption considered important oscillation behaviors of the coupled system, the features mentioned above were not examined and it is unclear whether the required destructive forces can be realized by a bistable attachment to suppress the forced vibrations of a host, linear structure.

To resolve these unknowns and identify effective means for vibration absorption with snap-through dynamics, the following sections present an analytical and experimental investigation of the effectiveness of a bistable oscillator as an absorber of vibrations via direct cancellation of the excitation. By drawing on a broader range of theoretical and applied evidence, the aim is to formulate a complete picture of the potential for harnessing snap-through dynamics as a means to reactively attenuate the oscillations of harmonically forced structures.

3.6.1 Model Formulation of a Bistable Vibration Absorber

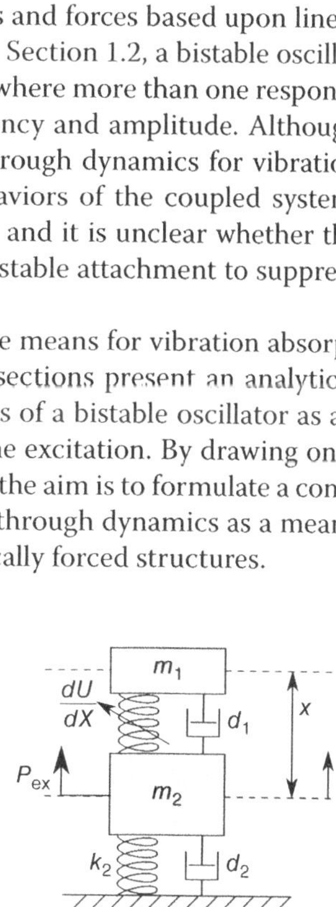

Figure 3.38 Excited host linear structure with bistable device attachment.

The schematic in Figure 3.38 shows the general coupled system dynamic framework under investigation: a bistable attachment connected to a primary linear host structure that is excited by a harmonic force. The aim is to attenuate the forced vibration of the host structure using the reaction forces induced via the motion of the bistable attachment. The motion of the bistable attachment is denoted using $x(t)$, which is the relative displacement of the bistable inertial mass with respect to the host structure mass as a deviation from the central unstable equilibrium position of the bistable attachment given to be $x=0$. The absolute displacement of the primary structure is described by $y(t)$. System parameters are as follows: m_1 and m_2 represent mass, parameters d_1 and d_2 damping constant, k_2 stiffness, $dU/dx=-k_1x+k_3x^3$ is the bistable spring force, t is time, and an overdot operator indicates a derivative with respect to time. A periodic external force $P_{ex}(t)=P_0\cos\Omega t$ is applied to the host structure. The governing equations of motion are therefore

$$m_2\ddot{y}+d_2\dot{y}+k_2y-d_1\dot{x}+k_1x-k_3x^3=P_0\cos\Omega t \tag{3.72a}$$

$$m_1\left(\ddot{x}+\ddot{y}\right)+d_1\dot{x}-k_1x+k_3x^3=0 \tag{3.72b}$$

Eq. (3.72a,b) are non-dimensionalized and rearranged to yield

$$y''+\gamma_2y'+y-\mu f\gamma_1x'+\mu f^2x-\mu f^2\beta x^3=p\cos\omega\tau \tag{3.73a}$$

$$x''+(1+\mu)f\gamma_1x'-(1+\mu)f^2x+(1+\mu)f^2\beta x^3-\gamma_2y'-y=-p\cos\omega\tau \tag{3.73b}$$

where $\omega_1^2=k_1/m_1$, $\omega_2^2=k_2/m_2$, $\tau=\omega_2t$ is non-dimensional time, $p=P_0/k_2$ is the normalized excitation amplitude, $\omega=\Omega/\omega_2$ is the non-dimensional excitation frequency, $\mu=m_1/m_2$ is the ratio of the added bistable attachment mass to the host structure mass, $f=\omega_1/\omega_2$ is the ratio of the linear natural frequencies of the bistable attachment to the host structure, $\beta=k_3/k_1$ is the nonlinearity strength, and $\gamma_1=d_1/m_1\omega_1$, $\gamma_2=d_2/m_2\omega_2$ are the damping factors of the subsystems. In Eq. (3.73), the apostrophes indicate derivatives with respect to non-dimensional time. The harmonic balance method is employed to solve for the steady-state responses of the bistable and linear structures, x and y, respectively. Thus, Eq. (3.74a,b) are the assumed Fourier series expansion of the displacements of the bistable and linear components, respectively.

$$x(\tau)=c_1(\tau)+a_1(\tau)\sin\omega\tau+b_1(\tau)\cos\omega\tau \tag{3.74a}$$

$$y(\tau)=a_2(\tau)\sin\omega\tau+b_2(\tau)\cos\omega\tau \tag{3.74b}$$

The amplitude coefficients c_1, a_1, b_1, a_2, and b_2 are assumed to vary slowly in time. Eq. (3.74) is then substituted into Eq. (3.73). Neglecting higher order harmonics and equating the harmonic coefficients to be zero due to the independence of the harmonics, the following system of equations is produced:

$$-(1+\mu)f\gamma_1c_1'=(1+\mu)f^2\left[-1+\beta\left(c_1^2+\frac{3}{2}r_1^2\right)\right]c_1 \tag{3.75a}$$

$$\begin{aligned}-(1+\mu)f\gamma_1a_1'+2\omega b_1'+\gamma_2a_2'=\left[(1+\mu)f^2\Lambda-\omega^2\right]a_1-(1+\mu)f\gamma_1\omega b_1\\-a_2+\gamma_2\omega b_2\end{aligned} \tag{3.75b}$$

$$-2\omega a_1' - (1+\mu) f\gamma_1 b_1' + \gamma_2 b_2' = (1+\mu) f\gamma_1 \omega a_1 + \left[(1+\mu) f^2\Lambda - \omega^2\right] b_1 - \gamma_2\omega a_2 - b_2 + p \tag{3.75c}$$

$$\mu f\gamma_1 a_1' - \gamma_2 a_2' + 2\omega b_2' = -\mu f^2 \Lambda a_1 + \mu f\gamma_1 \omega b_1 + \left(1-\omega^2\right) a_2 - \gamma_2\omega b_2 \tag{3.75d}$$

$$\mu f\gamma_1 b_1' - 2\omega a_2' - \gamma_2 b_2' = -\mu f\gamma_1 \omega a_1 - \mu f^2 \Lambda b_1 + \gamma_2 \omega a_2 + \left(1-\omega^2\right) b_2 - p \tag{3.75e}$$

where $\Lambda = -1 + \beta\,[3c_1^2 + \frac{3}{4}(a_1^2 + b_1^2)]$. The steady-state values of the amplitude coefficients are found by equating the time derivatives to be zero, and then proceeding through algebraic manipulations to reduce the equation to a single expression. Defining $r_1^2 = a_1^2 + b_1^2$ and $r_2^2 = a_2^2 + b_2^2$ to be the squared amplitudes of the bistable and linear oscillators, respectively, the steady-state solutions to Eq. (3.75a) are given by

$$c_1^2 = \begin{cases} \dfrac{1}{\beta} - \dfrac{3}{2} r_1^2 \\ 0 \end{cases} \tag{3.76}$$

Stated explicitly, if the offset coefficient $c_1 \neq 0$, then the bistable attachment is predicted to oscillate about one of the nonzero stable equilibria positions, which is the indicator for the intrawell vibration regime. Assuming $c_1 = 0$, then interwell motions are activated for the bistable attachment.

By employing one of the above assumptions in regards to the value of c_1, Eqs (3.75d,e) are solved in terms of a_2 and b_2 substituted into Eqs (3.75b,c). The latter are squared and summed to produce the following third order polynomial for the bistable attachment relative response amplitude r_1^2:

$$p^2\omega^4 = \left\{ \begin{array}{l} \left[\left(f^2 - (1+\mu) f^2\omega^2\right)^2 + \left(f^2\gamma_2\omega\right)^2\right]\Lambda^2 \\ + \left[-2f^2\omega^2\left\{\left(1-\omega^2\right)\left(1-(1+\mu)\omega^2\right) + \left(\gamma_2\omega\right)^2\right\}\right]\Lambda \\ + \omega^2 \left[\begin{array}{l} \left(1-\omega^2\right)^2\omega^2 + \left(f\gamma_1\right)^2\left(1-(1+\mu)\omega^2\right)^2 \\ + \left\{\left(f\gamma_1\gamma_2\right)^2 + \left(\gamma_2\omega\right)^2 + 2\mu f\gamma_1\gamma_2\omega^2\right\}\omega^2 \end{array}\right] \end{array} \right\} r_1^2 = \alpha r_1^2 \tag{3.77}$$

For a given set of system parameters, Eq. (3.77) is solved to determine the positive, real roots of r_1^2. Then, by considering Eq. (3.75), the amplitude coefficients are expressed in terms of Λ:

$$a_1 = \frac{p\omega^3}{\alpha}\left[f^2\gamma_2\Lambda - \gamma_2\omega^2 + f\gamma_1\left(1-(1+\mu)\omega^2\right)\right] \tag{3.78a}$$

$$b_1 = \frac{p\omega^2}{\alpha}\left[f^2\Lambda\left(1-(1+\mu)\omega^2\right) - f\gamma_1\gamma_2\omega^2 - \left(1-\omega^2\right)\omega^2\right] \tag{3.78b}$$

$$a_2 = \frac{1}{\rho}\left[\lambda_1 a_1 + \lambda_2 b_1 + (\gamma_2 \omega) p\right] \tag{3.78c}$$

$$b_2 = \frac{1}{\rho}\left[-\lambda_2 a_1 + \lambda_1 b_1 + \left(1-\omega^2\right) p\right] \tag{3.78d}$$

with the following definitions

$$\rho = \left(1-\omega^2\right)^2 + (\gamma_2 \omega)^2 \tag{3.79a}$$

$$\lambda_1 = \left(1-\omega^2\right)\left(\mu f^2 \Lambda\right) + (\gamma_2 \omega)(\mu f \gamma_1 \omega) \tag{3.79b}$$

$$\lambda_2 = (\gamma_2 \omega)\left(\mu f^2 \Lambda\right) - \left(1-\omega^2\right)(\mu f \gamma_1 \omega) \tag{3.79c}$$

The stability of the predicted stationary solutions is evaluated by linearizing the system response governed by Eq. (3.75) around a given solution set of coefficients. If all the eigenvalues of the linearized state matrix have negative real parts, then the corresponding solution is stable. See Section 3.4.3 for a more detailed description of the procedure. In the analytical predictions presented in this study, only stable responses are plotted in the figures. The absence of stable solutions predicted over a given parameter range indicates that no single-periodic motions are predicted via this solution strategy for the governing Eq. (3.73).

The individual responses of the bistable attachment and host linear structure, as expressed in Eq. (3.74), may be rewritten as

$$x(\tau) = c_1(\tau) + r_1(\tau)\cos\left[\omega\tau - \phi_1\right] \tag{3.80a}$$

$$y(\tau) = r_2(\tau)\cos\left[\omega\tau - \phi_2\right] \tag{3.80b}$$

where the phase lags of the responses with respect to the applied force are $\tan\phi_1 = a_1 / b_1$ and $\tan\phi_2 = a_2 / b_2$ for the bistable attachment and linear structure, respectively. The non-dimensional force applied by the bistable attachment to the structure is

$$P_{\text{att}}(\tau) = \mu f \gamma_1 x' - \mu f^2 x + \mu f^2 \beta x^3 \tag{3.81}$$

Equation (3.80a) is substituted into Eq. (11), higher order harmonics are neglected, to obtain

$$P_{\text{att}}(\tau) = P_{a1}\cos\left[\omega\tau - \Phi_1\right] \tag{3.82}$$

where the attachment force amplitude and phase are

$$P_{a1} = \mu f r_1 \sqrt{(f\Lambda)^2 + (\omega\gamma_1)^2} \tag{3.83a}$$

$$\tan\Phi_1 = \frac{f\Lambda a_1 - \omega\gamma_1 b_1}{\omega\gamma_1 a_1 + f\Lambda b_1} \tag{3.83b}$$

Equation (3.82) can be trigonometrically summed with the excitation force applied to the host structure, $p\cos\omega\tau$, to yield the total force applied to the host structure:

$$P_{tot}(\tau) = A_1 \cos[\omega\tau - \psi_1] \tag{3.84}$$

where the total force amplitude and phase are

$$A_1 = \sqrt{P_{a1}^2 + 2P_{a1}p\cos\Phi_1 + p^2} \tag{3.85a}$$

$$\tan\psi_1 = \frac{P_{a1}\sin\Phi_1}{P_{a1}\cos\Phi_1 + p} \tag{3.85b}$$

When A_1 is close to zero, the bistable attachment is predicted to supply a sufficiently destructive force to the host, linear structure so as to cancel the effects of the driving excitation. Thus, effective vibration absorption is achieved for the smallest values of A_1.

3.6.2 Analytical Investigation of Force Cancellation Performance

In this section, the displacement and force response of the host structure and bistable attachment are investigated with the goal of using these quantities to evaluate the effectiveness of the bistable attachment as a vibration absorption mechanism. Figure 3.39 plots the analytically predicted displacement responses of the bistable attachment and host structure for different frequency tuning ratios f when the remaining system and excitation parameters are $p = 0.2$, $\mu = 0.3$, $\beta = 1$, and $\gamma_1 = \gamma_2 = 0.05$. The selection of the excitation level p and nonlinearity strength β parameters are in accordance with previous examinations of buckled beam structural dynamics in which case a single-term harmonic balance was successfully employed to predict the fundamental behaviors [96]. The damping constants γ_1 and γ_2 are set to relatively small values, since maximum absorption of vibrations is achieved with very small damping [3], although realistic structural systems always include inherent dissipation, typically due to a hysteretic relation between the stress and strain of oscillations [5]. From several examinations, the selection of mass ratio μ is representative of typical results although a broader range of the value will be considered in the following studies below.

In the results shown in Figure 3.39, the solid curves show that the bistable attachment undergoes interwell dynamics, whereas dashed curves indicate the attachment exhibits intrawell oscillations around one of the stable equilibria. At lower frequencies, it is shown in Figure 3.39a that the intra- and interwell dynamics may coexist, and that there is a significant difference in the amplitude of bistable attachment displacement between the two response forms. Figure 3.39c shows that the interwell dynamics of the bistable attachment *amplify* the host structure responses, which is adverse in terms of vibration control performance. In some frequency regions, particularly close to the host structure resonance, $\omega = 1$, no responses are predicted by the analytical method. When such a result occurs from the analytical solution strategy it is an indicator that

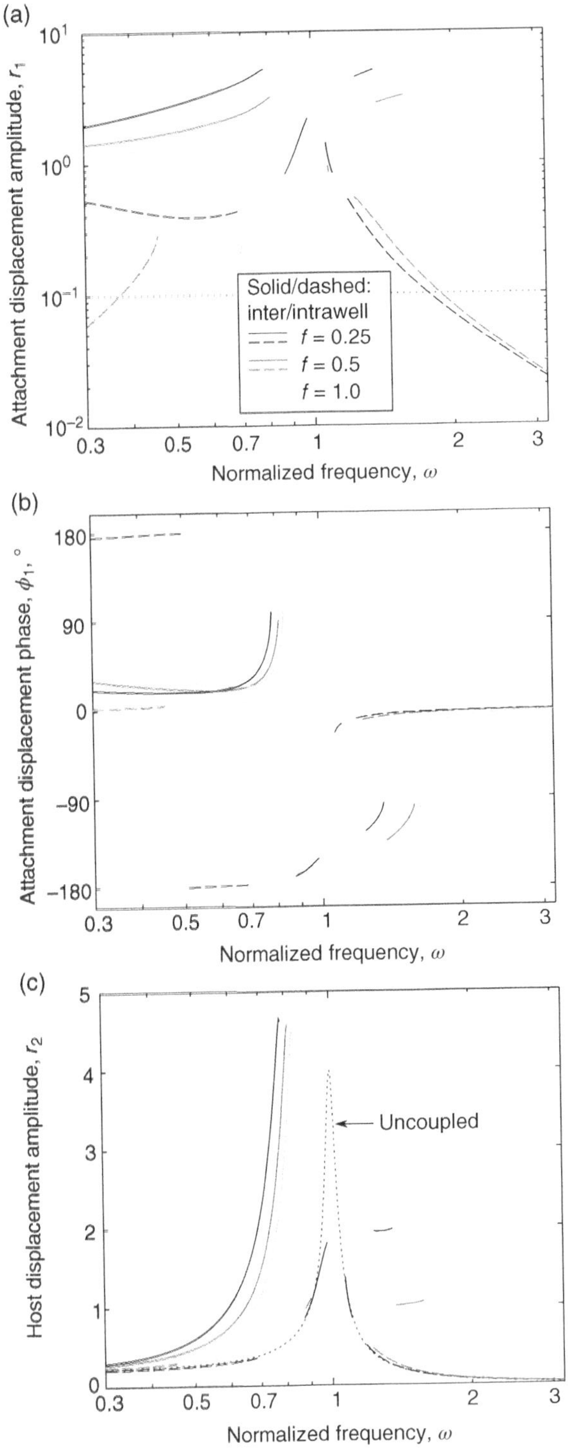

Figure 3.39 System dynamics as tuning ratio f varies. (a) Bistable attachment displacement amplitude and (b) phase lag, (c) host structure displacement amplitude.

single-periodic dynamics proportional to the excitation frequency are not anticipated. A similar absence of stable single-periodic dynamics has been predicted for other implementations of monostable nonlinear oscillators attached to excited host linear structures [74,110] where it was likewise concluded that multiharmonic or aperiodic motions would be the prevalent response. This analytical result for the present case of the bistable attachment as a vibration absorber encourages further examination.

At a glance, Figure 3.39c shows that the snap-through dynamic of the bistable attachment increases the response amplitude of the primary structure over a wide band of excitation frequencies, when compared to the uncoupled response, which is shown as the dotted line. Despite tailoring the tuning ratio f, the host structural vibrations are still adversely amplified. As indicated in Figure 3.39a, this effect at lower frequencies, such as $\omega < 0.8$, is associated with the bistable attachment undergoing large amplitude interwell dynamics. On the other hand, in a narrow frequency region near the uncoupled host structural resonance frequency $\omega = 1$, certain tuning frequency ratios such as $f = 0.5$ produce an interwell dynamic which is seen to suppress the vibrations of the main structure. Figure 3.39a indicates that the corresponding motions of the bistable attachment exhibit displacement amplitudes that are noticeably less than the snap-through responses. In addition, Figure 3.39b shows that the attachment oscillates with displacements that are mostly out-of-phase with respect to the excitation for the response regime around $\omega = 1$ that provides a degree of vibration suppression using the selection of tuning frequency ratio $f = 0.5$.

To further explore the results of this first study, the attachment force amplitude and phase, computed by Eq. (3.83), as well as the total force amplitude applied to the structure, determined from Eq. (3.85a), are presented in Figure 3.40. In the operation region below frequencies of approximately $\omega = 0.85$, Figure 3.40b shows that the attachment force phase is mostly less than 90°. As a result, the attachment force for the interwell dynamics below $\omega = 0.85$ provides *constructive* forces with the excitation, and thus amplifies the host structural vibrations. In contrast, near $\omega = 1$, the interwell dynamics of the bistable attachment induce forces which lag the excitation by typically more than 90°, which is adequate to destructively cancel the excitation force. Thus, in Figure 3.40c the total force applied to the host is suppressed around $\omega = 1$ as compared to the uncoupled case (shown by the dotted line). This presents a clear picture of the constructive and destructive force capacity provided by the bistable attachment and motivates further attention to design factors that might be utilized to ensure the destructive forces are activated for favorable absorption of the host structural vibrations.

Figure 3.39c shows that near $\omega = 1$ the consequence of the destructive bistable attachment force is to attenuate the vibrations from the uncoupled case. More interestingly, it does so without otherwise changing the resonant spectral characteristics of the host. In other words, with the bistable attachment supplying the attenuating force, the host structure vibration response appears to be that of a single DOF oscillator with an effective increase in damping near resonance, although the system actually constitutes two, distinct DOF. This is of course very unique as compared to the results obtained with a linear vibration absorber where the vibrational response of the host structure is "split" between two modal

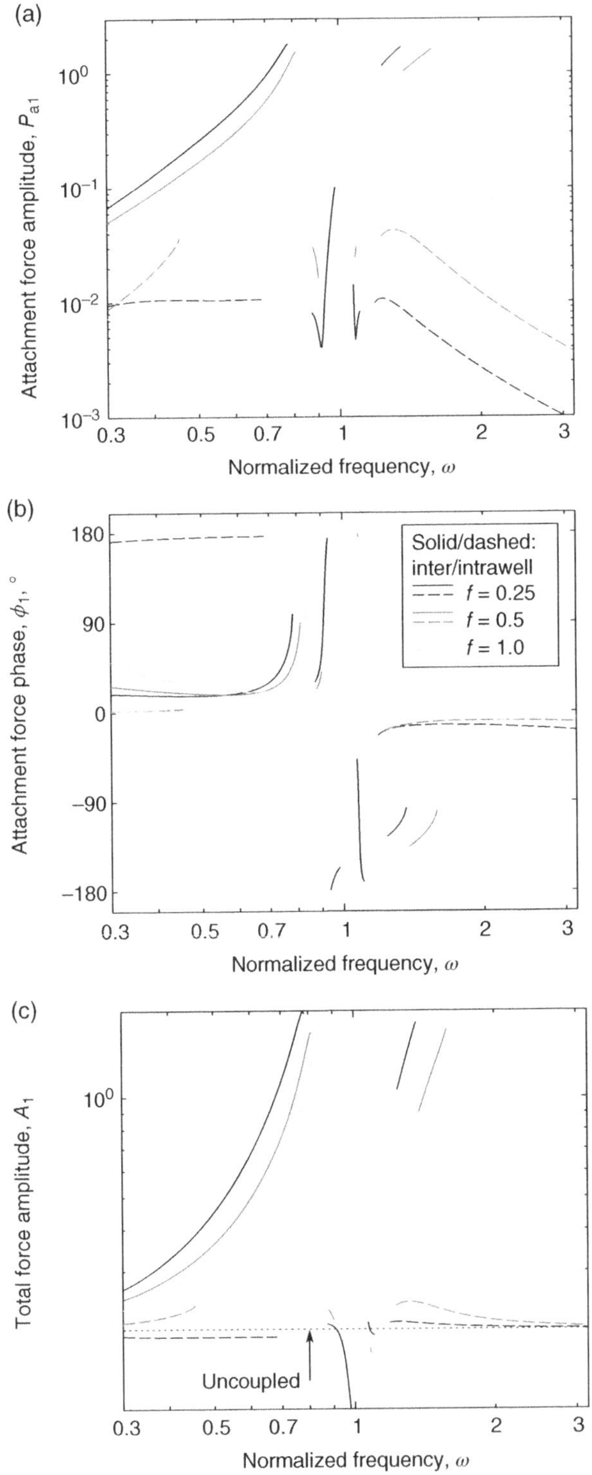

Figure 3.40 (a) Force amplitude and (b) force phase as applied by attachment. (c) Total force amplitude applied to the host structure for different tuning frequency ratios f.

behaviors above and below the frequency of the uncoupled resonance [3]. For the bistable attachment, the absence of spectral feature change of the host structure is beneficial when it is difficult to tailor the system damping without a corresponding shift in frequency characteristics.

Figure 3.41 examines the response of the system for different levels of the mass ratio μ, while the remaining parameters are $p = 0.2$, $f = 0.25$, $\beta = 1$, and $\gamma_1 = \gamma_2 = 0.05$. The mass ratio is varied from moderate to relatively large values, from 0.1 to 1.0, for the purposes of illustration as to the utility of the bistable attachment, since it is well-known that a larger linear absorber supplies a more

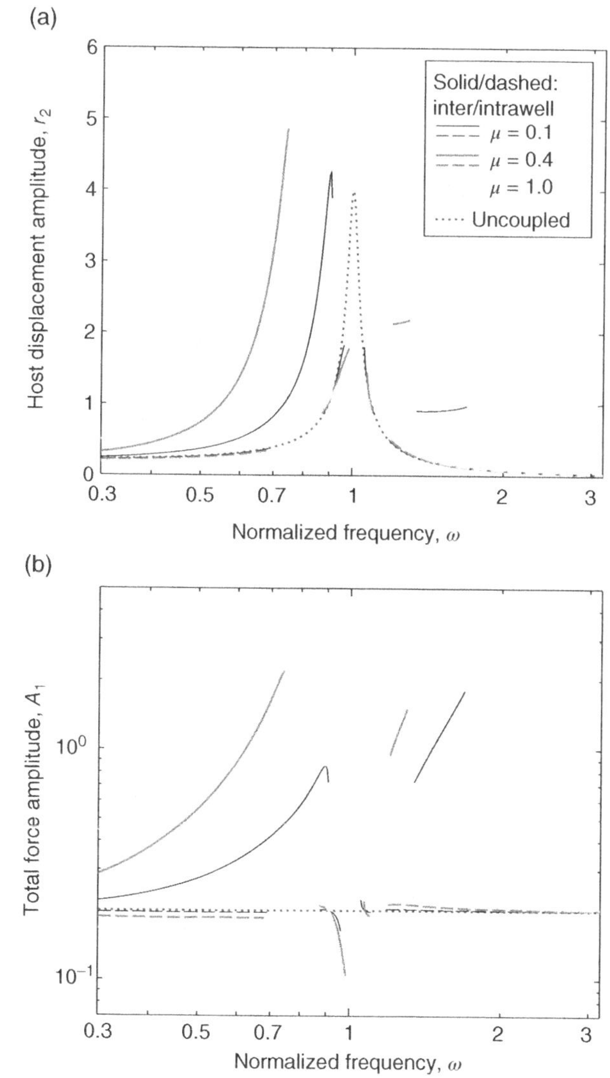

Figure 3.41 System dynamics as mass ratio μ varies. (a) Host displacement amplitude, (b) total force applied to host structure.

effective reactive force. As shown clearly in Figure 3.41a, the maximum frequency of stable constructive snap-through is tailored by adjusting μ: as mass ratio increases, the maximum stable frequency of this dynamic decreases. Near $\omega = 1$, regions of favorable destructive snap-through may be activated. Around this frequency, a variation in the mass ratio has less effect on the stable bandwidth of the response that provides the destructive force, as compared to the prominent influence by changing mass ratio observed in the constructive region for frequencies $\omega < 1$. Additionally, the greater added mass of the bistable attachment is found to yield greater cancellation of the total force, Figure 3.41b.

Figure 3.42 examines the influence of changing the bistable attachment damping γ_1 while the other parameters are $p = 0.2$, $f = 0.25$, $\beta = 1$, $\mu = 0.3$, and $\gamma_2 = 0.05$. As the damping constant of the attachment is increased, the undesirable constructive forces are predicted to occur over progressively narrower bandwidths of frequency. For the greatest level of damping considered here, $\gamma_1 = 0.5$, Figure 3.42 shows that the interwell dynamic that might otherwise induce a constructive force does not exist at low frequencies. At the same, the increase in bistable attachment damping γ_1 is found to broaden the frequency range of stable, destructive forces which may occur near $\omega = 1$, as exemplified via the host structure response in the inset of Figure 3.42a.

Figure 3.42d shows that as the damping of the bistable attachment increases, the phase of the attachment force approaches 180° out-of-phase in the frequency region around $\omega = 1$. The result is an improvement in effectiveness of the interwell dynamics for vibration control, since the host structural motions are more sufficiently suppressed near $\omega = 1$ for higher degrees of damping. Roughly speaking, this is the opposite effect of increasing the damping of a linear vibration absorber, in which case the phase of the attachment reactive force is progressively shifted away from the ideal of 180° out-of-phase with respect to the excitation. Comparable to the findings in Figure 3.41, Figure 3.42a also shows that the influence of the bistable attachment near $\omega = 1$ is more like an enhancement in the host structural damping than a multimodal behavior which is commonly associated with the two DOF dynamics of the primary structure and attachment.

3.6.3 Experimental Investigation of Force Cancellation Performance

The lack of experimental components in the previous research of bistable devices as absorbers of forced vibrations indicates that much remains to be uncovered in terms of putting the concept into practice. This section seeks to remedy such deficiency of insight. The analytical studies in Section 3.6.2 found that employing a bistable device as a vibration absorption implement is favorable only for some configurations of the device and only over certain bandwidths of frequency. On the other hand, by using key design parameters and certain excitation conditions, it was shown that a properly deployed bistable attachment may induce destructive forces with respect to the excitations, and lead to attenuation of the host structural oscillations. Interestingly, it did not appear that the energetic snap-through motions are the dynamic regime of the bistable attachment that provided this opportunity for vibration control. The experiments of this section thus seek to verify these analytical findings and present a first set of hands-on insights as to the utility of snap-through dynamics for vibration absorption.

The experimental system configuration is shown in Figure 3.43. A host linear oscillator (frame) is attached to an electrodynamic shaker platform via a low-friction translational bearing and coil spring. Another translational bearing mass is mounted on the host structure via an inclined spring and a bearing slide. The natural length of the inclined spring is greater than the distance between the

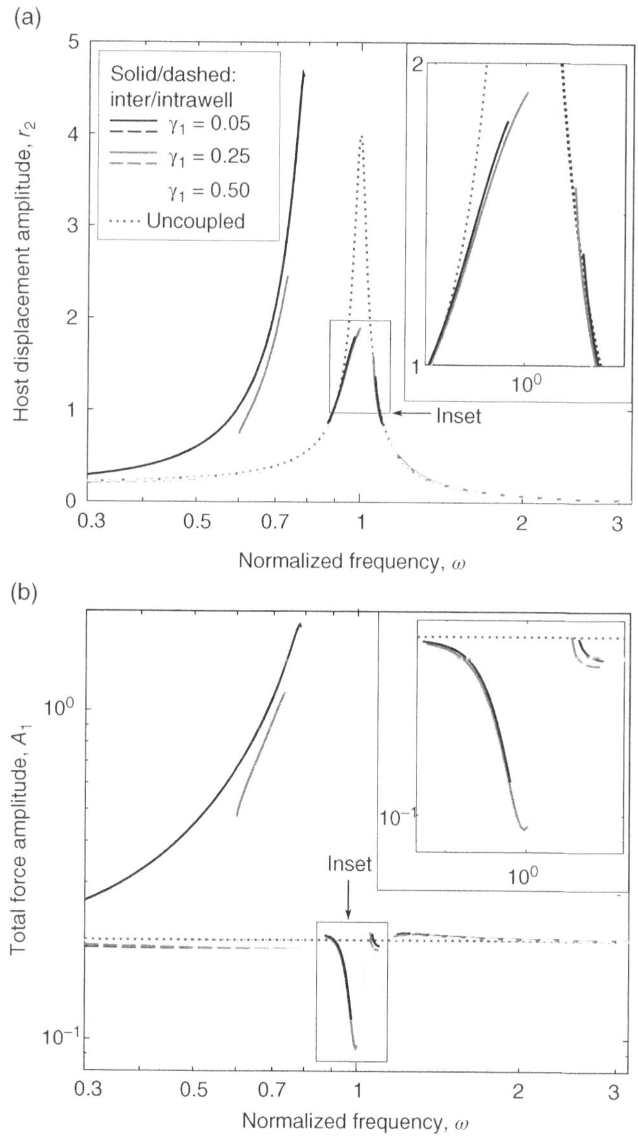

Figure 3.42 System dynamics as damping constant γ_1 varies. (a) Host structure displacement amplitude, (b) total force amplitude on host structure, (c) bistable attachment force amplitude and (d) phase.

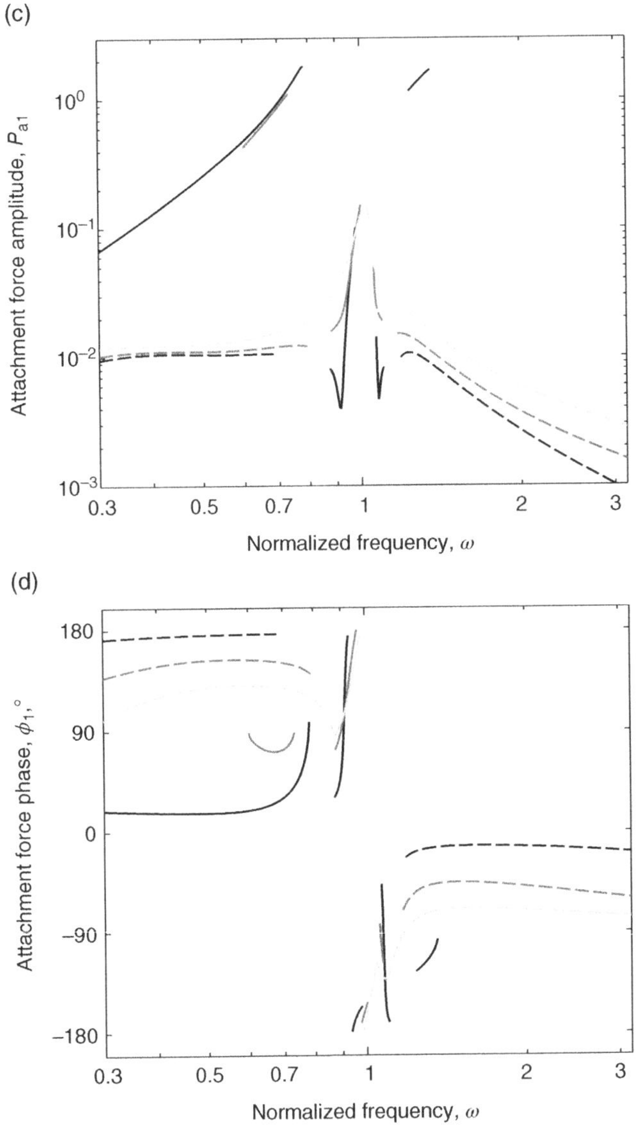

Figure 3.42 (Continued)

frame and the mounted bearing, which therefore induces bistability in the attachment. The configuration in Figure 3.43 shows the bistable attachment positioned at one of the stable equilibria, while the unstable configuration corresponds to the bearing position such that the inclined spring is perfectly vertical and is thus compressed.

In the experiments, the signal of excitation is a slowly swept sinusoid (+0.0092 Hz s^{-1}) provided to the shaker at various input levels. A potentiometer measures the displacement of the shaker platform and accelerometers measure

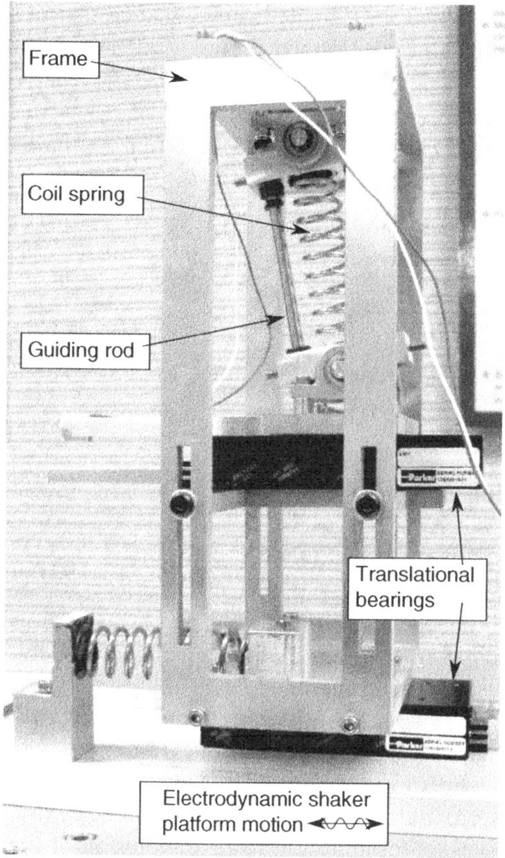

Figure 3.43 Experimental system of primary host structure and bistable attachment.

the vibration of the oscillator inertial masses. Experiments are conducted with and without the bistable attachment coupled to the host oscillator. The attachment has a mass ratio of $\mu \approx 0.39$ and tuning ratio of $f \approx 0.3$. The resonance frequency of the uncoupled linear oscillator is computed to be 5.20 Hz based on the known linear spring constant (960 N m^{-1}) and mass (0.90 kg).

A direct comparison of results between the *force-excited* system examined in the analysis and the *base-excited* system of the experiment is possible by correctly utilizing the acquired data. First, consider the governing equations for the system studied in Section 3.6.1, shown in Figure 3.38, excepting that the force excitation is replaced by ground acceleration, $\ddot{z}$, such as that employed in the experimentation.

$$m_2(\ddot{v}+\ddot{z})+d_2\dot{v}+k_2v-d_1(\dot{u}-\dot{v})+k_1(u-v)-k_3(u-v)^3=0 \tag{3.86a}$$

$$m_1(\ddot{u}+\ddot{z})+d_1(\dot{u}-\dot{v})-k_1(u-v)+k_3(u-v)^3=0 \tag{3.86b}$$

The displacement coordinates are defined such that v is the displacement of the host structure and u is the displacement of the bistable attachment, both relative to the moving ground. The bistable coordinate is defined such that the inertial mass is in the unstable equilibrium position when both $u=0$ and $v=0$. Rearranging terms in Eq. (3.86):

$$m_2\ddot{v}+d_2\dot{v}+k_2v=-m_2\ddot{z}+\left[d_1\left(\dot{u}-\dot{v}\right)-k_1\left(u-v\right)+k_3\left(u-v\right)^3\right] \tag{3.87a}$$

$$m_1\left(\ddot{u}+\ddot{z}\right)=-\left[d_1\left(\dot{u}-\dot{v}\right)-k_1\left(u-v\right)+k_3\left(u-v\right)^3\right]=-F_{\text{att}} \tag{3.87b}$$

Subsequently, Eq. (3.87a) becomes

$$m_2\ddot{v}+d_2\dot{v}+k_2v=-m_2\ddot{z}-m_1\left(\ddot{u}+\ddot{z}\right)=F_{\text{ex}}+F_{\text{att}} \tag{3.88}$$

In order to compare the base-excited system with the force-excited system, Eq. (3.72b) is rearranged using,

$$m_1\left(\ddot{x}+\ddot{y}\right)=-\left[d_1\dot{x}-k_1x+k_3x^3\right]=-P_{\text{att}} \tag{3.89}$$

and thereafter express Eq. (3.72a) as

$$m_2\ddot{y}+d_2\dot{y}+k_2y=P_0\cos\Omega t-m_1\left(\ddot{x}+\ddot{y}\right)=P_{\text{ex}}+P_{\text{att}} \tag{3.90}$$

Equations (3.88) and (3.90) show that the force-excited system of the analysis and the base-excited system of the experiment are mathematically equivalent. The summation of excitation and bistable attachment forces (F_{ex} and F_{att}, respectively) in the experimental system Eq. (3.88) are both exerted upon the *relative* vibration of the host structure v, whereas their counterpart forces (P_{ex} and P_{att}) in the analytical system expressed using Eq. (3.90) work upon the *absolute* vibration of the host system y. Therefore, to verify the analytical development of this work, the measurements are post-processed to compute the relative vibration of the host structure with respect to the input shaker accelerations. Finally, one should note that by comparing systems Eq. (3.88) and (3.90) and following a normalization of the equations as employed in the analytical formulation of Section 3.6.1, the parameters $\omega_{1,2}$, μ, f, $\gamma_{1,2}$, and β have identical interpretations between the systems, including in terms of their magnitudes. In other words, no reinterpretation or transformation is required to compare the dynamical changes observed by tailoring the system and excitation parameters in the analytical studies of Section 3.6.2 with those in the experimental system.

The host structure relative vibration v is plotted in Figure 3.44, evaluated by a frequency response function (frf) of host acceleration to input acceleration. Three levels of base excitation are considered having root-mean-square (RMS) values of 1.07, 1.35, and 1.60 m s^{-2}. The results measured for these excitation levels are shown in Figure 3.44a–c, respectively. Due to the nonlinearity in the system, the acceleration frf measure will vary based upon the excitation level. As anticipated, the uncoupled system exhibits a strong resonance around 5.20 Hz, denoted by a large peak in the frf at this frequency for each of the excitation levels presented in Figure 3.44. For the coupled system around 2.5 to 3 Hz for each case of base acceleration, the bistable attachment undergoes an energetic snap-through dynamic. It is seen that the displacements of the bistable attachment mass are primarily in-phase with those of the host structure when this dynamic occurs.

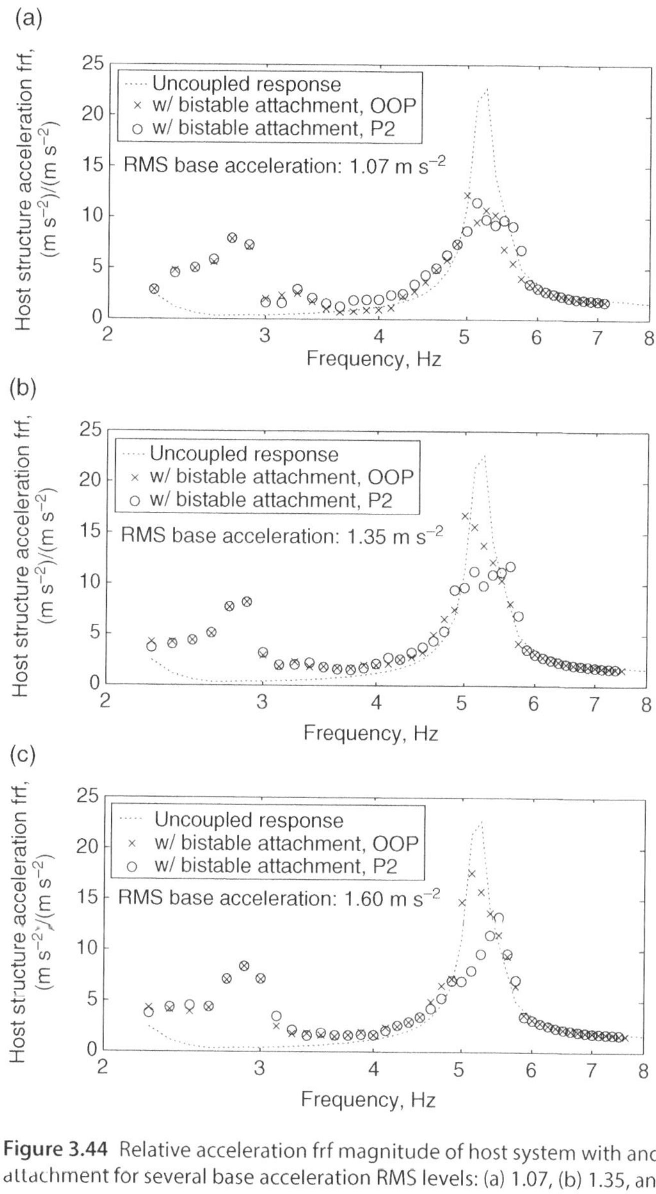

Figure 3.44 Relative acceleration frf magnitude of host system with and without the bistable attachment for several base acceleration RMS levels: (a) 1.07, (b) 1.35, and (c) 1.60 m s^{-2}.

Using the experimental data of base acceleration $\ddot{z}$ and the absolute acceleration of the bistable attachment $\ddot{u}+\ddot{z}$, the excitation and bistable attachment forces are determined and their phase relationships are compared to assess the degree of constructive or destructive superposition. A time series of the two forces is shown in Figure 3.45, for the base acceleration of 1.07 m s^{-2}, to verify the relative phases of the forces that work upon the host structure. Figure 3.45a shows that the bistable attachment force $F_{att} = -m_1(\ddot{u}+\ddot{z})$ is constructive with the excitation force $F_{ex} = -m_2\ddot{z}$ for the lower frequency excitations that lead to the

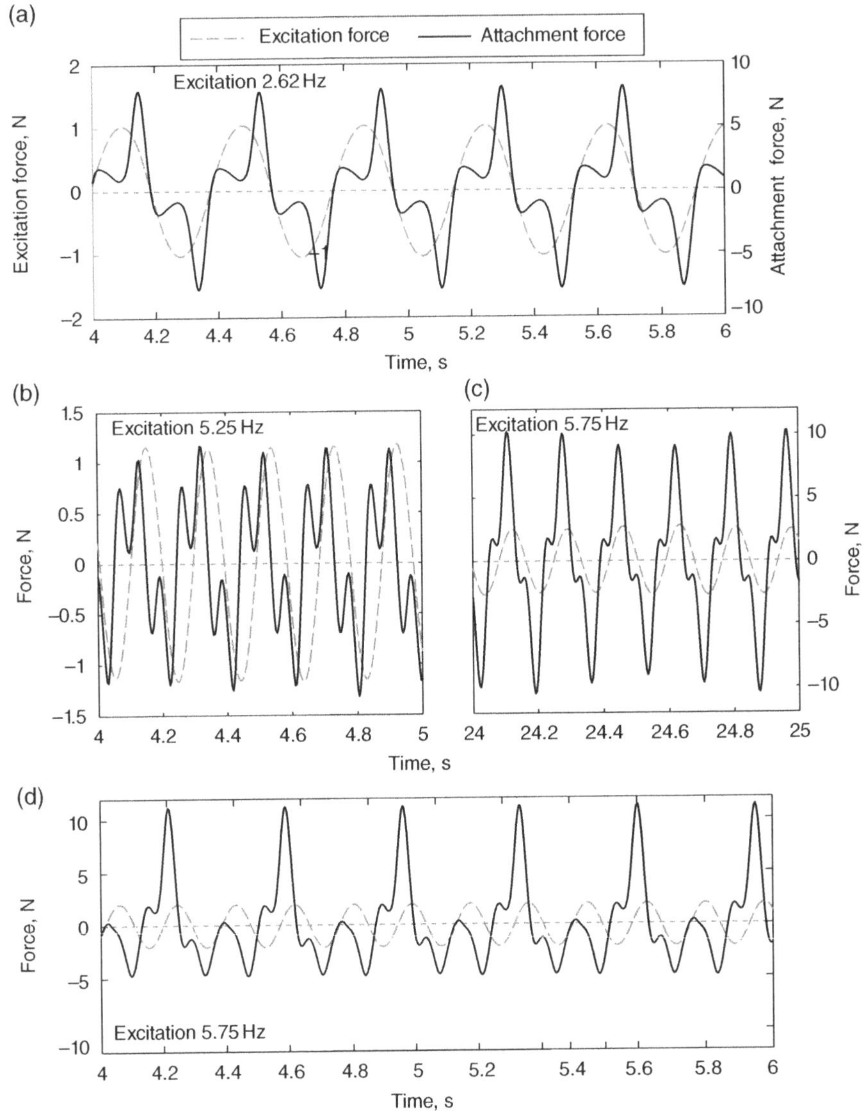

Figure 3.45 Time series of forces as computed via measured data. Forces plotted correspond to the following dynamic phenomena: (a) Constructive snap-through at 2.62 Hz. (b) Destructive snap-through during near-stabilization of unstable equilibrium at 5.25 Hz (OOP). (c) Destructive snap-through at 5.75 Hz (OOP). (d) P2 snap-through at 5.75 Hz.

energetic snap-through dynamics of the attachment. One must also consider the distinct vertical axes presented in Figure 3.45a to observe that the attachment forces produced by the snap-through dynamics are several times *larger* in magnitude than the excitation forces. The large constructive force explains the resulting detrimental amplification of the host oscillator response in the frequency range around 2.5 to 3 Hz. This also serves to validate the analytical formulation, since it was predicted that at frequencies less than the host structure uncoupled resonance, the snap-through dynamic may provide a constructive force leading to a large amplification in the host structure response.

The analytical results in Figure 3.39b,c indicate that as the forcing frequency is increased, the constructive snap-through forces may persist as the frequency approaches the uncoupled linear oscillator resonance and adversely amplify the host system response. However, the present experimental configuration was not capable of sustaining such high-energy dynamics for frequencies greater than about 3 Hz and instead returned to small intrawell oscillations, as observed for each excitation level shown in Figure 3.44. As Figure 3.42 shows through the analyses, increased damping in the bistable attachment may be one means to reduce the persistence of the snap-through dynamics that induce the adverse constructive forces. Indeed, the experimental system has a moderate amount of damping due to the bearing motions and this may be the explanation for the limited frequency band across which the unfavorable constructive forces are activated.

As the excitation frequency approaches 5 Hz, the coupled system undergoes a resonance-like behavior during which time the bistable attachment exhibits one of two very distinct responses, indicated in Figure 3.44 as squares (out-of-phase snap-through displacements with respect to host structure displacements, OOP) or circles (period-2 snap-through motions, P2). It is clear from a large number of experiments that the initial conditions determine which response form is produced. It is also observed from numerous trials that both responses would be activated with similar probabilities; in fact, physical impulses into the system around these frequencies were capable of consistently transitioning the system response from P2 to OOP and back again. Around the excitation frequencies of 5 to 5.25 Hz for the OOP dynamic, the bistable inertial mass has little absolute motion, although there is still relative motion between it and the surrounding frame. The attachment force which is induced by this behavior is of comparable magnitude with the applied excitation and mostly out-of-phase, as shown in Figure 3.45b. This is conducive for effective vibration cancellation, although Figure 3.45b indicates that the attachment force response is partially diffused into an order-3 harmonic such that a perfect cancellation of the exciting disturbances is not realized. Figure 3.44 shows that the primary influence provided by the OOP response is a damping-like effect upon the original host structure resonance with no shift in spectral characteristics. This suggests that the bistable spring enables the attachment to behave as a purely damping device instead of like an additional DOF, and is in good agreement with the findings of the analysis in regards to the consequences of the single-periodic motions that occur near the uncoupled host structure resonance frequency such that $\omega = 1$.

As the excitation frequency increases slightly to 5.75 Hz, the bistable attachment exhibits pronounced OOP behaviors. The forces induced by the bistable attachment, as shown in Figure 3.45c, are partially destructive with the excitation force. Approximately one-half of each excitation period in Figure 3.45c indicates opposing forces, and the remaining portion of the period suggests substantial reinforcement of the excitation load. However, the trend of the OOP response plotted in Figure 3.44 is in good agreement with the analytical findings in

Figure 3.39c for $f = 0.25$ near $\omega = 1$. The analyses predict that the coupled system undergoes a resonance feature in this bandwidth although at a reduced peak amplitude when compared to the uncoupled system. In Figure 3.44, the OOP dynamic in the bistable attachment leads to a similar reduction in peak amplitude of the host structure vibrations as compared with the resonance peak of the uncoupled system. The attenuation observed near the uncoupled host structure resonance peak during the OOP response is similar to the expected effect of directly adding damping to the host structure. This observation is in good agreement with analytical findings.

A second and repeatable dynamic that leads to attenuation of host structure motion is observed near the uncoupled linear system resonance frequency. Figure 3.45d shows the time series of the force induced by this dynamic during which the bistable oscillator vibrates primarily with a period that is twice that of the excitation (P2). In contrast, the host system is observed to vibrate mostly at the driving frequency. Figure 3.45d indicates several moments during a forcing period when the bistable attachment and excitation forces are destructive which provides an explanation for the reasonable degree of vibration suppression that it supplies, as shown in Figure 3.44, for all levels of base excitation measured here. Another explanation for this attenuation is that the net harmonic energy of the coupled system is more readily diffused from the host structure into the bistable attachment in the form of the strong P2 harmonic. This is a similar nonlinear vibration control methodology as employed by other researchers [9], sometimes classified as nonlinear energy pumping or energy transfer.

3.6.4 Summary

Previous reliance on analytical studies using nonlinear normal modes or only examining the transient dynamics has provided an inconclusive assortment of evidence as to the practical effectiveness of a bistable attachment when employed to be an absorber of forced vibrations. This study aimed to remedy the absence of needed insights and provided analytical and experimental assessments of the vibration absorption capabilities of a bistable device attached to an excited, host, linear structure. Contrary to a prevailing expectation in the literature, it is shown that the energetic snap-through dynamics uniformly *amplify* the vibration of the host structure. This is confirmed by observing that the forces which the snap-through behaviors induce are in-phase with the excitation forces. Analysis and experiments show that a unique out-of-phase dynamic, which is primarily of the same frequency as the excitation, can provide a reduction in the host structural vibrations around the uncoupled resonance via a certain degree of destructive forces. The effect of the opposing force dynamic has a similar consequence to a direct increase in the host structure damping. As a result, the bistable attachment effectively allows the structure to maintain the original spectral features near resonance, appearing like the host structure is dynamically uncoupled from other dynamics bodies, with the appearance of enhancing the damping characteristics. The experimentation verifies the analytical findings and also uncovers regions of excitation frequency that may induce other-order harmonic behaviors not predictable by the present analytical solution strategy, such as period-2 oscillations.

Since these latter dynamics are also found to provide a greater level of vibration attenuation than the out-of-phase dynamic, future studies should evaluate the efficacy of deliberately exploiting steady-state dynamics exhibiting a strong contribution of frequencies other than the excitation as a means to favorably attenuate the dynamics of a host structure having a bistable attachment.

References

[1] C.R. Fuller, S.J. Elliott, P.A. Nelson, *Active Control of Vibration* (Academic Press, London, 1996).

[2] K.W. Wang, J. Tang, *Adaptive Structural Systems with Piezoelectric Transducer Circuitry* (Springer, New York, 2008).

[3] J.P. Den Hartog, *Mechanical Vibrations* (McGraw-Hill, New York, 1947).

[4] D.J. Inman, *Vibration with Control* (John Wiley and Sons, Ltd., Chichester, 2006).

[5] L. Cremer, M. Heckl, B.A.T. Petersson, *Structure-Borne Sound: Structural Vibrations and Sound Radiation at Audio Frequencies* (Springer, Berlin, 2005).

[6] X. Gong, Y. Fan, S. Xuan, *et al.*, Control of the damping properties of magnetorheological elastomers by using polycaprolactone as a temperature-controlling component, *Industrial & Engineering Chemistry Research* **51**, 6395–6403 (2012).

[7] B. Zhu, B.D. Rahn, C.E. Bakis, 2013 Vibration damping of a cantilever beam utilizing fluidic flexible matrix composites. In *Proceedingsof SPIE vol. 8688, Active and Passive Smart Structures and Integrated Systems 2013*.

[8] C. Chen, W.H. Liao, A self-sensing magnetorheological damper with power generation, *Smart Materials and Structures* **21**, 025014 (2012).

[9] A.F. Vakakis, O.V. Gendelman, L.A. Bergman, *et al.*, *Nonlinear Targeted Energy Transfer in Mechanical and Structural Systems* (Springer, Dordrecht, 2008).

[10] G.S. Agnes, D.J. Inman, Nonlinear piezoelectric vibration absorbers, *Smart Materials and Structures* **5**, 704–714 (1996).

[11] R. Lakes, *Viscoelastic Materials* (Cambridge University Press, Cambridge, 2009).

[12] D.J. Warkentin, N.W. Hagood, 1997 Nonlinear piezoelectric shunting for structural damping. In *Proceedings of SPIEVol. 3041 Smart Structures and Materials 1997*, pp. 747–757.

[13] R.S. Lakes, Extreme damping in composite materials with a negative stiffness phase, *Physical Review Letters* **86**, 2897–2900 (2001).

[14] S.P. Timoshenko, *Theory of Elastic Stability* (McGraw-Hill, New York, 1936).

[15] L.N. Virgin, R. Wiebe, On damping in the vicinity of critical points, *Philosophical Transactions of the Royal Society A* **371**, 20120426 (2013).

[16] R.S. Lakes, W.J. Drugan, Dramatically stiffer elastic composite materials due to a negative stiffness phase?, *Journal of the Mechanics and Physics of Solids* **50**, 979–1009 (2002).

[17] W.J. Drugan, Elastic composite materials having a negative stiffness phase can be stable, *Physical Review Letters* **98**, 055502 (2007).

[18] R.S. Lakes, Extreme damping in compliant composites with a negative-stiffness phase, *Philosophical Magazine Letters* **81**, 95–100 (2001).

[19] R.S. Lakes, T. Lee, A. Bersie, Y.C. Wang, Extreme damping in composite materials with negative-stiffness inclusions, *Nature* **410**, 565–567 (2001).
[20] Y.C. Wang, R.S. Lakes, Stable extremely-high-damping discrete viscoelastic systems due to negative stiffness elements, *Applied Physics Letters* **84**, 4451–4453 (2004).
[21] Y.C. Wang, R.S. Lakes, Composites with inclusions of negative bulk modulus: extreme damping and negative Poisson's ratio, *Journal of Composite Materials* **39**, 1645–1657 (2005).
[22] Y.C. Wang, M. Ludwigson, R.S. Lakes, Deformation of extreme viscoelastic metals and composites, *Materials Science and Engineering A* **370**, 41–49 (2004).
[23] T. Klatt, M.R. Haberman, A nonlinear negative stiffness metamaterial unit cell and small-on-large multiscale material model, *Journal of Applied Physics* **114**, 033503 (2013).
[24] J. Matthews, T. Klatt, C.C. Seepersad, *et al.*, 2013 Hierarchical design composite materials with negative stiffness inclusions using a Bayesian network classifier. In *Proceedings of the ASME 2013 International Design Engineering Technical Conferences and Computers and Information in Engineering Conference*; Portland, Oregon, USA. DETC2013-13128.
[25] D.M. Kochmann, W.J. Drugan, Dynamic stability analysis of an elastic composite material having a negative-stiffness phase, *Journal of the Mechanics and Physics of Solids* **57**, 1122–1138 (2009).
[26] D.M. Kochmann, Stable extreme damping in viscoelastic two-phase composites with non-positive-definite phases close to the loss of stability, *Mechanics Research Communications* **58**, 36–45 (2014).
[27] F. Fritzen, D.M. Kochmann, Material instability-induced extreme damping in composites: a computational study, *International Journal of Solids and Structures* **51**, 4101–4112 (2014).
[28] C.S. Wojnar, D.M. Kochmann, A negative-stiffness phase in elastic composites can produce stable extreme effective dynamic but not static stiffness, *Philosophical Magazine* **94**, 532–555 (2014).
[29] G. Puglisi, L. Truskinovsky, Mechanics of a discrete chain with bi-stable elements, *Journal of the Mechanics and Physics of Solids* **48**, 1–27 (2000).
[30] A.M. Balk, A.V. Cherkaev, L.I. Slepyan, Dynamics of chains with non-monotone stress-strain relations. i. model and numerical experiements, *Journal of the Mechanics and Physics of Solids* **49**, 131–148 (2001).
[31] G. Puglisi, L. Truskinovsky, A mechanism of transformational plasticity, *Continuum Mechanics and Thermodynamics* **14**, 437–457 (2002).
[32] A. Cherkaev, L. Zhornitskaya, Protective structures with waiting links and their damage evolution, *Multibody System Dynamics* **13**, 53–67 (2005).
[33] A. Cherkaev, E. Cherkaev, L. Slepyan, Transition waves in bistable structures. I. Delocalization of damage, *Journal of Mechanics and Physics of Solids* **53**, 383–405 (2005).
[34] L. Slepyan, A. Cherkaev, E. Cherkaev, Transition waves in bistable structures. II. Analytical solution: wave speed and energy dissipation, *Journal of Mechanics and Physics of Solids* **53**, 407–436 (2005).

[35] T. Cohen, S. Givli, Dynamics of a discrete chain of bi-stable elements: a biomimetic shock absorbing mechanism, *Journal of the Mechanics and Physics of Solids* **64**, 426–439 (2014).

[36] L. Dong, R.S. Lakes, Advanced damper with negative structural stiffness elements, *Smart Materials and Structures* **21**, 075026 (2012).

[37] L. Dong, R. Lakes, Advanced damper with high stiffness and high hysteresis damping based on negative structural stiffness, *International Journal of Solids and Structures* **50**, 2416–2423 (2013).

[38] S. Barbarino, M.E. Pontecorvo, F.S. Gandhi, 2012 Energy dissipation of a bi-stable von-Mises truss under harmonic excitation. In *Proceedings of the 54th AIAA/ASME/ASCE/AHS/ASC Structures, Structural Dynamics, and Materials Conference*; Honolulu, Hawaii, USA. AIAA 2012-1712.

[39] S. Barbarino, M.E. Pontecorvo, F.S. Gandhi, 2013 Energy dissipation of a bi-stable von-Mises truss under impulsive excitation. In *Proceedings of the 54th AIAA/ASME/ASCE/AHS/ASC Structures, Structural Dynamics, and Materials Conference*; Boston, Massachusetts, USA. AIAA 2013-1794.

[40] M.E. Pontecorvo, S. Barbarino, F.S. Gandhi, *et al.*, 2013 A load-bearing structural element with energy dissipation capability under harmonic excitation. In *Proceedings of the ASME 2013 Conference on Smart Materials, Adaptive Structures and Intelligent Systems*; Snowbird, Utah, USA. SMASIS 2013-3060.

[41] M.E. Pontecorvo, S. Barbarino, F.S. Gandhi, *et al.*, 2013 Impact load mitigation with bi-stable von-Mises trusses. In *Proceedings of the ASME 2013 Conference on Smart Materials, Adaptive Structures and Intelligent Systems*; Snowbird, Utah, USA. SMASIS 2013-3061.

[42] G.J. Murray, F. Gandhi, 2011 The use of damping to mitigate violent snap-through of bistable systems. In *Proceedings of the ASME 2011 Conference on Smart Materials, Adaptive Structures and Intelligent Systems SMASIS2011*; Scottsdale, Arizona. SMASIS 2011-4997.

[43] D.J. Inman, *Engineering Vibration* (Prentice Hall, Saddle River, New Jersey, 2001).

[44] R.A. Ibrahim, Recent advances in nonlinear passive vibration isolators, *Journal of Sound and Vibration* **314**, 371–452 (2008).

[45] L.N. Virgin, *Vibration of Axially Loaded Structures* (Cambridge University Press, Cambridge, 2007).

[46] J. Winterflood, T. Barber, D.G. Blair, Using Euler buckling springs for vibration isolation, *Classical and Quantum Gravity* **19**, 1639–1645 (2002).

[47] J. Winterflood, D.G. Blair, B. Slagmolen, High performance vibration isolation using springs in Euler column buckling mode, *Physics Letters A* **300**, 122–130 (2002).

[48] E.J. Chin, K.T. Lee, J. Winterflood, L. Ju, D.G. Blair, Low frequency vertical geometric anti-spring vibration isolators, *Physics Letters A* **336**, 97–105 (2005).

[49] L.N. Virgin, R.B. Davis, Vibration isolation using buckled struts, *Journal of Sound and Vibration* **260**, 965–973 (2003).

[50] R.H. Plaut, J.E. Sidbury, L.N. Virgin, Analysis of buckled and pre-bent fixed-end columns used as vibration isolators, *Journal of Sound and Vibration* **283**, 1216–1228 (2005).

[51] R.H. Plaut, H.M. Favor, A.E. Jeffers, *et al.*, Vibration isolation using buckled or pre-bent columns - part 1: two-dimensional motions of horizontal rigid bar, *Journal of Sound and Vibration* **310**, 409–420 (2008).
[52] D.L. Platus. Minus K® Technology Inc. [Online].; 2014. Available from: http://www.minusk.com/.
[53] D.L. Platus, 1991 Negative-stiffness-mechanism vibration isolation systems. In *Proceedings of SPIE Vol. 1619*, pp. 44–54.
[54] D.L. Platus, 1999 Negative-stiffness-mechanism vibration isolation systems. In *Proceedings of SPIE Vol. 3786*, pp. 98–105.
[55] A. Carrella, M.J. Brennan, T.P. Waters, K. Shin, On the design of a high-static-low-dynamic stiffness isolator using linear mechanical springs and magnets, *Journal of Sound and Vibration* **315**, 712–720 (2008).
[56] A.D. Shaw, S.A. Neild, D.J. Wagg, *et al.*, A nonlinear spring mechanism incorporating a bistable composite plate for vibration isolation, *Journal of Sound and Vibration* **332**, 6265–6275 (2013).
[57] N. Zhou, K. Liu, A tunable high-static-low-dynamic stiffness vibration isolator, *Journal of Sound and Vibration* **329**, 1254–1273 (2010).
[58] A. Carrella, M.J. Brennan, I. Kovacic, *et al.*, On the force transmissibility of a vibration isolator with quasi-zero-stiffness, *Journal of Sound and Vibration* **322**, 707–717 (2009).
[59] A. Carrella, M.J. Brennan, T.P. Waters, *et al.*, Force and displacement transmissibility of a nonlinear isolator with high-static-low-dynamic-stiffness, *International Journal of Mechanical Sciences* **55**, 22–29 (2012).
[60] J. Yang, Y.P. Xiong, J.T. Xing, Dynamics and power flow behaviour of a nonlinear vibration isolation system with a negative stiffness mechanism, *Journal of Sound and Vibration* **332**, 167–183 (2013).
[61] X. Huang, X. Liu, J. Sun, *et al.*, Vibration isolation characteristics of a nonlinear isolator using Euler buckled beam as negative stiffness corrector: a theoretical and experimental study, *Journal of Sound and Vibration* **333**, 1132–1148 (2014).
[62] S.A. Emam, A.H. Nayfeh, On the nonlinear dynamics of a buckled beam subjected to a primary-resonance excitation, *Nonlinear Dynamics* **35**, 1–17 (2004).
[63] A.F. Arrieta, S.A. Neild, D.J. Wagg, On the cross-well dynamics of a bi-stable composite plate, *Journal of Sound and Vibration* **330**, 3424–3441 (2011).
[64] L. Kashdan, C.C. Seepersad, M. Haberman, *et al.*, Design, fabrication, and evaluation of negative stiffness elements using SLS, *Rapid Prototyping Journal* **18**, 194–200 (2012).
[65] B.A. Fulcher, D.W. Shahan, M.R. Haberman, *et al.*, Analytical and experimental investigation of buckled beams as negative stiffness elements for passive vibration isolation and shock isolation systems, *Journal of Vibration and Acoustics* **136**, 031009 (2014).
[66] L.N. Virgin, *Introduction to Experimental Nonlinear Dynamics* (Cambridge University Press, Cambridge, 2000).
[67] A. Sarlis, D. Pasala, M. Constantinou, *et al.*, Negative stiffness device for seismic protection of structures, *Journal of Structural Engineering* **139**, 1124–1133 (2013).

[68] D.T.R. Pasala, A.A. Sarlis, S. Nagarajaiah, *et al.*, Adaptive negative stiffness: new structural modification approach for seismic protection, *Journal of Structural Engineering* **139**, 1112–1123 (2013).
[69] H. Frahm. Means for damping the rolling motion of ships. U.S. Patent 970,368. U.S. Patent Office; 1910.
[70] H. Frahm. Device for damping vibration of bodies. U.S. Patent 989,958. U.S. Patent Office; 1911.
[71] J.P. Den Hartog, J. Ormondroyd, Theory of the dynamic vibration absorber, *Journal of Applied Mechanics* **50**, 11–22 (1928).
[72] A.H. Nayfeh, D.T. Mook, *Nonlinear Oscillations* (Wiley-VCH Verlag GmbH, Weinheim, 1995).
[73] R.E. Roberson, Synthesis of a nonlinear dynamic vibration absorber, *Journal of the Franklin Institute* **254**, 205–220 (1952).
[74] J. Shaw, S.W. Shaw, A.G. Haddow, On the response of the non-linear vibration absorber, *International Journal of Non-linear Mechanics* **24**(4), 281–293 (1989).
[75] S.S. Oueini, A.H. Nayfeh, J.R. Pratt, A nonlinear vibration absorber for flexible structures, *Nonlinear Dynamics* **15**, 259–282 (1998).
[76] V.A. Buzhinskii, Use of the snap-through membrane effect to limit dynamic loads, *Mechanics of Solids* **24**(4), 40–45 (1989).
[77] O.V. Gendelman, C.H. Lamarque, Dynamics of linear oscillator coupled to strongly nonlinear attachment with multiple states of equilibrium, *Chaos, Solitons and Fractals* **24**, 501–509 (2005).
[78] E. Gourdon, C.H. Lamarque, Energy pumping with various nonlinear structures: numerical evidences, *Nonlinear Dynamics* **40**, 281–307 (2005).
[79] K.V. Avramov, Y.V. Mikhlin, Forced oscillations of a system, containing a snap-through truss, close to its equilibrium position, *Nonlinear Dynamics* **35**, 361–379 (2004).
[80] K.V. Avramov, Y.V. Mikhlin, Snap-through truss as an absorber of forced oscillations, *Journal of Sound and Vibration* **290**, 705–722 (2006).
[81] K.V. Avramov, O.V. Gendelman, Interaction of elastic system with snap-through vibration absorber, *International Journal of Non-Linear Mechanics* **44**, 81–89 (2009).
[82] L.I. Manevitch, G. Sigalov, F. Romeo, *et al.*, Dynamics of a linear oscillator coupled to a bistable light attachment: analytical study, *Journal of Applied Mechanics* **81**, 041011 (2014).
[83] F. Romeo, G. Sigalov, L.A. Bergman, *et al.*, Dynamics of a linear oscillator coupled to a bistable light attachment: numerical study, *Journal of Computational and Nonlinear Dynamics* **10**, 011007 (2015).
[84] D.A. Bies, C.H. Hansen, *Engineering Noise Control: Theory and Practice* (Taylor & Francis, New York, 2003).
[85] R. Wiebe, L.N. Virgin, I. Stanciulescu, *et al.*, Characterizing dynamic transitions associated with snap-through: a discrete system, *Journal of Computational and Nonlinear Dynamics* **8**, 011010 (2013).
[86] W. Szemplińska-Stupnicka, J. Rudowski, Steady states in the twin-well potential oscillator: computer simulations and approximate analytical studies, *Chaos* **3**, 375–385 (1993).

[87] I. Kovacic, M.J. Brennan, eds, *The Duffing Equation: Nonlinear Oscillators and their Behaviour* 1st ed. (John Wiley & Sons, Ltd, Chichester, 2011).

[88] C.R. McInnes, D.G. Gorman, M.P. Cartmell, Enhanced vibrational energy harvesting using nonlinear stochastic resonance, *Journal of Sound and Vibration* **318**, 655–662 (2008).

[89] R.A. Ibrahim, Excitation-induced stability and phase transition: a review, *Journal of Vibration and Control* **12**(10), 1093–1170 (2006).

[90] Z. Wu, R.L. Harne, K.W. Wang, Excitation-induced stability in a bistable Duffing oscillator: analysis and experiment, *Journal of Computational and Nonlinear Dynamics* **10**, 011016 (2015).

[91] S.Y. Kim, Y. Kim, Dynamic stabilization in the double-well Duffing oscillator, *Physical Review E* **61**, 6517–6520 (2000).

[92] F.C. Moon, *Chaotic and Fractal Dynamics : an Introduction for Applied Scientists and Engineers* (John Wiley & Sons, Inc. New York, 1992).

[93] P. Holmes, A nonlinear oscillator with a strange attractor, *Philosophical Transactions of the Royal Society of London A* **292**(1394), 419–448 (1979).

[94] K.B. Blair, C.M. Krousgrill, T.N. Farris, Harmonic balance and continuation techniques in the dynamic analysis of Duffing's equation, *Journal of Sound and Vibration* **202**, 717–731 (1997).

[95] Y. Kim, S.Y. Lee, S.Y. Kim, Experimental observation of dynamic stabilization in a double-well Duffing oscillator, *Physics Letters A* **275**, 254–259 (2000).

[96] W.Y. Tseng, J. Dugundji, Nonlinear vibrations of a buckled beam under harmonic excitation, *Journal of Applied Mechanics* **38**(2), 467–476 (1971).

[97] R.L. Harne, M. Thota, K.W. Wang, Concise and high-fidelity predictive criteria for maximizing performance and robustness of bistable energy harvesters, *Applied Physics Letters* **102**, 053903 (2013).

[98] Z. Wu, R.L. Harne, K.W. Wang, Energy harvester synthesis via coupled linear-bistable system with multistable dynamics, *Journal of Applied Mechanics* **81**, 061005 (2014).

[99] M.I. Friswell, J.E.T. Penny, The accuracy of jump frequencies in series solutions of the response of a Duffing oscillator, *Journal of Sound and Vibration* **169**, 261–269 (1994).

[100] S.C. Stanton, B.A.M. Owens, B.P. Mann, Harmonic balance analysis of the bistable piezoelectric inertial generator, *Journal of Sound and Vibration* **331**, 3617–3627 (2012).

[101] H. Gunderson, H. Rigas, F.S. VanVleck, A technique for determining stability regions for the damped Mathieu equation, *SIAM Journal on Applied Mathematics* **26**, 345–349 (1974).

[102] D.W. Jordan, P. Smith, *Nonlinear Ordinary Differential Equations: An Introduction for Scientists and Engineers* (Oxford University Press, Oxford, 2007).

[103] F.C. Moon, Experiments on chaotic motions of a forced nonlinear oscillator: strange attractors, *Journal of Applied Mechanics* **47**, 638–644 (1980).

[104] A. Carrella, M.J. Brennan, T.P. Waters, Static analysis of a passive vibration isolator with quasi-zero-stiffness characteristic, *Journal of Sound and Vibration* **301**, 678–689 (2007).

[105] H.J. Ahn, Performance limit of a passive vertical isolator using a negative stiffness mechanism, *Journal of Mechanical Science and Technology* **22**, 2357–2364 (2008).

[106] Z. Lu, M.J. Brennan, T. Yang, *et al.*, An investigation of a two-stage nonlinear vibration isolation system, *Journal of Sound and Vibration* **332**, 1456–1464 (2013).

[107] L. Tang, Y. Yang, C.K. Soh, Improving functionality of vibration energy harvesters using magnets, *Journal of Intelligent Material Systems and Structures* **23**, 1433–1449 (2012).

[108] A. Preumont, K. Seto, *Active Control of Structures* (John Wiley & Sons, Ltd, Chichester, 2008).

[109] S.W. Shaw, C. Pierre, Normal modes for non-linear vibratory systems, *Journal of Sound and Vibration* **164**, 85–124 (1993).

[110] W. Szemplińska-Stupnicka, J. Bajkowski, Multi-harmonic response in the regions of instability of harmonic solution in multi-degree-of-freedom non-linear systems, *International Journal of Non-linear Mechanics* **15**, 1–11 (1980).

4

Vibration Energy Harvesting

This chapter details how the snap-through dynamics of base-excited, bistable, electromechanical structures may be harnessed to transform the energetic mechanical motions into electrical signals of considerable magnitude. The focus here is to understand the dynamical sensitivities of *bistable vibration energy harvesters* which are essential to elucidate for effective conversion of the ambient vibration resource into usable electrical power. The chapter begins with a broad review of the latest research activities in this area spanning novel device explorations in the laboratory to rigorous analytical constructions. Then, a series of detailed and comprehensive studies is presented that contribute to new understanding on the dynamics and design methodologies that guide development of bistable energy harvesters providing high and robust power generation performance.

4.1 Topic Review

Modern civilization is comprehensively established upon infrastructures, broadly speaking, which rely on energy resources. Whether to assist agricultural labors, enable wireless communications, or propel advanced land-, water-, and air-based transportation systems, humans draw from a massive number of energy pools to meet their various needs. Prior to the Industrial Revolution, a large proportion of the energy demands were satisfied using resources that were renewable on a human timescale [1], such as the motion of water or hitched animals to operate grinding mills, burning wood to cook food, and capturing wind to pilot watercraft. Since the Industrial Revolution, the energy needs to establish an ever-growing array of technologies have commonly been met using supplies which require hundreds of thousands of years (or longer) to replenish, such as coal, oil, and gas. Yet, based upon the rate of technological growth and projections of available non-renewable resources into the future, it has become plainly apparent that the sustainability of current-day civilization will be severely challenged unless our energy portfolio is substantially augmented with renewable resources [2].

Responding to this call, a revival of research, development, and deployment of energy capture and conversion methods based upon renewables has come about,

Harnessing Bistable Structural Dynamics: For Vibration Control, Energy Harvesting and Sensing, First Edition. Ryan L. Harne and K.W. Wang.

such that renewable energy supplies account for approximately 5.6% of the world's power generation as of 2014 [3]. While solar, hydroelectric, and wind power supplies may leverage a greater density of energy resource per capture area [4], there are many applications for which these sustainable solutions are insufficient. For example, small microelectronics for distributed sensors for structural condition and health monitoring, automotive tire air pressure monitors, and even some devices transported by or embedded in humans – these are a few representative applications for which the previous renewable energy approaches are often inadequate to meet the energy needs or are unable to be adapted to the respective operating environment, often due to the embedded characteristic of the electronics in application [5]. However, based upon the low power requirements for many modern microelectronics [6] and due to the presence of vibrations in such working environments, it has been found that co-located electromechanical devices may convert the kinetic energies into satisfactory power supplies for various functions of the microelectronics [7]. On a different scale and application setting, the interest in converting the oscillation energy of water wave heaves into grid-scale electrical power became a center of intense attention starting in the 1970s during the worldwide energy crisis spurred by sudden increase in the price of crude oil [8]. Thus, whether for low-power microelectronics or to supplement the needs of continental electric grids, harvesting the *vibrations* around us has developed into an extensive and engaging subject of research, with some solutions already arriving at stages of commercial penetration.

Electromechanical devices that convert vibrations into electrical power are commonly termed "vibration energy harvesters". The transduction mechanisms used to date include a broad range of strategies, including adaptive materials like piezoelectrics [9] and dielectric elastomers [10], as well as conventional devices like rotary induction motors [8]. These, and almost all other transduction methods, induce greater amounts of electrical current or voltage in proportion to the dynamic displacement amplitude (and/or rate thereof) of the harvester inertial mass when excited by the driving vibrations. The mass-scaling performance of energy harvesters is a critical factor that has established important design principles applicable to numerous operational scenarios [7,11]. Based on engineering intuition, it is therefore essential that the harvester exhibit a significant sensitivity to the spectrum and form of the environmental vibrations to effect the best energy conversion performance. To provide such sensitivity, many energy harvesters are designed as underdamped, harmonic oscillators, where a portion of the restoring forces is attributed to structural, mechanical, and/or material design, and the remainder come about by the electromechanical influences of connecting the transduction mechanism to an electrical circuit. While linear dynamics-based energy harvesting platforms are able to realize effective upper bounds on power generation performance [12,13], they possess numerous practical concerns such as excessively large stroke motions if resonating at low excitation frequencies, and narrow bandwidths of favorable operation. Thus, from a variety of investigations, researchers have recently discovered that deliberately leveraging *nonlinear* designs and dynamics is a promising and viable means to overcome the limitations evidenced in the linear counterparts [13].

Among the investigations on favorable nonlinear structural designs and dynamics that improve power generation capability in energy harvesting devices [14–18], architectures that leverage ***bistability*** have been found to readily induce large amplitude electromechanical dynamics under a wide variety of vibration excitation environments. The source of the advantageous power generation capability is the non-resonant nature of snap-through dynamics, which makes the phenomenon largely frequency-insensitive. This property uniquely makes the key benefit of bistable vibration energy harvesting devices independent of the input vibration *type* (i.e., input vibration that is harmonic, random, etc.) so long as the excitation *amplitude* exceeds a threshold. In this way, an effective deployment of bistable structures for energy harvesting purposes must employ a critical understanding of the ambient vibration resource with respect to the harvester system parameters that lead to the required amplitude-based sensitivity. When compared to the excitation frequency- and amplitude-based sensitivities typical of a large proportion of linear and monostable nonlinear energy harvesting architectures, the non-resonant property of snap-through has motivated considerable research efforts to develop understanding and favorable design methodologies for maintaining the performance benefits of bistable device platforms via continued activation of snap-through. The following sections survey the progression of the research, catalogued according to the types of findings which support the conclusions, namely those studies employing mostly experimental and numerical simulation components and those supported by (semi-) analytical efforts.

4.1.1 Experimental and Numerical Developments in Energy Harvesting with Bistable Devices

Research backed by evidence obtained in laboratory explorations and via computer simulations propelled early interest to exploit the snap-through dynamics of bistable structures to energy harvesting advantage. This is because such efforts easily accommodate the strongly nonlinear dynamics when compared with the analytical tools common to most investigators, and because experimental evidence of the performance benefits is a convincing demonstrator and serves to encourage later close theoretical inspections. Baker *et al.* [19] made important early steps to illustrate the advantages of snap-through, noting that the novel dynamic "relax[es] the constraint of frequency matching", alluding to the non-resonant property. When compared to the same simply-supported piezoelectric beam if uncompressed, the axially compressed, bistable harvester beam yielded greater steady-state power generation across the full range of excitation frequencies evaluated in the testing. Presuming that all of the dissipation in the device could be equated to be the energy harvested, Ramlan *et al.* [20] used simulations of the governing equations of a bistable device, similar in design to that in Figure 1.1, and found that the snap-through motions enhanced power generation over that achieved by the counterpart linear platform for harmonic excitation frequencies considerably lower than resonance. These studies came early in the timeline of development and helped to direct subsequent research efforts to examine bistable structures for vibration energy harvesting in ever-greater detail.

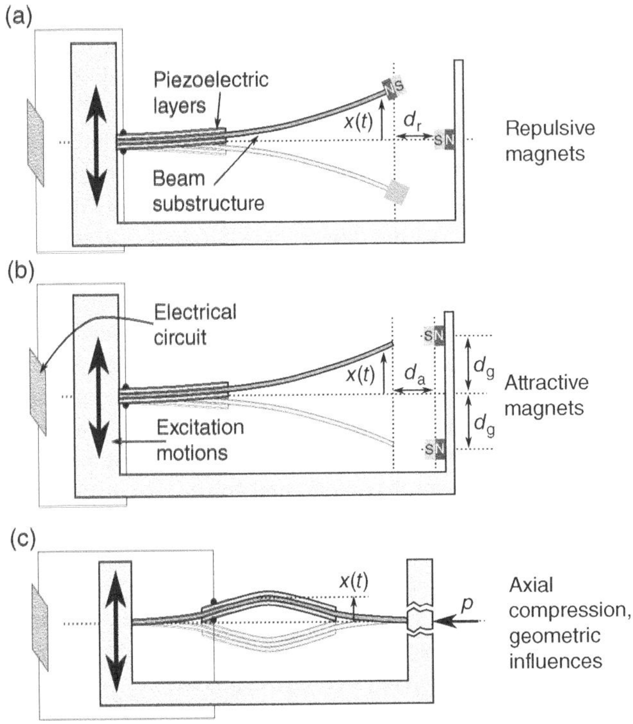

Figure 4.1 Common design strategies to induce bistability into vibration energy harvesting systems. (a) Two magnets repulse each other. (b) A ferromagnetic substructure is attracted to a pair of frame-mounted magnets. (c) Geometric influences or axial compressive forces are exerted on the energy harvesting structure. In all sub-figures, the lightly shaded bistable energy harvester represents the other statically stable equilibrium configuration. The piezoelectric layers are one representative example of including electromechanical transduction.

Among the many possibilities, the utilization of magnetic forces upon other magnetic or ferromagnetic inertial masses is a thoroughly considered design methodology to result in bistability for an energy harvester. The popularity of this multiphysics solution may be partially explained by the fact that magnets can be used in ways such that bistability is induced without physically contacting the underlying energy harvesting structure, where the contact approach is recognized to potentially introduce adverse loss, damping, or fatigue effects. One idea is for magnets to repel each other in the axis of motion of the harvester mass generalized displacement. A common implementation of this strategy is shown in Figure 4.1a where a piezoelectric beam with magnetic tip mass is repulsed by another frame-mounted magnet. According to the sensitivities imposed and the frequencies of excitation typically considered, the electroelastic dynamics of the beam are largely limited to displacement distributions comparable to the fundamental mode motions of the underlying substructure. Focusing upon this design, Cottone *et al.* [21] and Gammaitoni *et al.* [22] showed that the power generation potential was enhanced when bistable energy harvesters are excited with Gaussian noise. Using numerical and experimental efforts, the investigators demonstrated that stochastic resonance-like phenomena could lead to larger power harvesting by promoting snap-through motions, whereas the performance of linear harvesters is theoretically independent of design when excited with white noise. Lin and Alphenaar [23] arrived at similar insights when a comparable harvester platform excited with random input vibrations was attached to a variety of electrical circuits.

Via a series of meso- and microscale experimental systems, Ando *et al.* [24], Ferrari *et al.* [25], Meimukhin *et al.* [26], and De Paula *et al.* [27] examined the influence of such repulsive magnetic forces and found that preferred spacing exists between the beam tip-mounted magnet and an excitation frame-mounted magnet for the greatest enhancement of the bistable harvester sensitivity to random or harmonic excitations. Also considering harvester architectures leveraging repulsive magnetic interactions to induce bistability, Stanton *et al.* [28] and Mann and Owens [29] probed the influences of amplitude- and frequency-swept harmonic base excitations and concluded that excitation properties critically govern the possibility to activate favorable snap-through motions providing large electromechanical transduction. Karami and Inman [30] explored this bistable electromechanical platform in the context of energy harvesting from heartbeat vibrations; the researchers uncovered a considerable robustness of the favorable power-generating dynamics to a large range of heartbeat behaviors, and helped to establish a new concept that could lead to self-powered medical implants like pacemakers. Tang *et al.* [31] presented a comprehensive experimental study that explored the sensitivities of the gap between repulsive magnets mounted on a piezoelectric beam tip and the frame, in the process determining what gap spacing leads to the best power generation or storage from random or harmonic excitations. Rather than a cantilevered substructure, Zhu and Zu [32] explored a harmonically-excited and doubly-clamped energy harvesting beam where one of the clamped ends was free to translate axially due to the repulsive magnetic force effects. Simulations and experiments showed that such an implementation of bistability in tandem with the structural design provided the broadband frequency sensitivity reported for bistable harvesters when cantilevered from one end.

Similar to an idea initiated by Lin *et al.* [33], several research teams have explored how compliance in the (otherwise grounded) repulsive magnet mounting may lead to more pronounced and broadband performance in bistable energy harvesting systems. Wu *et al.* [34], Upadrashta and Yang [35], and Su *et al.* [36] proposed novel, inverted structural designs to create beam-in-beam energy harvesters wherein the clamped end at the exciting base became its own repulsive magnetic source with respect to the most suspended end of the cantilever system. By this implementation, the sensitivity of the devices is seen to be greatly intensified, compared to linear counterparts, whether under single-frequency sinusoidal or stochastic input excitations. Su and Zu [37] explored alternative implementations of the compliant mounting of the repulsive magnet so as to provide for a bi-directional excitation sensitivity of the energy harvesting system. Rather than employing one harvester beam as in Figure 4.1a, Zhou *et al.* [38] examined a design employing two such beams facing and repulsing each other and found that non-periodic motion regularly resulted when the two-beam system was excited with sinusoidal inputs. Zhu and Beeby [39] found that this architecture was a better choice than two uncoupled bistable devices or the two devices operating in a "linear" mode without magnets when the several platforms were excited with white noise excitations.

Extrapolating an idea drawn from historic, dynamical investigations [40,41], Erturk *et al.* [42] and Erturk and Inman [43] examined a piezomagnetoelastic energy harvesting beam where bistability was induced by the attraction between a ferromagnetic piezoelectric cantilever and two magnets affixed to the clamped ground frame. This platform is illustrated in Figure 4.1b and the general design approach has become another staple means to induce bistability in energy harvesting systems. Under single frequency harmonic base accelerations, the researchers [42,43] found that the resonant excitation of the underlying piezoelastic beam (having removed the attractive magnets from the frame) was the only condition in which the linear dynamics of the piezoelastic harvester outperformed the bistable counterpart. Thus, once steady-state snap-through dynamics were realized, the bistable piezomagnetoelastic beam consistently generated greater average power than the linear counterpart in the experimental and numerical evaluations for all cases of off-resonant excitation. Litak *et al.* [44] studied the same bistable harvester architecture and concluded that its sensitivities under Gaussian noise excitations were strongly tailored by the excitation variance and ratio of time constants, where the latter property relates the electrical to the mechanical natural frequencies. The favorable energy conversion capability of this device platform motivated efforts by Zhou *et al.* [45,46] and Cao *et al.* [47] to enhance performance further by incorporating a means to rotate the attractive magnets and by promoting the achievement of snap-through via external impact, a strategy originally suggested by Erturk and Inman [43].

Considering the baseline bistable harvester architecture of Figure 4.1b in greater detail, Zhao and Erturk [48] obtained further insight by critically exploring the relative advantages of the bistability compared to the underlying monostable piezoelastic beam when excited by Gaussian noise. The findings showed that bistability enhanced the long-time power generation only for intermediate levels of the excitation variance: for smaller variance, the harvester mass would not escape a local well of potential energy, whereas for higher variances the increased net stiffness associated with the magnetic fields inhibited greater dynamic motions and thus reduced the straining of the piezoelectric layers. Contradictory to some experimental observations, Halvorsen [13] used linear analytical arguments alongside nonlinear model simulations to show that there is no theoretical justification for why the nonlinear devices might outperform weakly coupled linear energy harvesters under Gaussian excitations, since the prior exhibit additional restoring forces that would limit mass motion. On the other hand, the study suggested that practical factors, such as limitations on mass motion or the unavoidable inducement of large harvester beam displacements, may lead to nonlinear behavior in the otherwise presumed linear device.

Apart from myriad methods to deploy magnets for generating bistability in vibration energy harvesting systems, another pool of research has been devoted to exploring design configurations based around mechanical, elastic, or geometric means to induce bistability. The platform shown in Figure 4.1c exemplifies one implementation of this approach, wherein an axial compressive force is directly applied to the clamped-clamped piezoelectric beam, such as the idea studied by Baker *et al.* [19]. An alternative method is to apply geometric displacement constraints, which Ramlan *et al.* [20] considered via an energy harvester model using lumped-parameter methods. For energy harvesters using beam substructures, the difference between force or geometry boundary conditions is primarily a matter of the exact harvester implementation and modeling, since the various beam theories are applicable only to certain boundary conditions [49,50]. Masana and Daqaq [51–53] have presented several studies to elucidate the performance advantages and disadvantages of clamped-clamped piezoelectric energy harvesting beams when buckled via compressive forces. The authors showed that under single-frequency harmonic excitations the bistable devices were favorable compared to a counterpart monostable implementation when the excitation level could activate snap-through dynamics at frequencies sufficiently lower than the linearized resonance, sometimes correlated to other forms of periodic dynamics, for instance the superharmonics [51]. Looking at the superharmonics in greater detail, the researchers found that by exploiting these behaviors the power generation could be just as large as when excited at frequencies closer to the linearized resonance [52]. Under bandlimited stochastic excitations, which may be more representative of ambient vibrations than Gaussian (white) noise, Masana and Daqaq [53] concluded that, in contrast to the results for harmonic excitations, there are many properties of bandlimited noise for which the bistable harvester could produce greater electrical output than the counterpart monostable device. These studies illustrate the high potential that bistability may cultivate in axially compressed, clamped-clamped energy harvesting beams.

Cottone *et al.* [54] studied numerically and experimentally the response of the clamped-clamped piezoelectric beam platform when the excitations were white noise; the data showed dramatic enhancement in the power generation due to the activation of cross-well behaviors for the buckled device when compared to the same harvester but not compressed. Considering different boundary conditions on geometrically-constrained energy harvesting beams, Van Blarigan *et al.* [55] and Xu *et al.* [56] found that these bistable implementations of the beams could consistently outperform the comparative linear designs, whether due to activating broadband aperiodic snap-through or based on those snap-through oscillations that are proportional to the single frequency of excitation.

Other researchers have proposed novel energy harvesters that leverage geometric constraints and uniquely designed components for enhanced performance from more conventional buckled beam-type structures. For example, Kim and Kim [57] developed a biologically-inspired piezoelectric energy harvesting system that included a four-bar linkage design which facilitated snap-through motions under harmonic excitation frequencies considerably less than the linearized natural frequency of the underlying piezoelectric beam. As a result, for excitation frequencies more than an order of magnitude less than the linear resonance, the bistable platform was still able to generate a viable electrical power for electronic functions that required a power supply on the order of μW.

Using motion magnifier mechanisms, Liu *et al.* [58] created a bistable structure which provided a greater power density than other bistable energy harvester designs considered in the study. A frequency transform analysis indicated that the sinusoidal excitation of the bistable harvester led to a considerable diffusion of energy to other harmonics in the resulting response spectrum, while an investigation of the device when attached to an advanced charging circuitry uncovered a notable electromechanical coupling influence that could dramatically modulate the resulting snap-through dynamics [59]. Porter and Berfield [60] explored a bistable energy harvesting structure that utilized torque arms to effectively constrain a clamped-clamped piezoelectric beam at its centerpoint and induce higher-order mode buckling dynamics. A broadening of the excitation frequency sensitivity was observed by applying the additional design components, but the peak power generation was ultimately reduced from a baseline comparative system [60], showing that neither all designs nor implementations of bistable energy harvesters may be favorable.

One strategy for inducing bistability into an electromechanical device for energy harvesting purposes exploits the elastic influences brought on by composite material lay-up and thermal forming. It is well-known that by using composite material layers, such as carbon-reinforced woven-fiber sheets, particular stacking sequences and relative stacking angles will lead to a state of elastic prestress for the final, free structure at room temperature [61]. Several research teams have exploited such bistable composite panels or plates as energy harvesting structures using piezoelectric materials bonded to the curved surfaces. Arrieta *et al.* [62] studied an early conception of this architecture where four piezoelectric PZT layers were adhered to the high curvature regions of the plate. Sinusoidal excitation of the plate center point with the four edges unrestrained showed the full range of anticipated dynamics from a bistable energy harvester: low amplitude intrawell dynamics, aperiodic cross-well motion, multiharmonic behaviors, and finally continuous snap-through dynamics at a frequency identical to the excitation. The persistent snap-through plainly led to the greatest average power generation although the input accelerations needed to sustain such oscillations was very high (around $20\,\mathrm{m\,s^{-2}}$) with respect to many ambient vibration levels. The researchers continued to explore the concept of energy harvesting from piezoelectric bistable composites and Arrieta *et al.* [63] later examined the plate when cantilevered and composed using a new stacking sequence. The asymmetry induced unique stable plate curvatures that led to a particularly broadband frequency sensitivity while fabrication experiences suggested design guidelines that greatly reduced the harmonic acceleration levels required to trigger snap-through oscillations (now only around $2.4\,\mathrm{m\,s^{-2}}$) when compared to the original implementation.

Betts *et al.* [64] determined optimal lay-up configurations and aspect ratios for bistable piezoelectric plates when centrally excited and pinned at the edges. It was found that square plates were the optimal lamina shape despite the greater out-of-plane deflections attainable by higher aspect ratios. The asymmetry induced by ordinary fabrication tolerances and limitations was explored and digital image correlation identified the dynamic curvatures and mode shapes evident in the snap-through oscillations that led to a usable average power generation for small microelectronics (order of mW) [65]. Li *et al.* [66] showed that the second mode of a bistable and piezoelectric composite plate could be favorably leveraged to induce snap-through dynamics for large voltage transduction at higher frequencies.

To summarize, a great proportion of early developments to explore the potential for leveraging the snap-through dynamics of bistable structures for energy harvesting purposes, as well as much of the research carried out to explore new bistable harvester device concepts, have largely relied upon numerical and experimental studies to derive their conclusions. These studies give important foundational understanding, but formal and generalized findings must be determined using more rigorous (semi-) analytical methods to take the exciting empirical observations an important step closer to confident, practical deployment.

4.1.2 Analytical Developments in Energy Harvesting with Bistable Devices

In contrast to the ease of fabricating experimental testbeds or numerically simulating a set of representative governing equations of motion, to date there has been less use of analytical tools to catalog and assess the dynamics of bistable vibration energy harvesters. In part this is due to the strongly nonlinear nature of the dynamics. Whether excited by steady-state, stochastic, or non-stationary inputs, the intrawell oscillations are by definition near to equilibria, whereas interwell vibrations are far from equilibria. This substantial qualitative distinction between the two dynamic regimes suggests that it may be challenging to identify an analytical tool which is comprehensively beneficial for the study of electromechanical bistable devices.

However, to arrive at broadly meaningful insights, analytical or approximate analytical, sometimes termed *semi-analytical*, methods must be employed[1]. An early step to derive greater understanding of the potential for exploiting the dynamics of bistable oscillators for energy harvesting enhancements was taken by Stanton *et al.* [67]. The research team used the method of harmonic balance, presuming all responses are periodic in proportion to the frequency of a harmonic excitation source. The study closely explored the role of load impedance properties in determining the resulting average power production, and gave analytical evidence, supported by numerical simulations, of the persistence of snap-through vibrations at very low frequencies, which agrees with previous experimental and numerical work [20,68]. Mann *et al.* [69] and Sneller *et al.* [70] also used this approximate analytical tool to uncover a pronounced robustness in bistable energy harvesting devices when compared to vulnerabilities exhibited by linear, softening monostable, or hardening monostable counterparts. For a prescribed potential of parametric uncertainty, the range of resulting performance

1 To assess the dynamics of bistable systems properly, semi-analytical techniques are almost always required. The remainder of "analytical" techniques described in this book are therefore specifically the *semi*-analytical type. Exceptional and *truly* analytical methods which exactly predict the dynamics of bistable systems under special conditions are detailed in Refs. [89] and [142].

measures was consistently greater for the bistable energy harvester than for the competitors [69].

Karami and Inman [71] used the method of multiple scales to predict the intrawell dynamics of a hybrid piezoelectric–electromagnetic bistable energy harvester. The phase portraits of intrawell displacements and velocities were in good agreement with trajectories determined through direct numerical integration of the governing equations of motion. The researchers pointed out the inadequacy of the method of multiple scales to predict the snap-through dynamics, and consequently employed numerical approaches to examine the interwell vibrations. To explore the design and excitation parameters critical for maintaining viable power generation capabilities, Stanton *et al.* [72] used Melnikov's method [41] to derive performance prediction tools. By the good agreement shown with representative simulations, the approach illustrated a valuable potential to estimate the relevant critical parameters when a bistable energy harvester was excited by single- and multi-frequency vibrations, or by stochastic inputs. Cohen and Bucher [73] deployed an approximate model of an electromagnetic bistable energy harvester and used a mapping approach, adapted from a strategy developed by Shaw and Holmes [74], to assess the stability of periodic orbits when the device was excited at frequencies well below a linearized resonance condition. Excellent agreement was found comparing the approximate analytical predictions of the stable periodic dynamics to numerical and experimental data sets, showing that integrating the model with the mapping technique facilitated a viable assessment of power generation capabilities of bistable harvesters excited at infrasonic frequencies [73].

Panyam *et al.* [75] combined rescaling procedures with the method of multiple scales to conduct detailed analyses on the bifurcations associated with the greatest extents of intra- and interwell steady-state oscillations, using a generalized model that accounted for either piezoelectric or electromagnetic transductions. The rescaling enabled the prediction of snap-through vibrations which the method of multiple scales is otherwise insufficient to achieve. Supported by numerical simulation examples, the study provided insights such as the relative insensitivity of power generation performance to change in the mechanical–electrical time constant ratio but the substantial performance sensitivity to change in the electromechanical coupling [75]. Panyam and Daqaq [76] later used this analytical foundation to determine and compare the optimal power generation capabilities of mono- and bistable energy harvesters under harmonic excitations. The results showed that bistable devices could produce greater steady-state electrical power than monostable energy harvesters when the two designs were prescribed the same normalized linear natural frequency, although the advantage of the bistable platforms would be best leveraged at lower frequencies of excitation.

In a series of efforts, a research team led by Daqaq has developed an extensive analytical foundation in the understanding of nonlinear vibration energy harvesters excited by random inputs. The promising potential of bistable devices under steady-state excitations motivated explorations of their performance capabilities under stochastic vibrations where it was found that when excited by white Gaussian noise, linear and bistable electromagnetic harvesters yield the same mean output power [77]. Yet many real-world stochastic vibration sources are not purely Gaussian and may be more realistically represented using exponentially correlated noise. For such low-pass filtered noise, Daqaq [77] found that the double-well potential may be designed – for example, by changing the structural characteristics with respect to the buckling method – so as to yield greater power output from the bistable harvester than the linear counterpart. Similar trends were observed for piezoelectric devices, where the sensitivity of bistable energy harvesters could either enhance or detract from the mean power generation under white noise excitations [78]. He and Daqaq [79,80] employed statistical linearization and the finite element method to reconsider the suitability of bistable energy harvesters under white noise excitations. The studies uncovered new insights, including the uniformly detrimental influence of potential energy asymmetries (design parameter deviations) on the power generation performance, and the unequivocal advantage of the symmetric double-well potential energy profile when optimized for a given noise excitation intensity. To validate the many findings, this research team led by Daqaq used numerical simulations that have also defined the limitations of the theoretical formulations and guiding assumptions.

4.1.3 Summary

The wide variety of research to develop, model, and understand bistable vibration energy harvesters has uncovered important characteristics of the platforms which enable electric power generation enhancements over other common baseline designs, such as linear and monostable nonlinear counterparts. The considerable body of experimental and numerical work has given fuel to the research activities, suggesting that bistable architectures could be a kind of golden standard energy harvesting platform due to the activation of the energetic snap-through dynamics. On the other hand, analytical works have offered counterexamples in which the energy barrier separating intra- from interwell oscillations can serve to make bistable harvesters less appealing in either steady-state or stochastic excitation environments. To put the outstanding questions to rest and develop pathways to promote continued innovations, it is clear that a conclusive assessment of the benefits of bistable vibration energy harvesters requires analytical treatment supported with a direct comparison of experimental and numerical results. The goal of the following sections of this chapter is to provide such a comprehensive approach for meaningful and rigorously verified insights on the potential performance enhancements that result from effectively deployed bistable energy harvesters.

4.2 Effective and Straightforward Design Guidelines for High Performance Operations[2]

A great compilation of research findings has conclusively demonstrated that successful deployment of bistability in energy harvesting platforms may result in the greatest electrical power production potential for a given environmental

2 Material from this section has been reproduced with permission from R.L. Harne, M. Thota, and K.W. Wang, Concise and high-fidelity predictive criteria for maximizing performance and robustness of bistable energy harvesters, Appl. Phys. Lett. 102, 053903 (2013); doi: 10.1063/1.4790381. Copyright 2013, AIP Publishing LLC.

vibration resource [21,48,81]. Despite the enthusiasm developed from the findings, a critical factor that inhibits the unrestrained acceptance of bistable energy harvesting devices for practical applications is the possibility that the favorable snap-through dynamics may coexist with an intrawell dynamic regime that yields considerably less electromechanical energy conversion. Such coexistence has been observed in the laboratory under steady-state excitation environments [51], as well as when bistable harvesters are under the influence of stochastic inputs, in which case snap-through behaviors are seen to occur only occasionally [24]. On the other hand, a careful understanding of where dynamic coexistence may be manifest, as related to design parameters and operational conditions, may enable bistable harvesters to be implemented with maximum confidence.

A recent study has shown that analytical methods are available to explicitly predict coexisting steady-state oscillation regimes. Stanton *et al.* [67] used the method of harmonic balance to approximate the displacement and average power amplitudes generated by a piezoelectric energy harvesting beam buckled due to the repulsion of magnetic forces. With this analytical formulation, the intra- and interwell dynamic behaviors were reconstructed for a prescribed set of design and harmonic excitation parameters, and coexistence of the dynamics was explicitly predicted in some cases. An analytical strategy such as the method of harmonic balance has broad merit, in that the steady-state dynamics may be reproduced using very low computing effort. Consequently, it enables a substantially greater range of the parameter space to be explored at a fraction of the cost needed by numerical simulations, since the analytical approach provides *direct* solution means in contrast to the *indirect* pathway to insights given by the numerical alternatives.

In fact, using experimentally-validated analytical techniques to predict the strongly nonlinear dynamics of bistable energy harvesters can be of further benefit to developing critical understanding if the approaches are leveraged for insights that the numerical strategies cannot provide. This is the methodology employed in this study, where the governing equations of a bistable energy harvester are examined with respect to *tracking the bifurcations* that demarcate intra- and interwell dynamic regimes as critical parameters are changed, an ability only possible using analytical methods. The following sections present the formulation of this analytical approach from which are derived concise and high-fidelity criteria to predict the regions of design and operation where a bistable harvester is most effective due to activation of the snap-through dynamics. Then, experiments validate the premise upon which the analysis is based, and the criteria are explored to elucidate the sensitivities of bistable harvester operation where the results are cross-checked using series of numerical simulations.

4.2.1 Analytical Formulation of Bifurcations Associated with Achieving Snap-through

To develop analytical expressions that explicitly detail the sensitivities encountered to maintain snap-through dynamics for a bistable energy harvester, a fundamental mathematical principle is used which corresponds to a critical case of bistable system dynamics. The governing equation of motion is adopted according to the assumption of a large ratio between the mechanical and electrical time constants of the electromechanical system; for example, this characteristic is common for electromagnetic energy harvesters [82,83] and piezoelectric systems operated at low frequencies [84]. The effect of this assumption is that the electrical influences are accurately approximated as a damping contribution to be added to the mechanical damping effects [77,83]. Then, the equation of motion using normalized notations is given to be

$$x'' + \gamma x' - x + \beta x^3 = p\cos\omega\tau \tag{4.1}$$

where $x(\tau)$ is the normalized, generalized relative displacement between the bistable harvester inertial mass and exciting ground (such as a beam tip motion with respect to the exciting clamped end), and is defined such that the central unstable equilibrium is $x = 0$; γ is the damping factor that integrates the influences of mechanical and electrical dissipations; β is a nonlinearity strength which is a ratio of nonlinear to linear stiffnesses; p is a normalized base acceleration level of excitation, while ω is a normalized excitation frequency. Finally, ( )' indicates differentiation with respect to a non-dimensional time τ. The dynamics associated with Eq. (4.1) include a variety of phenomena due solely to the single-frequency harmonic excitation: intrawell, interwell, aperiodic, and multiharmonic vibrations are all possible [43,75], although only one such behavior occurs at any one time. While distinct, these dynamics may in fact be triggered within a narrow range of excitation parameters, indicating a large sensitivity to excitation uncertainties. Thus, it is critical to effectively elucidate the relationships amongst excitation conditions and design parameters so that snap-through dynamics of a bistable energy harvester are sustained and power generation potential is maximized.

To solve Eq. (4.1), a fundamental Fourier series expansion is employed, according to the method of harmonic balance.

$$x(\tau) = k(\tau) + m(\tau)\sin\omega\tau + n(\tau)\cos\omega\tau \tag{4.2}$$

In this assumed response form, the coefficient k provides for the possibility that the harvester mass may vibrate locally within one well of potential energy: non-zero k is indicative of this condition. Equation (4.2) may be alternatively expressed using $x(\tau) = k(\tau) + r(\tau)\cos[\omega\tau - \phi(\tau)]$ where the amplitude of displacement determined according to $r = (m^2 + n^2)^{1/2}$ and the phase lag with respect to the excitation is computed from $\tan\phi = m/n$. Then, substituting Eq. (4.2) into Eq. (4.1), assuming slow-varying coefficients, steady-state response, and neglecting higher order terms from the trigonometric relations, two cubic characteristic polynomials are produced according to the displacement amplitude squared r^2. One polynomial corresponds to the intrawell dynamics ($k \neq 0$) whereas the latter characterizes the snap-through motions ($k = 0$).

$$\frac{225}{16}\beta^2 r^6 - \frac{15}{2}\beta\left(2-\omega^2\right)r^4 + \left[\left(2-\omega^2\right)^2 + (\gamma\omega)^2\right]r^2 - p^2 = 0 \tag{4.3}$$

$$\frac{9}{16}\beta^2 r^6 - \frac{3}{2}\beta\left(1+\omega^2\right)r^4 + \left[\left(1+\omega^2\right)^2 + (\gamma\omega)^2\right]r^2 - p^2 = 0 \tag{4.4}$$

Equations (4.3) and (4.4) are the characteristic cubic polynomials which enable one to predict the displacement amplitudes for intra- and interwell steady-state vibration, respectively, for a prescribed set of parameters. In other words, solving Eqs (4.3) or (4.4) yields three roots for r^{*2} that represent the possible, squared response amplitudes for a given set of system parameters (p,ω,γ,β). However, the meaningful and stable predictions must be extracted from the many results. First, negative real roots are eliminated, as these lead to complex amplitudes r^*, and, second, perturbations are applied to the response equations supplied with the predicted coefficients to resolve the stability of a given prediction of steady-state behavior r^* [85].

It is well known that the roots of cubic polynomials are expressible exactly [86]. To determine the roots of Eqs (4.3) and (4.4), a generic notation is employed according to the variable s using a polynomial form $as^3+bs^2+cs+d=0$. The coefficients (a,b,c,d) are related to the coefficients of Eqs (4.3) and (4.4) by observation. Then, the roots are computed from

$$s_1=-b/(3a)-G/(3a)-\left(b^2-3ac\right)/(3aG) \tag{4.5}$$

$$s_2=-b/(3a)+G\left(1+i\sqrt{3}\right)/(6a)-\left(1-i\sqrt{3}\right)\left(b^2-3ac\right)/(6aG) \tag{4.6}$$

$$s_3=-b/(3a)+G\left(1-i\sqrt{3}\right)/(6a)-\left(1+i\sqrt{3}\right)\left(b^2-3ac\right)/(6aG) \tag{4.7}$$

where $G=\left[(H+2b^3-9abc+27a^2d)/2\right]^{1/3}$ and $H=\left[(2b^3-9abc+27a^2d)^2-4(b^2-3ac)^3\right]^{1/2}$.

Figure 4.2a and b show the normalized displacement amplitude transfer function (with respect to the harmonic excitation amplitude) and the phase lag, respectively, of the bistable energy harvester determined for the set of parameters $(p,\gamma,\beta)=(0.16, 0.05, 1)$ and plotted as functions of changing excitation frequency ω. Intrawell and interwell dynamics are shown as dashed and dark-solid curves, respectively, while the unstable analytical predictions are light shaded curves. Over a portion of the frequency bandwidth, the intrawell and snap-through dynamics are analytically predicted to be theoretically coexistent. The coexistence is evident at frequencies less than about $\omega\approx 1.4$, although the extremely low frequency predictions are known to be an erroneous outcome of the single-term Fourier series expansion for the assumed solution [87]. The concern for maximizing the energy conversion performance of a given bistable energy harvester design is to determine the set of excitation parameters (p,ω) that allow snap-through without the susceptibility to yield coexisting dynamic behaviors. The threshold conditions that demarcate the sustainability of snap-through are presented by the points A_c and B_c in Figure 4.2. Point A_c indicates the highest frequency at which the persistent interwell oscillations may be obtained while point B_c indicates that intrawell dynamics are destabilized and lead to snap-through for further increase in excitation frequency [88].

Points A_c and B_c are also unique in Figure 4.2 as the locations where two polynomial roots coalesce, such that analytical predictions of a pair of stable and unstable roots vanish at a common frequency. Based upon the expressions for

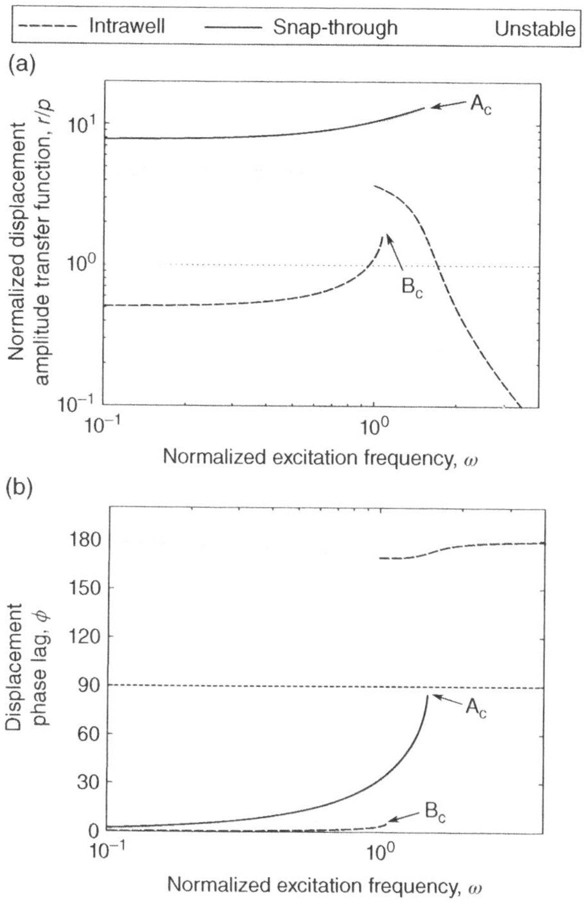

Figure 4.2 Analytical predictions of (a) normalized displacement amplitude transfer function and (b) displacement phase lag in degrees relative to the harmonic excitation. Parameters are $(p, \gamma, \beta)=(0.16, 0.05, 1)$.

the roots in Eqs (4.5)–(4.7), criteria may be derived where these critical transitions occur in parameter space. These critical cases are exploited here to develop simple expressions that provide design and operational requirements for the maintenance of snap-through oscillations in the bistable energy harvester. It is apparent that the coalescence of roots occurs when the "branch" of responses exhibits a vertical tangent. The tangency is exhibited in displacement growth for the critical point B_c in Figure 4.2a and phase growth to 90° for point A_c in Figure 4.2b. To give a firm foundation to an analytical prediction strategy based upon such critical phenomena, experimental verification of such trends as the vertical tangencies are required. While other studies have illustrated the vertical tangency in displacement amplitude that denotes point B_c spanning intra- to interwell dynamic regimes [29,51], it is needed to verify the other critical tangency event. The following section provides this validation by demonstrating the clear and apparent growth in phase lag that precedes the loss of persistent snap-through dynamics, representative of point A_c.

4.2.2 Experimental Validation of the Analytical Premise

Based upon the modeling formulation, the experimental system used in this study consists of a bistable oscillator with tunable damping provided by a variable orifice viscous damper. A schematic of the set-up is given in Figure 4.3a while b shows a photograph of the system. Bistability is induced using a spring that is pre-compressed from an upright position with respect to its top attachment point and a moving inertial mass at the bottom. The bistable oscillator is attached to an electrodynamic shaker table and the displacement and velocity of the inertial mass are measured using a laser interferometer. The sinusoidal displacement of the shaker table is measured using a potentiometer. The oscillator displacement

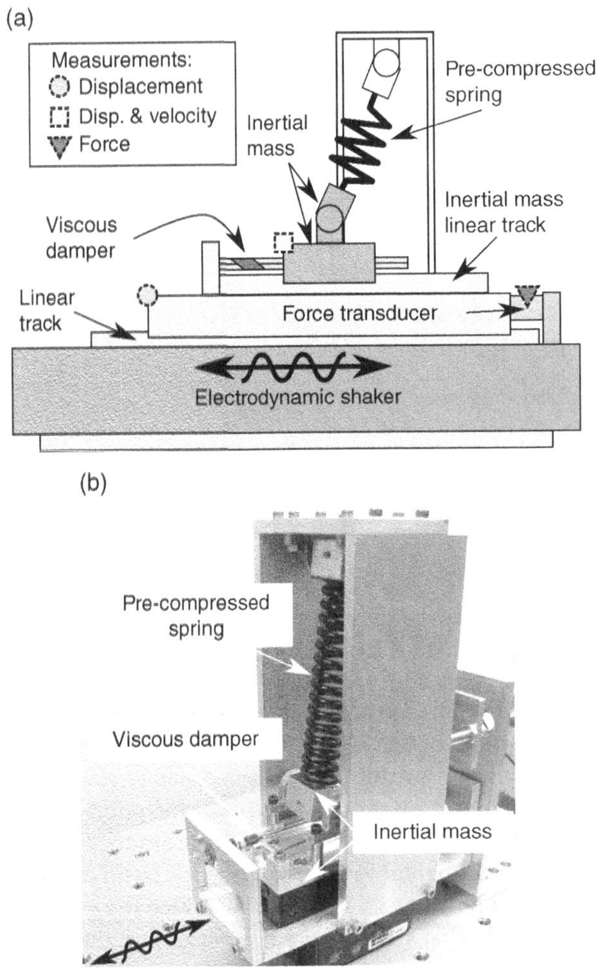

Figure 4.3 (a) Schematic and (b) photograph of experimental set-up.

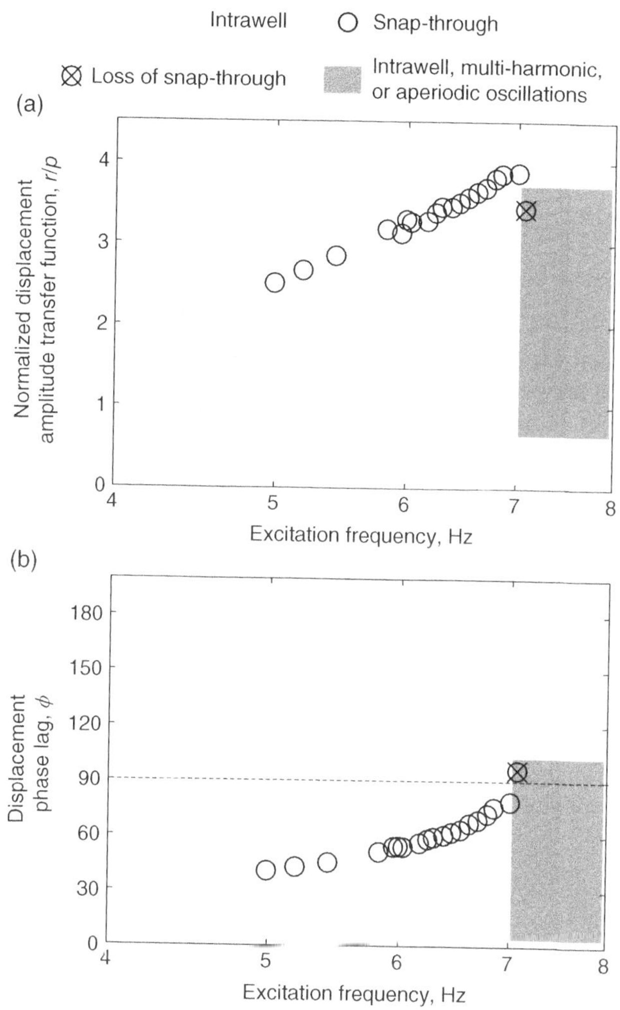

Figure 4.4 Measurements of (a) normalized displacement amplitude transfer function and (b) displacement phase lag in degrees relative to the harmonic excitation.

amplitude and phase lag are measured under a series of constant excitation conditions, such that the harmonic amplitude remains the same while the excitation frequency is incrementally changed, and the results are presented in Figure 4.4a and b, respectively.

Identically to point A_c in the analytical predictions of Figure 4.2b, the experimental results in Figure 4.4b show a phase lag approaching 90° prior to the loss of the snap-through vibrations. By connection to the analytical formulation, these experimental results verify that the mathematical coalescence of the polynomial roots is indicative of realistic boundaries of sustainable snap-through dynamics, and thus a measure of the viability of a given bistable harvester design. Moreover, while Figure 4.2a suggests that snap-through is achievable at all frequencies below a critical point, the experimental data plainly show in Figure 4.4a that such energetic dynamics are not possible down to vanishingly small excitation frequencies (unless the excitation amplitude is excessively large, as will be shown through the analysis in the following sections). These experimentally observed characteristics justify the importance of determining critical excitation parameters that genuinely induce snap-through for favorable power generation performance, since the original analytical predictions in Figure 4.2a are ambiguous in this regard.

4.2.3 Derivation of Criteria for Sustaining High Power Generation Performance

Then, the cubic polynomial expressions of Eqs (4.3) and (4.4) are exploited to determine the criteria that serve as bounds on the critical excitation amplitude–frequency $(p-\omega)$ plane to demarcate the potential for snap-through vibration. When two roots coalesce into one, the following criterion is satisfied:

$$\left[\left(2b^3-9abc+27a^2d\right)^2-4\left(b^2-3ac\right)^3\right]^{1/2}=0 \tag{4.8}$$

The cubic polynomial form in Eq. (4.4) is first employed. The appropriate terms making up the polynomial coefficients (a,b,c,d) are inserted, and the excitation amplitude p is solved. Eq. (4.9) results and is here called Criterion A. The expression defines the critical (squared) amplitude of excitation p above which snap-through oscillations are achieved for a given set of remaining excitation and design parameters (ω,γ,β). Based upon the nature of the bifurcation involved, this criterion provides the boundary in the high frequency regime of the $(p-\omega)$ plane.

$$p^2=\frac{8}{81}\frac{1}{\beta}\left\{\left(1+\omega^2\right)\left[\left(1+\omega^2\right)^2+\left(3\gamma\omega\right)^2\right]-\left[\left(1+\omega^2\right)^2-3\left(\gamma\omega\right)^2\right]^{3/2}\right\} \tag{4.9}$$

A second bound is defined from the cubic polynomial in Eq. (4.3). The double root expression is implemented, $s_d=(bc-9ad)/\left[2(3ac-b^2)\right]$. It is seen that the denominator is always negative for $\omega<1$ and for typical degrees of net mechanical and electrical damping, $\gamma<0.75$. Since this boundary will occur at lower frequencies and it is pertinent to consider underdamped oscillators in the context of energy harvesting investigations, these parameter constraints are easily accommodated. Should the numerator of the double root expression become positive, $bc-9ad>0$, then the double root r^2 is negative which means the response amplitude r is imaginary and non-physical. In terms of the dynamical behaviors, the intrawell oscillations escape into a more energetic interwell regime. To find the critical condition, the numerator of the root s_d is set to zero, and appropriate coefficients (a,b,c,d), supplied by the polynomial of Eq. (4.3), are inserted, and then again the excitation amplitude p is solved. Eq. (4.10), Criterion B, represents the minimum level of (squared) harmonic excitation required to theoretically trigger the snap-through vibrations in the lower frequency regime of the $(p-\omega)$ plane.

$$p^2 = \frac{8}{135}\frac{1}{\beta}\left(2-\omega^2\right)\left[\left(2-\omega^2\right)^2+\left(\gamma\omega\right)^2\right] \tag{4.10}$$

Equations (4.9) and (4.10) set boundaries on the excitation levels p at which the snap-through oscillations of bistable energy harvesters are achievable. The derivations of Criteria A and B have followed the insight provided by phenomenological root-coalescence. This was observed analytically and verified in experimentation, to show conclusively that roots of the cubic polynomials Eqs (4.3) and (4.4) are reduced to a double root and real root under conditions distinct to the original polynomials and contexts (intra- and interwell vibrations, respectively). Past attempts to determine similar criteria have led to measures of varying accuracy and rarely to as straightforward equations [88] as Eqs (4.9) and (4.10). Melnikov theory provides the most tractable expression elsewhere in the literature and was recently used in the study of a bistable energy harvester [72]. The Melnikov criterion is $p=\left[\gamma\sqrt{2}\cosh(\omega\pi/2)\right]/\left[3\sqrt{\beta}(\omega\pi/2)\right]$ [89]. Melnikov predictions are known to be conservative in terms of achieving well escape [72,88], suggesting it could underestimate the excitation level required to sustain the large amplitude interwell vibrations.

4.2.4 Evaluation of the Criteria Accuracy

To assess the fidelity of Criteria A and B, the predictions for two cases of system parameters are plotted in Figure 4.5, for a: $(\gamma,\beta)=(0.01,1)$ and b: $(\gamma,\beta)=(0.05,0.1)$. Included in Figure 4.5 are Melnikov predictions (solid curves) and results of numerically integrating governing Eq. (4.1) 50 times for each point in the parameter space (p,ω) with unique initial conditions for every long-time evaluation. The simulations are conducted on the $p-\omega$ grid and Figure 4.5 indicates regions where persistent, single-periodic snap-through was maintained for 10, 50, and 90% of the 50 simulations by light, medium, and dark gray shading, respectively.

From Figure 4.5, it is clear that Criteria A and B (denoted using dashed and dash-dot curves, respectively) much more closely bound the genuine cases of persistent snap-through where maximum harvesting performance is achieved, as compared to Melnikov predictions. This is evident due to the better consolidation of the numerically-determined snap-through results (shadings) provided by the new Criteria A and B when compared to the Melnikov equation. The unshaded regions below A and B led primarily to the intrawell oscillations, while the unshaded regions above Criteria A and B often yielded multiharmonic or aperiodic responses. Figure 4.5a,b shows that the new criteria lead to substantially more accurate predictions of achieving snap-through vibrations for two contrasting examples of damping and nonlinearity. As a result, the improved predictive capabilities of Criteria A and B indicate they are substantially more effective as design and implementation tools for bistable energy harvesters. The non-resonant nature of snap-through is also exemplified by the parameter demarcation given by the criteria: so long as the harmonic excitation amplitude p is great enough, the steady-state snap-through vibrations may be achieved for nearly any frequency of excitation ω.

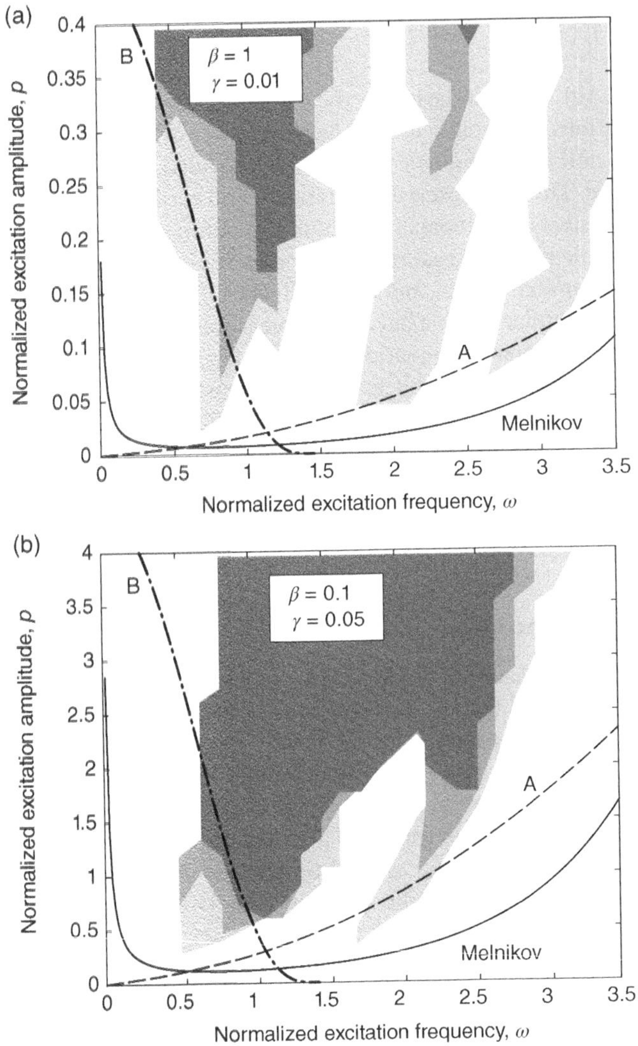

Figure 4.5 Comparison of critical normalized excitation frequency and amplitudes which activate snap-through dynamics for high power generation. Criteria A (dashed) and B (dash-dot) and Melnikov's equation (solid) are labeled, denoting the boundaries described in the text. Parameters (a) $(\gamma, \beta) = (0.01, 1)$ and (b) $(\gamma, \beta) = (0.05, 0.1)$. Light, medium, and dark gray shadings, respectively, refer to simulation results for which 10, 50, and 90% of the numerical trials resulted in steady-state snap-through vibrations of the bistable harvester.

4.2.5 Summary

In conclusion, this study addressed a critical need to create effective means for successfully developing and deploying bistable vibration energy harvesters. Maximizing the power generation performance and enhancing device robustness by proper design necessitates accurate and comprehensible criteria to demarcate parameter regimes wherein snap-through vibrations are ensured.

Here, new criteria are derived which delineate such operational and design-based boundaries using a bifurcation-tracking strategy based upon the bistable harvester governing equation. Compared to a classical method, the new criteria are shown to substantially enhance the fidelity of the predictions when compared to directly simulating the dynamics of the governing equations. Looking forward, it is therefore clear that analytical tools may serve to greatly enhance the understanding of bistable energy harvester design and development so as to maximize the effectiveness of their energy conversion potential.

4.3 Understanding Superharmonic Energy Diffusion in Bistable Energy Harvesters[3]

For nonlinear systems driven by a continuous, single-frequency excitation source, it is common for the induced dynamics to be composed of steady-state motions that occur at frequencies other than the drive motion [90]. When the system contains a restoring force described using a polynomial series expansion of terms, such as $F = k_1x + k_2x^2 + \ldots + k_nx^n$ where x is the generalized coordinate and k_i are stiffness coefficients, then steady-state dynamics of x are possible at all integer and rational combinations of the polynomial orders. For example, when the force is given by $F = k_1x + k_2x^2$ then prominent contributions of dynamics at frequencies Ω, 2Ω, and $1/2\Omega$ may be activated when the nonlinear system is excited at frequency Ω. In the context of harvesting the energy from oscillating nonlinear electromechanical systems, the diffusion of the dynamics is critical to elucidate since practical energy harvesting circuitry must be switched at the primary frequency of oscillation in order to maximize the stored power [91]. The benefit of such frequency switching strategy in comparison to ordinary AC–DC conversion (like a diode bridge) is almost an order of magnitude power improvement [92,93], so a decisive understanding of the harmonic energy contributions in the harvester dynamics is required.

A recent investigation by Masana and Daqaq [52] showed experimentally and numerically that harmonically driven bistable energy harvesters may be particularly susceptible to inducing large amplitude dynamics attributed to lower-order dynamics, for example the second or third order superharmonics. The researchers gave evidence that such dynamical features could be exploited for enhanced performance when the harvester is excited at sub-resonant frequencies. For snap-through vibrations, it is well-known that the symmetry of the oscillations is adequately characterized by two frequency components [94], namely one proportional to the excitation frequency Ω and one proportional to three times the excitation 3Ω (the third order superharmonic). To study such vibration behaviors, the method of harmonic balance is an ideal analytical tool for valuable and

3 Material from this section has been reprinted by permission of SAGE Publications, Inc., from R.L. Harne and K.W. Wang, On the fundamental and superharmonic effects in bistable energy harvesting. Journal of Intelligent Material Systems and Structures 26(8) pp. 937–950, copyright 2013 by The Authors. doi: 10.1177/1045389X13502856.

generalized insights. In their original demonstration of the dynamics of a post-buckled beam, Tseng and Dugundji [94] demonstrated the quantitative and qualitative accuracy of the harmonic balance method in characterizing the intra- and interwell response regimes. The Fourier series expansion

$$r(t)=c(t)+\sum_{k=1}^{N}a_k(t)\sin[k\omega t]+b_k(t)\cos[k\omega t] \tag{4.11}$$

was employed, where in general an increasing number of terms N improves the fidelity of the prediction. The study presented in Section 4.2 used a fundamental Fourier series expansion, retaining only the lowest order harmonic components proportional to the excitation frequency and the leading constant or bias term. This reduced expansion therefore assumes that all dynamic energy is concentrated at forced motions occurring at the driving frequency. Yet, in fact, the previous dynamical knowledge [94] and the recent experimental and numerical study in the context of energy harvesting [52] have both shown that bistable structures may readily activate dynamics occurring at other-order harmonics, particularly the third-order superharmonic. However, the specific conditions under which the response energy is diffused, and to what extent, are important factors to understand before bistable harvesters may be effectively deployed in practical situations.

This study seeks to provide such insights about bistable vibration energy harvesters by developing and examining an experimentally-validated analytical formulation that may explicitly characterize the relevant energy diffusion sensitivities of the favorable snap-through dynamics. In the following sections, the method of harmonic balance is used in a way to account for the third order superharmonic contributions with respect to those dynamics proportional to the frequency of excitation. Numerical and experimental data verify the analytical predictions and the relative advantages and disadvantages of the superharmonic dynamics that pertain to a large range of harmonic excitations conditions are carefully investigated.

4.3.1 Bistable Energy Harvester Modeling: Electromechanical Governing Equations

For extensible application of the findings derived from the present study, the bistable energy harvester platform of interest is generalized to be a base-excited bistable structure which has either an electromagnetic or piezoelectric transduction mechanism, where the electrical outputs from the transductions are connected to corresponding external load resistances, as shown in Figure 4.6, representing generic energy harvesting storage circuits. The governing equations for the coupled electromechanical system are expressed using

$$m\ddot{\bar{x}}+b\dot{\bar{x}}-k_1\bar{x}+k_3\bar{x}^3=-m\ddot{\bar{z}}-\Gamma_1\bar{v}-\Gamma_2\bar{c} \tag{4.12}$$

$$C_{\mathrm{p}}\dot{\bar{v}}+\frac{1}{R_{\mathrm{p}}}\bar{v}=\Gamma_1\dot{\bar{x}};\ \text{for piezoelectric coupling} \tag{4.13}$$

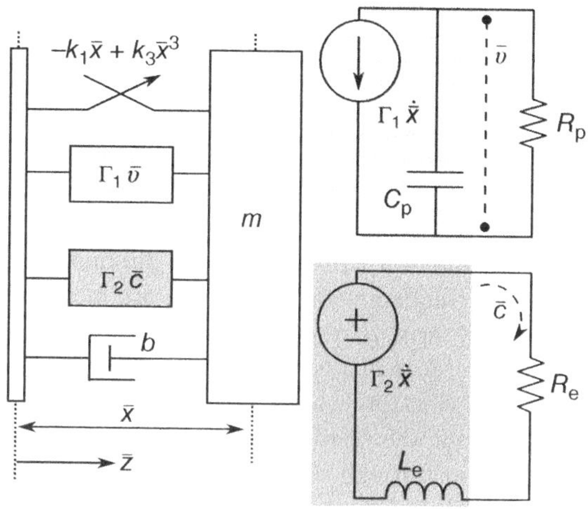

Figure 4.6 Bistable energy harvester driven by base excitation having either piezoelectric or electromagnetic transduction mechanisms. The respective harvesting circuits of the transduction mechanisms are shown on the right.

$$L_e\dot{\bar{c}} + R_e\bar{c} = \Gamma_2\dot{\bar{x}}; \text{ for electromagnetic coupling} \tag{4.14}$$

where $\ddot{\bar{z}}(t)$ is the exciting base acceleration given by $\ddot{\bar{z}} = -P_0 \cos\Omega t$ with Ω being the frequency of the excitations; $\bar{x}(t)$ is the relative displacement of the harvester inertial mass, such as a beam tip motion, with respect to the base motion; $\bar{v}(t)$ is the voltage across the load resistance R_p; $\bar{c}(t)$ is the current flow through resistance R_e; m is the mass; b is a mechanical damping constant; k_1 and k_3 are linear and nonlinear stiffness constants, respectively; C_p is the piezoelectric material capacitance; L_e is the inductance of the electromagnetic converter; Γ_1 and Γ_2 are piezoelectric and electromagnetic coupling constants, respectively; and the overdot is differentiation with respect to time t. Equations (4.12)–(4.14) are effectively a hybrid construction of equations previously developed by individual research teams to investigate bistable piezoelectric beams [52] and electromagnetic oscillators [69]. Following non-dimensionalization and simplification, the governing equations are reduced in a manner that identically represents either transduction mechanism, with a change of coefficient interpretations as necessary:

$$x'' + \gamma x' - x + \beta x^3 = p\cos\omega\tau - \varepsilon\theta i \tag{4.15}$$

$$i' + \rho i = \theta x' \tag{4.16}$$

The following parameters are defined

$$\tau = \omega_0 t;\ \omega = \Omega/\omega_0;\ \omega_0 = \sqrt{k_1/m};\ \gamma = b/m\omega_0;\ \beta = k_3 x_0^2/m\omega_0^2;\ p = P_0/x_0\omega_0^2 \tag{4.17}$$

$$i = v;\ \varepsilon = \frac{C_p}{m}\left(\frac{v_0}{x_0\omega_0}\right)^2;\ \theta = \frac{x_0}{v_0}\frac{\Gamma_1}{C_p};\ \rho = \frac{1}{R_p C_p \omega_0};\ \text{for piezoelectric coupling} \tag{4.18}$$

$$i = c;\ \varepsilon = \frac{L_e}{m}\left(\frac{c_0}{x_0\omega_0}\right)^2;\ \theta = \frac{x_0}{c_0}\frac{\Gamma_2}{L_e};\ \rho = \frac{R_e}{L_e\omega_0};\ \text{for electromagnetic coupling} \tag{4.19}$$

where x_0, v_0, and c_0 are characteristic length, voltage, and current, respectively, such that the non-dimensional generalized displacement and transduced quantities are $x = \bar{x}/x_0$, $v = \bar{v}/v_0$, and $c = \bar{c}/c_0$, respectively. Here, $(\ )'$ indicates differentiation with respect to non-dimensional time τ. The non-dimensional excitation frequency and amplitude are ω and p, respectively. In Eqs (4.15) to (4.19), the term ρ represents the ratio of electrical to mechanical natural frequencies, θ is analogous to a transducer constant, and ε characterizes the strength of electromechanical coupling.

To examine the relative importance of the third order superharmonic 3ω on the net snap-through vibration response of the harvester platform with respect to those induced dynamics which are proportional to the frequency of excitation ω, the following Fourier series expansions are assumed as solutions to Eqs (4.15) and (4.16)

$$x(\tau) = c(\tau) + a_1(\tau)\sin\omega\tau + b_1(\tau)\cos\omega\tau + a_3(\tau)\sin 3\omega\tau + b_3(\tau)\cos 3\omega\tau \tag{4.20}$$

$$i(\tau) = g_1(\tau)\sin\omega\tau + h_1(\tau)\cos\omega\tau + g_3(\tau)\sin 3\omega\tau + h_3(\tau)\cos 3\omega\tau \tag{4.21}$$

The coefficients are allowed to slowly vary in time, which provides a means to assess the stability of the dynamic predictions. The constant term c is necessary to describe the bistable harvester dynamics since the system permits oscillations around one of the statically stable equilibria which leads to a bias in the non-dimensional generalized displacement.

4.3.2 Amplitude Response Equations

Equations (4.20) and (4.21) are substituted into Eqs (4.15) and (4.16) and the coefficients of the constant, $\sin\omega\tau$, $\cos\omega\tau$, $\sin 3\omega\tau$, and $\cos 3\omega\tau$ terms are equated. This strategy yields the following set of nine, ordinary differential equations.

$$-\gamma c' = \Lambda_c c \tag{4.22}$$

$$-\gamma a_1' + 2\omega b_1' = \Lambda_1 a_1 - \omega\gamma b_1 + \frac{3}{4}\beta\left(-a_1^2 + b_1^2\right)a_3 - \frac{3}{2}\beta\left(a_1 b_1\right)b_3 + \varepsilon\theta g_1 \tag{4.23}$$

$$-2\omega a_1' - \gamma b_1' = \omega\gamma a_1 + \Lambda_1 b_1 + \frac{3}{2}\beta\left(a_1 b_1\right)a_3 + \frac{3}{4}\beta\left(-a_1^2 + b_1^2\right)b_3 + \varepsilon\theta h_1 - p \tag{4.24}$$

$$-\gamma a_3' + 6\omega b_3' = \frac{1}{4}\beta\left(-a_1^2 + 3b_1^2\right)a_1 + \Lambda_3 a_3 - 3\omega\gamma b_3 + \varepsilon\theta g_3 \tag{4.25}$$

$$-6\omega a_3' - \gamma b_3' = \frac{1}{4}\beta\left(-3a_1^2 + b_1^2\right)b_1 + 3\omega\gamma a_3 + \Lambda_3 b_3 + \varepsilon\theta h_3 \tag{4.26}$$

$$\theta a_1' - g_1' = \omega\theta b_1 + \rho g_1 - \omega h_1 \tag{4.27}$$

$$\theta b_1' - h_1' = -\omega\theta a_1 + \omega g_1 + \rho h_1 \tag{4.28}$$

$$\theta a_3' - g_3' = 3\omega\theta b_3 + \rho g_3 - 3\omega h_3 \tag{4.29}$$

$$\theta b_3' - h_3' = -3\omega\theta a_3 + 3\omega g_3 + \rho h_3 \tag{4.30}$$

The following terms are defined

$$\Lambda_c = -1 + \beta\left(c^2 + \frac{3}{2}r_1^2 + \frac{3}{2}r_3^2\right) \tag{4.31}$$

$$\Lambda_1 = -1 - \omega^2 + \beta\left(3c^2 + \frac{3}{4}r_1^2 + \frac{3}{2}r_3^2\right) \tag{4.32}$$

$$\Lambda_3 = -1 - 9\omega^2 + \beta\left(3c^2 + \frac{3}{4}r_3^2 + \frac{3}{2}r_1^2\right) \tag{4.33}$$

$$r_1^2 = a_1^2 + b_1^2 \tag{4.34}$$

$$r_3^2 = a_3^2 + b_3^2 \tag{4.35}$$

The terms r_1 and r_3 are the displacement amplitudes proportional to the normalized frequency of excitation, ω, and proportional to three times the excitation frequency, 3ω, respectively. The steady-state dynamics represented by the Eqs (4.22) through (4.30) are predicted by assuming the time derivatives are zero. With this assumption, the following results are obtained.

$$g_1 = \omega\Phi_1 a_1 - \rho\Phi_1 b_1 \tag{4.36}$$

$$h_1 = \rho\Phi_1 a_1 + \omega\Phi_1 b_1 \tag{4.37}$$

$$g_3 = 3\omega\Phi_3 a_3 - \rho\Phi_3 b_3 \tag{4.38}$$

$$h_3 = \rho\Phi_3 a_3 + 3\omega\Phi_3 b_3 \tag{4.39}$$

where additional terms are given by

$$f_1^2 = g_1^2 + h_1^2 \tag{4.40}$$

$$f_3^2 = g_3^2 + h_3^2 \tag{4.41}$$

$$\Phi_1 = \frac{\omega\theta}{\rho^2 + \omega^2} \tag{4.42}$$

$$\Phi_3 = \frac{3\omega\theta}{\rho^2 + 9\omega^2} \tag{4.43}$$

Equations (4.40) and (4.41) define the non-dimensional electrical response amplitudes associated with the fundamental and third harmonics, respectively. Equations (4.36)–(4.39) are then substituted into Eqs. (4.25) and (4.26), thereby providing means to solve for the coefficients a_3 and b_3.

$$a_3 = -\frac{1}{4\Delta_3}\beta\left[\left(\Lambda_3 + 3\varepsilon\theta\Phi_3\omega\right)\left(-a_1^2 + 3b_1^2\right)a_1 + \left(3\gamma\omega + \varepsilon\theta\Phi_3\rho\right)\left(-3a_1^2 + b_1^2\right)b_1\right] \quad (4.44)$$

$$b_3 = \frac{1}{4\Delta_3}\beta\left[\left(3\gamma\omega + \varepsilon\theta\Phi_3\rho\right)\left(-a_1^2 + 3b_1^2\right)a_1 - \left(\Lambda_3 + 3\varepsilon\theta\Phi_3\omega\right)\left(-3a_1^2 + b_1^2\right)b_1\right] \quad (4.45)$$

given that

$$\Delta_3 = \left(\Lambda_3 + 3\varepsilon\theta\Phi_3\omega\right)^2 + \left(3\gamma\omega + \varepsilon\theta\Phi_3\rho\right)^2 \quad (4.46)$$

Finally, Eqs (4.36), (4.37), (4.44) and (4.45) are substituted into Eqs (4.23) and (4.24). The resulting equations are squared and summed to obtain the first frequency response Eq. (4.47):

$$256\Delta_3 p^2 = \left(9\beta^4 r_1^8 + 96\Sigma\beta^2 r_1^4 + 256\Delta_1\Delta_3\right)r_1^2 = \kappa r_1^2 \quad (4.47)$$

$$\Delta_1 = \left(\Lambda_1 + \varepsilon\theta\Phi_1\omega\right)^2 + \left(\gamma\omega + \varepsilon\theta\Phi_1\rho\right)^2 \quad (4.48)$$

$$\Sigma = \left(\gamma\omega + \varepsilon\theta\Phi_1\rho\right)\left(3\gamma\omega + \varepsilon\theta\Phi_3\rho\right) - \left(\Lambda_1 + \varepsilon\theta\Phi_1\omega\right)\left(\Lambda_3 + 3\varepsilon\theta\Phi_3\omega\right) \quad (4.49)$$

Unlike 1-term solution formulations using the method of harmonic balance, the frequency response Eq. (4.47) contains two unknowns to be computed: the two non-dimensional displacement amplitudes of the harmonics, r_1 and r_3. Thus, a second equation is required. This is determined by squaring and summing Eqs (4.44) and (4.45).

$$16\Delta_3 r_3^2 = \beta^2 r_1^6 \quad (4.50)$$

Employing the harmonic balance method in this way, the frequency response equations are two coupled, nonlinear Eqs (4.47) and (4.50) which must be simultaneously solved for r_1^2 and r_3^2. Then, the electrical response amplitudes may be computed from Eqs (4.40) and (4.41). Average non-dimensional electrical power is determined from Eqs (4.51) and (4.52) for power proportional to the fundamental and third-order harmonics, respectively.

$$P_1 = \frac{1}{2}\rho f_1^2 = \frac{1}{2}\rho\Phi_1^2\left(\rho^2 + \omega^2\right)r_1^2 = \frac{1}{2}\rho\frac{\omega^2\theta^2}{\rho^2 + \omega^2}r_1^2 \quad (4.51)$$

$$P_3 = \frac{1}{2}\rho f_3^2 = \frac{1}{2}\rho\Phi_3^2\left(\rho^2 + 9\omega^2\right)r_3^2 = \frac{1}{2}\rho\frac{9\omega^2\theta^2}{\rho^2 + 9\omega^2}r_3^2 \quad (4.52)$$

Lastly, for completeness, the final two coefficients are presented explicitly.

$$a_1 = \frac{16}{\kappa}p\left[3\beta^2\left(3\gamma\omega + \varepsilon\theta\Phi_3\rho\right)r_1^4 + 16\Delta_3\left(\gamma\omega + \varepsilon\theta\Phi_1\rho\right)\right] \quad (4.53)$$

$$b_1 = \frac{16}{\kappa}p\left[-3\beta^2\left(\Lambda_3 + 3\varepsilon\theta\Phi_3\omega\right)r_1^4 + 16\Delta_3\left(\Lambda_1 + \varepsilon\theta\Phi_1\omega\right)\right] \quad (4.54)$$

Thus, having solved for the non-dimensional response amplitudes r_1 and r_3, the set of coefficients may be computed from Eqs (4.36) through (4.39), (4.44), (4.45), (4.53) and (4.54). In this study, the stability of a given set of predictions is assessed by perturbing the system of Eqs (4.22) to (4.30) around the predicted set of coefficients [85,95].

Due to the present motivation towards evaluating the contribution of the third order superharmonic behaviors which are more pronounced for the interwell dynamic regime and because of the greater importance of understanding these snap-through motions towards the overall power generation potential of bistable energy harvesters, the following examinations focus on the case in which the constant term is zero, $c = 0$. In spite of the current focus, the analytical formulation developed here is equally able to probe the multiharmonic characteristics of the intrawell oscillations, by carrying out similar computations provided that the c coefficient employed is the non-trivial stationary solution to Eq. (4.22).

To solve the two nonlinearly-coupled polynomials, Eqs (4.47) and (4.50), the MATLAB command fsolve is employed. The algorithm utilizes a trust-region numerical optimization routine; readers interested in greater detail in such computational routines are referred to Ref. [96]. This and similar methods have been commonly employed in recent research in dynamics and vibrations where comparable systems of nonlinear equations are simultaneously solved [97–99]. As a result, the *analytical* method developed here requires the implementation of a *numerical* approach to obtain the final predictions of the fundamental and third order superharmonic response amplitudes of the bistable energy harvester. Despite the dual analytical/numerical nature of the formulation, the numerical calculations require little more computational effort to converge to final solutions using reasonably accurate initial guesses than what is required for the solution procedures used elsewhere in this book which resolve only one cubic polynomial in a purely analytical way. In other words, the additional effort to conduct the numerical component of the analytical/numerical solution strategy does not deter the investigations herein or future implementation of the proposed prediction method.

4.3.3 Selection of System Parameters for Investigation

Throughout the following studies, unless noted otherwise, the design-based parameters employed are identical to those used in a previous investigation on an electromagnetic bistable energy harvester [69]: $\gamma = 0.01$, $\beta = 0.09$, $\theta = 10$, $\rho = 2500$, and $\varepsilon = 0.8$. To put these non-dimensional parameters into a meaningful context, the damping factor γ is set to a value comparable to a harvester exhibiting mild, hysteretic material losses. The selection of β, as a ratio of nonlinear to linear restoring forces in the double-well potential, can be realistically varied over a wide range based on design parameters like axial load on a clamped-clamped beam [100] or repulsive magnet distance [28]. The electrical and electromechanical parameters θ, ρ, and ε are characteristic of inertial generators having very high electrical natural frequency with respect to mechanical resonance and a mild degree of coupling, which is common for many electromagnetic harvester realizations [83]; typical piezoelectric harvester designs may exhibit smaller natural frequency ratios ρ but comparable-order coupling terms θ and ε [101]. In the following sections, the primary focus is on the influence of excitation level on the superharmonic dynamics of the harvester within certain bands of excitation frequencies. The interest is justified because a bistable energy harvester deployed in a moderately well-understood vibration environment is more likely to be exposed to a greater range of excitation levels and frequencies around an anticipated operating region, than the degree of deviation experienced in terms of the change in system design parameters.

4.3.4 Comparison to 1-Term Harmonic Balance Solution

To first evaluate the relevance of accounting for the third order superharmonic energy diffusion, the results of the 2-term harmonic balance analytical prediction developed here are compared to a recently derived 1-term solution of the dynamics of a bistable energy harvester [67,69]. Figure 4.7a and b present the results of 1- and 2-term harmonic balance solutions for the normalized harvester displacement amplitudes $r_{1,3}$ and average electrical powers $P_{1,3}$, respectively, for a normalized excitation amplitude of $p = 0.2$ which represents a mild excitation level according to the normalization scheme. In light of the experimental results presented in Section 4.3.7, remarks on the qualitative excitation amplitude classification will be reserved for later sections below.

As shown in Figure 4.7, as the excitation frequency nears $\omega \approx 0.15$, several stable interwell dynamics are predicted using the 2-term harmonic balance response expansion. (Recall that only the interwell, snap-through dynamics are presented in the plots due to the focus of this study on the fundamental and superharmonic energy diffusions of snap-through). When the excitation frequency approaches $\omega \approx 0.3$, the average power proportional to the superharmonic, which is only predicted by the 2-term analytical approach, begins to rise dramatically above the power generated by the fundamental harmonic motions. This trend was numerically and experimentally observed over a large range of excitation amplitudes by Masana and Daqaq [52] and is explicitly predicted by the present approach using both harmonic terms in the analytical method. The enhancement of power proportional to the amplified dynamics around $\omega \approx 0.3$ is suggested by the mathematical expression for the average power proportional to the third harmonic, Eq. (4.52). It is seen that the superharmonic power term is multiplied by a factor $9\omega^2$ as opposed to ω^2 for the average power directly proportional to the excitation frequency and given by Eq. (4.51). As a result, one may be led to ask why the contribution to the total average power is not *always* enhanced for the superharmonic dynamics. The answer is observed via the displacement amplitudes which show that the motions of the bistable harvester proportional to the superharmonic are diminished at higher frequencies, such that the power generated at high frequencies likewise reduces. (For completeness, it is also recognized that the denominator of Eq. (4.52) increases slightly faster than that of Eq. (4.51) for $\omega > 1$ but both denominators include the term $\rho^2 = 2500^2$ which dwarfs the influence of the other denominator terms in this respect).

Although the 1- and 2-term harmonic balance formulations are in fair agreement at frequencies less than $\omega \approx 0.3$ in terms of the fundamental displacement

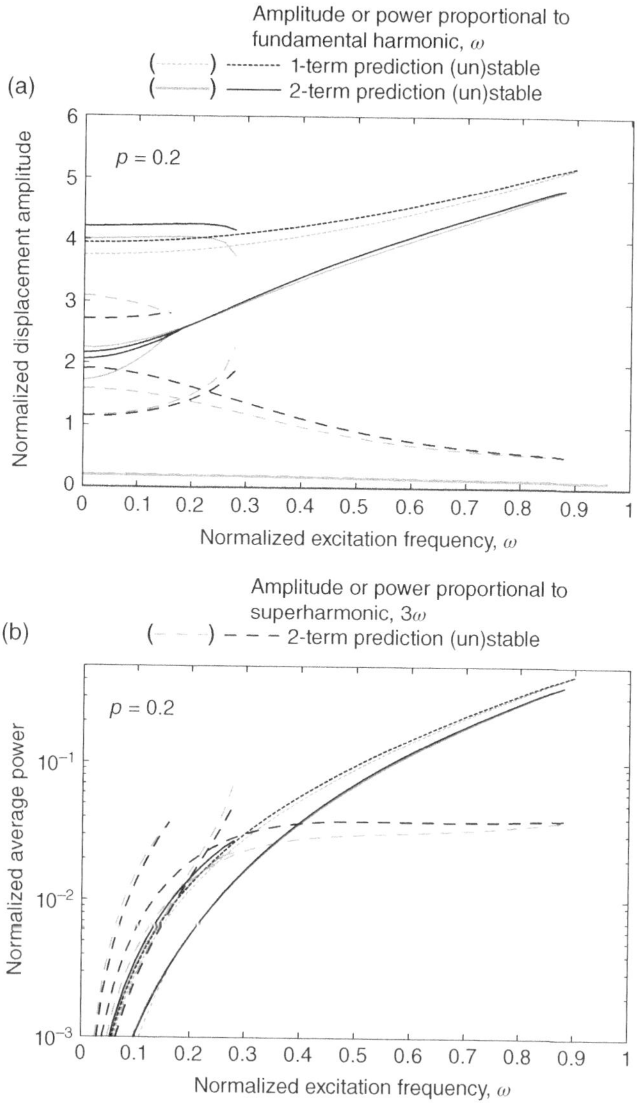

Figure 4.7 Comparison of 1-term and 2-term harmonic balance predictions for an excitation amplitude $p = 0.2$, showing those responses proportional to the excitation frequency ω and three times the excitation frequency 3ω. Excitation frequency dependence of the (a) displacement amplitudes $r_{1,3}$ and (b) average powers $P_{1,3}$.

amplitude, as the excitation frequency nears $\omega \approx 0.3$ the displacement response proportional to the fundamental harmonic computed with the 2-term formulation drops significantly due to the crossing of a saddle-node bifurcation which is not predicted by the 1-term formulation. A comparably large drop in response was observed in experiments and numerical simulations using a mild amplitude of base acceleration, $2.0\,\mathrm{m\,s^{-2}}$ [52], but such a finding has yet to be explicitly predicted by an analytical tool. After the drop in amplitude near $\omega \approx 0.3$, Figure 4.7

shows that the subsequent stable response amplitude gradually follows similar trends to the 1-term solution when the frequency ω increases. This comparison reveals that even for relatively mild amplitudes of excitation, the 1-term harmonic balance solution is incapable of providing complete insight into the refined dynamics and resulting harvested power induced by the superharmonic response. It should be emphasized that comparison between the 1- and 2-term predictions is best made in terms of general response trends and phenomena instead of between the values of the displacement and power amplitudes themselves. This is because the 2-term formulation "balances" more of the spectral energy, so that the net 2-term response given by $|x| = \left(r_1^2 + r_3^2\right)^{1/2}$ will inherently be distinct from the 1-term displacement amplitude $|x| = r_1$.

It is also seen that both 1- and 2-term harmonic balance solutions predict stable snap-through behaviors having finite values in the limit that the excitation frequency vanishes, $\omega \to 0$. In fact, these dynamics are not necessarily realizable but are the consequence of the strictly odd-order Fourier series solution expansion [87]. Thus, to determine boundaries which demarcate realistically obtainable interwell dynamics from the intrawell oscillations in the low frequency regime, a Fourier series solution expansion additionally including the second-order contribution is necessary [75,87]. Finally, as shown in Figure 4.7a some of the displacement amplitudes predicted by the 2-term analytical formulation have individual amplitudes less than the distance between a stable equilibria position x^* and the central unstable position: $r_{1,3} < x^* = 1/\sqrt{\beta} = 3.3\bar{3}$. This may lead one to believe the bistable harvester is not snapping through. Yet if one recalls that the net response amplitude is the vector sum of r_1 and r_3, then it is apparent that the net displacement amplitude is always greater than x^*. Unique exceptions to this rule will be discussed in Section 4.3.5.

4.3.5 Effects of Varying Excitation Amplitude

Previous research efforts have shown that the activation of superharmonic behaviors in a sinusoidally-forced nonlinear system is amplified as a result of increasing excitation amplitude [52,90,102]. Since the excitation amplitude p is normalized with respect to the linear stiffness of the bistable device, which is likely to be small so as to better facilitate snap-through vibrations for useful energy harvesting purposes, it is plausible that a bistable energy harvester will regularly encounter "high" excitation levels. Furthermore, the above comparison with the 1-term harmonic balance analysis shows that substantial deviation in the predicted responses occurs when incorporating the superharmonic contribution, even for mild excitation levels. Together, these factors suggest that it is essential to evaluate the relative importance of the third-order superharmonic dynamics of the bistable energy harvester when driven with greater amplitudes of excitation, so as to characterize the energy contributions between the two most important harmonics.

Figure 4.8 plots the displacement amplitudes and average harvested powers when the amplitude of excitation is $p = 1.6$. For clarity, this level $p = 1.6$ will be referred to as "moderate", those lower as "low" or "mild" levels, and those greater as "high" levels. The remaining system parameters employed in the analytical

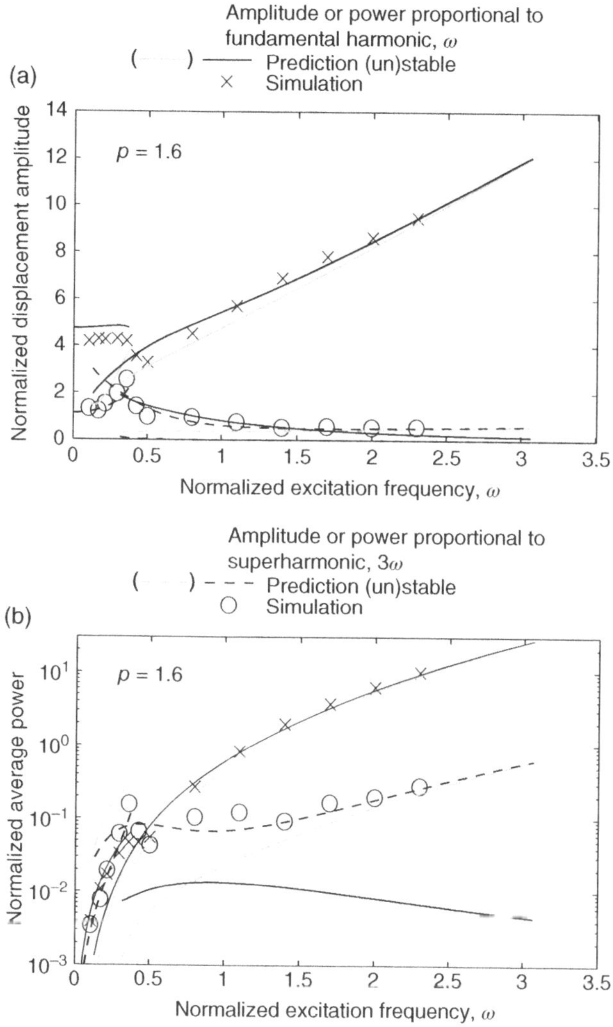

Figure 4.8 Excitation frequency dependence of the (a) displacement amplitudes $r_{1,3}$ and (b) average powers $P_{1,3}$ for an excitation amplitude $p = 1.6$.

predictions are the same as those used in the prior section. Figure 4.8 also presents the displacement and power responses of the snap-through dynamics as determined from long-time numerical simulations of the governing Eqs (4.15) and (4.16). For each frequency of excitation ω, the numerical data points are the fundamental ω and superharmonic 3ω spectral components from a fast Fourier transform of the steady-state response x, which is the analogous presentation of information generated by the 2-term harmonic balance method.

As evidenced in Figure 4.8, inclusion of the superharmonic terms in the analysis results in an intricate variety of potential electrodynamic behaviors in the excitation frequency range of $0.15 < \omega < 0.3$. At $\omega \approx 0.15$, a superharmonic response appears due to saddle-node bifurcation, producing a sudden increase in electrical power, seen in Figure 4.8b and verified by the simulations. As the frequency is increased towards $\omega \approx 0.3$, the superharmonic power continues to increase until a second saddle-node bifurcation eliminates its dominance. This effect simultaneously induces a sudden drop in the system response for the responses directly proportional to the excitation frequency. As the frequency increases still further, the dynamics of the fundamental harmonic become dominant. As shown in Figure 4.8a, the numerical simulations are in good agreement with these overall trends and quantitative values of the non-dimensional displacement amplitudes and average powers, which verifies the importance of the superharmonic spectral contribution and response-altering phenomena in the bandwidth around $\omega \approx 0.3$. On the other hand, some of the dynamics predicted analytically, particularly those yielding smaller amplitudes of displacement and power for both harmonic terms, were not uncovered by the simulations, suggesting that the feasibility of realizing such behaviors in practice is likewise diminished and thus not a major drawback of the analytical tools developed here.

When the excitation frequency $\omega > 0.3$, the electrical power attributed to the superharmonic spectral component is seen to gradually increase in Figure 4.8b, despite the fact that the superharmonic displacement amplitude asymptotically approaches a constant value, as observed in Figure 4.8a. This is due to the role of the superharmonic power multiplicative factor $9\omega^2$ mentioned in the prior section. However, the substantial reduction in the corresponding superharmonic displacement r_3 at higher frequencies explains why the power component P_3 does not exceed power proportional to the fundamental response P_1.

To provide greater clarity to the appearance of the superharmonic responses, Figure 4.9 presents the results using the excitation level of $p = 1.6$ but focused specifically on the low frequency regime that has been observed to more readily activate the larger amplitude superharmonic snap-through behaviors. As seen in Figure 4.9a in the bandwidth around $0.07 \le \omega \le 0.3$, a bifurcation is predicted to suddenly activate a response pair characterized by fundamental and superharmonic spectra having comparable amplitudes ranging from around $1.5 \le r_{1,3} \le 3.5$. However, Figure 4.9a shows that the numerical simulations do not detect the existence of such dynamics in spite of employing a large number of trials using various initial conditions. When the excitation frequency nears $\omega \approx 0.4$, the harvester is predicted to undergo an apparent saddle-node bifurcation whereby the amplitudes at frequencies just greater than $\omega \approx 0.4$ are notably lower than those before the bifurcation. The simulations corroborate this analytically predicted trend, although deviating slightly in the exact frequency and response amplitudes at which the critical event occurs. In fact, the sudden decrease in the principal displacement response (a reduction of about 20–25%) predicted analytically in Figure 4.9a was likewise observed experimentally by Masana and Daqaq [52]. Only by accounting for the higher-order superharmonic dynamics is the analytical strategy developed here capable of reconstructing such intricate dynamics and energy diffusion in the bistable harvester. It is clear from the results presented above that understanding the roles of the excitation amplitude and frequency on triggering these trends in the energy conversion capabilities is needed for the effective utilization of bistable harvesters, even in environments where the sinusoidal excitations have mild or moderate amplitudes.

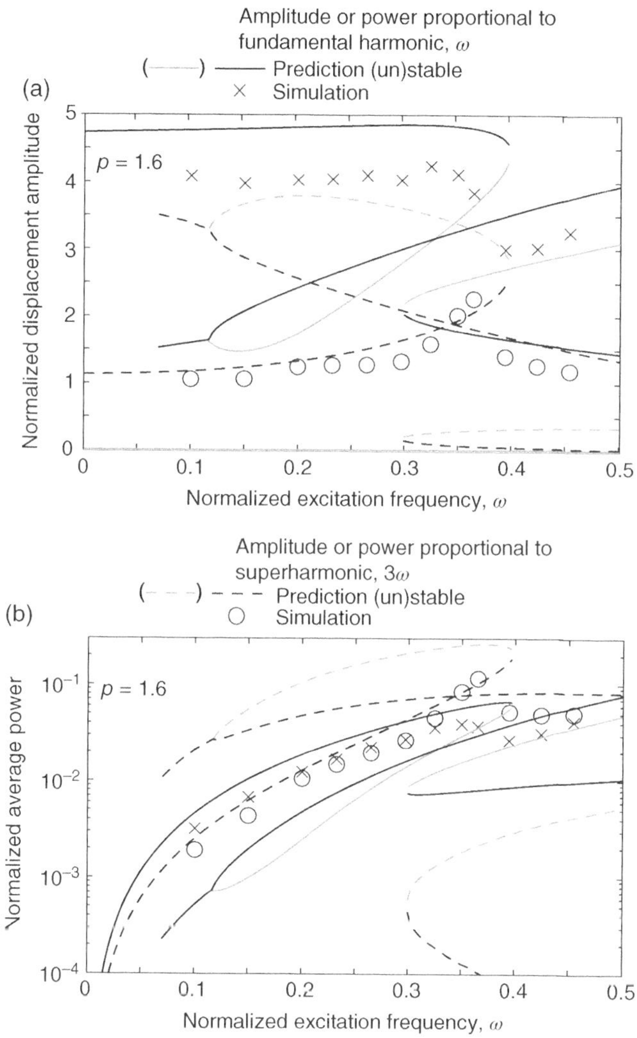

Figure 4.9 Zoom on Figure 4.8 in the excitation frequency bandwidth of significant activation of superharmonic dynamics.

Figure 4.10 plots the predicted responses for an excitation amplitude of $p = 4.5$; the remaining system parameters are the same as those employed above. Logarithmic scales are used in both axes due to the range of displacement amplitudes and average powers predicted by the analysis under this greater amplitude of harmonic excitation. As shown for the case of $p = 1.6$ in Figure 4.9, the results in Figure 4.10 uncover several bifurcations that are activated using the greater excitation level $p = 4.5$ in the lower frequency regime. Likewise, these bifurcations induce sudden enhancement of the dynamics proportional to the superharmonic. The steady-state oscillations of the harvester, such as those around a frequency $\omega \approx 0.4$, may produce a greater average power than those directly proportional to the excitation frequency, suggesting that the higher-order

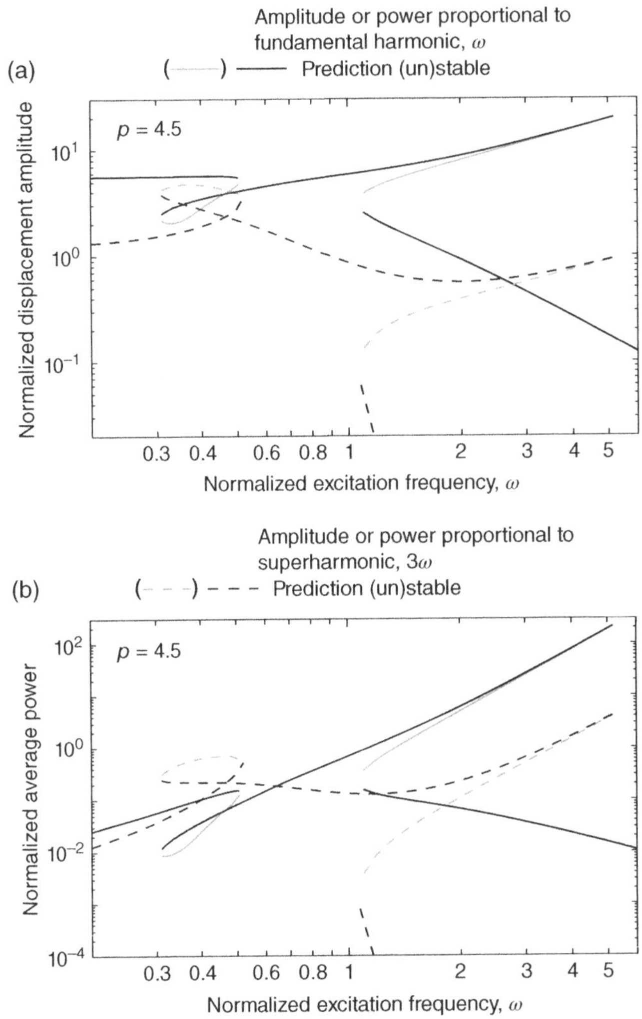

Figure 4.10 Excitation frequency dependence of the (a) displacement amplitudes $r_{1,3}$ and (b) average powers $P_{1,3}$ for an excitation amplitude $p = 4.5$.

harmonic phenomenon could be exploited for improved performance from the bistable harvester. At higher excitation frequencies, similar trends are observed using this greater amplitude of excitation $p = 4.5$: the principal dynamic contribution to the harvested energy is that proportional to the driving frequency ω although a non-trivial amount of the power is generated at frequencies three times this excitation, 3ω.

Also evident in Figure 4.10 is a predicted response of considerably less amplitude in displacement and power which occurs at higher frequencies for this large amplitude of excitation. In fact, accounting for both harmonic displacement amplitudes reveals that their vector sum is less than the vibration magnitude between the two stable equilibria, $|x| = \left(r_1^2 + r_3^2\right)^{1/2} < 1/\sqrt{\beta} = 3.3\bar{3}$. Although not shown in the figure, the phase of this particular displacement response is seen to lag the excitation by approximately 180°, where the phase lag of the *ith* harmonic $\phi_i(i=1,3)$ is computed from $\tan\phi_i = a_i / b_i$. Given that the dynamic represents an interwell oscillation in the sense that the response mean is zero over a period of the harmonic forcing, the predicted dynamic represents interwell-type oscillations of small amplitude that remain close to the unstable equilibrium at $x=0$. Indeed, this represents a dynamic stabilization of an otherwise unstable periodic orbit [103,104], due to the high frequency and amplitude of excitation on the bistable harvester. The likelihood of realizing this *excitation-induced stability* (EIS) response is subject to the bistable harvester initial condition sensitivities. These latter factors are strongly governed by design parameters and excitation conditions, and may be characterized using basin of attraction mapping techniques (see Ref. [105] or Section 3.3 of this book). However the lack of published reports of EIS for bistable energy harvesters and the authors' own absence of numerical and experimental evidence (given in Section 4.3.7) suggest that bistable harvesters may not often be employed in parametric or excitation environments which are favorable to realizing EIS oscillations. In fact, this is a favorable result since, as Figure 4.10 shows, the EIS behaviors lead to poor power harvesting performance.

To summarize the observed influences on the amplitude of excitation across a wide range of excitation frequencies, it is clear that the concentration of vibrational energy in harmonically-excited bistable energy harvesters is particularly intricate in the frequency band around $\omega \approx 0.3$. The significance of the dynamics which are proportional to three times the excitation frequency 3ω is closely governed by the excitation conditions, but greater diffusion of the response energy to the higher harmonic motions is more likely around a frequency $\omega \approx 0.3$. Since the results analytically predicted and numerically corroborated in Figure 4.7 through Figure 4.10 consistently find that the average harvested power proportional to the third-order superharmonic in this frequency band is greater than that generated by dynamics directly proportional to the excitation, there is strong motivation to explore this trend in greater detail to examine the range of conditions for which such power enhancement is possible by strategic exploitation.

4.3.6 Superharmonic Energy Harvesting Analysis

To follow the motivation from the previous section, Figure 4.11 presents the analytically predicted role of change in excitation amplitude p on the bistable harvester displacement and average power when the excitation frequency is held constant at $\omega = 0.32$. All other system parameters remain the same as in the prior studies. From the findings in Figure 4.11b, it is clear that the average power generated at frequencies three times the excitation, 3ω, is greater than that generated at the excitation frequency ω for all amplitudes of excitation greater than the lower limit of the plot and less than approximately $p \approx 4.6$. This represents over an order of magnitude in excitation level range across which the superharmonic dynamics of bistable energy harvesters may be deliberately leveraged for performance improvements. As shown in Figure 4.11b, across this span of excitation levels the average power improvement may be

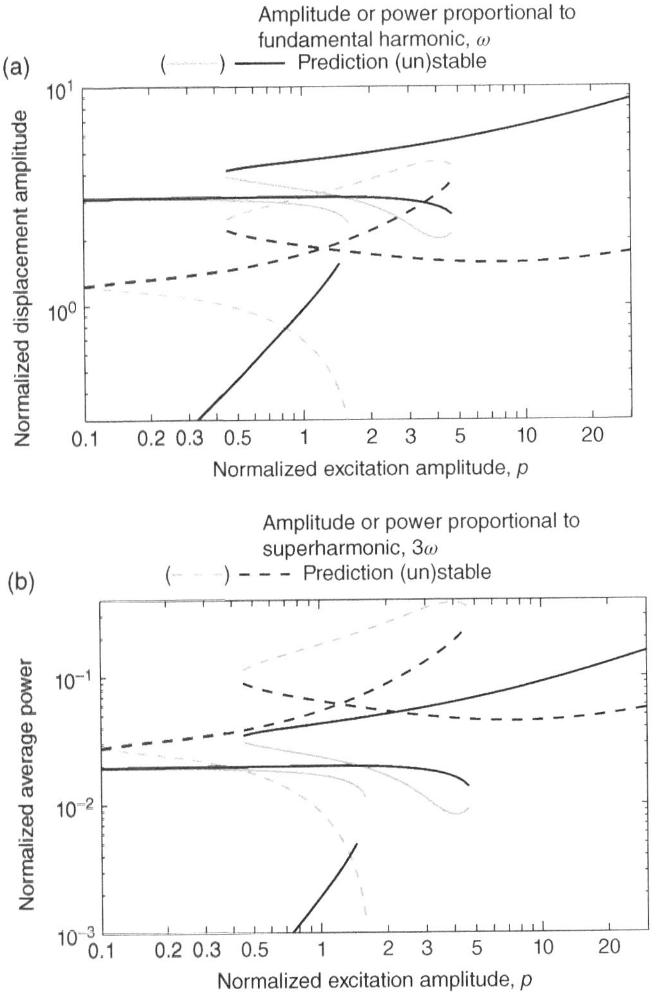

Figure 4.11 Excitation amplitude dependence of (a) displacement amplitudes $r_{1,3}$ and (b) average harvested powers $P_{1,3}$. $\omega = 0.32$.

from several times to almost an order of magnitude when compared to the power generated at the excitation frequency. By accounting for these behaviors via the theoretical formulation, the analytical tools developed here uncover this important opportunity that promotes a more effective utilization of bistable harvesters in excitation environments having a diverse range of excitation levels. On the other hand, Figure 4.11 also indicates that excessive amplitudes of excitation, at least in relation to the non-dimensionalization of parameter p, will inhibit the means to exploit harvesting the energy generated at the superharmonic; for excitation levels greater than $p \approx 4.6$ the analysis predicts that the favorable high-amplitude superharmonic motions are eliminated due to a bifurcation in the global dynamic behaviors.

4.3.7 Experimentally Investigating the Contribution of Total Harvested Energy by the Superharmonic Component

The predictions from the present analysis are in good agreement with the corresponding numerical data as well as with the trends observed throughout the experimental and numerical study on energy harvesting in the low excitation frequency regime carried out by Masana and Daqaq [52]. On the other hand, the previous experimental work did not examine how the vibrational energy relatively diffuses between the fundamental and superharmonic oscillations; instead observations were made as to the overall response envelopes as being more or less excited in the excitation frequency regime which most easily activates the superharmonic dynamics [52]. Yet, it is the specific *energy diffusion* which critically determines the quantity of collected power possible from a vibration energy harvester by way of employing storage circuitry and necessary switching strategies [91–93]. Therefore, new experimentation is carried out here to validate the insights thus far obtained from the analytical and numerical studies of the previous sections and to confirm the specific energy diffusion characteristics of the bistable energy harvester when excited across a large range of base acceleration amplitudes.

A clamped, ferromagnetic spring steel beam buckled via magnetic attraction is employed, Figure 4.12. The beam has length L = 134 mm, width 12.7 mm, and thickness 0.508 mm. The magnets are stood off from the free beam tip by a distance δ = 6 mm with their centers equidistant to the beam tip equilibrium

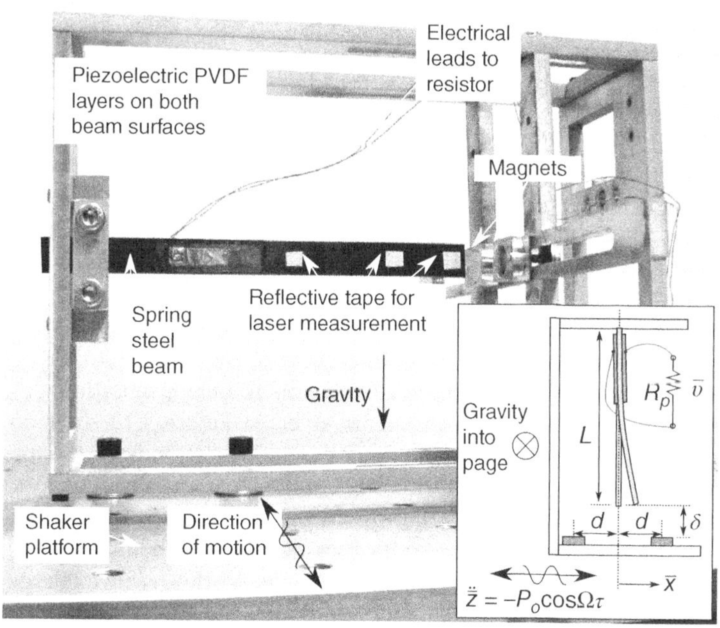

Figure 4.12 Experimental set-up. Bistable ferromagnetic cantilever with piezoelectric PVDF patches, attached to shaker platform, with PVDF electrode leads connected to external resistance (not shown).

position when the magnets are absent: d = 14 mm. In this configuration, the two intrawell natural frequencies (and damping ratios) are identified as 14.94 Hz (0.016) and 15.06 Hz (0.019) with stable equilibria at $\bar{x} = \pm 5.06$ mm. These results show that the experimental platform exhibits a near symmetry between the intrawell oscillation characteristics, which is in agreement with the modeling strategy and assumptions employed in this study. Piezoelectric PVDF layers of 28 μm thickness are bonded to both beam surfaces near the clamped end with poled axes facing outwards and electrode leads connected in series to an external load resistance $R_p = 4.5$ MΩ. The frame to which the beam was clamped is firmly attached to an electrodynamic shaker in an orientation such that gravitational influences do not affect cantilevered beam bending deflections of interest. The measured responses during testing are the base acceleration (measured by an accelerometer), beam tip velocity and displacement (measured by a laser interferometer), and the voltage across the load resistor.

A series of experiments is conducted to validate the trends observed analytically that the amplitude of the exciting base acceleration critically governs the energy diffusion to the third-order superharmonic dynamics, which can enhance the power generation possible in such a low frequency regime, near $\omega \approx 0.3$. For each experiment, the shaker base acceleration level is fixed while increasing frequency sweeps of harmonic base motion are carried out, particularly around the excitation frequency regime that corresponds to the third-order superharmonics of the harvester. A very slow sweeping rate of +0.01 Hz s^{-1} is employed to ensure that the excitation properties are nearly stationary over a given portion of time (thus providing an opportunity for the oscillations of the bistable harvester to obtain steady-state conditions, if possible). The slow sweep strategy also enables the data to be post-processed with short time fast Fourier transforms to extract the spectral data proportional to the excitation frequency, ω, and proportional to three times the driving frequency, 3ω. Over these intervals of sampling, the mean excitation frequency is determined and the two spectral components of the average power harvested across the load resistance are plotted in the left column of Figure 4.13, showing the responses proportional to the excitation frequency ω using black crosses while the gray circles denote the proportion of net harvested power that occurs at the superharmonic frequency 3ω. The instantaneous voltage envelopes corresponding to these power results are given in the right-hand column of Figure 4.13. The horizontal axes in both columns are the mean instantaneous excitation frequency normalized with respect to the linear natural frequency of the harvester, to recover the non-dimensional frequency notation ω as used in the analyses above. The measurements given in Figure 4.13 show an increasing root-mean-square (rms) amplitude of base acceleration from the top row to the bottom, spanning 2.53 to 4.39 m s^{-2}. From the plots, the absolute average powers are very low (order of nW), yet this is to be anticipated due to the small piezoelectric PVDF layers used as the transduction mechanisms. Such power levels should not detract from the current focus on the *relative* energy diffusion that exists between the fundamental and superharmonic dynamics. More strongly coupled piezoelectric materials, such as lead zirconate titanate (PZT), would enable greater harvested power with comparable roles observed as relate to the energy diffusion.

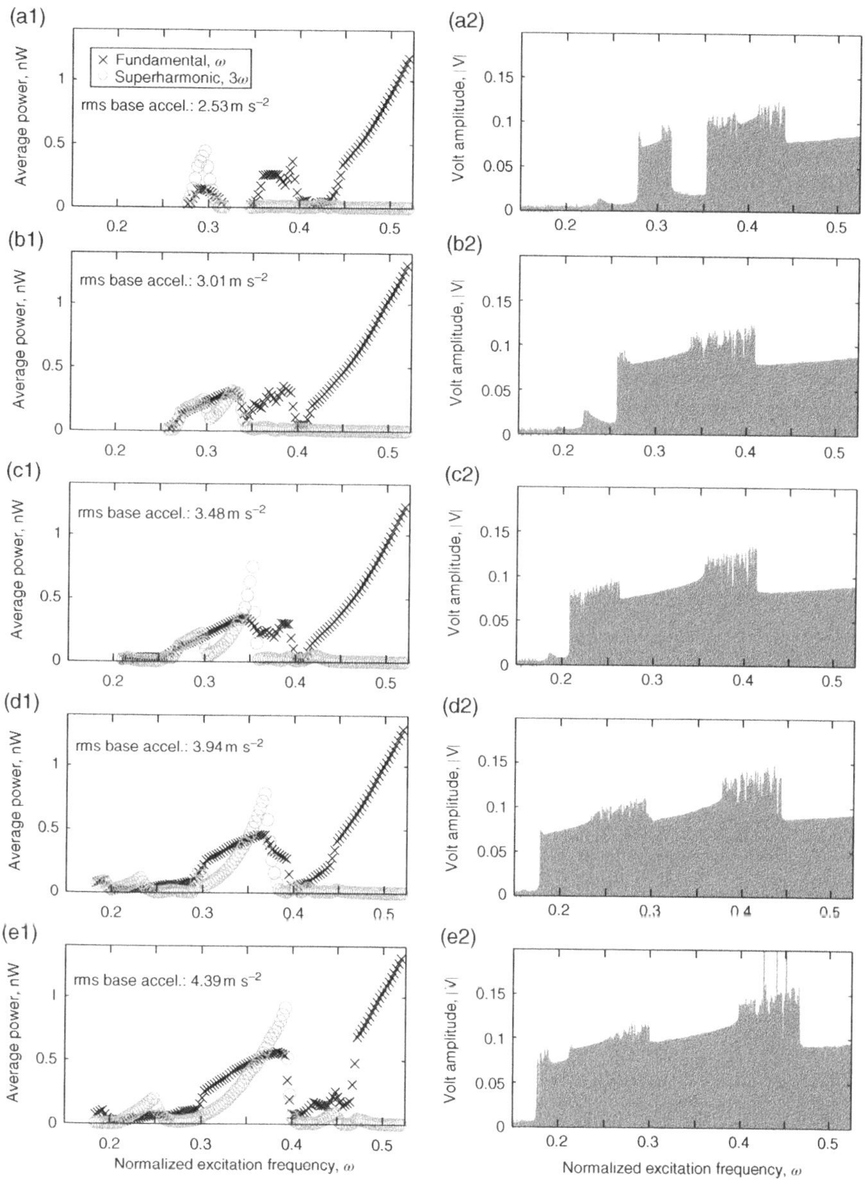

Figure 4.13 Experimentally measured bistable harvester responses. Left-hand column (1), average power across the load resistance showing spectral components of fundamental ω and superharmonic 3ω in consequence to excitation at frequency ω. Right-hand column (2), piezoelectric voltage amplitude during excitation frequency sweep. Using base acceleration levels of (a) 2.53, (b) 3.01, (c) 3.48, (d) 3.94, and (e) 4.39 m s^{-2}.

Throughout the experimental results presented in Figure 4.13, and in agreement with the analytical predictions presented in Sections 4.3.4 to 4.3.6, dynamics proportional to the third-order superharmonic are triggered in the frequency regime near $0.3 \leq \omega \leq 0.4$. This is evident in the left-hand column of

Figure 4.13 where the superharmonic average power contributions (circle data points) are obviously activated and rise to increasing levels as the excitation frequency approaches the end of this bandwidth near $\omega \approx 0.4$. In very good agreement with the analytical findings shown in Figure 4.7 to Figure 4.10, the proportion of average power produced by the bistable harvester which is concentrated at a frequency three times the excitation rises to levels considerably *greater* than that proportional to the excitation frequency itself. This verifies that the particular sensitivities of bistable harvesters may be deliberately exploited to enhance the power generation when the devices are excited with lower frequency input accelerations. Importantly, such knowledge guides the necessary storage circuit switching strategies to operate at the superharmonic 3ω rather than at the excitation frequency itself, so as to collect the greater amount of harvested power generated by the device.

For continued increase in the excitation frequency in excess of this key bandwidth, the superharmonic power contributions are seen to drop suddenly, in very good agreement with the analytical predictions, such as in Figure 4.9b. The measurements in this frequency band just above $\omega \approx 0.4$ are primarily aperiodic in nature, giving very low average power levels for both the fundamental and superharmonic spectral components. Previous investigations of bistable structures for energy harvesting likewise observed that aperiodic motions are unfavorable for providing useful *average* power when compared to dynamics which induce persistent snapping of the buckled harvester from one stable state to the other [43]. The voltage measurements in the right-hand column of Figure 4.13 corroborate the aperiodicity of these dynamics above the favorable excitation frequency regime for exploiting the superharmonic behaviors (about $0.3 \leq \omega \leq 0.4$), since the voltage envelopes lose a consistent structure characteristically observed for the steady-state oscillations.

Finally, another trend observed via the analytical predictions has a direct counterpart to the experimental findings. Figure 4.13 shows that the frequency at which the favorable amplified superharmonic dynamics suddenly lose their stability is progressively increased as the amplitude of the base acceleration increases. For instance, from Figure 4.13b1 to e1, the highest normalized excitation frequency at which the beneficial amplification of the superharmonic power contribution occurs extends from about $\omega \approx 0.33$ to $\omega \approx 0.4$, considering the base acceleration range of 3.01 to 4.39 m s^{-2}. In comparison, from Figure 4.7 to Figure 4.10, the increase in the comparable frequency is about $\omega \approx 0.28$ to $\omega \approx 0.5$, spanning the normalized excitation amplitudes of $p = 0.2$ (the "mild" level) to $p = 4.5$ (the "high" level). This qualitative agreement between the experiments and analysis suggests that the experimentation is conducted using a regime of base accelerations that may be considered as "moderate". Considering these results and the analytical formulation in detail, recall that the nondimensional excitation level p is normalized with respect to the linear natural frequency ω_0 of the bistable harvester, $p = P_0 / x_0 \omega_0^2$. This frequency is often minimized so as to better facilitate persistent snap-through oscillations under sinusoidal excitations, effectively yielding a "shallower" double-well potential energy profile. Thus, it is apparent from the analytical and experimental work presented in this study that the diffusion of energy to superharmonic dynamics should be readily observed in bistable energy harvesters in practice, namely because designing the structures to have shallower potential energy profiles reduces the linear natural frequency and correspondingly increases the normalized amplitude of excitation experienced by the device. From the findings reported here, it is clear that this property is not to be avoided but rather *exploited* since, in the low frequency regime, the triggering of the superharmonic dynamics leads to a harvested energy greater than that which is attributed to the harmonic excitation frequency itself. Therefore, suitable switching strategies can be devised to leverage this unique feature of the bistable energy harvester architecture for collecting and storing maximum converted energy from the excitation environment.

4.3.8 Summary

Recent experimental and numerical investigations of a bistable energy harvesting beam observed that a dramatic enhancement of the net harvested power could be obtained if the device was activated in a low frequency regime such that superharmonic dynamics are triggered. To more clearly elucidate this phenomenon and uncover the conditions for which it may be exploited, this study developed an experimentally-validated analytical model of a generalized bistable energy harvester excited by harmonic motions. Because practical energy harvesting and collecting circuits must employ switching technologies for greatly enhanced electrical power capture, the energy diffusion between the fundamental and third-order superharmonics is essential to explain so that one may identify the best frequency for the switching protocol. By using a higher-order harmonic balance solution strategy, the analytical formulation presented here opens the door to the direct prediction of energy diffusion amongst the multiharmonic oscillations and therefore characterizes the conditions for which the superharmonic vibrations may be exploited for enhancement of power generation capabilities. The experiments agree with all salient trends predicted by the analysis and help to illustrate the ease with which such amplified superharmonic dynamics are exhibited in the bistable harvester. The findings of this study draw the important conclusion that due to the need to reduce the linear natural frequency of bistable harvesters to facilitate persistent snap-through oscillations, the relative level of excitation may be "high", which induces a significant diffusion of energy to the superharmonic dynamics when the harvesters are excited in a frequency regime around one-third of the linear natural frequency. Therefore, this particular triggering of snap-through dynamics at the superharmonic frequency may be leveraged to enhance the collected power in energy harvesting storage circuitry, giving bistable devices a particular merit and distinction for such an important performance-enhancing capability.

4.4 Optimal and Robust Energy Harvesting from Realistic Stochastic Excitations Using the Dynamics of Structures Designed Near the Elastic Stability Limit[4]

The successful *implementation* of a vibration energy harvesting system relies on *designing* the device so as to undergo large amplitude oscillations in consequence to the vibration spectra and form which excite it. Consequently, device development is best accomplished following careful assessment of the input excitation source. Although some vibration sources possess a considerable concentration of energy at one or a few tones to effectively characterize the motions as sinusoidal inputs, in fact many common ambient vibration sources in environments important for energy harvesting applications are stochastic and have peak spectral distributions in the very low frequency range of a few Hz or less [8,106–108]. Above the frequencies associated with peak acceleration amplitudes, the power spectra roll off rapidly, which is a common feature of exponentially-correlated (colored) noise sources [109]. These are the key vibrational sources of interest which motivate the investigations in this study.

As evident in the survey given in Section 4.1, the exploitation of bistability is a widely successful means to enhance energy harvester sensitivities to such challenging, diffused, and low frequency exciting vibration spectra. By gradually changing a design parameter proportional to an axial compressive force (e.g., such as those techniques depicted in Figure 4.1), a variety of nonlinear energy harvesting beam designs transition from a monostable state to a statically bistable state. A critical axial load occurs at the transition, denoted as the *limit of elastic stability*, and the linear natural frequency vanishes [110]. Exactly at the transition, the beam is called *essentially nonlinear* because the linear restoring forces are eliminated from the corresponding governing equation of motion. Quinn and colleagues [111,112] have investigated essentially nonlinear energy harvesters for harmonic and impulsive loading, but conclusions on their performance in stochastic environments have yet to be made. Moreover, several research efforts have found clues that energy harvester design at or around the elastic stability limit is an optimal design space for stochastic excitation environments. Tang *et al.* [31] observed experimentally that when excited by white noise a harvester beam having magnetic repulsion to buckle the system achieved "optimal performance.... near the monostable-to-bistable transition region". Zhao and Erturk [48] cautioned that the performance enhancement of a bistable energy harvester beam under stochastic excitations was limited to the case in which the interwell oscillations could be persistently maintained, thus encouraging a bistable design but one with shallow wells of potential energy to surmount via snap-through motions. In contrast to these experimental observations,

4 Material from this section has been reprinted by permission of ASME, from R.L. Harne and K.W. Wang, Prospects for nonlinear energy harvesting systems designed near the elastic stability limit when driven by colored noise. Journal of Vibration and Acoustics 136:021009, 2014. doi: 10.1115/1.4026212. Copyright © 2014 by ASME.

Daqaq [77,78] found through numerically-verified analyses that bistability cannot improve power generation over a counterpart linear energy harvesting architecture in white noise environments when the ratio of mechanical to electrical time constants τ_m/τ_e is large, which is representative of all harvesters designed near the stability limit because this ratio grows asymptotically $\tau_m / \tau_e \rightarrow \infty$ at the limit. These findings suggest that the merit of deploying harvesters designed near the elastic stability in colored noise vibration environments (which are more representative of ambient vibration sources than white noise) appears to be warranted but remains incompletely understood since definitive conclusions on best energy harvester design practices are outstanding.

One critical factor not addressed by the previous investigations is the possibility that an energy harvesting beam may possess asymmetry, for instance due to fabrication-related imperfections. Careful studies of this inevitability have yet to be carried out although several authors have reported asymmetric biases in their respective bistable experimental harvesters [28,68,70]. Very subtle factors can cause such asymmetries for mono- or bistable architectures, including the straightness of the harvester beam, eccentricity of the buckling force, and even gravity.

Considering the lack or incompleteness of the understanding as to the advantages of energy harvesters designed near the elastic stability limit when excited by colored noise, including the roles of asymmetry, this study seeks to discover new knowledge that will enlighten optimal and robust energy harvesting development. Because this design space is already recognized to enable substantial performance enhancements under sinusoidal excitations by leveraging slightly *bistable* energy harvesters, an aim here is to uncover how this performance benefit translates to colored noise excitations which also might readily activate the sensitive, low frequency snap-through oscillations of bistable structures. The following sections present a study which scrutinizes these questions and concerns through analytical, numerical, and experimental results for decisive answers and insights on the relative advantages of exploiting the energy harvesters designed near the elastic stability limit, whether before this transition (monostable), beyond it (bistable), or at the limit itself (essentially nonlinear).

4.4.1 Modeling of Nonlinear Energy Harvester Platform

The nonlinear energy harvester architecture considered here is depicted in Figure 4.14 using the convention of lumped parameter elements to represent the fundamental modal contributions of a nonlinear energy harvesting beam, including its modal mass m, modal damping constant b, and nonlinear restoring force $f(k_0, k_1, k_3)$ which includes three contributions representative of a bias (asymmetry) and linear and nonlinear stiffnesses. The harvester is excited with ground acceleration $g(t)$ such that the relative displacement between the ground and inertial mass of the harvester is $z(t)$. Here, the electromechanical transduction mechanisms are generalized to include either piezoelectric or electromagnetic induction, which is a matter only of changing the interpretation of the electrical transduction, similarly as shown in Section 4.3.1. For instance, considering piezoelectric coupling, Kirchhoff's laws represent the transduction as an equivalent

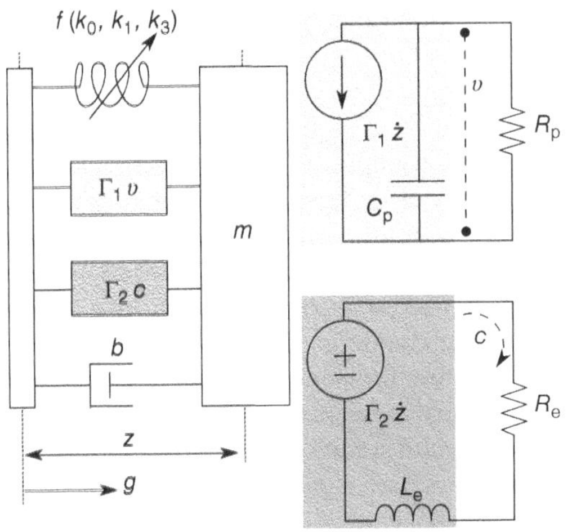

Figure 4.14 Nonlinear energy harvester with piezoelectric and electromagnetic induction mechanisms and corresponding circuitry.

current source such that the state variable is the voltage $v(t)$ across the piezoelectric material leads, whereas for an electromagnetic harvester the device is equivalent to a voltage source described by a current flow state $c(t)$. The governing equations for the platform are derived and given in Eq. (4.55).

$$m\ddot{z}+b\dot{z}+\bar{V}'(z)=-m\ddot{g}-\Gamma_1 v-\Gamma_2 c \tag{4.55a}$$

$$\bar{V}(z)=\int f(z)\mathrm{d}z=k_0 z-\frac{1}{2}k_1 z^2+\frac{1}{4}k_3 z^4 \tag{4.55b}$$

$$C_\mathrm{p}\dot{v}+\frac{1}{R_\mathrm{p}}v=\Gamma_1\dot{z};\ \text{for piezoelectric coupling} \tag{4.55c}$$

$$L_\mathrm{e}\dot{c}+R_\mathrm{e}c=\Gamma_2\dot{z};\ \text{for electromagnetic coupling} \tag{4.55d}$$

The derivative operators indicate $(\cdot)=\mathrm{d}/\mathrm{d}t$ and $(\,)'=\mathrm{d}/\mathrm{d}z$. Although asymmetries have also been considered using cubic potential energy contributions [79], in this study the restoring potential $\bar{V}(z)$ neglects a cubic term because prior investigations demonstrate that probabilistic outcomes for systems having quartic potentials are substantially less influenced by asymmetries tending from cubic potential terms (or quadratic forces) as compared to linear terms (or static, bias forces) [113,114]. Biases, included here by coefficient k_0, are commonly encountered in practical energy harvesting system development, including biases imposed by gravitational loading or due to manufacturing imperfections.

In the light of the context of the present study, an important modeling assumption is applied that accounts for the energy harvester design space of interest. To target high power generation performance under ambient vibration sources with greatest energy density in the sub-Hz range, the interest is to realize a harvester having a linear natural frequency as low as feasibly possible. Compared to available piezoelectric materials and electromagnetic induction harvester realizations, this implies that the mechanical time constant will be significantly greater than the electrical time constant. In this context, the primary influence of either transduction is a dissipation of energy in the external circuit [16,25]. With this assumption in mind, Eqs (4.55) are reduced and rearranged.

$$\ddot{z} + \mu\dot{z} + V'(z) = \zeta(t) \tag{4.56a}$$

$$V(z) = \alpha z - \frac{1}{2}\kappa z^2 + \frac{1}{4}\beta z^4 \tag{4.56b}$$

$$\alpha = k_0/m;\ \kappa = k_1/m;\ \beta = k_3/m;\ \mu = (b + \phi)/m \tag{4.56c}$$

$$\phi = R_p\Gamma_1^2;\ \text{for piezoelectric coupling} \tag{4.56d}$$

$$\phi = \Gamma_2^2/R_e;\ \text{for electromagnetic coupling} \tag{4.56e}$$

The result of this assumption is that the system dissipation coefficient μ is due to a combination of mechanical b and electrical ϕ influences. Power generated by the harvester is proportional to the velocity $\dot{z}$ via

$$P = \left|v^2\right|/R_p = \phi\left|\dot{z}^2\right|;\ \text{for piezoelectric coupling} \tag{4.57a}$$

$$P = R_e\left|c^2\right| = \phi\left|\dot{z}^2\right|;\ \text{for electromagnetic coupling} \tag{4.57b}$$

The colored stochastic exciting base acceleration $\zeta(t) = -\ddot{g}$ has a correlation function given by $\langle\zeta(t)\zeta(t^*)\rangle = rD\exp\{-r|t - t^*|\}$, where r is the correlation bandwidth (inverse correlation time) and D is the noise intensity. The power spectral density of this stochastic base excitation form is

$$S_{\zeta\zeta}(\omega) = 2D/\left[1 + (\omega/r)^2\right] \tag{4.58}$$

such that the motion is governed by a first-order Langevin equation

$$\dot{\zeta} + r\zeta = \rho(t) \tag{4.59a}$$

$$\langle\rho(t)\rangle = 0;\ \langle\rho(t)\rho(t^*)\rangle = \delta(t - t^*) \tag{4.59b}$$

where ρ is a Gaussian white noise process. Similar reformulations may be employed to incorporate and study other types of stochastic processes; by enlarging the state space, a stochastic process of arbitrary correlation function may be related to a Wiener process [115]. Then equation system (4.56) is expressed in Itô stochastic differential form as presented by equation system (4.60) [116,117].

$$\mathrm{d}\boldsymbol{x} = \begin{bmatrix} x_2 \\ -V'(x_1) - \mu x_2 + x_3 \\ -rx_3 \end{bmatrix}\mathrm{d}t + \begin{bmatrix} 0 \\ 0 \\ r\sqrt{2D} \end{bmatrix}\mathrm{d}B \tag{4.60a}$$

$$\boldsymbol{x} = \left[x_1, x_2, x_3\right]^{\mathrm{T}} = \left[z, \dot{z}, \zeta\right]^{\mathrm{T}};\ \mathrm{d}B/\mathrm{d}t = \rho(t) \tag{4.60b}$$

In the present work, directly simulated results are obtained by integrating equation system (4.60) using the stochastic differential equation (SDE) toolbox in MATLAB with the Euler–Maruyama (E–M) algorithm. The corresponding Fokker–Planck equation governing the pdf $\bar{p}(\boldsymbol{x}, t)$ is given by [116,117]

$$\frac{\partial \bar{p}}{\partial t} = -x_2 \frac{\partial \bar{p}}{\partial x_1} + V'(x_1)\frac{\partial \bar{p}}{\partial x_2} + \mu \frac{\partial (x_2 \bar{p})}{\partial x_2} - x_3 \frac{\partial \bar{p}}{\partial x_2} + r \frac{\partial (x_3 \bar{p})}{\partial x_3} + r^2 D \frac{\partial^2 \bar{p}}{\partial x_3^2} \quad (4.61)$$

satisfying $\bar{p}(\boldsymbol{x}, t) \to 0$ as $|\boldsymbol{x}| \to \infty$. Stationary solutions to Eq. (4.61) are analytically possible only for quadratic potentials $V(x_1)$, that is, linear systems. Other potential forms require evaluation of the bandwidth r to facilitate investigation of the stationary pdf because various approximation methods, each suited to particular assumptions and accuracy requirements, are available. Since r is a parameter normalized to the mechanical system's characteristic frequency (i.e. $r = 1$ indicates the noise level begins a 20 dB/decade roll-off at the system's linear natural frequency), one must select the appropriate method to determine an approximate stationary pdf based upon the narrowness or breadth of r with respect to prominent mechanical responses. The practical interest of this study encourages focus on bandwidths around 1 or less, $r \leq 1$, since maximum ambient vibration energy of target excitation sources is concentrated at very low frequencies and because it is envisioned that the energy harvesters are designed around the point of elastic instability at which linear natural frequency asymptotically approaches zero.

A preferred scheme to determine pdf response is therefore a decoupling approximation approach set forth by Hänggi [118], later extended and utilized by collaborators [119,120], and recently employed by Daqaq [77] in a study on the bistable inductive generator. For brevity, only the outcomes of the analytical derivations are presented here since the technique has been formalized in full by Moss and McClintock [116] and Hänggi and Jung [117]. Past studies have used this pdf approximation technique to investigate chiral symmetry breaking in chemical reactions, laser instability, and crystal formation [113,117,121], but it is utilized in this study for new insight into nonlinear energy harvester design synthesis near the elastic stability limit. The key result is an approximate, decoupled stationary pdf $p(x_1, x_2)$ expressed as

$$p(x_1, x_2) = Z^{-1} \exp\{-V(x_1)/\sigma_1\} \exp\{-x_2^2/2\sigma_2\} \quad (4.62a)$$

$$Z = \int_{-\infty}^{\infty}\int_{-\infty}^{\infty} \exp\{-V(x_1)/\sigma_1\} \exp\{-x_2^2/2\sigma_2\} \mathrm{d}x_1 \mathrm{d}x_2 \quad (4.62b)$$

$$\sigma_1 = (D/\mu)/\{1 + \langle V''(x_1)\rangle/[\mu r + r^2]\} \quad (4.62c)$$

$$\sigma_2 = (D/\mu)/\{1 + \mu/r + \langle V''(x_1)\rangle/r^2\} \quad (4.62d)$$

$$\langle x_1^2 \rangle = \int_{-\infty}^{\infty}\int_{-\infty}^{\infty} x_1^2 p(x_1, x_2) \mathrm{d}x_1 \mathrm{d}x_2 \quad (4.62e)$$

$$\left\langle x_2^2 \right\rangle = \int_{-\infty}^{\infty}\int_{-\infty}^{\infty} x_2^2 p(x_1, x_2) \mathrm{d}x_1 \mathrm{d}x_2 \tag{4.62f}$$

Here, Z is a normalization constant. In applying the above approximation scheme for an electromagnetic bistable energy harvester, Daqaq [77] noted the challenges involved in determining the decoupled pdf $p(x_1, x_2)$ since it necessitates computation of σ_1 which is a function of the mean-square displacement $\left\langle x_1^2 \right\rangle$, via $\langle V''(x_1) \rangle$, which is itself dependent on the pdf. Consequently, Daqaq [77] demonstrated the use of an iterative computational method to resolve the circular-dependence concern; the method was shown to be accurate with respect to direct integration of the SDE. Likewise, the computational method is employed here via the MATLAB command fsolve to determine solutions to Eqs (4.62). The performance metric of interest is the generated power of the harvester, Eq. (4.57), which is proportional to the square of the device velocity for both transductions. Therefore, a straightforward means to evaluate the relative advantages and disadvantages of exploiting nonlinear energy harvesting system design near the elastic stability limit when the devices are excited by colored noise is the mean-square velocity $\left\langle x_2^2 \right\rangle$, Eq. (4.62f), following the observation that $\left\langle P \propto \left\langle x_2^2 \right\rangle \right\rangle$.

4.4.2 Preliminary Remarks on Accuracy, Comparisons, and Experimentation

The accuracy of the decoupling approximation in analytically predicting the mean-square responses is excellent so long as the bandwidth of the colored noise r is not too small and while the noise intensity D is not too large [117,120]. However, accuracy is maintained even for small r so long as noise level is limited to approximately $D \leq 0.3$ [120]. Predictions for high bandwidth $r >> 1$ (i.e., the white noise limit) are seen to correctly converge to Gaussian response estimates, while the predictions accounting for quadratic potentials (linear systems) also lead to accurate mean-square responses in that limiting case. The direct simulation of the Itô stochastic differential in Eqs (4.60) is complicated by the extremely small time steps required to maintain E–M integration algorithm stability and the need to integrate for long times to obtain meaningful mean-square values. To accommodate these concerns, this study carries out 80 simulations for each instance considered (e.g., a given set of excitation and design parameters), where each simulation lasts 4000 natural periods. The results from these simulations are then averaged to yield one numerical validation data point. To similarly account for the need to obtain convergence to the results, the experimentation was undertaken using long evaluation times to attain consistent mean-square responses. These numerous challenges simply to obtain reliable data sets strongly suggest that the greatest number of comparisons amongst results helps ensure that insights obtained are valid to the highest degrees of confidence.

The experimental set-up used in this study is identical to that employed in Section 4.3. Here, the cantilevered ferromagnetic spring steel beam has length $L = 132$ mm, width 12.7 mm, and thickness 0.51 mm, and is oriented such that gravity plays no role in the axis of motion. Piezoelectric PVDF layers of length 38 mm, width 11 mm, and 28 μm thickness are bonded to both sides of the beam

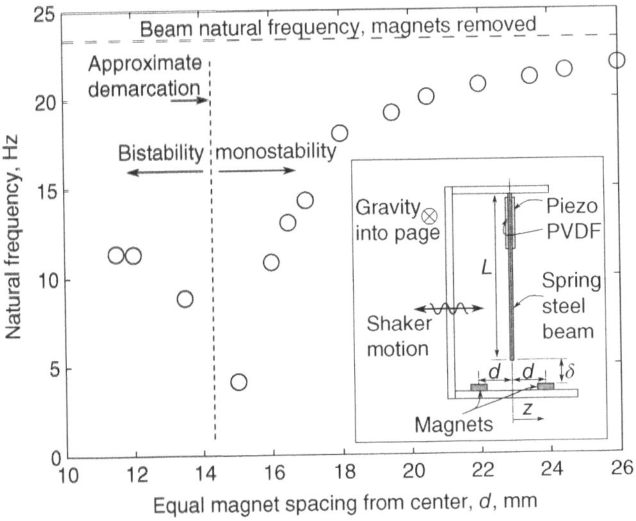

Figure 4.15 Measured natural frequencies of beam as equal magnet spacing d is adjusted. Inset shows experimental platform schematic.

near the clamped end – with poling directions extending away from the beam surfaces – and with electrode leads connected in series to a load resistance of 5.4 MΩ. The permanent magnets used to exert nonlinear restoring forces on the beam's generalized displacement are neodymium with diameter 12.7 mm, thickness 6.35 mm, and flux density 1.42 T (as quoted by the distributor). Throughout experimentation the magnets are positioned a distance $\delta = 8$ mm away from the beam free end, according to the schematic shown in the inset of Figure 4.15, while the magnet spacing distance d is adjusted to tailor the net restoring forces on the harvester beam. In the course of changing magnet position in the range of elastic instability, large changes in natural frequency are observed, as shown in Figure 4.15 and as anticipated according to the fundamental theory of elastic stability [110]. In the absence of the two attractive magnets, the beam natural frequency is 23.38 Hz (defined here to be the linear design). The nearest to a genuine design case of essential nonlinearity obtained in this study was when the magnets were positioned in such a way as to lead to a nonlinear monostable beam having natural frequency 4.141 Hz ($d = 15$ mm); hereafter, this experimental design is loosely called the essentially nonlinear case. A ring-down response of the essentially nonlinear harvester is shown in Figure 4.16 to exemplify the extremely low frequency oscillations. In the following section, the essentially nonlinear harvester design is compared to a very weakly bistable device. For this bistable configuration, the magnets are positioned equally apart from the beam center position, $d = 13.5$ mm (8.853 Hz intrawell natural frequency), which was the "weakest" and most symmetric bistable configuration achieved. The stable equilibria of this harvester configuration displaced the beam free end by just +/− 3 mm when compared to the full 132 mm length. Hereafter, this experimental case is referred to as the bistable harvester.

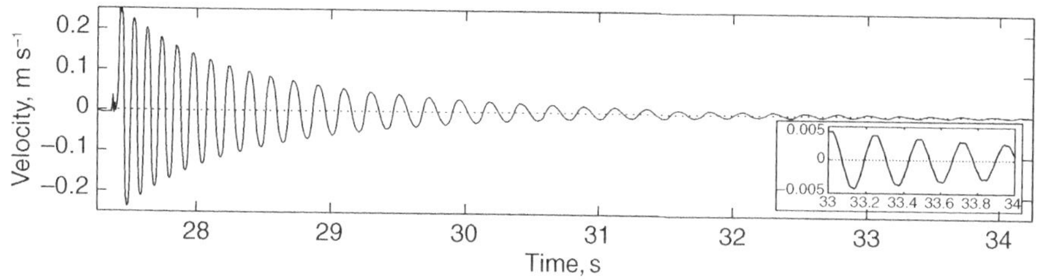

Figure 4.16 Piezoelectric cantilever free tip velocity vibrations due to initial displacement imposed on beam tip. Here, the magnet spacing is $d = 15$ mm, near point of essential nonlinearity.

When generating colored noise, the bandwidth r refers to a value normalized with respect to the time constant of the underlying structural system time constant. Thus, in the event of a bandwidth $r = 1$ for the linear harvester, this means the noise bandwidth extends to 23.38 Hz, whereas for the experimental essentially nonlinear case the comparable $r = 1$ band extends to just 4.141 Hz. Employing similarly non-dimensionalized bandwidths ensures that the most challenging performance comparisons amongst the various harvester designs are made. Final remarks on this stringent comparison method are given in the summary of this study. For experimentation, after selecting the appropriate bandwidth, colored noise is generated for a 300-second time period by solving Eq. (4.59) using a fourth-order routine of Runge–Kutta numerical integration. The result is recorded in an uncompressed audio file and played back as input to the electrodynamic shaker amplifier using various amplitudes of amplifier gain to explore the role of the mean-square excitation level. Following each experiment, the noise intensity level D is determined by fitting a power spectral density curve, using Eq. (4.58), to the measured input acceleration spectrum. These values of the noise intensity D are thereafter employed in analysis and simulation for direct comparisons.

4.4.3 Noise Bandwidth and Level Influences on Ideal Designs

Figure 4.17 presents a collection of analytical, numerical, and experimental findings of harvester mean-square velocity – proportional to generated power according to Eq. (4.57a) – normalized to noise correlation maxima rD as a function of bandwidth r. In Figure 4.17a,b, numerical validations are the filled data points on the plots. Figure 4.17a,b present the analytical and numerical results while 4.17c,d show experimental findings with legends indicating the mean-square acceleration input noise levels to provide an intuitive interpretation of the excitation levels employed. Results for the essentially nonlinear harvester are given in Figure 4.17a,c, while 4.17b,d present bistable harvester data. Results for the linear systems are given on each plot in Figure 4.17. In analysis the linear results, determined using $\kappa = -1$, $\alpha = \beta = 0$, are independent of excitation level (thus, no numerical verification data are required) whereas the linear harvester experimental data represent values averaged over three measured excitation levels which were found to yield very similar ratios of mean-square velocity to noise

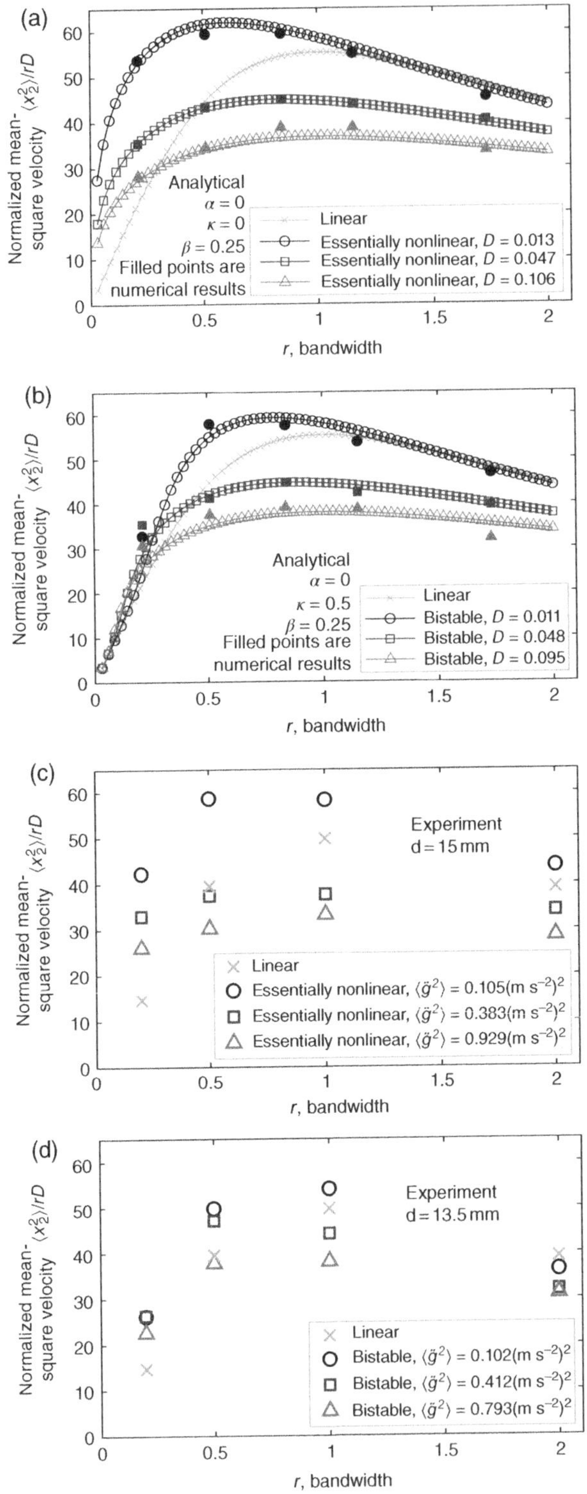

Figure 4.17 Energy harvester mean-square velocity normalized to noise correlation maxima. Part (a,b) shows the analytical results and simulated data (filled data points). Part (c,d) shows the experimental results with legends indicating mean-square noise input acceleration values.

characteristics rD. Damping of the system is experimentally determined to be $\mu = 0.009$ (combined mechanical and electrical dissipation) and the value is correspondingly used in all subsequent analysis and simulation. The restoring potential parameters κ and β employed for the analytical and numerical assessments represent energy harvesters designed near the elastic stability limit and are given in the corresponding sub-figure. However, there is no specific justification for the exact values selected apart from the agreement obtained from the theoretical findings with respect to experimental observations using comparative energy harvester designs.

Figure 4.17a,b show that the peak normalized responses of essentially nonlinear and bistable harvesters are very similar for each level of noise, although they occur for slightly different bandwidths of the colored noise excitation. For low noise levels, it is apparent that both nonlinear harvester configurations perform better than the linear counterpart by inducing greater mean-square velocity, particularly for noise bandwidths less than the individual systems' natural frequencies, $r < 1$. The peak of the linear harvester velocity responses occurs at $r = 1$, which indicates that maximum performance of a linear harvester in colored noise environments is obtained when the natural frequency of the system is within the noise bandwidth. Interestingly, as shown in Figure 4.17, the essentially nonlinear and bistable harvesters do not require that the excitation bandwidth extends to their respective linear resonance frequencies. Optimal energy conversion from the nonlinear platforms is in fact achieved for bandwidths less than the linear natural frequencies. The specific bandwidths at which the optima occur are seen to shift as functions of noise level. When compared to both the linear and weakly bistable energy harvester configurations, the essentially nonlinear harvester is found to provide improved power generation when excited by almost any noise level environment having low bandwidths, particularly evident in Figure 4.17a,c for $r \leq 0.5$. In comparison, the bistable platform shows a degradation of performance at low excitation bandwidths for all excitation levels, such as in Figure 4.17b for $r \leq 0.3$. Experimentally, this is observed to be due to a confinement of the stochastic oscillations to one local well of potential energy: thus, an intrawell dynamic regime. On the other hand, as observed in Figure 4.17b,d, for low noise bandwidths and for smaller intensities of the noise excitation, the bistable harvester plainly outperforms the linear counterpart by way of triggering interwell oscillations. Yet, as the excitation level and bandwidth increase, the performance improvements of the essentially nonlinear and bistable harvesters are diminished and the linear design leads to greater mean-square velocity and thus mean-square harvested power. This result is intuitively explained by the cubic restoring forces of the nonlinear configurations that stem from the large beam bending displacements and magnetic forces, which provide greater resistance to beam motions than the restoring forces of the linear harvester possessing no such higher-order contribution.

The numerically simulated results in Figure 4.17a,b show excellent quantitative agreement with analysis. Moreover, all of the key trends predicted analytically are plainly observed in the experimental measurements, Figure 4.17c,d. Although imperfect, there is also a good quantitative agreement between the analytical and experimental findings, which shows that the theoretical formulation assumptions are practically justified and suggests that useful insights may be derived from further analytical study with the model.

4.4.4 Criticality of Design at the Elastic Stability Limit

Although the results throughout Figure 4.17 show that the greatest power conversion is obtained by the essentially nonlinear harvester under colored noise excitations, the experimental realization of such harvester architecture used here is not precisely "essentially nonlinear". As described in Section 4.4.2, given its non-zero linear natural frequency, the essentially nonlinear experimental platform is actually very slightly monostable, albeit very nonlinear (see Figure 4.16). This fact should be considered alongside the observation by Tang *et al.* [31] that the "monostable-to-bistable" transition region experimentally led to best harvesting performance in white noise environments. Yet, if the transition itself appears to be a high-performance design space and there remains difficultly in realizing ideal essential nonlinearity in practice, a question is naturally raised as to whether it makes a difference to aim for a harvester design that is slightly monostable or slightly bistable if the essentially nonlinear "ideal" cannot be practically obtained. Therefore, the attention of the investigations turns to probing how the stochastic dynamic behaviors of the harvester vary based upon the linear stiffness term κ which characterizes the degree of monostability ($\kappa < 0$) or bistability ($\kappa > 0$) with respect to essential nonlinearity ($\kappa = 0$). Due to very good agreement between the experimental measurements and analytical predictions, the following evaluation employs the analytical model for new insights.

Figure 4.18 plots the influence of the linear stiffness term κ on the excitation-normalized harvester mean-square velocity predicted via the analytical model. Note that ideal essential nonlinearity corresponds to the region between the left and right sub-figures. On the whole, the results show that performance of the nonlinear energy harvester is not considerably different between designs which induce a weak monostability or weak bistability, such as when $|\kappa| < 0.1$. This corroborates similar experimental findings by Tang *et al.* [31] who employed white noise excitations. However, the logarithmic scaling of the horizontal axes may be misleading because adjustment of the buckling force is difficult to finely control in close proximity to the elastic stability limit, whether the means is by attractive magnets, repulsive magnets, geometric constraints, and so on. As shown in Figure 4.15, the system may rapidly switch from mono- to bistability due to common limitations on the resolution achievable in applying such a force. Then, considering Figure 4.18 near values $\kappa = \pm 1$, it is seen that bistable harvester performance is greatly reduced when the noise intensity is low. The explanation for this performance drop is the confinement of the beam oscillations to a single well of potential energy and the subsequent imposition of additional intrawell stiffnesses which raise the linearized natural frequency to around $\sqrt{2\kappa}$, thus increasing the local resistance to the stochastic oscillations. In contrast, such a vulnerability does not occur for slightly monostable harvesters since the linear natural frequency remains $\sqrt{|\kappa|}$. On the other hand, Figure 4.18 shows that very weakly bistable harvesters are able to achieve *optimal performance*, denoted by the circle data points in the figure. As the excitation levels increase, the linear

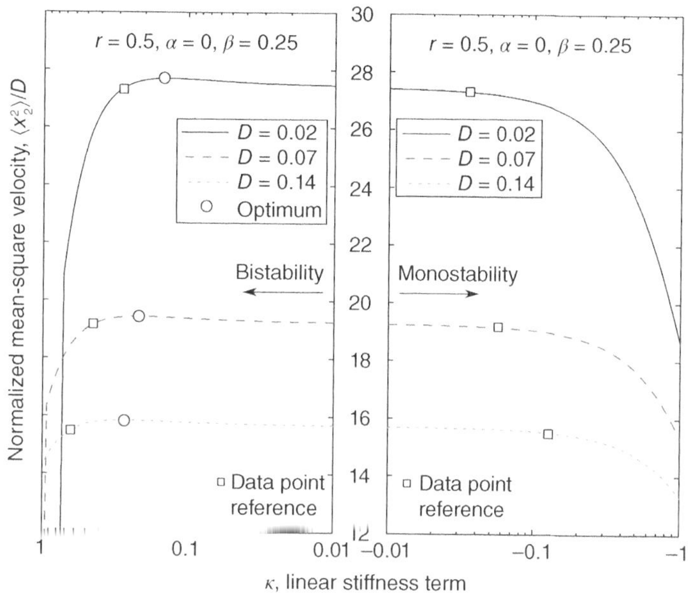

Figure 4.18 Influence of linear stiffness term κ on harvester mean-square velocity normalized to excitation level.

stiffness κ required for the bistable harvesters to achieve such best power generation is found to reduce. This is the result of achieving an optimum rate of well escape and may be similar to the stochastic resonance phenomenon [122]. Comparatively, as the magnitude of the linear stiffness increases for monostable harvesters for a given intensity of the colored noise excitation, the mean-square responses steadily and uniformly reduce. To synthesize these analytical findings for practical interpretation, in many cases it is challenging to realize perfectly an essentially nonlinear energy harvester which shows best performance over a wide range of noise bandwidths (Figure 4.17). Consequently, for a given bandwidth of the stochastic excitation and an expectation that the harvester design will inevitably be slightly mono- or bistable, it is discovered here that monostable harvesters possess an improved robustness to deviation in parameters although bistable harvesters are able to realize peak performance.

4.4.5 Impact of Asymmetry on Energy Harvesting Performance

In fabricating energy harvesters with restoring potentials configured near the elastic stability limit, numerous implementation issues may lead to asymmetric potentials, such that restoring forces do not equally and oppositely work on the harvester beam to bring it back to a resting position. Therefore, for the experimental results in Figure 4.17, both samples may not be precisely symmetric. Following assessment of the comparable importance of obtaining exact essential nonlinearity in Section 4.4.4, a similarly practical question is now addressed: to what degree should asymmetry concern energy harvesting design in the region of elastic instability?

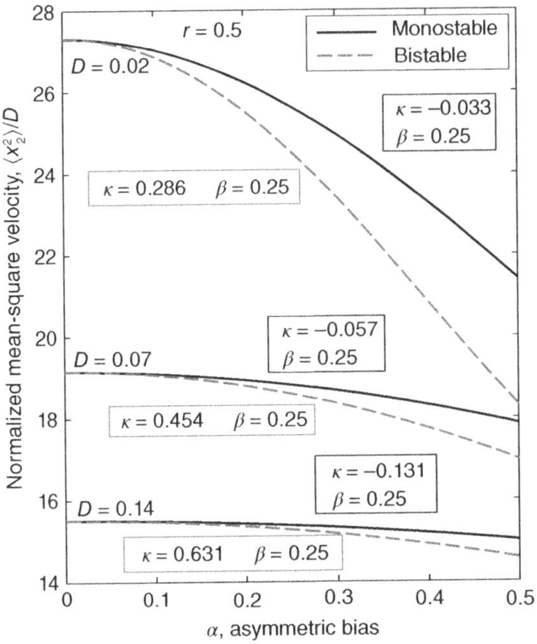

Figure 4.19 Analytically-predicted influence of bias α on harvester mean-square velocity normalized to excitation level.

Figure 4.19 presents analytical results of the excitation-normalized harvester velocity response levels in consequence to changing asymmetric bias α and when considering several excitation levels D. Given that the experimental sample classified as "essentially nonlinear" is actually slightly monostable, the comparisons begin using baseline symmetric harvesters with very minor mono- and bistability and examine how asymmetric biases from these symmetric designs affect the energy harvesting performance. Due to findings in Figure 4.18 where linear stiffness κ was shown to influence the ideal symmetric responses, the system responses in Figure 4.19 are computed with different linear stiffness κ for a given excitation level, such that the normalized response levels of the monostable and bistable harvesters are identical for the case of no bias terms $\alpha = 0$. These baseline starting values are denoted in Figure 4.18 using the square points and the corresponding values are given in Figure 4.19. In this way, a more meaningful sensitivity comparison is provided since baseline response levels of the mono- and bistable systems are the same.

Figure 4.19 reveals that an asymmetric bias is uniformly more detrimental to a bistable energy harvester's power generation capacity than it is for a slightly monostable system. In perturbing both systems from ideal configurations, the vulnerability of the bistable system to become confined to an individual intrawell response, and the subsequent additional stiffnesses associated with this deviation, suggest that bistable energy harvesting systems have a distinct disadvantage unless designed and deployed in ways that ensure the symmetry of the restoring forces. The influence of asymmetry is far more pronounced at low excitation levels, indicating well-escape is inhibited for the bistable configuration under such conditions. In contrast, the asymmetric nonlinear monostable system is less susceptible to performance degradation due to the same degree of bias.

To validate these predictions, gravitational forces are exploited as a means to apply controlled restoring force biases upon the experimental systems. The test fixture is arranged with the cantilevered harvester free end hanging downward but inclined from the vertical direction (gravitational force direction), shown schematically in Figure 4.20a. The shaker direction of motion remains along the beam's axis of transverse motion, inclined from the horizontal plane with angle θ. By changing the inclination angle θ of the whole experimental environment, the linear and cubic restoring force terms proportional to κ and β are held constant since magnet spacing remains fixed, and the amplitude of the exciting noise motions from the shaker will not deviate. Using this protocol, the responses of a nonlinear monostable harvester with equal magnet spacing $d = 16$ mm are compared to a bistable configuration with magnet spacing $d = 13.5$ mm. Both experimental samples represent nonlinear harvesters having restoring potentials slightly perturbed from essential nonlinearity (Figure 4.15). The piezoelectric PVDF voltage output serves as the response signal and these values are normalized with respect to measurements taken without inclination, $\theta = 0°$; thus, data points for both systems coincide at $\theta = 0°$.

From top to bottom, Figure 4.20b,c,d show experimental results for steadily increasing colored noise base acceleration levels when the correlation bandwidth is $r = 0.5$. Mean-square excitation levels for each test are provided in the legends and each test acquired 500 seconds of data. As seen from the experimental results, the first inclination of 2.6° leads to uniform reductions in the nonlinear harvester responses, regardless of noise excitation level, and the influence of the force bias is more apparent for the bistable harvester design. At low noise levels and for greater inclinations, Figure 4.20b, the reduced response levels are most pronounced for both systems. These findings are in good qualitative agreement with the analytical predictions in Figure 4.19. It was confirmed that for an inclination of 5.9° the bistable device still exhibited static bistability; yet, only occasional cross-well escape events were observed during the experiments carried out at 5.9°. Intuitively, the results demonstrate that the addition of a potential barrier for the bistable harvester deters power generation capability under colored-noise excitation environments when asymmetries become manifest, whether due to, for example, design imperfection or unsatisfactory attachment to the oscillating excitation system such that gravitational forces lead to bias effects.

4.4.6 Summary

Adopting a practical approach to vibrational energy harvesting from realistic ambient vibration sources characterized as colored noise, this study sought to answer several outstanding questions in the literature as to the suitability of energy harvesters designed near the elastic stability limit, which is the design space denoting the transition between mono- and bistability. By a combination of analytical, numerical, and experimental approaches showing very good agreement amongst each other, a number of new and important conclusions may be

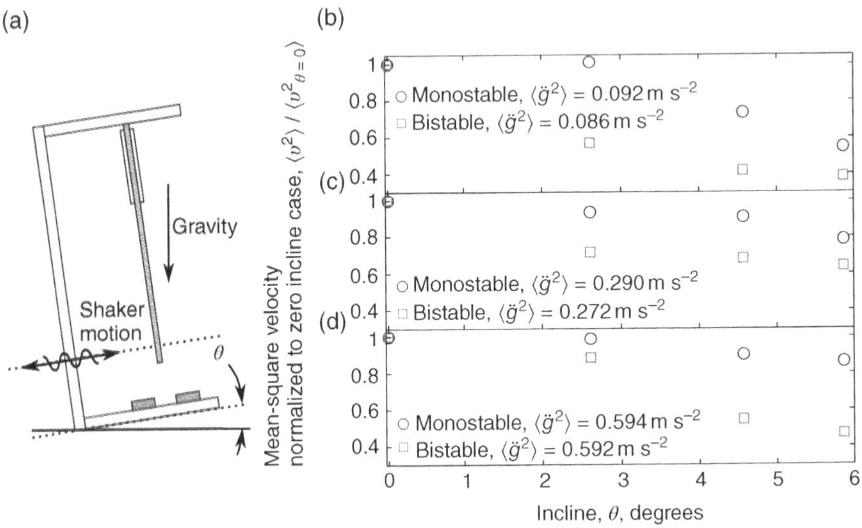

Figure 4.20 (a) Experimental set-up for evaluating influence of gravitational bias. (b,c,d) Experimental harvester mean-square voltage responses normalized to data for inclination $\theta = 0°$ using increasing noise excitation levels from (b) to (d). Legends indicate mean-square base acceleration levels for each test.

drawn. A first finding of relevance is the advantage realized by exploiting nonlinear harvesters in lower intensity noise environments when compared to the linear design. The low frequency sensitivities produced by the nonlinear restoring forces promote larger amplitude vibrations and an enhancement of the mean-square harvested power. Since many realistic ambient vibration sources are stochastic with energy concentrations at very low frequencies, this finding justifies the motivation to research and exploit the nonlinear design platforms. Comparing a variety of nonlinear harvester designs under such excitation conditions, it is discovered that essentially nonlinear energy harvesters are the best choice when the noise bandwidth is extremely small (Figure 4.17). These devices are those designed precisely at the elastic stability limit such that no linear restoring forces remain.

Yet, such an ideal design is challenging to achieve in practice based upon available means to exert influences on energy harvesting beams to eliminate the linear restoring forces. Consequently, the role of the linear restoring forces, which assess the degree of mono- or bistability when compared to the essentially nonlinear platform, are examined for fixed noise bandwidth (of reasonable extent $r = 0.5$) but varying noise intensity. The results draw a slightly different conclusion in that optimal harvested mean-square power is achieved by a weakly bistable energy harvester, possibly explained by the occurrence of a stochastic resonance-like phenomenon. On the other hand, the "slightness" of the bistability is a relative term, and too great bistability is a clear detriment. Therefore, it is seen that the weakly monostable harvesters benefit from a robustness to parameter uncertainty by a performance that is less inhibited for the same magnitude of change in linear stiffness (Figure 4.18). When asymmetry in the restoring forces is included, these insights are reiterated: bistable harvesters are more susceptible in lower excitation intensity environments due to the inconsistent triggering of snap-through oscillations, which is a property not evident in the monostable counterpart.

Bringing these findings together, it is tempting to conclude that bistable energy harvesters have distinct disadvantages in colored noise vibration environments when compared to essentially nonlinear designs, which conflicts with recent experimental observations that considered white noise excitations [31,48]. However, the results of this study indicate that with sufficient *a priori* knowledge of the noise intensity and bandwidth properties (factors which may readily be obtained through site monitoring), one may develop a weakly bistable harvester sensitive to such colored noise motions for optimal harvesting performance. Accounting for reasonable deviation in the excitation properties by some margins of safety in the design is one way to ensure that the necessary interwell oscillations are persistently triggered. In fact, the experimentally-validated analytical model developed here facilitates such a design procedure. A final remark taken from this study is the critical nature of the bistable device's snap-through oscillations towards providing a performance benefit when compared to other nonlinear monostable or linear energy harvesting platforms. Stated concisely, it is essential to heighten the sensitivity of the snap-through dynamics for energy harvesting advantage, whether the bistable device is excited by stochastic or harmonic vibration sources.

4.5 Amplifying the Snap-through Dynamics of a Bistable Energy Harvester Using an Appended Linear Oscillator[5]

A recurring theme in the conclusions drawn from the prior studies is that promoting snap-through dynamics is critical to maximize the energy harvesting success of bistable structures. Such high amplitude interwell oscillations induce large displacements of the inertial mass and lead to substantial transduction of voltage via the embedded electromechanical mechanism. With such goals in mind, the prior studies of this chapter have elucidated the inherent sensitivities of bistable harvesters, validating the analytical findings with numerical and experimental evidence.

Alternatively, another popular approach for best realizing the snap-through vibrations is to develop novel *designs* that assist to trigger favorable large amplitude oscillations in energy harvesting systems. Section 4.1 surveyed a variety of designs that have been explored to date for bistable harvesters having one degree-of-freedom (DOF), where the primary motions are often proportional to

5 Material from this section has been reprinted by permission of IOP Publishing, from R.L. Harne, M. Thota and K.W. Wang, Bistable energy harvesting enhancement with an auxiliary linear oscillator. Smart Materials and Structures 22:125028, 2013. doi: 10.1088/0964-1726/22/12/125028.

the fundamental mode dynamics of a cantilevered beam relative to the base excitation. In tandem with these investigations, other researchers have explored two- or multiple-DOF *linear* energy harvesters. In such cases, an excited linear energy harvester (e.g., piezoelectric cantilever) serves as a host for one or more linear oscillators, which are hoped to be as lightly damped as possible. Liu *et al.* [123] examined the influences of attaching a single, lumped linear oscillator to a harmonically-excited piezoelectric cantilever. The attached oscillator's mass and placement along the cantilever length were found to intricately govern the amplification of the piezoelectric cantilever's vibrations and thus enhance the efficiency of the energy conversion. Tang *et al.* [124,125] investigated experimentally and analytically an appended cantilever on an excited host piezoelectric cantilever. In spite of the term employed, the "parasitic" attachment was shown to be an effective means to broaden the resonant bandwidth of the energy harvesting system under sinusoidal excitation, since multiple modes could be tuned in close spectral proximity without an excessive addition of mass to the primary harvester. This benefit was absent from the reversed design concept: that of an excited, host linear cantilever with piezoelectric cantilever harvester thereafter attached. Qi *et al.* [126] assessed the use of multiple cantilevers attached to the primary, excited, host cantilever harvester. Although the "multiresonant" nature of the global dynamics was favorable in promoting a very broadband excitation frequency sensitivity, the authors noted that the power density is reduced by the addition of oscillators which do not contribute to power generation. Therefore, one must strike a balance between the advantages of broadband frequency sensitivity and a peak power density which is realized at an individual excitation frequency [126].

The investigations [123–126] illustrate that an auxiliary linear oscillator appended to an energy harvesting platform is a promising design strategy for improving the overall system performance. The multimodal dynamics activated in such multi-DOF structures broaden the range of operating frequencies over which large, resonance-based dynamics are attainable, which gives rise to an enhancement in energy harvesting effectiveness. Based on the prior studies surveyed in Section 4.1 and those detailed in the previous sections of this chapter, it is a logical progression from the discoveries about multi-DOF linear energy harvesters to question how might the dynamics of a bistable harvester be enhanced by attaching an auxiliary linear oscillator? Yet, there is an important distinction to draw between the linear and bistable platforms that relates to the dynamical behaviors involved. The enhancement of multi-DOF linear energy harvesting systems results from multimodal (i.e., resonance-based) vibration properties. In contrast, snap-through is not a resonant phenomenon: excitations of any frequency can induce the unique behavior, given sufficient excitation amplitude. Therefore, whether or not an attached linear oscillator will heighten the sensitivity of snap-through dynamics in a bistable harvester is a matter that requires close investigation.

The following study pursues such insight towards understanding how to effectively append an additional, linear dynamic body to a bistable energy harvester so as to enhance the energy conversion capability of the resulting system. The explorations utilize insights obtained in prior sections of this chapter, including the potential for multiharmonic dynamics and the benefit of a comprehensive collection of evidence, to probe deeply into the relative advantages and disadvantages of an auxiliary linear oscillator on the resulting performance of a harmonically-excited bistable energy harvester.

4.5.1 Coupled Energy Harvesting System Governing Equations

The energy harvesting architecture of interest is modeled using equivalent, lumped-parameter components. This modeling approach is applicable for understanding the lowest-order dynamics of a variety of distributed-parameter systems such as piezoelectric harvesters [67,75,127,128] in addition to designs utilizing electromagnetic transduction which often result in lumped-parameter platforms [29,129]. Such an approach is particularly relevant in the context of employing bistable energy harvesters which are typically intended to strictly undergo snap-through motions proportional to the fundamental mode dynamics of the underlying structure, such as an axially-compressed beam. As previous research shows, even if the attached oscillator is an appended cantilever [124], so long as only the lowest-order dynamics of the additional body are activated, the auxiliary oscillator is also able to be expressed as an equivalent lumped-parameter device. The lumped-parameter modeling approach is likewise applicable when explicitly accounting for the third-order superharmonic dynamics, subject to the constraint that this frequency is sufficiently less than higher-order structural modes of the actual distributed harvester under consideration. Indeed, the second natural frequency of a piezoelectric cantilever is approximately six times that of the fundamental frequency [130], so a considerable design space is available to develop two-DOF nonlinear energy harvesters using continuous sub-systems whereby the lumped-parameter modeling and analysis approach developed below is insightful.

Consequently, the architecture studied here is given by the schematic of Figure 4.21. An auxiliary linear oscillator of mass m_2 is attached to a bistable energy harvester of mass m_1 which is subjected to a base acceleration $\ddot{z}(t) = -W\cos\Omega t$ of acceleration amplitude W and angular frequency Ω. The transduction mechanism considered is electromagnetic induction between the relative displacement $x(t)$ of the harvester mass and base motion, such as by using a magnetic mass that passes through the axis of a ground-fixed coil [29]. The low inductance of the bistable electromagnetic transduction mechanism is shunted by a resistive load R to induce a flow of current $i(t)$ in the harvesting circuit when the relative velocity $\dot{x}(t)$ is non-zero. The double-well potential energy profile of the bistable device is

$$U(x) = -\frac{1}{2}k_1x^2 + \frac{1}{4}k_3x^4 \tag{4.63}$$

where position $x = 0$ is defined as the central, unstable equilibrium position of the double-well, as illustrated in Figure 4.21, since the curvature of the energy is negative at this location. The governing equations for the system are then expressed by

$$m_1(\ddot{x} + \ddot{z}) + d_1\dot{x} - k_1x + k_3x^3 - d_2\dot{y} - k_2y + \psi i = 0 \tag{4.64}$$

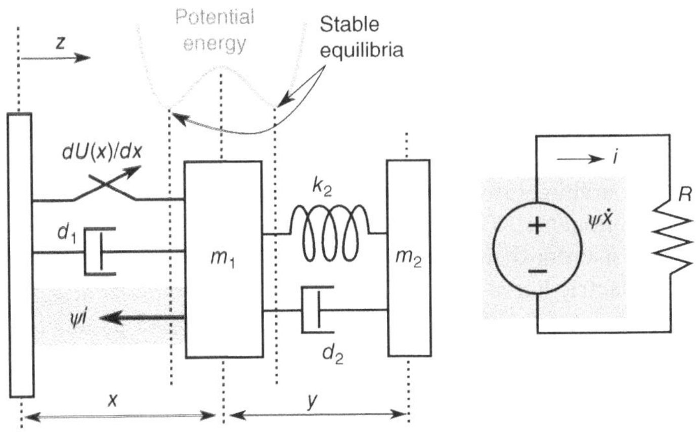

Figure 4.21 Schematic of base-excited, inductive bistable energy harvester to which is attached a linear oscillator. The corresponding external harvesting circuit is shown on the right.

$$m_2\left(\ddot{x}+\ddot{y}+\ddot{z}\right)+d_2\dot{y}+k_2y=0 \tag{4.65}$$

$$iR=\psi\dot{x} \tag{4.66}$$

where $y(t)$ is the relative displacement between the bistable inertial mass and the linear oscillator mass and ψ is an electromechanical coupling coefficient. For energy harvesters having an electrical time constant much smaller than the mechanical natural period, it is well-known that the effects of external circuitry upon the mechanical response of the harvester are primarily additional dissipative influences [5,83,84] and this assumption is employed in the derivation of Eq. (4.66) from Kirchhoff's Law. Following substitution and non-dimensionalization, the governing equations are then given as follows.

$$x''+\gamma_1x'-x+\beta x^3-\mu f\gamma_2y'-\mu f^2y=p\cos\omega\tau \tag{4.67}$$

$$y''+(1+\mu)f\gamma_2y'+(1+\mu)f^2y-\gamma_1x'+x-\beta x^3=0 \tag{4.68}$$

where the following constants are defined

$$\omega_1^2=k_1/m_1;\ \omega_2^2=k_2/m_2;\ \mu=m_2/m_1;\ f=\omega_2/\omega_1;\ \omega=\Omega/\omega_1;\ p=W/\omega_1^2 \tag{4.69}$$

$$\begin{aligned}&\gamma_1=\left(d_1+\varepsilon^2\right)/m_1\omega_1;\ \gamma=d_1/m_1\omega_1;\ \varepsilon^2=\psi^2/R;\\&\gamma_2=d_2/m_2\omega_2;\ \beta=k_3/k_1;\ \tau=\omega_1t\end{aligned} \tag{4.70}$$

and ( )′ indicates a derivative with respect to non-dimensional time τ.

4.5.2 Analytical Formulation by the Harmonic Balance Method: 1-Term Prediction

The dynamics of the two-DOF harvester which are proportional to the harmonic excitation frequency may be approximated using fundamental (1-term) Fourier series expansions for the displacements of each body. This strategy has limitations, as will be uncovered below, but it sheds first light on the overall dynamics behaviors activated by the auxiliary linear oscillator on the underlying bistable harvester and usefully leads to the direct analytical solution to the resulting modulation equation. Therefore, the 1-term harmonic balance response solutions assumed here are

$$x(\tau)=c_1(\tau)+a_1(\tau)\sin\omega\tau+b_1(\tau)\cos\omega\tau \tag{4.71}$$

$$y(\tau)=g_1(\tau)\sin\omega\tau+h_1(\tau)\cos\omega\tau \tag{4.72}$$

where c_1 is required due to the possibility that the bistable oscillator may exhibit intrawell oscillations which have a non-zero displacement bias, according to the potential energy profile of Eq. (4.63). The attached linear oscillator does not have a constant term in its respective fundamental Fourier series expansion due to the notation that y represents a relative coordinate. Then, Eqs (4.71) and (4.72) are substituted into (4.67) and (4.68), the higher-order terms are dropped, and coefficients of the constant, $\sin\omega\tau$, and $\cos\omega\tau$ terms are collected together. Due to the independence of these terms, solution of the resulting set of equations requires that the coefficients are independently zero, which yields five, algebraic governing equations for the slow-varying coefficients c_1, a_1, b_1, g_1, and h_1.

$$-\gamma_1 c_1' = \Lambda_c c_1 \tag{4.73}$$

$$-\gamma_1 a_1' + 2\omega b_1' + \mu f\gamma_2 g_1' = \left(\Lambda-\omega^2\right)a_1 - \omega\gamma_1 b_1 - \mu f^2 g_1 + \mu f\omega\gamma_2 h_1 \tag{4.74}$$

$$-2\omega a_1' - \gamma_1 b_1' + \mu f\gamma_2 h_1' = \omega\gamma_1 a_1 + \left(\Lambda-\omega^2\right)b_1 - \mu f\omega\gamma_2 g_1 - \mu f^2 h_1 - p \tag{4.75}$$

$$\gamma_1 a_1' - (1+\mu) f\gamma_2 g_1' + 2\omega h_1' = -\Lambda a_1 + \omega\gamma_1 b_1 + \Sigma g_1 - \sigma h_1 \tag{4.76}$$

$$\gamma_1 b_1' - 2\omega g_1' - (1+\mu) f\gamma_2 h_1' = -\omega\gamma_1 a_1 - \Lambda b_1 + \sigma g_1 + \Sigma h_1 \tag{4.77}$$

The following variables are defined

$$\Lambda_c = -1+\beta\left(c_1^2+\frac{3}{2}r_1^2\right) \tag{4.78}$$

$$\Lambda = -1+\beta\left(3c_1^2+\frac{3}{4}r_1^2\right) \tag{4.79}$$

$$\Sigma = (1+\mu)f^2-\omega^2 \tag{4.80}$$

$$\sigma = (1+\mu)f\omega\gamma_2 \tag{4.81}$$

$$r_1^2 = a_1^2 + b_1^2 \tag{4.82}$$

$$r_2^2 = g_1^2 + h_1^2 \tag{4.83}$$

Steady-state responses are determined by solving the coupled Eqs (4.73)–(4.77) after eliminating the time-varying terms. A first result is that Eqs (4.76) and (4.77) are solved explicitly in terms of a_1 and b_1 to determine coefficients g_1 and h_1.

$$g_1 = \frac{\Sigma\Lambda + \sigma\omega\gamma_1}{\Sigma^2 + \sigma^2} a_1 + \frac{\sigma\Lambda - \Sigma\omega\gamma_1}{\Sigma^2 + \sigma^2} b_1 \tag{4.84}$$

$$h_1 = -\frac{\sigma\Lambda - \Sigma\omega\gamma_1}{\Sigma^2 + \sigma^2} a_1 + \frac{\Sigma\Lambda + \sigma\omega\gamma_1}{\Sigma^2 + \sigma^2} b_1 \tag{4.85}$$

Then, Eqs (4.84) and (4.85) are substituted into (4.74) and (4.75) which are thereafter squared and summed to yield one equation for the steady-state displacement amplitude squared r_1^2 of the bistable harvester:

$$p^2 = r_1^2\left[\left(\kappa\Lambda + \delta\omega\gamma_1 - \omega^2\right)^2 + \left(\sigma\Lambda - \kappa\omega\gamma_1\right)^2\right] \tag{4.86}$$

given the following terms

$$\delta = \omega\gamma_2\Sigma\Theta - f\sigma\Theta \tag{4.87}$$

$$\kappa = 1 - f\Sigma\Theta - \omega\gamma_2\sigma\Theta \tag{4.88}$$

$$\Theta = \frac{\mu f}{\Sigma^2 + \sigma^2} \tag{4.89}$$

It is seen that the variable Λ contains the unknowns c_1 and r_1^2. Therefore Eq. (4.73) is first solved under steady-state conditions and a choice of the parameter c_1 is selected, based upon whether one considers the dynamics to be related to the intrawell oscillations of the bistable oscillator $(c_1 \neq 0)$ or related to the interwell vibrations $(c_1 = 0)$. Then, c_1 is substituted into Λ such that Eq. (4.86) is expressed fully in terms of a cubic polynomial of parameter r_1^2. The roots of Eq. (4.86) are then determined and the final coefficients are computed from their expressions as determined by simultaneously solving Eqs (4.74) and (4.75) with (4.84) and (4.85) inserted.

$$a_1 = \frac{-\delta\Lambda + \kappa\omega\gamma_1}{\left(\kappa\Lambda + \delta\omega\gamma_1 - \omega^2\right)^2 + \left(\delta\Lambda - \kappa\omega\gamma_1\right)^2} p \tag{4.90}$$

$$b_1 = \frac{\kappa\Lambda + \delta\omega\gamma_1 - \omega^2}{\left(\kappa\Lambda + \delta\omega\gamma_1 - \omega^2\right)^2 + \left(\delta\Lambda - \kappa\omega\gamma_1\right)^2} p \tag{4.91}$$

Having computed all five of the coefficients, the stability of the solutions r_1^2 may be determined via a perturbation method, such as by the means detailed in Section 3.4.3.

To investigate the phase relationships amongst the dynamic elements, the system responses are expressed alternatively by

$$x = c_1 + r_1\cos(\omega\tau - \phi_1) \tag{4.92}$$

$$y = r_2\cos(\omega\tau - \phi_2) \tag{4.93}$$

$$\tan\phi_1 = a_1/b_1;\ \tan\phi_2 = g_1/h_1 \tag{4.94}$$

where ϕ_1 and ϕ_2 are phase lags between the base excitation and the bistable harvester and linear oscillator displacements, respectively.

4.5.3 Analytical Formulation by the Harmonic Balance Method: 2-Term Prediction

The advantage of the 1-term assumed solution is the direct analytic expression – the cubic polynomial given by Eq. (4.86) – used to determine the responses. But as was discovered in Section 4.3, even a one-dimensional bistable energy harvester diffuses the vibrational energy into integer and fractional order harmonics of the driving frequency. In light of the fact that multiple DOF nonlinear systems are recognized to readily exhibit multiharmonic dynamics [90,131,132], it is logical to anticipate that other-order harmonics may be generated within the collective dynamical responses of the energy harvesting system composed of the bistable harvester and auxiliary linear oscillator attachment. The importance of the third-order superharmonic was previously examined in Section 4.3 due to its considerable dynamic contribution, and this particular energy diffusion feature is therefore accounted for in the 2-term Fourier series expansion of the assumed response solutions in the following analytical formulation.

$$x(\tau) = c_1(\tau) + a_1(\tau)\sin\omega\tau + b_1(\tau)\cos\omega\tau + a_3(\tau)\sin 3\omega\tau + b_3(\tau)\cos 3\omega\tau \tag{4.95}$$

$$y(\tau) = g_1(\tau)\sin\omega\tau + h_1(\tau)\cos\omega\tau + g_3(\tau)\sin 3\omega\tau + h_3(\tau)\cos 3\omega\tau \tag{4.96}$$

Equations (4.95) and (4.96) are substituted into (4.67) and (4.68) and coefficients of the constant, $\sin\omega\tau$, $\cos\omega\tau$, $\sin 3\omega\tau$, and $\cos 3\omega\tau$ terms are collected.

$$-\gamma_1 c_1' = \Lambda_s c_1 \tag{4.97}$$

$$\begin{aligned} -\gamma_1 a_1' + 2\omega b_1' + \mu f \gamma_2 g_1' = {} & \left(\Lambda_1 - \omega^2\right) a_1 - \gamma_1 \omega b_1 + \frac{3}{4}\beta\left(-a_1^2 + b_1^2\right) a_3 \\ & - \frac{3}{2}\beta\left(a_1 b_1\right) b_3 - \mu f^2 g_1 + \mu f \gamma_2 \omega h_1 \end{aligned} \tag{4.98}$$

$$\begin{aligned} -2\omega a_1' - \gamma_1 b_1' + \mu f \gamma_2 h_1' = {} & \gamma_1 \omega a_1 + \left(\Lambda_1 - \omega^2\right) b_1 + \frac{3}{2}\beta\left(a_1 b_1\right) a_3 \\ & + \frac{3}{4}\beta\left(-a_1^2 + b_1^2\right) b_3 - \mu f \gamma_2 \omega g_1 - \mu f^2 h_1 - p \end{aligned} \tag{4.99}$$

$$\begin{aligned} -\gamma_1 a_3' + 6\omega b_3' + \mu f \gamma_2 g_3' = {} & \frac{1}{4}\beta\left(-a_1^2 + 3b_1^2\right) a_1 + \left(\Lambda_3 - 9\omega^2\right) a_3 \\ & - 3\gamma_1 \omega b_3 - \mu f^2 g_3 + 3\mu f \gamma_2 \omega h_2 \end{aligned} \tag{4.100}$$

$$\begin{aligned} -6\omega a_3' - \gamma_1 b_3' + \mu f \gamma_2 h_3' = {} & \frac{1}{4}\beta\left(-3a_1^2 + b_1^2\right) b_1 + 3\gamma_1 \omega a_3 \\ & + \left(\Lambda_3 - 9\omega^2\right) b_3 - 3\mu f \gamma_2 \omega g_3 - \mu f^2 h_3 \end{aligned} \tag{4.101}$$

$$\gamma_1 a_1' - (1+\mu) f\gamma_2 g_1' + 2\omega h_1' = -\Lambda_1 a_1 + \gamma_1 \omega b_1 - \frac{3}{4}\beta\left(-a_1^2 + b_1^2\right)a_3 + \frac{3}{2}\beta\left(a_1 b_1\right)b_3 + \Psi_1 g_1 - \Psi_2 h_1 \quad (4.102)$$

$$\gamma_1 b_1' - 2\omega g_1' + (1+\mu) f\gamma_2 h_1' = -\gamma_1 \omega a_1 - \Lambda_1 b_1 - \frac{3}{2}\beta\left(a_1 b_1\right)a_3 - \frac{3}{4}\beta\left(-a_1^2 + b_1^2\right)b_3 + \Psi_2 g_1 + \Psi_1 h_1 \quad (4.103)$$

$$\gamma_1 a_3' + (1+\mu) f\gamma_2 g_3' + 6\omega h_3' = -\frac{1}{4}\beta\left(-a_1^2 + 3b_1^2\right)a_1 - \Lambda_3 a_3 + 3\gamma_1 \omega b_3 + \Psi_{1s} g_3 - \Psi_{2s} h_3 \quad (4.104)$$

$$\gamma_1 b_3' - 6\omega g_3' + (1+\mu) f\gamma_2 h_3' = -\frac{1}{4}\beta\left(-3a_1^2 + b_1^2\right)b_1 - 3\gamma_1 \omega a_3 - \Lambda_3 b_3 + \Psi_{2s} g_3 + \Psi_{1s} h_3 \quad (4.105)$$

where the following are defined

$$\Lambda_s = -1 + \beta\left(c_1^2 + \frac{3}{2}r_1^2 + \frac{3}{2}r_3^2\right) \quad (4.106)$$

$$\Lambda_1 = -1 + \beta\left(3c_1^2 + \frac{3}{4}r_1^2 + \frac{3}{2}r_3^2\right) \quad (4.107)$$

$$\Lambda_3 = -1 + \beta\left(3c_1^2 + \frac{3}{4}r_3^2 + \frac{3}{2}r_1^2\right) \quad (4.108)$$

$$r_1^2 = a_1^2 + b_1^2;\ r_3^2 = a_3^2 + b_3^2 \quad (4.109)$$

$$\Psi_1 = (1+\mu) f^2 - \omega^2 \quad (4.110)$$

$$\Psi_2 = (1+\mu) f\gamma_2 \omega \quad (4.111)$$

$$\Psi_{1s} = (1+\mu) f^2 - 9\omega^2 \quad (4.112)$$

$$\Psi_{2s} = (1+\mu) 3 f\gamma_2 \omega \quad (4.113)$$

Following the same procedure outlined in Section 4.5.2, the nine equations (4.97)–(4.105) are solved considering steady-state conditions. In this case, ultimately two, nonlinearly-coupled equations are determined with respect to the squared displacement amplitudes of the fundamental and third-order superharmonic motions of the bistable harvester, r_1^2 and r_3^2, respectively. Equations (4.114) and (4.115) are the two, coupled response equations.

$$256\Delta^2 p^2 = r_1^2 K \quad (4.114)$$

$$16\Delta^2 r_3^2 = \beta^2\left(\Gamma_1^2 + \Gamma_2^2\right)r_1^6 \quad (4.115)$$

where K, Δ, Γ_1, and Γ_2 are provided below.

$$
\begin{aligned}
K &= 9\beta^4 r_1^8\left(\Gamma_1^2+\Gamma_2^2\right)\left[\left(\Psi_1-\mu f^2\right)^2+\left(\Psi_2-\mu f\gamma_2\omega\right)^2\right] \\
&+96\Delta\beta^2 r_1^4\left\{\begin{array}{l}\left(\gamma_1\omega\Gamma_2-\Lambda_1\Gamma_1\right)\left[\left(\Psi_1-\mu f^2\right)^2+\left(\Psi_2-\mu f\gamma_2\omega\right)^2\right] \\ +\left(-\omega^2\right)\left[\mu f\gamma_2\omega\left(\Gamma_2\Psi_1+\Gamma_1\Psi_2\right)\right. \\ \left.+\mu f^2\left(\Gamma_1\Psi_1-\Gamma_2\Psi_2\right)-\Gamma_1\left(\Psi_1^2+\Psi_2^2\right)\right]\end{array}\right\} \\
&+256\Delta^2\left\{\begin{array}{l}\left(\Psi_1^2+\Psi_2^2\right)\left(\Lambda_1-\omega^2\right)^2+\left(\gamma_1\omega\right)^2\left[\left(\Psi_1-\mu f^2\right)^2\right. \\ \left.+\left(\Psi_2-\mu f\gamma_2\omega\right)^2\right] \\ +\mu f^2\left(\mu f^2\Lambda_1^2-\left(-\omega^2\right)2\Psi_2\gamma_1\omega\right) \\ -2\Lambda_1\left(\mu f^2\Psi_1+\mu f\gamma_2\omega\Psi_2\right)\left(\Lambda_1-\omega^2\right) \\ +\left(\mu f\gamma_2\omega\right)^2\Lambda_1^2-2\mu f\gamma_1\gamma_2\Psi_1\left(-\omega^2\right)^2\end{array}\right\}
\end{aligned}
\tag{4.116}
$$

$$
\begin{aligned}
\Delta &= \left[\left(\Lambda_3-9\omega^2\right)\left(f^2-9\omega^2\right)+\left(\mu f^2\right)\left(-9\omega^2\right)\right]^2 \\
&+\left(3f\gamma_2\omega\right)^2\left[\left(3\gamma_1\omega\right)^2+\left(\Lambda_3-(1+\mu)9\omega^2\right)^2\right] \\
&+\left[\left(3\gamma_1\omega\right)\left(f^2-9\omega^2\right)\right]^2+2\mu\left(3\gamma_1\omega\right)\left(3f\gamma_2\omega\right)\left(-9\omega^2\right)^2
\end{aligned}
\tag{4.117}
$$

$$
\begin{aligned}
\Gamma_1 &= \left(\Lambda_3-9\omega^2\right)\left(f^2-9\omega^2\right)^2+\left(\mu f^2\right)\left(-9\omega^2\right)\left(f^2-9\omega^2\right) \\
&+\left(\Lambda_3-(1+\mu)9\omega^2\right)\left(3f\gamma_2\omega\right)^2
\end{aligned}
\tag{4.118}
$$

$$
\Gamma_2 = \left(3\gamma_1\omega\right)\left(f^2-9\omega^2\right)^2+\left(3\gamma_1\omega\right)\left(3f\gamma_2\omega\right)^2+\mu\left(3f\gamma_2\omega\right)\left(-9\omega^2\right)^2 \tag{4.119}
$$

The numerical procedure employed in Section 4.3 to solve the two, nonlinear equations for the superharmonic dynamics of the individual bistable energy harvester is also utilized here to solve Eqs (4.114) and (4.115). Having solved the equations for a given set of system and excitation parameters, the individual coefficients are computed from the expressions provided below, and the stability of each given coefficient solution set is evaluated via a perturbation method as in Section 4.5.2.

$$
a_1 = \frac{16\Delta p}{K}\left\{\begin{array}{l}3\beta^2 r_1^4\left[\left(\Psi_1^2+\Psi_2^2\right)\Gamma_2-\mu f^2\left(\Gamma_2\Psi_1+\Gamma_1\Psi_2\right)\right. \\ \left.-\mu f\gamma_2\omega\left(-\Gamma_1\Psi_1+\Gamma_2\Psi_2\right)\right] \\ +16\Delta\left[\left(\Psi_1^2+\Psi_2^2\right)\gamma_1\omega-\mu f^2\left(\gamma_1\omega\Psi_1-\Lambda_1\Psi_2\right)\right. \\ \left.-\mu f\gamma_2\omega\left(\Lambda_1\Psi_1+\gamma_1\omega\Psi_2\right)\right]\end{array}\right\}
\tag{4.120}
$$

$$b_1 = \frac{16\Delta p}{K}\left\{\begin{array}{l} 3\beta^2 r_1^4\left[\left(\Psi_1^2+\Psi_2^2\right)\Gamma_1 - \mu f^2\left(\Gamma_1\Psi_1 - \Gamma_2\Psi_2\right)\right. \\ \left. -\mu f\gamma_2\omega\left(\Gamma_2\Psi_1+\Gamma_1\Psi_2\right)\right] \\ +16\Delta\left[-\left(\Psi_1^2+\Psi_2^2\right)\left(\Lambda_1-\omega^2\right) - \mu f^2\left(\Lambda_1\Psi_1+\gamma_1\omega\Psi_2\right)\right. \\ \left. -\mu f\gamma_2\omega\left(\gamma_1\omega\Psi_1 - \Lambda_1\Psi_2\right)\right] \end{array}\right\} \tag{4.121}$$

$$a_3 = -\frac{1}{4\Delta}\beta\left[\Gamma_1\left(-a_1^2+3b_1^2\right)a_1 + \Gamma_2\left(-3a_1^2+b_1^2\right)b_1\right] \tag{4.122}$$

$$b_3 = \frac{1}{4\Delta}\beta\left[\Gamma_2\left(-a_1^2+3b_1^2\right)a_1 - \Gamma_1\left(-3a_1^2+b_1^2\right)b_1\right] \tag{4.123}$$

$$g_1 = \frac{1}{\Psi_1^2+\Psi_2^2}\left\{\begin{array}{l} \left(\Lambda_1\Psi_1+\gamma_1\omega\Psi_2\right)a_1 + \left(\Lambda_1\Psi_2 - \gamma_1\omega\Psi_1\right)b_1 \\ -\frac{3}{16\Delta}\beta^2 r_1^4\left[\left(\Gamma_1\Psi_1-\Gamma_2\Psi_2\right)a_1 + \left(\Gamma_2\Psi_1+\Gamma_1\Psi_2\right)b_1\right] \end{array}\right\} \tag{4.124}$$

$$h_1 = \frac{1}{\Psi_1^2+\Psi_2^2}\left\{\begin{array}{l} \left(-\Lambda_1\Psi_2+\gamma_1\omega\Psi_1\right)a_1 + \left(\Lambda_1\Psi_1 + \gamma_1\omega\Psi_2\right)b_1 \\ +\frac{3}{16\Delta}\beta^2 r_1^4\left[\left(\Gamma_2\Psi_1+\Gamma_1\Psi_2\right)a_1 - \left(\Gamma_1\Psi_1-\Gamma_2\Psi_2\right)b_1\right] \end{array}\right\} \tag{4.125}$$

$$g_3 = \frac{1}{\Psi_{1s}^2+\Psi_{2s}^2}\left\{\begin{array}{l} \left[\frac{1}{4}\beta\left(-a_1^2+3b_1^2\right)a_1 + \Lambda_3 a_3 - 3\gamma_1\omega b_3\right]\Psi_{1s} \\ +\left[\frac{1}{4}\beta\left(-3a_1^2+b_1^2\right)b_1 + 3\gamma_1\omega a_3 + \Lambda_3 b_3\right]\Psi_{2s} \end{array}\right\} \tag{4.126}$$

$$h_3 = \frac{1}{\Psi_{1s}^2+\Psi_{2s}^2}\left\{\begin{array}{l} -\left[\frac{1}{4}\beta\left(-a_1^2+3b_1^2\right)a_1 + \Lambda_3 a_3 - 3\gamma_1\omega b_3\right]\Psi_{2s} \\ +\left[\frac{1}{4}\beta\left(-3a_1^2+b_1^2\right)b_1 + 3\gamma_1\omega a_3 + \Lambda_3 b_3\right]\Psi_{1s} \end{array}\right\} \tag{4.127}$$

4.5.4 Computing the Fundamental and Superharmonic Average Power and Power Density

Average power across the harvesting circuit resistive load R is computed from $P = R|i|^2/2$. By Eq. (4.66), which relates the current flow to the bistable harvester inertial mass velocity, and according to the solution forms Eqs (4.71) and (4.95), the amplitude of the generated power may be expressed as a sum of the respective harmonic displacement amplitudes, $P = P_1 + P_3$, where $P_1 \propto r_1^2$ and $P_3 \propto r_3^2$. This practically represents the guiding assumptions of the method of harmonic balance that the total system response is the summation of unique harmonic contributions [127]. Consequently, depending on whether one considers the fundamental or superharmonic response magnitude, expressed in Eq. (4.109), unique spectral components of harvested power are computed. To simplify the resulting average power expressions, the terms of the bistable harvester mass m_1

and linear natural frequency ω_1 are set to unity. Then, the average, generated powers related to the fundamental and third-order superharmonic are computed, respectively, from Eqs (4.128) and (4.129).

$$P_1 = (\omega \varepsilon r_1)^2 / 2 \tag{4.128}$$

$$P_3 = (3\omega \varepsilon r_3)^2 / 2 \tag{4.129}$$

In the following analytical investigations, the parameters $\gamma = 0.01$ and $\varepsilon^2 = 0.04$ are used to respectively represent a bistable harvester with mild mechanical viscous losses and mild electromagnetic coupling strength typical of many inductive platforms [77,83]. It is evident from Eqs (4.128) and (4.129) that the average power is directly proportional to the squared displacement amplitudes of the bistable energy harvester harmonics. In terms of evaluating the viability of the auxiliary linear oscillator as relates to the total energy harvesting *system*, a more meaningful assessment may be made by considering the *average power density*, which is the average power generated by each harmonic motion normalized to the net system mass. The following investigations therefore consider the power density, $P_{1,3}/(1+\mu)$, as the primary metric to consider to weigh the relative advantages and disadvantages of the auxiliary linear oscillator as a means to enhance the energy harvesting capability of the total system.

4.5.5 Roles of the Auxiliary Linear Oscillator

The coupling of lumped and distributed linear energy harvesting platforms was seen to be particularly enhanced when a simple linear oscillator was attached to the free-displacing mass of the host harvester [123–125]. Using the 1-term harmonic balance formulation, a first examination considers the roles of the auxiliary linear oscillator on the dynamics and power density of the bistable energy harvester. Here, the remaining system and excitation parameters are taken to be $p = 0.2$, $\beta = 1$, $f = 1$, $\gamma_2 = 0.001$, and considering a range of mass ratios, $\mu = [0.003, 0.01, 0.1, 0.3, 1, 2]$. These values are selected to be reflective of typical bistable energy harvester designs and implementation scenarios. For instance, the strength of the nonlinearity β may realistically be varied over a large range of values by tailoring the severity of the post-buckled state, such as was examined in Section 4.4. The amplitude of the harmonic excitation p represents a "mild" level with respect to the normalized definition in Eq. (4.69), such as was similarly found in Section 4.3. The damping ratio of the auxiliary attachment γ_2 is set at a reasonably low and achievable value (e.g., a music wire coil spring and mass, or spring–steel beam cantilever). The frequency, or tuning, ratio f can easily be adjusted across a wide range of values by changing the net harvester restoring stiffnesses or by changing the auxiliary attachment properties; however, only the latter directly adjusts the frequency ratio since the prior would tend to also tailor the nonlinearity strength parameter β. As has been a recurring theme from the prior studies of this chapter, the greatest benefit derived from bistable energy harvesters is when the snap-through oscillations are triggered. Consequently, the following plots of this section omit the harvester dynamics and generated

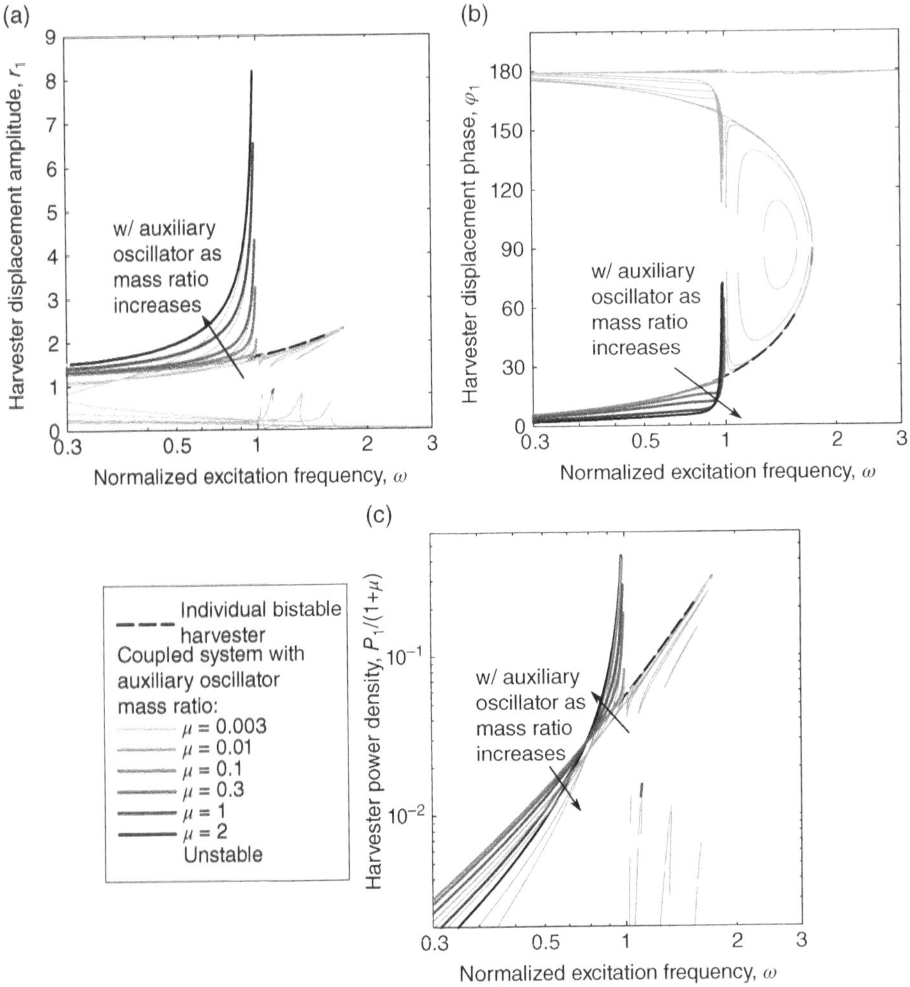

Figure 4.22 Analytically predicted responses of the bistable harvester as the mass ratio μ and excitation frequency ω are varied, corresponding to the snap-through dynamic regime. (a) Fundamental displacement amplitude r_1 and (b) phase lag ϕ_1 (degrees). (c) Average power density determined from $P_1/(1+\mu)$.

powers corresponding to the intrawell oscillation regime. In cases for which the large-amplitude interwell vibrations are not predicted by the analysis, the designer may be warned that the operating regime has vulnerability to the loss of the advantageous snap-through motions.

Using the above parameters, the displacement amplitude, phase, and the power density of the bistable inertial generator with and without the attached linear oscillator are plotted in Figure 4.22. The responses of the individual bistable harvester are computed for comparison by setting the attachment mass ratio to zero, $\mu = 0$, which reduces the polynomial in Eq. (4.86) to the expression given in Eq. (4.4) for the individual bistable harvester; these results are shown by the

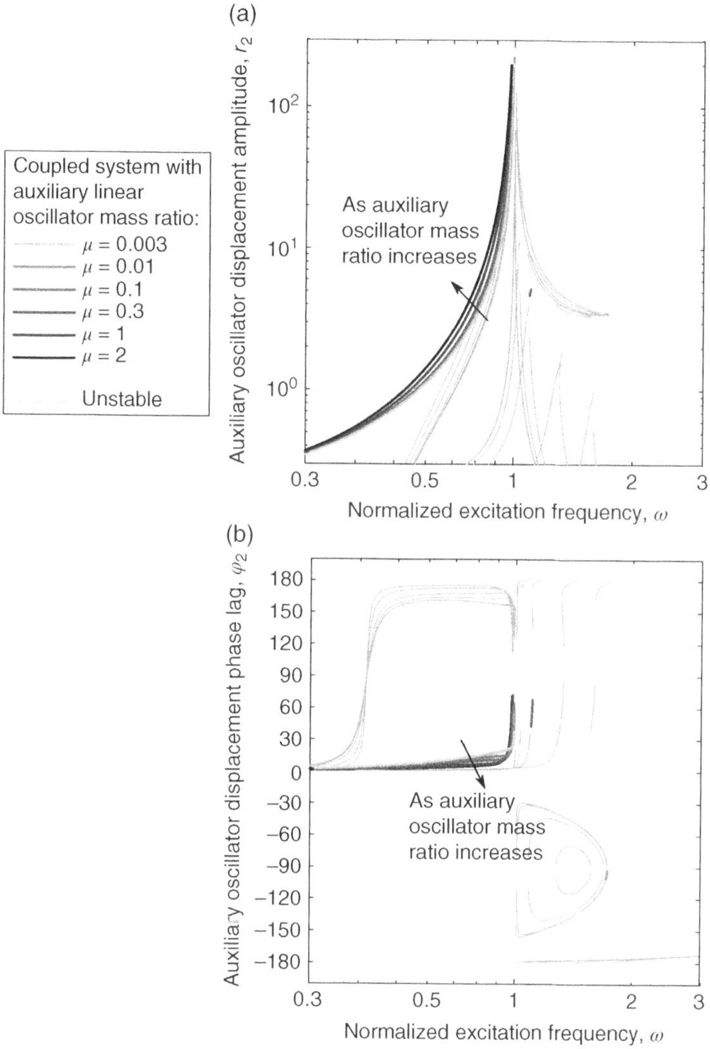

Figure 4.23 Analytically predicted responses of the auxiliary linear harvester as the mass ratio μ and excitation frequency ω are varied, corresponding to the snap-through dynamic regime of the harvesting system. (a) Fundamental displacement amplitude r_2 and (b) phase lag ϕ_2 (degrees).

dashed curves in Figure 4.22. For the coupled system, the displacement magnitude and phase lag of the linear oscillator are presented in Figure 4.23. The results of Figure 4.22 show that for the smallest mass ratios, the trends of the coupled energy harvesting system intuitively converge to those which occur when the individual bistable harvester is excited by the same harmonic base acceleration. As the attachment mass is increased, the displacement of the harvester snap-through oscillations is uniformly amplified so long as the excitation frequency remains less than the natural frequency of the auxiliary linear oscillator. According to the frequency ratio used for these predictions, $f=1$, this indicates that the decoupled auxiliary oscillator natural frequency is at $\omega_2 = 1$ so that the amplification of the bistable harvester displacement amplitudes always occurs when the excitation frequency $\omega < 1$. Admittedly, such magnification of the harvester displacement is only realized so long as the snap-through motions persist. For example, the analytical predictions show a trade-off to such performance gain: the highest frequency of excitation at which snap-through occurs is reduced due to increase in the mass ratio of the auxiliary linear oscillator when considering this frequency ratio $f = 1$. Similarly, the greater mass of the auxiliary oscillator magnifies the vibrations of the attached device itself, as seen in Figure 4.23a.

Because the average power generated by the harvester is directly proportional to the displacement amplitude by Eq. (4.128), the average power increases in the same manner as the displacement amplitude in Figure 4.22a. In contrast, the results of Figure 4.22c show that the *power density* is enhanced, with respect to the individual bistable harvester, in a particular bandwidth of excitation frequencies. As seen in Figure 4.22c, an inexact inflection point appears to serve as a lower bound on the excitation frequency at which the power density increase occurs for increasing auxiliary oscillator mass. The highest frequency of the performance improvement is that at which the snap-through vibrations are maximally amplified, as shown in Figure 4.22a, in other words near to the linear oscillator natural frequency (here $\omega_2 = 1$, so that maximal power density improvement occurs as the excitation frequency approaches $\omega \rightarrow 1$). Since the power density enhancement describes the advantage of the coupled harvester platform when compared to the individual bistable energy harvester, increasing the power density is the genuine means to justifying the two-DOF harvester architecture.

There are important insights on the dynamics involved towards the performance enhancement to be derived by close consideration of the auxiliary linear oscillator responses plotted in Figure 4.23. Recall that in this example, the resonant excitation of the appended oscillator occurs when the excitation frequency is $\omega = 1$ (considering the decoupled natural frequency of the auxiliary oscillator). Even though the system is composed of two dynamic bodies which are individually governed by second-order differential equations, and assuming $\mu > 0$, the displacement amplitude and phase trends shown in Figure 4.23 suggest the system is a *single* DOF device excited at frequencies around resonance, excepting the frequencies at which no stable snap-through-related responses occur. As a result, it may be stated that from the perspective of the auxiliary linear oscillator, the snap-through dynamics of the bistable harvester are equivalent to a base excitation, albeit an equivalent excitation source that is vulnerable to the motions of the attachment. This interpretation is supported by several points, including the fact that snap-through is a non-resonant phenomenon and thus would not be expected to contribute to modal-type behaviors in a multi-DOF dynamic system. Once the linear attachment is excited above resonance, such that the excitation phase lag approaches 180°, the auxiliary oscillator vibrates primarily out-of-phase with respect to the bistable harvester, thus inhibiting its ability to maintain snap-through vibrations. The prevention to maintain snap-through appears to come about by two effects. First, the linear oscillator effectively acts as a "vibration absorber" when excited at frequencies above its decoupled natural frequency, since it vibrates mostly out-of-phase compared to the snap-through displacements of the harvester. Second, the bistable harvester is driven abruptly towards a 90° phase lag as the linear oscillator is excited towards resonance. As detailed in Section 4.2, sustainable snap-through vibrations necessitate no more than a 90° phase lag. From these observations, and from those obtained by considering the results of Figure 4.22, it is concluded that the mass ratio μ appears to govern the lower bound on the frequency bandwidth at which the enhancement of the average power density occurs while the frequency ratio f, corresponding to the decoupled resonant frequency of the auxiliary linear oscillator, appears to govern the upper frequency of this desirable performance-improving bandwidth.

To study the role of the frequency ratio, Figure 4.24 presents the displacement amplitude of the bistable harvester and power density generated by the system for the case in which the auxiliary oscillator mass ratio is $\mu = 1$. The values of the frequency ratio are taken as irrational numbers ($f = \pi / i$ where $i = 1, 2, \ldots, 6$) to avoid misleading results due to potential internal resonance phenomena [90]. As postulated in the preceding paragraph based on the observed dynamical trends, the results of Figure 4.24 demonstrate that the displacements of the harvester with the appended oscillator are uniformly amplified for excitation frequencies less than the appendage oscillator natural frequency. In fact, the frequency ratio may be tailored to *extend* the range of frequencies at which the harvester maintains snap-through motions when compared to the individual bistable harvester, such as that shown for the greatest selection of frequency ratio in Figure 4.24, $f = \pi$ (lightest shading of solid, thick curves). A comparable trend is observed in relation to the power density enhancement as found for the adjustment of the mass ratio: the improvement of power density is manifest in a finite bandwidth approximately upper-bounded in frequency by the linear oscillator natural frequency (if the excitation level is sufficient to sustain high-energy response to this frequency). These observations propose an initial framework for the designer seeking improvement upon the performance achieved with an individual bistable energy harvester. Enhanced power harvesting density may be obtained with the addition of the linear oscillator by *a priori* knowledge of the amplitude of the base excitation p (or a sufficient knowledge combined with a conservative margin of safety). Then, the frequency ratio and mass ratio are set by adjustment of the auxiliary oscillator parameters to tailor the frequency bandwidth of interest for the power density improvement as well as to meet design constraints, such as peak stroke amplitudes or absolute added mass limitations, while weighing such factors in relation to the desired power harvesting goals.

4.5.6 Roles of the Superharmonic Dynamics in the Energy Harvesting Performance

The influences of the auxiliary linear oscillator are found to substantially modulate the overall dynamic and power harvesting trends when compared to those realized by the individual bistable harvester. Taking the investigation further, this section evaluates the intricate roles of the appended oscillator with respect to the multiharmonic energy diffusion, in which case the energy diffuses to snap-through vibrations having contributions from the frequency proportional

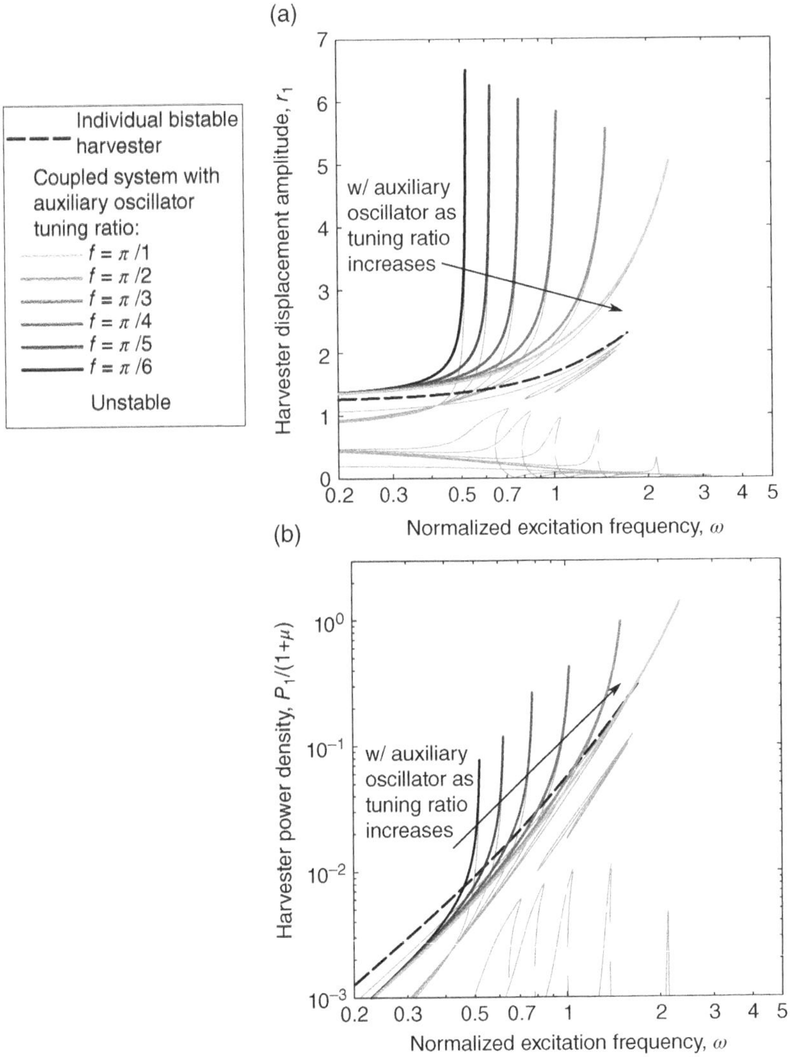

Figure 4.24 Analytically predicted responses of the bistable harvester as the frequency ratio f and excitation frequency ω are varied, corresponding to the snap-through dynamic regime. (a) Fundamental displacement amplitude r_1 and (b) average power density $P_1/(1+\mu)$.

to the excitation ω and from the frequency three times the excitation 3ω. The nonlinearly coupled Eqs (4.114) and (4.115) are solved using parameters $p=3$, $\beta=0.1$, $\mu=1$, and $f=1.67$ while the remaining parameters are the same as those employed in Section 4.5.5. The excitation amplitude $p=3$ is increased by approximately an order of magnitude from that level considered in Section 4.5.5 since greater excitation amplitudes are known to be one means of more readily inducing multiharmonic nonlinear phenomena [127]. Figure 4.25 presents the displacement amplitudes of the bistable harvester and the average powers generated, with respect to the fundamental and third-order superharmonic behaviors.

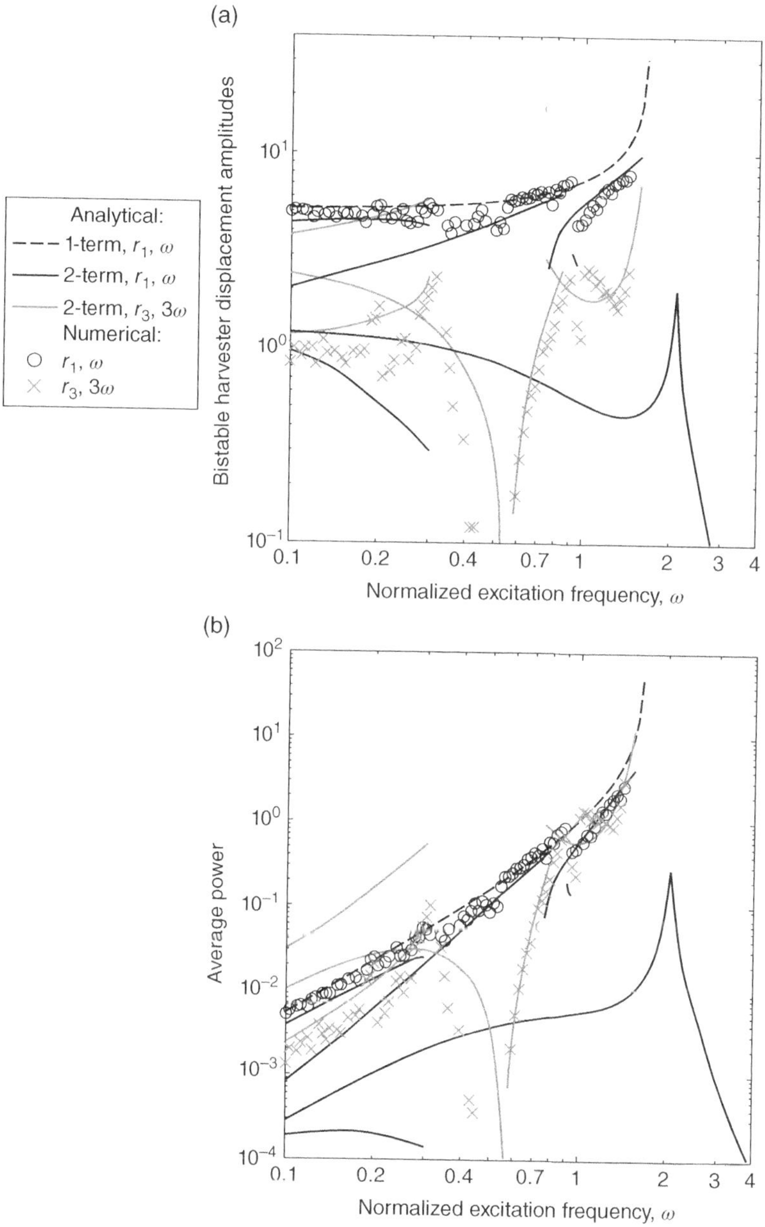

Figure 4.25 Comparison of numerically- and analytically-predicted snap-through (a) displacement amplitudes of the bistable harvester $r_{1,3}$ and (b) average powers $P_{1,3}$, considering contributions from the fundamental and third-order superharmonic. The auxiliary linear oscillator mass and frequency ratios are $\mu = 1$ and $f = 1.67$, respectively.

For each excitation frequency, 50 long-time simulations of governing Eqs (4.67) and (4.68) are carried out; then the fast Fourier transforms of the results are taken for the second half of the time series so as to extract the spectral components related to the fundamental and third-order superharmonics; the numerical data points shown in Figure 4.25 are the output of the simulation data to be compared with the analytical predictions. For comparison between the 1- and 2-term analytical predictions, the thick dashed curves in Figure 4.25 are the analytical results computed using the 1-term Fourier series solution expansion which cannot take into account the multiharmonic energy diffusion characteristics.

Across the full bandwidth of excitation frequencies considered, the third-order dynamics are shown to play important total roles in the resulting system behaviors. This is evident in Figure 4.25 based on the notable differences between the 1- and 2-term analytical predictions of the fundamental harmonic motions (those related to ω). Moreover, the numerical data points exhibit considerably greater agreement with the 2-term analytical results than with the 1-term estimations. When the excitation frequency is less than approximately $\omega \approx 0.3$, the 2-term formulation predicts several fundamental and superharmonic response pairs are possible. However, the numerical simulations identified only one predominant pair of the dynamics in this frequency regime which suggests that the basins of attraction for the other steady-state dynamics are appreciably small. Near to the frequency $\omega \approx 0.3$, a sudden increase in the harvested power proportional to the superharmonic 3ω is predicted to occur in Figure 4.25b; the fundamental harmonic motions correspondingly reduce with the onset of this saddle-node bifurcation. This phenomenon is comparable to the discoveries in Section 4.3 that the third-order superharmonic contributions may heighten bistable harvester sensitivity and thereby yield enhanced performance in this low frequency regime near $\omega \approx 0.3$. The numerical findings plotted in Figure 4.25 corroborate the respective analytical predictions in these regards. In the bandwidth around $\omega \approx 0.4$ to $\omega \approx 0.7$, the 2-term analytical predictions suggest that the bulk of the oscillation energy is captured in the fundamental harmonic motions, such that the superharmonic contribution is negligibly small. Likewise, this prediction is plainly observed in the numerical data for verification of the trend. Such a finding suggests an advantageous operating condition if the energy harvesting storage circuit is to switch at a well-defined frequency to best convert the amplified, transduced voltage obtained for the coupled harvester. Thus, for frequencies within $\omega \approx 0.4$ to $\omega \approx 0.7$, advanced energy harvesting circuits [92,133] may be set to operate at the fundamental harmonic ω without excess concern of converted energy loss due to other-harmonic diffusion.

Unlike the trends for the fundamental and superharmonic dynamics characterized in Section 4.3, a *second* sudden drop in the fundamental harmonic dynamics is analytically-predicted in Figure 4.25 as the excitation frequency approaches $\omega \approx 0.8$. The numerical simulation results are in excellent agreement qualitatively and quantitatively with this finding. At the bifurcation itself, just prior to the drop in the fundamental harmonic response, the average power proportional to the third-order superharmonic rises to a level comparable to that extracted proportional to the driving frequency ω. For frequencies above this saddle-node bifurcation, the fundamental and superharmonic power levels have similar levels. In the context of extracting the most energy via the available switching circuit technologies, the equi-distribution of the harvested energy at these higher excitation frequencies is unfavorable since a considerable proportion of energy is lost by switching at either the frequency ω or 3ω. In contrast, in the bandwidth of $\omega \approx 0.4$ to $\omega \approx 0.7$, the unambiguous concentration of the energy conversion within the fundamental harmonic dynamics indicates that the switching circuits will operate to great advantage. Thus, from the application perspective, the best utilization of the auxiliary linear oscillator is to tailor the mass and frequency ratios so that the power density enhancement for the bistable harvester system is realized within an excitation frequency regime where either the fundamental or the superharmonic dynamics predominate. If one alternatively wishes to usee the standard rectifying energy harvesting circuit, such a strategy does not dictate the operating regime of greatest performance benefit since strictly amplitude of the transduced voltage (and current) is enough to improve the storage circuit charging performance [92], regardless of the spectral distribution. Overall, the numerical data in Figure 4.25 show that the multiharmonic dynamics of the bistable energy harvester with auxiliary linear oscillator are non-trivial and ultimately govern the operating conditions under which enhancements in the average power generation are obtained, thus justifying the need for more accurate analytical tools such as the 2-term formulation presented here.

4.5.7 Experimental Investigations of the Multiharmonic Dynamics Enhancement via the Auxiliary Linear Oscillator

By the redistribution of the net vibrational energy between the fundamental and third-order harmonics and the amplification of the dynamics under certain instances, power generation benefits are realized for the bistable harvester by the simple design alteration effected via the auxiliary linear oscillator. To experimentally explore such features as uncovered by the numerical and analytical investigations, an equivalent lumped-parameter bistable device and linear oscillator system is fabricated in the laboratory. A photograph of the set-up is given in Figure 4.26. Recall that the model composition assumes the electrical

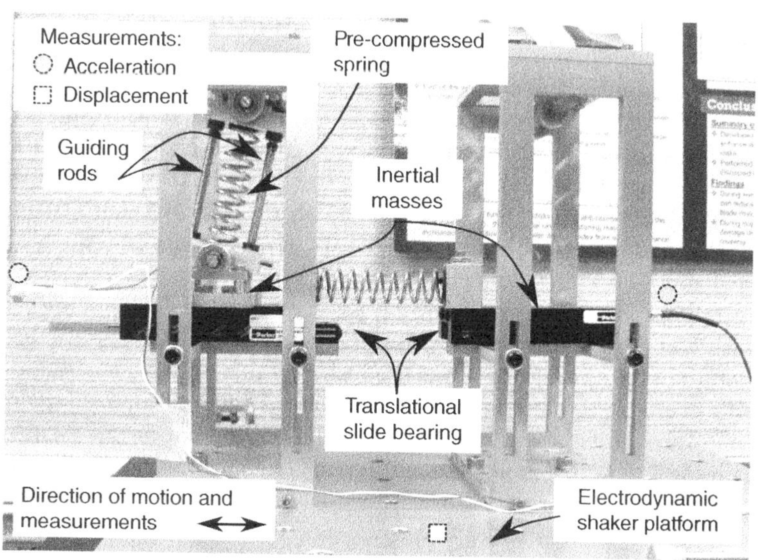

Figure 4.26 Experimental set-up of bistable device coupled to auxiliary linear oscillator.

time constant of the transduction mechanism is much smaller than the mechanical natural period, such that the exact electrodynamic influences may be effectively approximated as pure dissipation. This approximation enables the experimental system to be a strictly mechanical two-DOF device composed of the bistable and linear oscillators, as shown in Figure 4.26. The bistable device is constructed by attaching a translational slide bearing to a coil spring which is pre-compressed in a vertical, upright orientation. Guiding rods inhibit bending or twisting of the compressed coil spring during snap-through. The linear oscillator is realized by attaching a coil spring from the bistable device slide bearing to a separate slide bearing that is guided within a nearby frame (right in the figure). Here, two linear coupling springs are employed such that the resulting tuning ratios are $f_1 \approx 4$ and $f \approx 2$ given that the uncoupled bistable oscillator is observed to have a linear natural frequency $\omega_1/2\pi \approx 4$ Hz. The mass of the appended oscillator leads to a mass ratio of $\mu \approx 0.75$. The system is mounted on an electrodynamic shaker platform in a way that isolates the linear oscillator from the excitation, as is needed to replicate the analytical model schematic of Figure 4.21. Accelerometers are mounted on the inertial masses while a potentiometer measures the shaker displacements in the same horizontal axes of motion. To incorporate an electromechanical transduction mechanism to this architecture for future exploration, one approach is to add an inductive generator to the rotating pivot that connects the pre-compressed spring to the frame upon which the bistable device bearing mass moves. Such an electromechanical integration would convert the relative motion between the bistable harvester bearing mass and the exciting base motion to a flow of current, in agreement with the model framework.

The results of Section 4.5.5 indicated that the ratio between the linear natural frequencies of the attachment and host bistable harvester, parameter f, provides a means to tailor the excitation frequency bandwidth and extent of the dynamics enhancement, observed in terms of the displacement, power, and power density. To explore such important trends, the experimental platform is excited by base acceleration using slow frequency sweeps ($+0.15\,\mathrm{Hz\,s^{-1}}$) beginning at 2 Hz, which is the lowest feasible frequency of excitation at which the snap-through dynamics are achieved. The magnitude of the acceleration frequency response function (frf) for the bistable oscillator is plotted in Figure 4.27, where the frf is computed as the ratio of the bistable device acceleration to shaker input acceleration. Trials are conducted without the auxiliary linear oscillator (data plotted as crosses) and then with the attachment using the two cases of frequency ratio described above (square, f_1, and circle, $f_2 \Leftrightarrow f_2 < f_1$, data points).

The measurements in Figure 4.27 are those acquired for frequencies of excitation lasting from the start of the forward frequency sweep at 2 Hz until the point at which the snap-through motions are lost. The data give evidence that the bistable device with the auxiliary linear oscillator is able to realize larger amplitude snap-through vibrations than without the attachment for all excitation frequencies less than that at which the interwell dynamics are destabilized. This trend is in very good agreement with the analytical results discovered in Section 4.5.5. In addition, the smaller frequency ratio f_2 leads to a greater acceleration frf amplitude prior to the loss of snap-through, an event which moreover

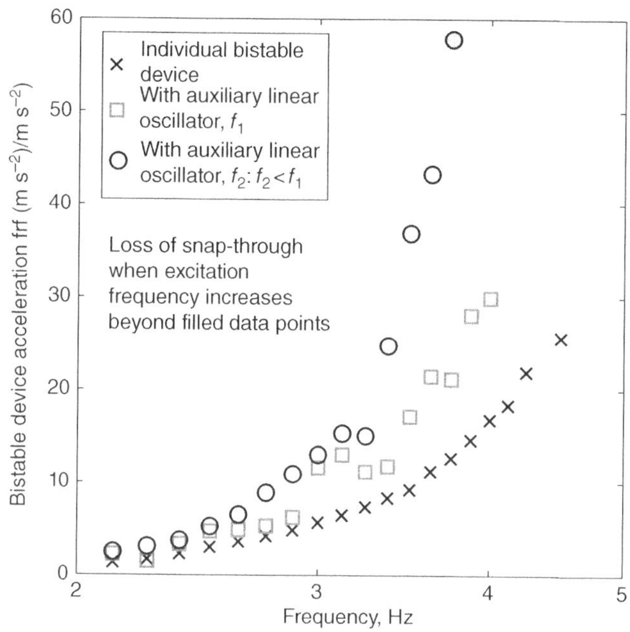

Figure 4.27 Snap-through acceleration frf amplitude of the bistable device with and without the linear oscillator attachment.

occurs at a lower excitation frequency for the smaller value of f. These results are also in good agreement with the analytical predictions shown in Figure 4.24. Such useful and satisfactory agreement in the qualitative nature of the dynamics of the strongly nonlinear two-DOF platform between analysis and experiment show the advantage of the 1-term analytical formulation in efficiently reconstructing important global dynamic behaviors.

During experimentation, coexisting dynamic regimes are found for certain constant amplitudes and frequencies of the base acceleration excitation, whether one considers the bistable device individually or when the linear oscillator is appended. When the excitation is at 4 Hz, the measurements shown in Figure 4.28 and Figure 4.29 exemplify the prevalence of the relative energy diffusion between the fundamental and third-order superharmonic frequencies. In Figure 4.28, the dynamics of the individual bistable device and of the coupled system (using frequency ratio $f_1 = 4$) yield the most significant energy concentration at the same frequency as the 4 Hz excitation. The acceleration autospectra in Figure 4.28a confirm this relative distribution of the vibrational energy, and that the next largest harmonic components are at least an order of magnitude less in amplitude than the fundamental. Figure 4.28a shows that the addition of the linear oscillator leads to a small diffusion of vibrational energy to the oscillator natural frequency, in this case around 4 times the excitation frequency due to the frequency ratio $f_1 = 4$. A portion of the acceleration time series, from which the autospectra are computed, is shown in Figure 4.28b; the data verify that the dynamics occur at a steady-state oscillation period which is the same as the exciting base acceleration.

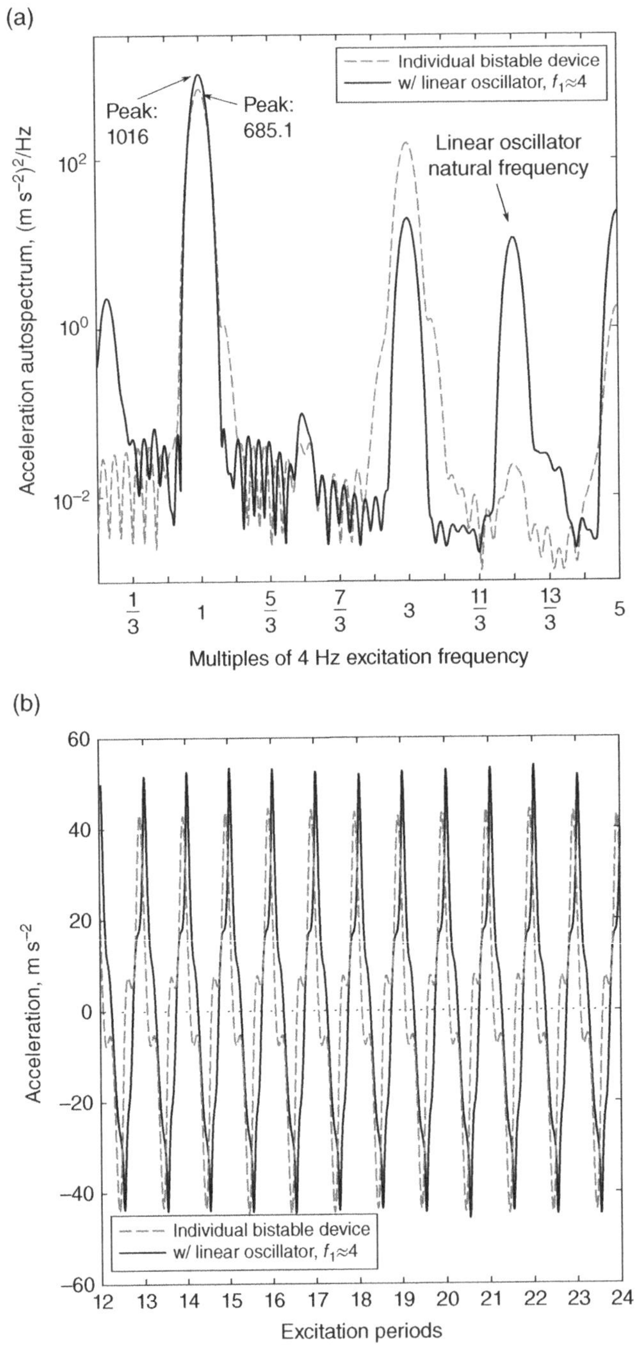

Figure 4.28 Snap-through due to base acceleration excitation at 4 Hz. (a) Acceleration autospectra and (b) time series for the dynamic regime dominated by motions proportional to the frequency of excitation.

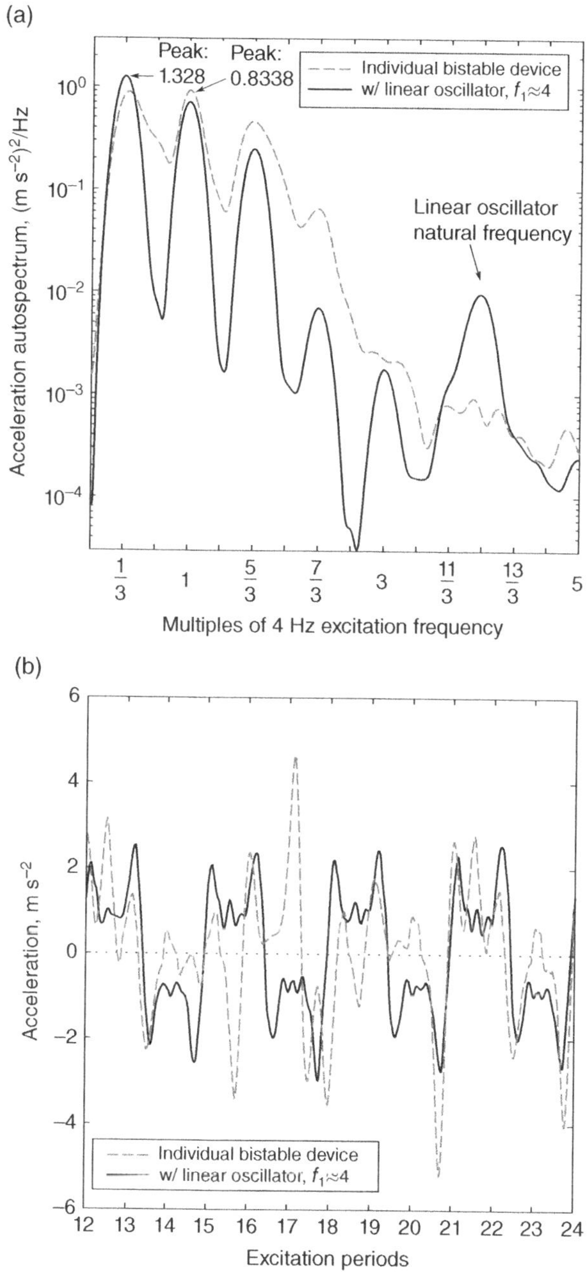

Figure 4.29 Snap-through due to base acceleration excitation at 4 Hz. (a) Acceleration autospectra and (b) time series for the dynamic regime dominated by motions proportional to the third-order superharmonic.

The alternative set of coexisting dynamic regimes, for the individual bistable device and for the coupled system, are shown in Figure 4.29. As has been described throughout this book, the alternative dynamic regime is obtained using different initial conditions. In these cases, the different initial conditions are effected via fast manual impact perturbations that transitioned the systems from the steady-state dynamic regimes shown in Figure 4.28 to those plotted in Figure 4.29, all the while the excitation amplitude and frequency remain unchanged. Regardless of whether or not the auxiliary linear oscillator is considered, the coexisting dynamic regime leads to a considerable redistribution of the vibration energy to sub- and superharmonics in the frequency spectrum. But, distinctively, the individual bistable oscillator exhibits a substantially greater aperiodicity in its dynamics, which is evident via the more broadly diffused autospectrum in Figure 4.29a and the time series in Figure 4.29b which shows poor consistency of the oscillations over many periods of excitation. In contrast to these trends, the bistable device with linear oscillator attachment exhibits far more concentration of the vibration energy to a frequency which is one-third of the excitation. Indeed, so much energy is diffused that the autospectrum in Figure 4.29a reveals 49.6% greater concentration of the energy at one-third of the excitation frequency than at the driving frequency itself. The multiharmonic susceptibilities observed in Figure 4.29 suggest a unique benefit of the coupled bistable–linear oscillator system. By capturing a periodic steady-state, even one not primarily concentrated at the excitation frequency itself, a coupled bistable–linear energy harvesting platform may realize a periodic dynamic regime that is well-suited for the desirable energy harvesting storage circuitry that employs switching regimes at a prescribed switching period [91,92,133]. Comparably, the aperiodic dynamics of the individual bistable harvester would result in poor energy capture and storage because the broadly diffused energy of the vibrations presents no ideal frequency at which the switching strategies may be carried out [43].

4.5.8 Summary

The recent discoveries that simple, lumped oscillator attachments may enhance the performance of a linear energy harvester motivate the consideration of a comparable strategy to amplify the utility of bistable energy harvesting platforms which operate in the desirable snap-through dynamic regime. At the same time, the incorporation of additional dynamic bodies to a strongly nonlinear system is well-known to exacerbate the possibility for multiharmonic vibrations. Taking these factors into close consideration, this study probed the numerous roles that an auxiliary linear oscillator plays towards the global electromechanical dynamics of a bistable vibration energy harvester. Through numerically- and experimentally-verified analytical modeling, it is seen that the fundamental snap-through dynamics are uniformly amplified in a bandwidth of frequencies less than the decoupled natural frequency of the attached oscillator, resulting in power density enhancement within this bandwidth. From the perspective of such an appendage, the snap-through behaviors of the bistable harvester are an equivalent base excitation, which leads to a feedback-like mechanism for system dynamics enhancement at frequencies less than the auxiliary oscillator's natural frequency and leads to a strong inhibition of snap-through dynamics above this frequency. Thus, for the purposes of improving energy harvesting performance, the utilization of the coupled system is evidently beneficial only when excited at frequencies less than the linear oscillator natural frequency. Furthermore, the diffusion of energy amongst other-harmonic behaviors is found to be an interesting benefit of the coupled bistable harvester and linear oscillator system, in contrast to the bistable harvester on its own. Due to the additional periodic, steady-state snap-through vibrations (having different harmonic component contributions) that occur for the coupled platform, a bistable energy harvester is less vulnerable to undergoing poor-performing aperiodic dynamics by the addition of the auxiliary linear oscillator. Thus, the snap-through dynamics of bistable harvesters may be facilitated and promoted using a strategically designed appendage oscillator, and the power density enhancement uncovered in this study provides strong justification for the design concept.

4.6 A Linear Dynamic Magnifier Approach to Bistable Energy Harvesting[6]

Whereas the investigations motivating the prior study explored linear energy harvesters with appendage linear oscillators [123–126], other researchers have recently considered designs which leverage a linear "dynamic magnifier" stage. The concept centers around an oscillatory interface between the energy harvester and the exciting base motion: attached to the excitation source is a linear oscillator to which the harvester is then attached. This dynamic magnifier design strategy is contrasted against the auxiliary oscillator approach in Figure 4.30

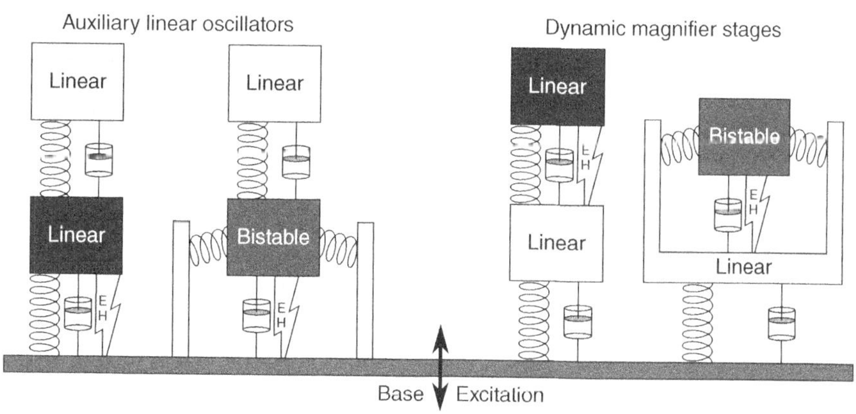

Figure 4.30 Comparison of linear and bistable energy harvesting system architectures which leverage auxiliary linear oscillators (left) compared to linear dynamic magnifier stages (right). The stage linearity is denoted by the text on the respective inertial mass. The components labeled by EH are the energy harvesting transduction mechanisms.

6 Material from this section has been reprinted by permission of ASME, from Z. Wu, R.L. Harne and K.W. Wang, Energy harvester synthesis via coupled linear-bistable system with multistable dynamics. Journal of Applied Mechanics 81:061005, 2014. doi: 10.1115/1.4026555.

in the context of employing a *bistable* energy harvester, unlike the linear harvesters previously considered [123–126]. Depending on the placement of the energy harvesting transduction mechanism, labeled as the "EH" components in the figure, the relative advantages of the linear device inclusions vary. For purely linear systems, the auxiliary oscillator approach is a lightweight means to realize closely-spaced resonances, broaden the effective frequency bandwidth of operation, and improve the power production [125]. The peak power generation of the dynamic magnifier approach can be substantially enhanced when the electromechanical coupling of the linear harvester is large and the linear, amplifying stage is lightly damped [125,134,135]. The dynamic magnifier approach also provides considerably broadened bandwidths of effective operation by strategic design of the magnifier mass and the stage's decoupled natural frequency [136,137].

These advantages of the dynamic magnifier design strategy are critically tied to the multimodal nature of the dynamics involved for the all-linear systems. As was made clear in Section 4.5, the non-resonant feature of snap-through dynamics discourages the assumption that a concept developed for a multi-DOF linear energy harvester will be similarly favorable when applied to a harvester possessing bistability. At the same time, it was also shown in the prior study that great benefit may be derived by the appropriate utilization of the auxiliary linear oscillator approach even though the method was originally devised to augment linear energy harvesters. Inspired by the favorable outcomes from careful study of the auxiliary oscillator approach to enhance power generation of the bistable harvester in Section 4.5 and by the positive performance enhancements enabled by the dynamic magnifier design for linear harvesters, this study explores the application of a dynamic magnifier stage to a bistable energy harvester and examines how to effectively leverage the additional dynamic body for encompassing improvements in the energy conversion utility. The right-most schematic in Figure 4.30 illustrates the system considered here, where the base-excited linear magnifier stage supports the bistable energy harvester. In this configuration, the electromechanical transduction mechanism interfaces the inertial masses of the magnifier stage and bistable harvester. The following sections explore this novel design through analytical, numerical, and experimental efforts, and thoroughly examine how the power generation capabilities may be uniquely promoted as compared to the energy harvesting system with auxiliary oscillator studied in Section 4.5.

4.6.1 Governing Equations for the Bistable Harvester with Linear Dynamic Magnifier

The energy harvesting system architecture considered here is depicted in Figure 4.31. The bistable harvester utilizes an electromagnetic induction mechanism such that motions of the magnetic mass m_1 along the axis of a coil fixed to the magnifier stage induce a flow of current in the external power harvesting circuitry, as shown in the right part of the figure. The magnetic mass is coupled to the linear dynamic magnifier mass m_2 (the outer frame in the figure) through several dissipative and restorative force mechanisms. The electromagnetic

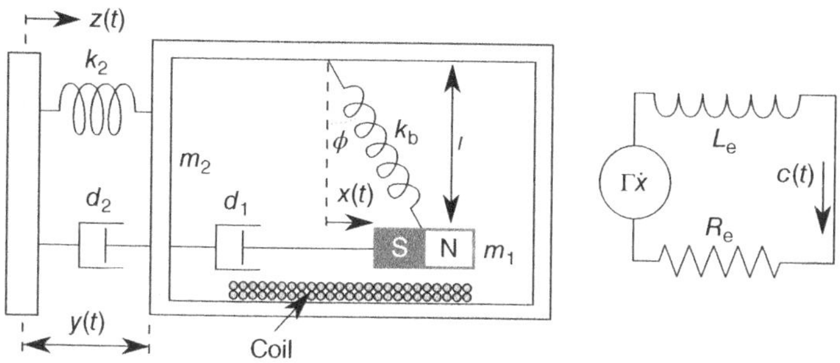

Figure 4.31 Base-excited linear oscillator with attached bistable energy harvester having electromagnetic transduction mechanism, with corresponding resistive load power harvesting circuit shown on the right right.

induction coupling leads to both types of coupling forces. The restoring force of the spring k_b in the direction of motion of the masses is given by

$$F(x)=k_b\left(\sqrt{x^2+l^2}-l_0\right)\sin\phi \tag{4.130}$$

where $x(t)$ is the relative displacement between the linear oscillator mass and magnetic (bistable) inertial mass; k_b and l_0 are the stiffness and free length of the spring, respectively; l is the vertical distance between the linear frame and bistable mass; and ϕ is the inclination of the spring as measured from the central, unstable equilibrium position of the bistable harvester ($x=0$), where $\sin\phi = x/\sqrt{x^2+l^2}$. Assuming the oscillations remain close to the unstable equilibrium, $x/l << 1$ – which may alternatively be ensured by consideration of "weakly bistable" designs where the stable equilibria are close together – a binomial series expansion is then employed to simplify the restoring force expression to be

$$F=k_b\left(1-\frac{l_0}{l}\right)x+\frac{k_b l_0}{2l^3}x^3+O\left(x^5\right) \tag{4.131}$$

The governing equations for the system are found to be

$$m_1\left(\ddot{x}+\ddot{y}+\ddot{z}\right)+d_1\dot{x}-k_1x+k_3x^3+\Gamma c=0 \tag{4.132}$$

$$m_2\left(\ddot{z}+\ddot{y}\right)+d_2\dot{y}+k_2y-d_1\dot{x}+k_1x-k_3x^3-\Gamma c=0 \tag{4.133}$$

$$L_e\dot{c}+R_ec=\Gamma\dot{x} \tag{4.134}$$

where $y(t)$ is the relative displacement between the linear oscillator and base motion $z(t)$; $c(t)$ is the current flow through the energy harvesting circuit load resistance R_e; L_e is the inductance of the electromagnetic generator; Γ is an electromagnetic coupling constant; d_1 and d_2 are mechanical damping constants; $k_1=k_b\left[\left(l_0/l\right)-1\right]$ and $k_3=k_bl_0/2l^3$ are linear and nonlinear stiffness parameters due to the interfacing spring k_b, respectively; k_2 is the spring stiffness of the dynamic magnifier stage; and the overdot is differentiation with respect to time t. The linear magnifying stage is harmonically excited by a uniform base acceleration $\ddot{z}(t) = -Z_0 \cos\Omega t$.

In the following, it is assumed that $l_0 > l$. Then, Eqs (4.132)–(4.134) are simplified and non-dimensionalized to yield

$$x'' + (1+\mu) f\gamma_1 x' - (1+\mu) f^2 x + (1+\mu) f^2 \beta x^3 - \gamma_2 y' - y + (1+\mu)\varepsilon\theta c = 0 \tag{4.135}$$

$$y'' + \gamma_2 y' + y - \mu f \gamma_1 x' + \mu f^2 x - \mu f^2 \beta x^3 - \mu\varepsilon\theta c = p\cos\omega\tau \tag{4.136}$$

$$c' + \rho c = \theta x' \tag{4.137}$$

where the parameters are defined as

$$\omega_1^2 = k_1/m_1;\ \omega_2^2 = k_2/m_2;\ \mu = m_1/m_2;\ f = \omega_1/\omega_2;\ p = Z_0/k_2;\ \gamma_1 = d_1/m_1\omega_1$$
$$\gamma_2 = d_2/m_2\omega_2;\ \theta = \Gamma/L_e;\ \rho = R_e/L_e\omega_2;\ \varepsilon = L_e/m_1\omega_2^2;$$
$$\omega = \Omega/\omega_2;\ \beta = k_3/k_1;\ \tau = \omega_2 t$$

and the operator $(\ )'$ represents a derivative with respect to non-dimensional time τ. Similarly as with the parameters utilized in Section 4.5, here μ is a ratio of mass between the bistable and dynamic magnifier oscillators, and f is a ratio of the linear natural frequencies between the same respective dynamic bodies. Also, the excitation frequency ω is normalized with respect to the natural frequency of the magnifier stage itself ω_2, which is the comparable normalization applied to define the non-dimensional time τ.

4.6.2 Approximate Solution by the Method of Harmonic Balance

To capture the most fundamental dynamics of the nonlinear energy harvester, the relative displacements of the bistable and linear oscillators and the current flow in the harvesting circuit are approximated using fundamental Fourier series expansions as follows

$$x(\tau) = c_1(\tau) + a_1(\tau)\sin\omega\tau + b_1(\tau)\cos\omega\tau \tag{4.138}$$

$$y(\tau) = a_2(\tau)\sin\omega\tau + b_2(\tau)\cos\omega\tau \tag{4.139}$$

$$c(\tau) = a_3(\tau)\sin\omega\tau + b_3(\tau)\cos\omega\tau \tag{4.140}$$

The coefficients are assumed to be slowly varying such that the second derivatives $(\ )''$ are negligible. Substituting Eqs (4.138)–(4.140) into (4.135)–(4.137), eliminating higher-order terms, and grouping the constant, $\sin\omega\tau$, and $\cos\omega\tau$ terms yields seven equations for the coefficients.

$$-\gamma_1 c_1' = f\Lambda_c c_1 \tag{4.141}$$

$$-\gamma_2 a_2' + 2\omega b_2' + \mu f \gamma_1 a_1' = -\mu\varepsilon\theta a_3 + (1-\omega^2) a_2 - \gamma_2\omega b_2 - \mu f^2 \Lambda a_1 + \mu f \gamma_1 \omega b_1 \tag{4.142}$$

$$-2\omega a_2' - \gamma_2 b_2' + \mu f \gamma_1 b_1' = -\mu\varepsilon\theta b_3 + \gamma_2\omega a_2 + \left(1-\omega^2\right)b_2 - \mu f \gamma\omega a_1 - \mu f^2 \Lambda b_1 - p \tag{4.143}$$

$$\gamma_2 a_2' - (1+\mu) f\gamma_1 a_1' + 2\omega b_1' = (1+\mu)\varepsilon\theta a_3 - a_2 + \gamma_2\omega b_2 + \sigma a_1 - (1+\mu) f\gamma_1\omega b_1 \tag{4.144}$$

$$\gamma_2 b_2' - 2\omega a_1' - (1+\mu) f\gamma_1 b_1' = (1+\mu)\varepsilon\theta b_3 - \gamma_2\omega a_2 - b_2 + (1+\mu) f\gamma_1\omega a_1 + \sigma b_1 \tag{4.145}$$

$$-a_3' + \theta a_1' = \rho a_3 - \omega b_3 + \theta\omega b_1 \tag{4.146}$$

$$-b_3' + \theta b_1' = \omega a_3 + \rho b_3 - \theta\omega a_1 \tag{4.147}$$

where the terms are defined as

$$\Lambda_c = -1 + \beta\left(c_1^2 + \frac{3}{2}r_1^2\right); \Lambda = -1 + \beta\left(3c_1^2 + \frac{3}{4}r_1^2\right); \sigma = (1+\mu)f^2\Lambda - \omega^2;$$
$$r_1^2 = a_1^2 + b_1^2$$

The displacement amplitude of the relative motion of the bistable harvester magnetic mass is given by r_1. The steady-state dynamics of the coupled energy harvesting system are determined by solving Eqs (4.141)–(4.147). First, from Eqs (4.146) and (4.147), a_3 and b_3 are expressed in terms of a_1 and b_1.

$$a_3 = \Phi(\omega a_1 - \rho b_1) \tag{4.148}$$

$$b_3 = \Phi(\rho a_1 + \omega b_1) \tag{4.149}$$

where $\Phi = \theta\omega/(\rho^2 + \omega^2)$. Substituting Eqs (4.148) and (4.149) into (4.144) and (4.145), a_2 and b_2 are also written in terms of a_1 and b_1.

$$a_2 = \frac{1}{\Delta}\left[\Gamma_1 a_1 + \Gamma_2 b_1 + \gamma_2\omega p\right] \tag{4.150}$$

$$b_2 = \frac{1}{\Delta}\left[-\Gamma_2 a_1 + \Gamma_1 b_1 + \left(1-\omega^2\right)p\right] \tag{4.151}$$

where the terms are defined as

$$\Delta = \left(1-\omega^2\right)^2 + (\gamma_2\omega)^2$$

$$\Gamma_1 = \mu\left[\left(1-\omega^2\right)\left(f^2\Lambda + \varepsilon\theta\omega\Phi\right) + (\gamma_2\omega)\left(f\gamma_1\omega + \varepsilon\theta\rho\Phi\right)\right]$$

$$\Gamma_2 = \mu\left[(\gamma_2\omega)\left(f^2\Lambda + \varepsilon\theta\omega\Phi\right) - \left(1-\omega^2\right)\left(f\gamma_1\omega + \varepsilon\theta\rho\Phi\right)\right]$$

After substituting Eqs (4.148)–(4.151) into (4.142) and (4.143) and then squaring and summing the resulting equations, a characteristic equation for the square of the displacement amplitude of the bistable inertial magnetic mass, r_1^2, is derived.

$$p^2\left[1+(\gamma_2\omega)^2\right]=\kappa r_1^2 \tag{4.152}$$

where κ is defined as

$$\begin{aligned}\kappa=&\left[\left(1-(1+\mu)\omega^2\right)^2+(\gamma_2\omega)^2\right](f^2\Lambda+\varepsilon\theta\omega\Phi)^2\\&-2\omega^2\left[\left(1-\omega^2\right)\left(1-(1+\mu)\omega^2\right)+(\gamma_2\omega)^2\right](f^2\Lambda+\varepsilon\theta\omega\Phi)\\&+\left\{\begin{aligned}&\left[\left(1-(1+\mu)\omega^2\right)^2+(\gamma_2\omega)^2\right](f\gamma_1\omega+\varepsilon\theta\rho\Phi)^2\\&+2\mu\gamma_2\omega^5(f\gamma_1\omega+\varepsilon\theta\rho\Phi)+\omega^4\left[\left(1-\omega^2\right)^2+(\gamma_2\omega)^2\right]\end{aligned}\right\}\end{aligned}$$

Eq. (4.152) contains two unknowns c_1 and r_1^2 based upon the functional dependences of the term Λ.

The constant term c_1 may be obtained by solving the steady-state response of Eq. (4.141) with a pre-conceived expectation as to the resulting dynamics. For example, if one wishes to consider the potential for snap-through (interwell) vibrations of the energy harvesting platform, the appropriate selection of c_1 is zero according to a stationary solution of Eq. (4.141). In contrast, intrawell oscillations are characterized and predicted by Eq. (4.152) by taking the non-trivial stationary solution to Eq. (4.141):

$$c_1^2=\frac{1}{\beta}-\frac{3}{2}r_1^2 \tag{4.153}$$

Therefore, depending on the dynamics of interest, a unique c_1 is substituted into Λ such that Eq. (4.152) is a function only of the bistable harvester displacement amplitude squared r_1^2. The roots of Eq. (4.152) are then computed. If the roots are positive, real numbers, the analytical predictions are considered to be meaningful. Then, a full set of solution coefficients are determined starting from a computation of a_1 and b_1:

$$\begin{aligned}a_1=\frac{p}{\kappa}&\left\{(1+\mu)\gamma_2\omega^3(f^2\Lambda+\varepsilon\theta\omega\Phi)\right.\\&\left.+\left[1-(1+\mu)\omega^2+(\gamma_2\omega)^2\right](f\gamma_1\omega+\varepsilon\theta\rho\Phi)-\gamma_2\omega^5\right\}\end{aligned} \tag{4.154}$$

$$b_1=\frac{p}{\kappa}\left\{\begin{aligned}&\left[1-(1+\mu)\omega^2+(\gamma_2\omega)^2\right](f^2\Lambda+\varepsilon\theta\omega\Phi)\\&-(1+\mu)\gamma_2\omega^3(f\gamma_1\omega+\varepsilon\theta\rho\Phi)-\left[1-\omega^2+(\gamma_2\omega)^2\right]\omega^2\end{aligned}\right\} \tag{4.155}$$

To determine the stability of physically meaningful solutions, Eqs (4.141)–(4.147) are first cast into the form $A\boldsymbol{x}'=F(\boldsymbol{x})$ where $\boldsymbol{x}$ is defined by $\boldsymbol{x}=[a_1, b_1, c_1, a_2, b_2, a_3, b_3]^{\mathrm{T}}$. Then, stable sets of the mathematical solutions are those for which the eigenvalues of the Jacobian of $\mathrm{G}=A^{-1}F(\boldsymbol{x})\big|_{\boldsymbol{x}=\boldsymbol{x}^*}$ possess only negative real roots, where $\boldsymbol{x}^*$ is a given set of coefficients as determined by solving Eq. (4.152).

For a comprehensive measure of benefit for the multi-DOF energy harvesting system, the average power density generated by the bistable harvester is determined by

$$P_1 = \frac{1}{1+\mu}\left[\frac{1}{2}\rho\left(a_3^2+b_3^2\right)\right] = \frac{\rho}{2(1+\mu)}\frac{\theta^2\omega^2}{\omega^2+\rho^2}r_1^2 \tag{4.156}$$

To take phase relationships into consideration, the assumed displacement expression for the bistable harvester may be alternatively expressed by

$$x = c_1 + r_1\cos\left[\omega\tau - \varphi_1\right];\ \tan\varphi_1 = a_1/b_1 \tag{4.157}$$

4.6.3 Analytical and Numerical Investigations on the Roles of the Linear Dynamic Magnifier Stage

For linear energy harvesting platforms, the use of a dynamic magnifier stage opens up means to broaden the operating frequency bandwidth as well as to enhance the peak power generation [134,137]. Of course, these are desirable outcomes for incorporating a linear magnifier stage with a bistable energy harvester. Yet, Section 4.5 found that purely linear multi-DOF counterparts do not accurately enlighten one as to the result of incorporating an auxiliary oscillator attachment with a *bistable* harvester. Therefore, the following investigations probe the roles of the linear dynamic magnifier stage in the resulting power generating performance of the bistable energy harvester snap-through dynamics.

4.6.3.1 Effect of the Mass Ratio

From Eq. (4.152), the stable, snap-through dynamics are determined using the parameters $p=0.2$, $f=0.2$, $\beta=1$, and $\gamma_1=\gamma_2=0.001$. The results corresponding to the intrawell oscillation regime are omitted from consideration due to the undesirable power harvesting performance of such dynamics in comparison to snap-through vibrations. The non-dimensional electromechanical coupling parameters are selected to be $\rho=2500$, $\theta=10$, and $\varepsilon=0.6$. These values are representative of an electromagnetic generator design having small inductance and moderate transducer coupling constant [69,127]. The ratio of bistable harvester to linear oscillator masses $\mu=m_1/m_2$ is first explored for its parametric roles. Previous studies on dynamic magnifier stages for linear energy harvesting systems found that the magnifier mass may be required to be on the order of or greater than that of the harvesting element for appreciable performance enhancement [135–137]. Thus, the following examination considers a large range of mass ratios $\mu=[0.5,1.5]$ that reflect the design principles of the earlier work ($\mu<1$) as well as the alternative design approach ($\mu>1$) that represents a smaller linear magnifier mass added into the bistable harvesting system. For comparison to the benchmark of the individual, base-excited bistable harvester, the dynamics of this single oscillator are predicted by solution of the corresponding governing equation non-dimensionalized using the time of the *coupled* system $\tau=\omega_2 t$

$$x'' + f\gamma_1 x' - f^2 x + f^2\beta x^3 + \varepsilon\theta c = p\cos\omega\tau \tag{4.158}$$

The corresponding electrical response equation for the single bistable harvester is the same as Eq. (4.137). Power density is thus determined from

$$P_1 = \frac{\rho}{2\mu}\frac{\theta^2\omega^2}{\omega^2+\rho^2}r_1^2 \tag{4.159}$$

In this and in the following sub-section, numerically-computed results are provided for several of the parameter cases studied. The values are obtained by long-time numerical integrations of the governing Eqs (4.135)–(4.137) and (4.158) using randomly generated initial conditions. The steady-state displacement amplitudes of the bistable harvester are then computed from the fast Fourier transform of the last 5% of the simulations, from which are also computed the power densities according to Eqs (4.156) and (4.159). The results of 20 simulations yielding single periodic snap-through responses are averaged for each data point and shown in Figure 4.32 and Figure 4.33 (later). When more than one single periodic steady-state is uncovered at a given frequency of excitation, separate averages are made to distinguish the dynamic regimes.

Figure 4.32 shows the displacement amplitude and phase between the bistable harvester and linear magnifier stage in a,b, respectively, and in part c the average power density generated by the bistable harvester component. These are plotted as solid curves. The corresponding predictions for the individual bistable harvester, that is, without the linear oscillator as determined by analytical solution to Eq. (4.158), are provided for comparison as the dashed curves. Since the mass ratio μ does not contribute in the formulation of Eq. (4.158), a change in μ does not affect the relative displacements of the individual bistable harvester. Accordingly, the findings in Figure 4.32a reveal no change in the displacement amplitudes of the individual harvester, although the power densities are plainly influenced due to the normalization given in Eq. (4.159). In comparison, Figure 4.32a shows that the coupled system leads to greater bistable harvester relative displacements than the harvester alone, so long as the excitation frequency is less than $\omega^* = 1/\sqrt{1+\mu}$ (see Section 4.6.4 for detailed discussion). As the mass ratio μ increases from 0.5 to 5, the amplification of the bistable harvester dynamics for the coupled system becomes more substantial with the clear trade-off that the snap-through motions are sustainable at progressively smaller frequencies of excitation. The numerical results are in very good agreement with these observations.

The destabilization of snap-through is correspondingly manifest in the phase relationships plotted in Figure 4.32b. It is seen that as the driving frequency nears the point at which the interwell oscillations are destabilized in the lower frequency regime, the bistable harvester approaches a 90° phase lag relative to the linear oscillator. However, recent investigations have found that large-amplitude snap-through dynamics are not sustainable beyond a 90° phase lag to a sinusoidal excitation [138], which verifies the source of the dynamics destabilization occurring in Figure 4.32 at the lower frequency regime for each mass ratio μ considered. In addition, the results in Figure 4.32(a) verify that the relative amplitude of displacement between the harvester and linear magnifier remains greater than 1, that is, greater than the amplitude between the two stable equilibria $1/\sqrt{\beta}$.

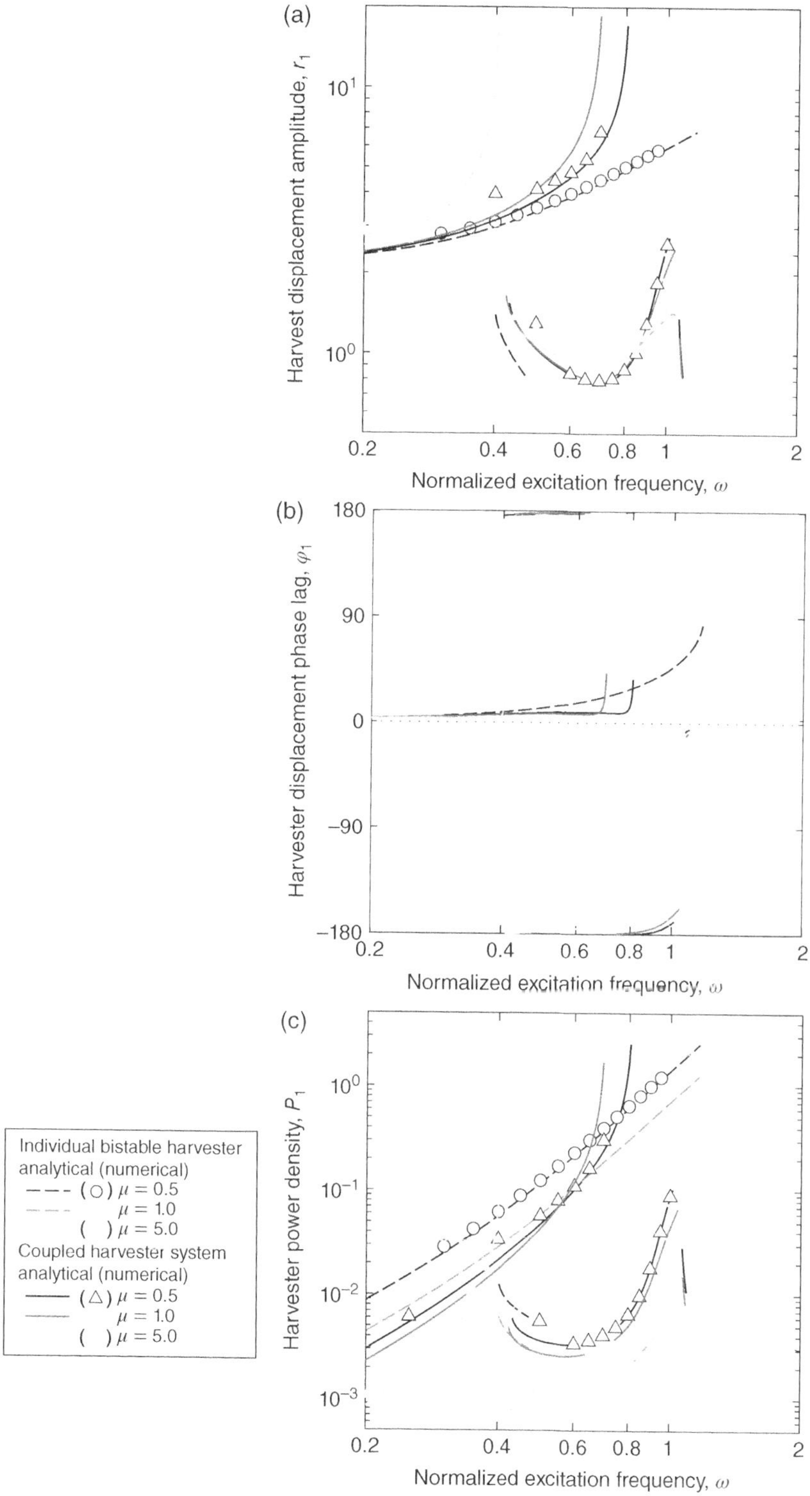

Figure 4.32 Snap-through dynamics of the bistable harvester as a function of excitation frequency ω, for a variety of mass ratios μ. Shown are the bistable harvester (a) displacement amplitude r_1; (b) phase lag φ_1; and (c) average power density P_1.

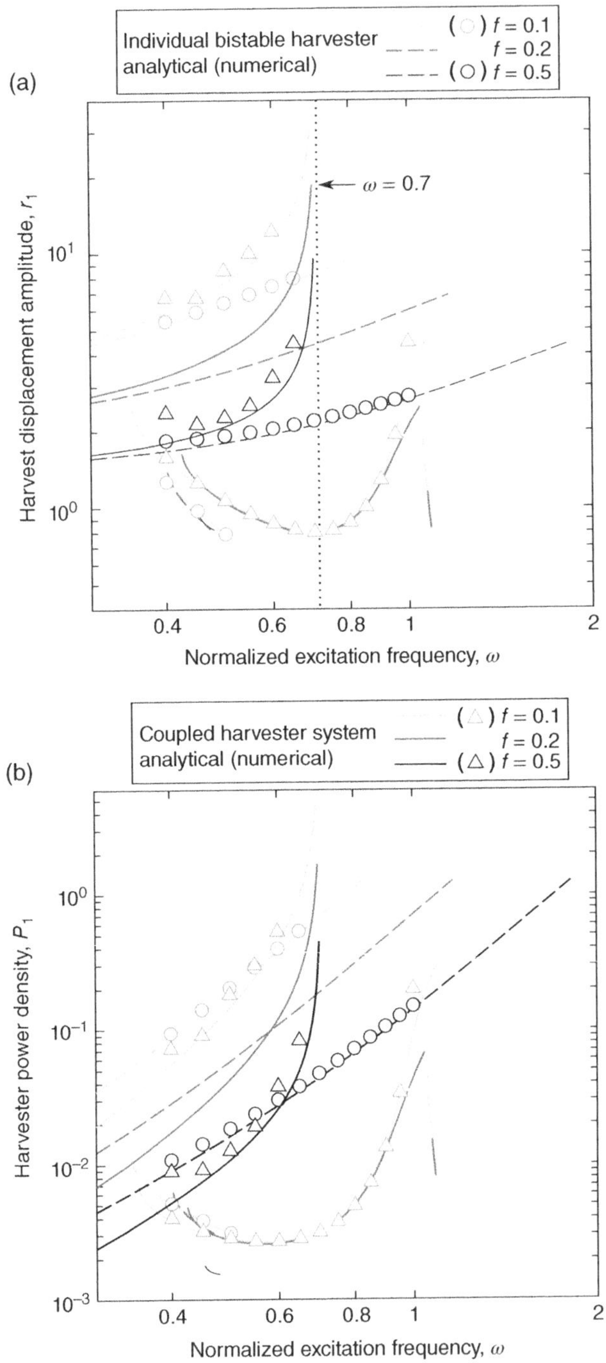

Figure 4.33 Snap-through dynamics of the bistable harvester as a function of excitation frequency ω, for a variety of frequency ratios *f*. Shown are the bistable harvester (a) displacement amplitude r_1 and (b) average power density P_1.

This indicates that the steady-state dynamics are legitimately cataloged as snap-through actions as opposed to an ordinary in-phase vibration of the same magnitude as the linear oscillator. Figure 4.32c reveals that as the mass ratio μ increases, the power density for coupled system is correspondingly increased when compared to the individual bistable harvester. Interestingly, this trend is the opposite of that observed for *linear* energy harvesting systems having a dynamic magnifier stage [135–137] where greater magnifier stage masses (here characterized by $\mu < 1$) are more beneficial to power harvesting performance. On the other hand, as seen in Figure 4.32c, even when the mass ratio is small, the average power density of the coupled harvester system still exceeds that achieved by the individual bistable harvester across a small but finite bandwidth of excitation frequencies. The corresponding numerical data verify these relative performance enhancements although the absolute peaks of the given "branches" of the dynamics are difficult to capture in the simulations due to heightened initial condition sensitivity closer to the frequencies at which the large amplitude snap-through dynamics are destabilized.

As the excitation frequency approaches the linear oscillator natural frequency ($\omega = 1$), another "branch" of the steady-state snap-through dynamics is analytically predicted and numerically uncovered for the coupled systems. Yet, according to the phase relationships plotted in Figure 4.32b, these oscillations are out-of-phase with respect to the motions of the dynamic magnifier stage. An increasing value of mass ratio μ reduces the displacement amplitudes of these responses but slightly broadens the frequency range across which they occur. Interestingly, such a dynamic regime is counter to the traditional understanding of snap-through which dictates that it is not sustainable for phase lags in excess of 90°. However, the displacements of these interwell dynamics are seen to have extremely small amplitudes. In some cases the bistable harvester displacements are near to the amplitude between the stable and unstable equilibria $r_1 = 1$, for example as seen in Figure 4.32a in the low amplitude branches around $\omega = 0.8$. Since the bistable harvester undergoes such small interwell vibrations, these dynamics are representative of the *excitation-induced stability* phenomenon described in detail in Section 3.3. Recall that this dynamic regime was not uncovered in the explorations of Section 4.5 where the auxiliary linear oscillator was considered as a means to promote the snap-through oscillations of the bistable harvester. This notable distinction in the achievable steady-state dynamics justifies the rationale for close study on each implementation of a bistable structure in the full framework of an energy harvesting system so as to first understand the various dynamic sensitivities and thereafter to best leverage the design parameters for performance enhancements. It is clear from the results of Figure 4.32 that the mass ratio μ plays intricate and novel roles towards governing these dynamics.

4.6.3.2 Effect of the Frequency Ratio

Considering the influences of the frequency ratio f, Figure 4.33 presents analytical and numerical results using a fixed mass ratio of $\mu = 1$ and range of frequency ratios $f = [0.1, 0.2, 0.5]$. The remaining parameters are the same as those employed in the previous computations. Similarly to change in the mass ratio, when the frequency ratio is tailored the results in Figure 4.33 show that two distinct interwell dynamic regimes can be identified: amplified bistable harvester snap-through dynamics in a frequency region less than $\omega^* = 1/\sqrt{1+\mu}$ and an out-of-phase response regime around $\omega = 1$. Regardless of the frequency ratio considered in Figure 4.33, the magnified snap-through motions are greatest as the excitation frequency approaches $\omega \leq 0.7$ from below, such that the dynamics are destabilized approximately at this critical frequency. In each of these respects, the numerical simulations are in good agreement with the analytical predictions. Together with the results shown in Figure 4.32, the findings suggest that the frequency at which in-phase snap-through dynamics of the coupled harvester system are destabilized is primarily controlled by the mass ratio while the frequency ratio principally determines the displacement amplitude and the rate at which the response grows asymptotically prior to destabilization. Also, as seen in Figure 4.33, the bandwidth of enhanced average power density for the coupled system is only slightly affected by the change in frequency ratio.

Figure 4.33a indicates that the frequency ratio also governs the existence of the out-of-phase dynamic regime for the coupled system around $\omega = 1$. It is evident that the relative displacement amplitude between the bistable harvester and dynamic magnifier decreases as the frequency ratio f increases, so much so that stable dynamics may eventually be impossible to realize should f become too large. With the aim of providing for broadened excitation frequency sensitivity in the coupled energy harvesting system, the achievement of the out-of-phase dynamic regime is important to ensure. Although the absolute power generated by such oscillations is relatively low, Figure 4.33 shows that the individual bistable harvester is not excited to snap-through oscillations up to such high frequencies. Consequently, while small in the absolute sense, the power generated by the out-of-phase dynamics of the bistable harvester with magnifier is still infinitely greater than the individual harvester which possesses no such response features at such high frequencies of excitation.

4.6.3.3 Effect of the Electromechanical Coupling

The influence of electromechanical coupling θ on the power harvesting performance is investigated using the system parameters $p = 0.2$, $f = 0.2$, $\beta = 1$, $\mu = 1$, $\gamma_1 = \gamma_2 = 0.001$, $\rho = 2500$, and $\varepsilon = 0.6$. The coupling factor is varied over a broad range $\theta = [5, 10, 20]$ that practically takes into consideration electromagnetic induction generator designs of low to high transducer constants [139], more commonly realized for meso-/macroscale energy harvesting platforms. The analytical predictions in Figure 4.34 show that a drastic degradation of the sustainable frequency bandwidth of snap-through vibrations is encountered for the individual bistable harvester as the electromechanical coupling factor θ is increased. For example, snap-through for the individual device is sustainable up to approximately $\omega = 2$ when the coupling factor is just $\theta = 5$; as the coupling factor is increased four-fold to $\theta = 20$, snap-through is possible only up to a maximum excitation frequency of about $\omega = 0.6$. In contrast, the bistable harvester with dynamic magnifier stage exhibits remarkable robustness of performance with increasing electromechanical coupling. For substantial levels of electromechanical coupling, such as is the case for larger-sized electromagnetic systems [139], a bistable harvester with linear dynamic magnifier maintains

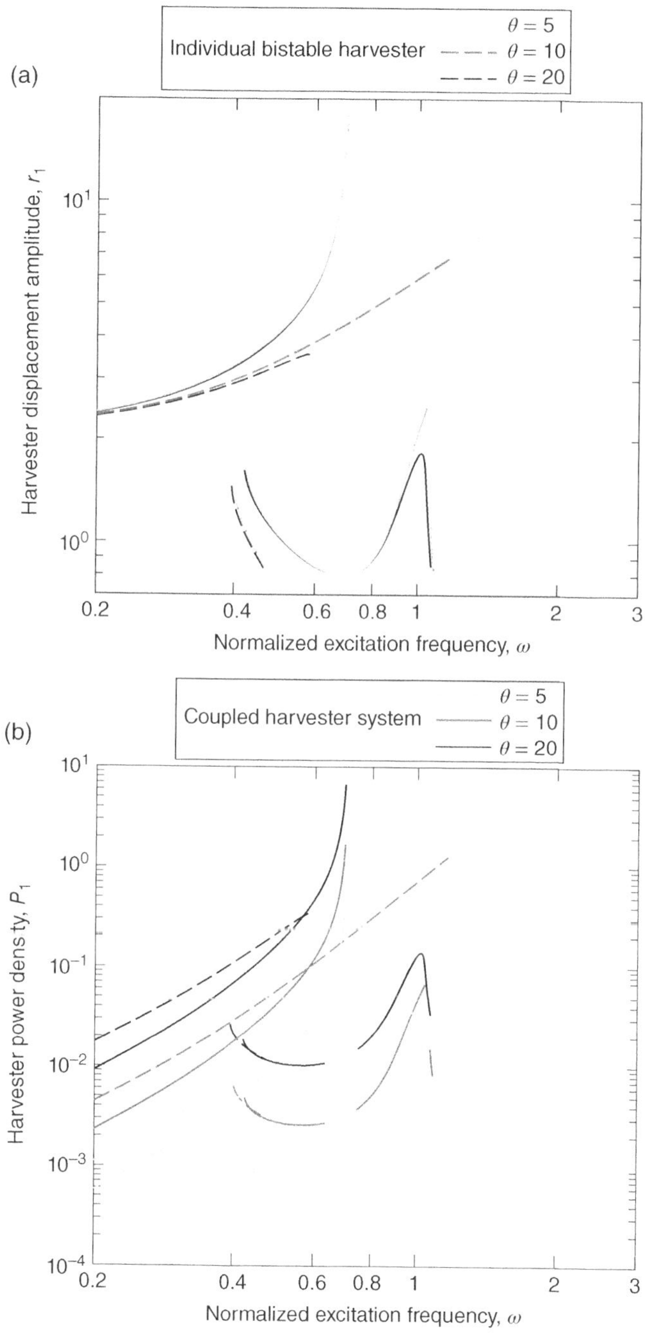

Figure 4.34 Snap-through dynamics of the bistable harvester as a function of excitation frequency ω, for a variety of electromechanical coupling coefficients θ. Shown are the bistable harvester (a) displacement amplitude r_1 and (b) average power density P_1.

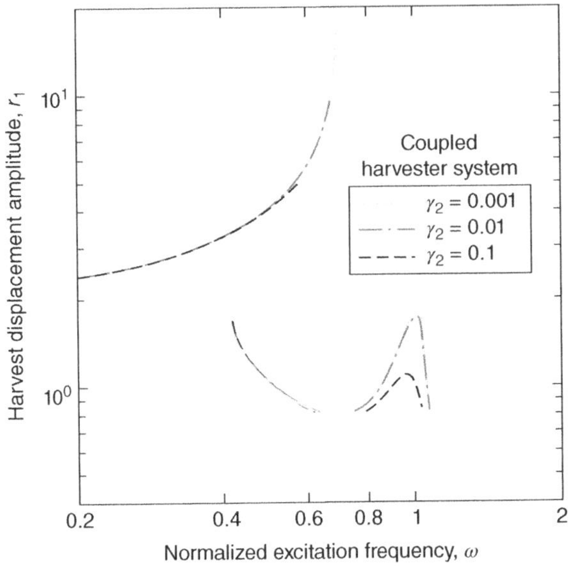

Figure 4.35 Snap-through dynamics of the bistable harvester in the coupled system as a function of excitation frequency ω, for a variety of linear dynamic magnifier stage damping factors γ_2.

high-performing snap-through oscillations across a considerably broader range of frequencies than the individual bistable harvester. This is a second case of an infinite performance improvement by the strategic utilization of the linear dynamic magnifier. Another observation is made that the greater coupling strengths θ assist in stabilizing the out-of-phase dynamics at higher frequencies. For example, Figure 4.34 shows that this response regime is maintained throughout the entire range of frequencies around $\omega = 1$ for the larger coupling factor $\theta = 20$. In comparison, the individual bistable harvester possesses no such broadband frequency sensitivity and great vulnerability to increase in the electromechanical coupling factor. It is worth noting that the latter finding is also seen for the purely linear energy harvesting platform which incorporates a dynamic magnifier stage: greater levels of electromechanical coupling do not significantly degrade the coupled linear system's energy harvesting performance [125,134].

In the light of the experimental findings presented in Section 4.6.5.1, the consequences of combined increases in electromechanical coupling (here using $\theta = 20$) and in mechanical dissipation of the linear oscillator γ_2 are explored. The dissipation factor γ_2 is varied from 0.001 to 0.1. Figure 4.35 plots the bistable harvester displacement amplitude as predicted by the analysis for the coupled system. The lower frequency snap-through dynamics are seen to destabilize with reducing frequencies of excitation as the damping of the linear magnifier stage γ_2 increases. Consequently, the in- and out-of-phase responses at lower and higher frequencies, respectively, become closer in amplitude. In terms of the dissipative effects in this bistable, electromagnetic energy harvesting system, the magnitude of lower frequency snap-through vibrations is governed by the damping influences of the magnifying stage while the low-amplitude, out-of-phase oscillations are primarily influenced by dissipative effects evident in the bistable harvester itself, whether they represent mechanical or electromechanical influences.

4.6.4 Interpreting Frequency Response Characteristics of the Coupled Energy Harvester System

The analytical investigations carried out above indicate that the coupled linear–bistable energy harvesting system yields particularly unique frequency response characteristics, even when considered with respect to the findings of Section 4.5 using a platform (bistable harvester and linear oscillator attachment) having the dynamic bodies in "reversed" order. However, like a linear two-DOF system, two resonance-like phenomena are observed when the system possesses bistability, in this case corresponding to the in- and out-of-phase interwell oscillation regimes (i.e., the snap-through and excitation-induced stability regimes, respectively). Figure 4.32 suggests that the mass ratio μ plays an important role in determining the frequency at which the first resonance feature appears and that the second resonance effect seems to occur around the natural frequency of the uncoupled linear system. Here, this observation is analytically confirmed, and guidelines are thereafter devised for the design of frequency response characteristics desirable for the coupled linear–bistable harvester system. In tandem with these efforts below, a new interpretation of the snap-through phenomenon is elucidated.

As has been reiterated throughout this book, the snap-through dynamics of bistable structures are a non-resonant phenomenon because *any form or frequency* of excitation may induce the characteristic switching of the mass from one stable state to the other, so long as a satisfactory level of excitation is applied. For a linear two-DOF system, a non-resonant oscillator of finite mass coupled to an excited sub-system may be realized in one of two ways: when the natural frequency of the non-resonant body approaches zero or infinity, given by $f \to 0$ or $f \to \infty$, respectively, according to the parametric definition provided here. For the case that the frequency ratio approaches zero, the non-resonant oscillator has an infinitesimally small natural frequency while in the case of an infinitely large frequency ratio the oscillator is infinitely stiff such that it appears to be "welded" to the excited sub-system. Therefore, one may interpret the manner in which the bistable device interacts with the linear oscillator during snap-through as being analogous to a linear device of $f \to 0$ and $f \to \infty$.

These perspectives are supported by considering the limiting cases of the undamped linear two-DOF system frequency response [140] in comparison to the present results. The characteristic equation governing the resonances of the linear two-DOF system is

$$\omega^4 - \left[1 - f^2(1+\mu)\right]\omega^2 + f^2 = 0 \tag{4.160}$$

where f and μ are the frequency and mass ratios, respectively, defined with respect to the excited linear sub-system. Using Eq. (4.160), a linear body of

$f \to 0$ splits the original system resonance into two normalized frequencies $\omega^* = 0$ and 1; of course, the frequency $\omega^* = 1$ corresponds to the uncoupled natural frequency of the underlying linear oscillator according to the normalized frequency definition. Thus, the out-of-phase interwell dynamic can be interpreted as the response of a linear oscillator with zero natural frequency coupled to an excited linear oscillator, since it does not shift the resonant frequency from the uncoupled resonance of the linear system, which is plainly observed in the results such as the resonance features around $\omega = 1$ shown in Figure 4.32. For the limiting case $f \to \infty$, only terms multiplied by f^2 in Eq. (4.160) are appreciable, such that only one resonance associated with the composite mass at frequency $\omega^* = 1/\sqrt{1+\mu}$ remains. Hence, the in-phase snap-through dynamics of the coupled system may be interpreted as the response of a linear oscillator with infinite natural frequency coupled to the excited linear oscillator since the resonance phenomenon occurs at a frequency associated with the net mass of the system: $1+\mu$. This defines the frequencies at which the two resonant features of the undamped coupled linear–bistable harvester system occur. To summarize, one resonant peak is seen at the uncoupled linear natural frequency $\omega^* = 1$ associated with out-of-phase interwell vibrations (the excitation-induced stability phenomenon) and a second resonant-like peak exists at a frequency associated with the combined mass of the two oscillators $1+\mu \to \omega^* = 1/\sqrt{1+\mu}$.

This perspective is verified by computing the undamped system dynamics using the model formulation detailed in Section 4.6.1 with parameters $p = 0.01$, $f = 0.25$, $\beta = 1$, $\theta = 0$. Figure 4.36 plots the resulting analytical predictions of bistable harvester displacement amplitude as the mass ratio μ is varied across several orders of magnitude. As plainly observed in the results, the frequencies at

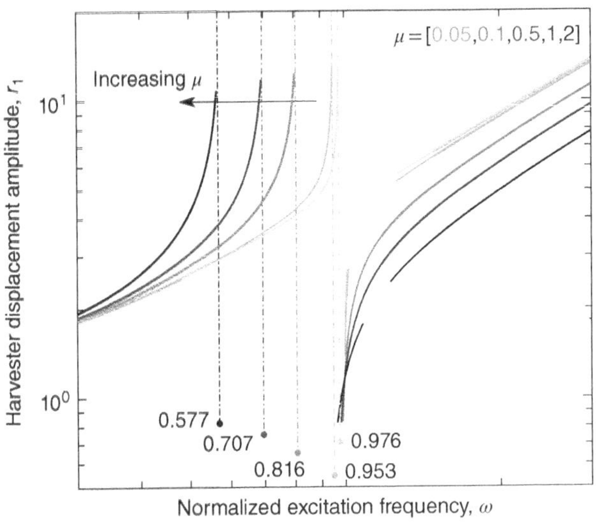

Figure 4.36 Snap-through dynamics of the undamped coupled bistable energy harvesting system with linear magnifier stage as a function of excitation frequency ω for a variety of mass ratios μ. The dashed lines are the respective computations of the limiting frequency $\omega^* = 1/\sqrt{1+\mu}$.

which the lower resonant-like characteristics occur are exactly those determined from the expression $\omega^* = 1/\sqrt{1+\mu}$. This confirms the observation in Section 4.6.3.1 that the mass ratio predominantly governs the frequency at which this critical dynamic feature occurs, critical in the sense that greatest power density enhancements are strongly correlated with this feature, for example as evident in Figure 4.32c. These perspectives and validations show that the frequency sensitivity of snap-through vibrations in the coupled bistable energy harvester platform studied here may be deliberately tailored for energy harvesting advantage by adjustment of the mass ratio μ. In this way, the power density enhancements over the individual bistable harvester may be changed to suit the excitation frequencies of the environmental vibrations. In contrast, no specific design changes are required to realize the out-of-phase resonance-like feature in the coupled bistable harvester system because this frequency is intricately related to the characteristics of the linear dynamic magnifier stage. However, tailoring the absolute frequency at which this feature occurs is obviously as straightforward as adjusting the uncoupled natural frequency of the underlying linear oscillator. Such enlightening design guidelines are not as apparently evident in the bistable harvester-auxiliary oscillator system considered in Section 4.5. This distinction highlights the importance of elucidating the source of the performance enhancement of the present approach which utilizes a linear dynamic magnifier stage along with the bistable energy harvester: with the discovery of the principles facilitating the enhancements, designers may best leverage the unique bistable harvester dynamics for the application at hand.

4.6.5 Experimental Validations and Investigations

Several important factors are uncovered in the analytical and numerical investigations of Section 4.6.3, such as (i) the amplification of a given base excitation level that results in a more intense in-phase snap-through dynamic and (ii) an out-of-phase snap-through response around the uncoupled linear oscillator natural frequency. These two features may enhance energy harvesting performance and are shown to be tailored for a prescribed excitation level by adjusting the mass ratio μ, electromechanical coupling factor θ, and frequency ratio f. These key analytical trends are next validated through a series of experiments.

For electromagnetic vibration energy harvesters possessing small inductance, which is common for many meso-/macroscale platforms [139], the governing equations of motion are reducible by substituting the electrical flow of current in Eq. (4.137) into Eqs (4.135) and (4.136) using the relation $c = \theta x'/\rho$ [77,84,141]. Defining a coupling factor $\kappa_c = \theta/\rho$ the effects of electromechanical coupling on the bistable harvester dynamics are then modeled as damping influences, where the net dissipation factor is $\gamma^* = \gamma_1 + \kappa_c^2$. Consequently, in the following experiments, an "all-mechanical" platform is considered without loss of generality.

A photograph of the experimental platform is presented in Figure 4.37. The linear oscillator mass is composed of the machined aluminum frame, bottom bearing mass, and bistable oscillator bearing guide rail. The inertial mass of the bistable device is composed of the bearing mass in the middle of the photograph along with the hardware used to affix the mass to the attached spring. Bistability

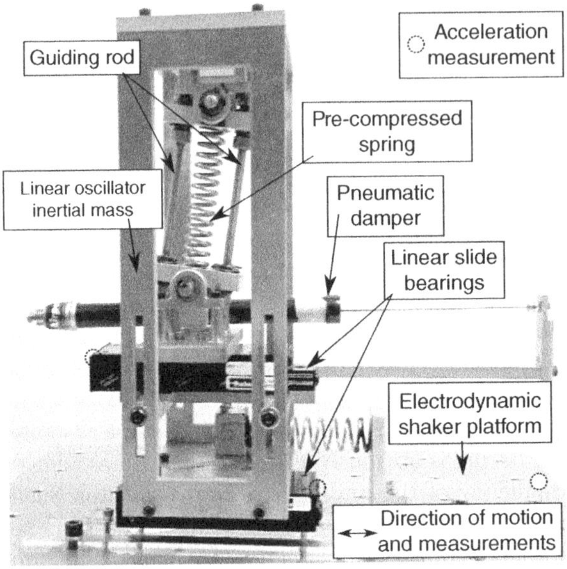

Figure 4.37 Photograph of coupled linear dynamic magnifier and bistable device system.

in this central device is generated by pre-compressing the spring in its upright, vertically-oriented configuration between the central, sliding inertial mass and a rotating pivot affixed to the linear oscillator. Therefore, the two stable equilibria for the bistable oscillator mass are to the right and left of the central unstable position. Guide rods prevent the compressed spring from bending or twisting as the inertial mass vibrates from one stable state to the other. Low-friction linear slide bearings are used while an adjustable, translational pneumatic damper modifies the net dissipative forces which occur due to relative motion between the linear oscillator and bistable device masses. The coupled system is mounted on an electrodynamic shaker table activated in the horizontal direction. The linear oscillator natural frequency is identified to be approximately 6 Hz. The shaker excitation during tests is a slowly forward swept sinusoid from 1 to 9 Hz ($+0.025\,\mathrm{Hz\,s^{-1}}$). To compare the dynamics of the individual bistable oscillator to the same driving excitation conditions, the linear oscillator is locked to the shaker platform, thereby only allowing relative motions between the bistable device mass and the shaker. Three accelerometers measure the dynamics of the shaker platform and linear and bistable oscillators, as indicated in Figure 4.37.

4.6.5.1 Effect of Bistable Harvester Electromechanical Coupling

Figure 4.38 plots the measured magnitudes of the acceleration frequency response functions (frf) for varying excitation frequency and for several cases of bistable device damping γ^* effected by adjusting the pneumatic damper. Recall that change in the damping for the electromagnetic bistable harvester may be equivalent to change in the electromechanical coupling factor κ_c according to the net damping definition $\gamma^* = \gamma_1 + \kappa_c^2$. The corresponding frfs for the individual bistable oscillator are the ratios of relative accelerations between the bistable device inertial mass to the driving shaker acceleration. For the coupled system,

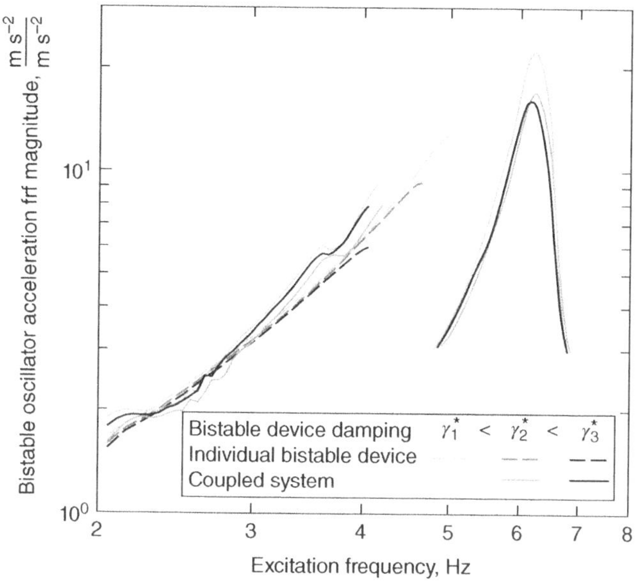

Figure 4.38 Relative acceleration frf magnitude of the bistable oscillator as a function of the base excitation frequency, for a variety of damping factors.

the frf is determined by the ratio of the relative acceleration between the bistable and linear oscillator masses to the shaker acceleration. Similarly to the previous reporting of the analytical and numerical results, only the dynamics related to the snap-through regime are plotted. The sweep test is conducted using an rms base acceleration of 2.83 m s^{-2}. As depicted in Figure 4.38, the large amplitude oscillations of the single bistable device destabilize at progressively lower frequencies when the net damping increases. In contrast, the snap-through dynamics of the coupled system are far less sensitive to the same increase in the damping, in very good agreement with the analytical trends uncovered in Figure 4.34.

While the individual bistable device exhibits only intrawell oscillations when the excitation frequency increases beyond that at which the snap-through motions are sustained, the coupled systems each undergo another energetic series of dynamics around the natural frequency of the linear dynamic magnifier. As the net damping in the bistable sub-system increases, the peak vibrational amplitude of this out-of-phase dynamic decreases, which agrees with the analytical predictions made in Section 4.6.3.3. Lastly, the results in Figure 4.38 show that the magnitudes of in- and out-of-phase snap-through dynamics are on the same order of magnitude. As examined in Section 4.6.3.3, this indicates that the levels of damping are relatively high in both the linear and bistable oscillator sub-systems, which may be practically attributed to bearing friction losses.

4.6.5.2 Effect of Bistable Harvester Mass

As observed in the analytical investigations, the mass ratio μ determines the frequency at which the favorable in-phase snap-through vibrations destabilize. This trend is probed experimentally by changing the net mass of the bistable oscillator. Yet, a direct change of the mass in the laboratory inherently also influences the linear natural frequency of the bistable sub-system and consequently changes other non-dimensional parameters such as f, γ_1, and also ε. Thus, to facilitate comparing trends between analysis and experiment, Figure 4.39a plots the analytically-predicted response amplitude of the bistable energy harvester having the dynamic magnifier for a variety of cases in which only the bistable mass is varied. As observed from the results in Figure 4.39a, the displacement amplitude of the bistable harvester in the coupled system is always enhanced for larger masses in the bistable harvester sub-system, when compared to the influence of the same mass change for the individual bistable harvester. For an increase in the harvester mass, the desirable in-phase snap-through destabilizes at lower frequencies, regardless of whether or not the bistable energy harvesting system includes the dynamic magnifier. In contrast, the out-of-phase dynamics for the system having the magnifier are not very sensitive to change in the absolute inertial mass of the bistable harvester.

Having an understanding of the specific influences of bistable harvester mass as predicted by the analytical formulation, Figure 4.39b plots the measured acceleration frfs for several cases of the bistable device mass while the base acceleration remains at an rms level of $2.07\,\mathrm{m\,s^{-2}}$. Each of the prior analytical trends is verified by the data. First, a frequency range is consistently found where the frf amplitudes for the coupled system are amplified above that of the counterpart individual bistable device. Next, the in-phase snap-through motions for both coupled and single bistable systems destabilize at progressively lower frequencies of excitation as the bistable device mass increases. Lastly, the out-of-phase oscillations exhibited by the coupled system are obviously insensitive to the change in the bistable device mass. These findings thoroughly support the analytical predictions so as to verify the efficacy of using the model for effective harvester system development and also demonstrate the advantages of the dynamic magnifier and bistable harvester configuration for robust energy conversion performance.

4.6.6 Summary

The recent discovery that a linear dynamic magnifier stage may intensify and multiply the vibration regimes of a linear energy harvester motivates this study to explore how the dynamic magnifier approach may favorably influence the performance of harmonically-excited bistable energy harvesters. Using a comprehensive collection of evidence, it is seen that the in-phase snap-through oscillations are significantly amplified when the excitation frequency is less than the resonant frequency of the uncoupled linear magnifier stage having an equivalent mass equal to the total dynamic mass of the two-DOF system. Average power density enhancements are greatest near this critical frequency, suggesting useful design guidelines in the light of known excitation spectral features. The mass ratio μ is found to primarily determine the frequency at which the coupled harvester system loses the capability to sustain in-phase snap-through vibrations, while the frequency ratio f mainly governs the magnitude and rate at which the amplified dynamics grow prior to losing sustainability. When the frequency of the base excitation is close to the resonant condition of the uncoupled linear magnifier stage, a second out-of-phase interwell oscillation regime is induced,

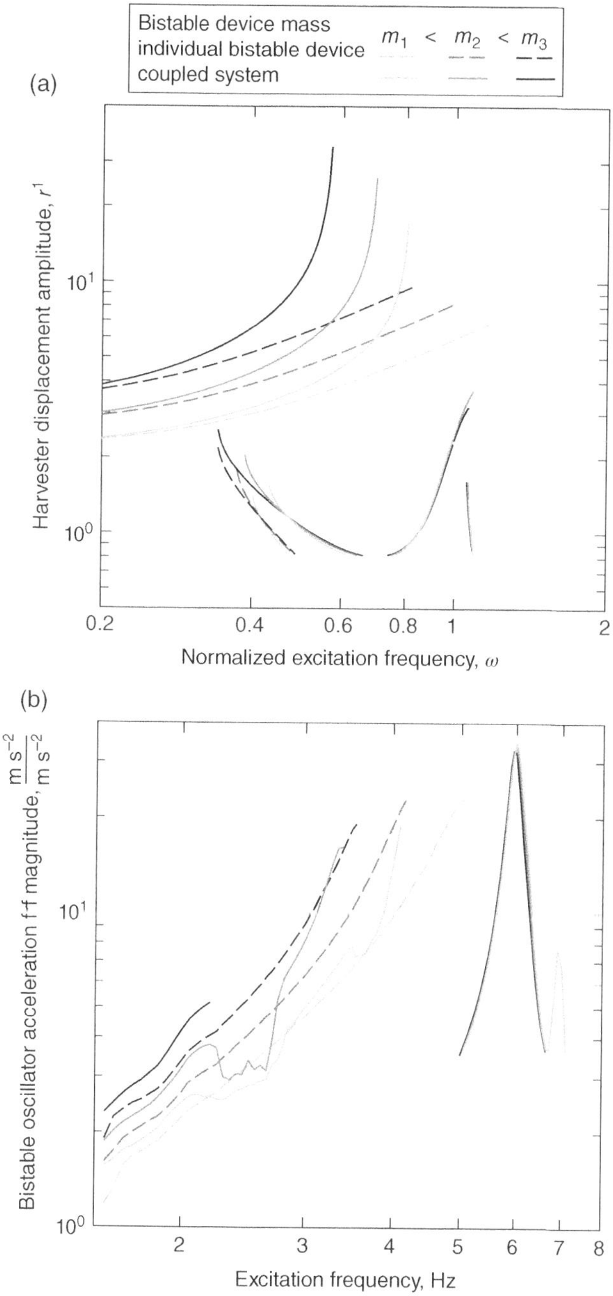

Figure 4.39 Snap-through dynamics of the bistable oscillator of the coupled system as functions of the excitation frequency for a variety of attached bistable device mass ratios. Shown in (a) is the bistable harvester displacement amplitude r_1 predicted analytically in comparison to the corresponding (b) measurements of relative acceleration frf of the bistable device.

an energy conversion opportunity unique to the multi-DOF bistable harvester system. Additionally, the multi-DOF platform exhibits a notable robustness in energy conversion capability for greater values of the electromechanical coupling factor θ where, in contrast, the individual bistable harvester shows considerable performance vulnerability to increases in the coupling. Taken together, the discoveries of Sections 4.5 and 4.6 indicate that intelligently deployed strategies to intensify or amplify the snap-through oscillations of bistable energy harvesters may be avenues to not only enhance the absolute power generation but also improve the more challenging measure of power density. The findings therefore justify the holistic design approach of a coupled linear–bistable system for advancing the aims of vibration-based energy harvesting.

References

[1] D.S. Landes, *The Unbound Prometheus: Technological Change and Industrial Development in Western Europe from 1750 to Present* (Cambridge University Press, Cambridge, 2003).
[2] World Energy Outlook 2014. Paris: International Energy Agency; 2014.
[3] BP Statistical Review of World Energy. BP; 2014.
[4] J. Falnes, A review of wave-energy extraction, *Marine Structures* **20**, 185–201 (2007).
[5] S.P. Beeby, M.J. Tudor, N.M. White, Energy harvesting vibration sources for microsystems applications, *Measurement Science and Technology* **17**, R175–R195 (2006).
[6] K. Baert, *et al.*, Technologies for highly miniaturized autonomous sensor networks, *Microelectronics Journal* **37**, 1563–1568 (2006).
[7] N.G. Elvin, N. Lajnef, A.A. Elvin, Feasibility of structural monitoring with vibration powered sensors, *Smart Materials and Structures* **15**, 977–986 (2006).
[8] J. Cruz, *Ocean Wave Energy: Current Status and Future Perspectives* (Springer, Berlin, 2008).
[9] A. Erturk, D.J. Inman, *Piezoelectric Energy Harvesting* (John Wiley & Sons, Ltd, Chichester, 2011).
[10] R.D. Kornbluh, *et al.*, From boots to buoys: promises and challenges of dielectric elastomer energy harvesting. In L. Rasmussen, ed. *Electroactivity in Polymeric Materials* (New York: Springer); 2012.
[11] S.D. Moss, O.R. Payne, G.A. Hart, *et al.*, Scaling and power density metrics of electromagnetic vibration energy harvesting devices, *Smart Materials and Structures* **24**, 023001 (2015).
[12] M.F. Daqaq, R. Masana, A. Erturk, *et al.*, On the role of nonlinearities in vibratory energy harvesting: a critical review and discussion, *ASME Applied Mechanics Review*, doi: 10.1115/1.4026278 (2014).
[13] E. Halvorsen, Fundamental issues in nonlinear wideband-vibration energy harvesting, *Physical Review E* **87**, 042129 (2013).
[14] S.D. Nguyen, E. Halvorsen, G.U. Jensen, Wideband MEMS energy harvester driven by colored noise, *Journal of Microelectromechanical Systems* **22**, 892–900 (2013).

[15] B.P. Mann, N.D. Sims, Energy harvesting from the nonlinear oscillations of magnetic levitation, *Journal of Sound and Vibration* **319**, 515–530 (2009).
[16] S. Leadenham, A. Erturk, M-shaped asymmetric nonlinear oscillator for broadband vibration energy harvesting: harmonic balance analysis and experimental validation, *Journal of Sound and Vibration* **333**, 6209–6223 (2014).
[17] L. Tang, Y. Yang, A nonlinear piezoelectric energy harvester with magnetic oscillator, *Applied Physics Letters* **101**, 094102 (2012).
[18] S.C. Stanton, C.C. McGehee, B.P. Mann, Reversible hysteresis for broadband magnetopiezoelastic energy harvesting, *Applied Physics Letters* **95**, 174103 (2009).
[19] J. Baker, S. Roundy, P. Wright, 2005 Alternative geometries for increasing power density in vibration energy scavenging for wireless sensor networks. In *Proceedings of 3rd International Energy Conversion Engineering Conference*; San Francisco. AIAA 2005–5617.
[20] R. Ramlan, M.J. Brennan, B.R. Mace, *et al.*, Potential benefits of a non-linear stiffness in an energy harvesting device, *Nonlinear Dynamics* **59**, 545–558 (2010).
[21] F. Cottone, H. Vocca, L. Gammaitoni, Nonlinear energy harvesting, *Physical Review Letters* **102**, 080601 (2009).
[22] L. Gammaitoni, I. Neri, H. Vocca, Nonlinear oscillators for vibration energy harvesting, *Applied Physics Letters* **94**, 164102 (2009).
[23] J.T. Lin, B. Alphenaar, Enhancement of energy harvested from a random vibration source by magnetic coupling of a piezoelectric cantilever, *Journal of Intelligent Material Systems and Structures* **21**, 1337–1341 (2010).
[24] B. Andò, S. Baglio, C. Trigona, *et al.*, Nonlinear mechanism in MEMS devices for energy harvesting applications, *Journal of Micromechanics and Microengineering* **20**, 125020 (2010).
[25] M. Ferrari, V. Ferrari, M. Guizzetti, *et al.*, Improved energy harvesting from wideband vibrations by nonlinear piezoelectric converters, *Sensors and Actuators A: Physical* **162**, 425–431 (2010).
[26] D. Meimukhin, N. Cohen, I. Bucher, On the advantage of a bistable energy harvesting oscillator under bandlimited stochastic excitation, *Journal of Intelligent Material Systems and Structures* **24**, 1736–1746 (2013).
[27] A.S. De Paula, D.J. Inman, M.A. Savi, Energy harvesting in a nonlinear piezomagnetoelastic beam subjected to random excitation, *Mechanical Systems and Signal Processing* **54**, 405–416 (2015).
[28] S.C. Stanton, C.C. McGehee, B.P. Mann, Nonlinear dynamics for broadband energy harvesting: investigation of a bistable piezoelectric inertial generator, *Physica D* **239**, 640–653 (2010).
[29] B.P. Mann, B.A. Owens, Investigations of a nonlinear energy harvester with a bistable potential well, *Journal of Sound and Vibration* **329**, 1215–1226 (2010).
[30] M.A. Karami, D.J. Inman, Powering pacemakers from heartbeat vibrations using linear and nonlinear energy harvesters, *Applied Physics Letters* **100**, 042901 (2012).
[31] L. Tang, Y. Yang, C.K. Soh, Improving functionality of vibration energy harvesters using magnets, *Journal of Intelligent Material Systems and Structures* **23**, 1433–1449 (2012).

[32] Y. Zhu, J. Zu, A magnet-induced buckled-beam piezoelectric generator for wideband vibration-based energy harvesting, *Journal of Intelligent Material Systems and Structures* **25**, 1890–1901 (2014).
[33] J.T. Lin, B. Lee, B. Alphenaar, The magnetic coupling piezoelectric cantilever for enhanced energy harvesting efficiency, *Smart Materials and Structures* **19**, 045012 (2010).
[34] H. Wu, L. Tang, Y. Yang, *et al.*, Development of a broadband nonlinear two-degree-of-freedom piezoelectric energy harvester, *Journal of Intelligent Material Systems and Structures* **25**, 1875–1889 (2014).
[35] D. Upadrashta, Y. Yang, Finite element modeling of nonlinear piezoelectric energy harvesters with magnetic interaction, *Smart Materials and Structures* **24**, 045042 (2015).
[36] W.J. Su, J. Zu, Y. Zhu, Design and development of a broadband magnet-induced dual-cantilever piezoelectric energy harvester, *Journal of Intelligent Material Systems and Structures* **25**, 430–442 (2014).
[37] W.J. Su, J.W. Zu, Design and development of a novel bi-directional piezoelectric energy harvester, *Smart Materials and Structures* **23**, 095012 (2014).
[38] S. Zhou, J. Cao, W. Wang, *et al.*, Modeling and experimental verification of doubly nonlinear magnet-coupled piezoelectric energy harvesting from ambient vibration, *Smart Materials and Structures* **24**, 055008 (2015).
[39] D. Zhu, S.P. Beeby, A broadband electromagnetic energy harvester with a coupled bistable structure, *Journal of Physics: Conference Series* **476**, 012070 (2013).
[40] P. Holmes, A nonlinear oscillator with a strange attractor, *Philosophical Transactions of the Royal Society of London A* **292**(1394), 419–448 (1979).
[41] F.C. Moon, P.J. Holmes, A magnetoelastic strange attractor, *Journal of Sound and Vibration* **65**, 275–296 (1979).
[42] A. Erturk, J. Hoffmann, D.J. Inman, A piezomagnetoelastic structure for broadband vibration energy harvesting, *Applied Physics Letters* **94**, 254102 (2009).
[43] A. Erturk, D.J. Inman, Broadband piezoelectric power generation on high-energy orbits of the bistable Duffing oscillator with electromechanical coupling, *Journal of Sound and Vibration* **330**, 2339–2353 (2011).
[44] G. Litak, M.I. Friswell, S. Adhikari, Magnetopiezoelastic energy harvesting driven by random excitations, *Applied Physics Letteres* **96**, 214103 (2010).
[45] S. Zhou, J. Cao, A. Erturk, J. Lin, Enhanced broadband piezoelectric energy harvesting using rotatable magnets, *Applied Physics Letters* **102**, 173901 (2013).
[46] S. Zhou, J. Cao, D.J. Inman, S. Liu, W. Wang, J. Lin, Impact-induced high-energy orbits of nonlinear energy harvesters, *Applied Physics Letters* **106**, 093901 (2015).
[47] J. Cao, S. Zhou, D.J. Inman, J. Lin, Nonlinear dynamic characteristics of variable inclination magnetically coupled piezoelectric energy harvesters, *Journal of Vibration and Acoustics* **137**, 021015 (2015).
[48] S. Zhao, A. Erturk, On the stochastic excitation of monostable and bistable electroelastic power generators: relative advantages and tradeoffs in a physical system, *Applied Physics Letters* **102**, 103902 (2013).

[49] D.J. Inman, *Vibration with Control* (John Wiley and Sons, Ltd, Chichester, 2006).

[50] L.N. Virgin, *Vibration of Axially Loaded Structures* (Cambridge University Press, Cambridge, 2007).

[51] R. Masana, M.F. Daqaq, Relative performance of a vibratory energy harvester in mono- and bi-stable potentials, *Journal of Sound and Vibration* **330**, 6036–6052 (2011).

[52] R. Masana, M.F. Daqaq, Energy harvesting in the super-harmonic frequency region of a twin-well oscillator, *Journal of Applied Physics* **111**, 044501 (2012).

[53] R. Masana, M.F. Daqaq, Response of duffing-type harvesters to band-limited noise, *Journal of Sound and Vibrations* **332**, 6755–6767 (2013).

[54] F. Cottone, L. Gammaitoni, H. Vocca, *et al.*, Piezoelectric buckled beams for random vibration energy harvesting, *Smart Materials and Structures* **21**, 035021 (2012).

[55] L. Van Blarigan, P. Danzl, J. Moehlis, A broadband vibrational energy harvester, *Applied Physics Letters* **100**, 253904 (2012).

[56] C. Xu, *et al.*, Bi-stable energy harvesting based on a simply supported piezoelectric buckled beam, *Journal of Applied Physics* **114**, 114507 (2013).

[57] G.W. Kim, J. Kim, Compliant bistable mechanism for low frequency vibration energy harvester inspired by auditory hair bundle structures, *Smart Materials and Structures* **22**, 014005 (2013).

[58] W.Q. Liu, A. Badel, F. Formosa, *et al.*, Novel piezoelectric bistable oscillator architecture for wideband vibration energy harvesting, *Smart Materials and Structures* **22**, 035013 (2013).

[59] W.Q. Liu, A. Badel, F. Formosa, *et al.*, Wideband energy harvesting using a combination of an optimized synchronous electric charge extraction circuit and a bistable harvester, *Smart Materials and Structures* **22**, 125038 (2013).

[60] D.A. Porter, T.A. Berfield, A bi-stable buckled energy harvesting device actuated via torque arms, *Smart Materials and Structures* **23**, 075003 (2014).

[61] A.W. Leissa. Advances in vibration, buckling and postbuckling studies on composite plates. In I.H. Marshall, ed. *Composite Structures*, pp. 312–334. (Applied Science Publishers, Essex, 1981).

[62] A.F. Arrieta, P. Hagedorn, A. Erturk, *et al.*, A piezoelectric bistable plate for nonlinear broadband energy harvesting, *Applied Physics Letters* **97**, 104102 (2010).

[63] A.F. Arrieta, T. Delpero, A.E. Bergamini, *et al.*, Broadband vibration energy harvesting based on cantilevered piezoelectric bi-stable composites, *Applied Physics Letters* **102**, 173904 (2013).

[64] D.N. Betts, H.A. Kim, C.R. Bowen, *et al.*, Optimal configurations of bistable piezo-composites for energy harvesting, *Applied Physics Letters* **100**, 114104 (2012).

[65] D.N. Betts, C.R. Bowen, H.A. Kim, *et al.*, Nonlinear dynamics of a bistable piezoelectric-composite energy harvester for broadband application, *The European Physical Journal* **222**, 1553–1562 (2013).

[66] H. Li, F. Dai, S. Du, Broadband energy harvesting by exploiting nonlinear oscillations around the second vibration mode of a rectangular piezoelectric bistable laminate, *Smart Materials and Structures* **24**, 045024 (2015).

[67] S.C. Stanton, B.A.M. Owens, B.P. Mann, Harmonic balance analysis of the bistable piezoelectric inertial generator, *Journal of Sound and Vibration* **331**, 3617–3627 (2012).

[68] N. Cohen, I. Bucher, M. Feldman, Slow-fast response decomposition of a bi-stable energy harvester, *Mechanical Systems and Signal Processing* **31**, 29–39 (2012).

[69] B.P. Mann, D.A.W. Barton, B.A.M. Owens, Uncertainty in performance for linear and nonlinear energy harvesting strategies, *Journal of Intelligent Material Systems and Structures* **23**, 1451–1460 (2012).

[70] A.J. Sneller, P. Cette, B.P. Mann, Experimental investigation of a post-buckled piezoelectric beam with an attached central mass used to harvest energy, *Proceedings of the Institution of Mechanical Engineers, Part I: Journal of Systems and Control Engineering* **225**, 497–509 (2011).

[71] M.A. Karami, D.J. Inman, Equivalent damping and frequency change for linear and nonlinear hybrid vibrational energy harvesting systems, *Journal of Sound and Vibration* **330**, 5583–5597 (2011).

[72] S.C. Stanton, B.P. Mann, B.A.M. Owens, Melnikov theoretic methods for characterizing the dynamics of a bistable piezoelectric inertial generator in complex spectral environments, *Physica D* **241**, 711–720 (2012).

[73] N. Cohen, I. Bucher, On the dynamics and optimization of a non-smooth bistable oscillator: application to energy harvesting, *Journal of Sound and Vibration* **333**, 4653–4667 (2014).

[74] S.W. Shaw, P. Holmes, A periodically forced piecewise linear oscillator, *Journal of Sound and Vibration* **90**, 129–155 (1983).

[75] M. Panyam, R. Masana, M.F. Daqaq, On approximating the effective bandwidth of bi-stable energy harvesters, *International Journal of Non-Linear Mechanics* **67**, 153–163 (2014).

[76] M. Panyam, M.F. Daqaq, A comparative performance analysis of electrically optimized nonlinear energy harvesters, *Journal of Intelligent Material Systems and Structures*, doi: 10.1177/1045389X15573344 (2015).

[77] M.F. Daqaq, Transduction of a bistable inductive generator driven by white and exponentially correlated Gaussian noise, *Journal of Sound and Vibration* **330**, 2554–2564 (2011).

[78] M.F. Daqaq, On intentional introduction of stiffness nonlinearities for energy harvesting under white Gaussian excitations, *Nonlinear Dynamics* **69**, 1063–1079 (2012).

[79] Q. He, M.F. Daqaq, Influence of potential function asymmetries on the performance of nonlinear energy harvesters under white noise, *Journal of Sound and Vibration* **333**, 3479–3489 (2014).

[80] Q. He, M.F. Daqaq, New insights into utilizing bistability for energy harvesting under white noise, *Journal of Vibration and Acoustics* **137**, 021009 (2015).

[81] M.F. Daqaq, R. Masana, A. Erturk, *et al.*, On the role of nonlinearities in vibratory energy harvesting: a critical review and discussion, *Applied Mechanics Reviews* **66**, 040801 (2014).

[82] C.B. Williams, R.B. Yates, Analysis of a micro-electric generator for microsystems, *Sensors and Actuators A* **52**, 8–11 (1996).

[83] B.P. Mann, N.D. Sims, On the performance and resonant frequency of electromagnetic induction energy harvesters, *Journal of Sound and Vibration* **329**, 1348–1361 (2010).

[84] R.L. Harne, K.W. Wang, Prospects for nonlinear energy harvesting systems designed near the elastic stability limit when driven by colored noise, *ASME Journal of Vibration and Acoustics* **136**, 021009 (2014).

[85] A.H. Nayfeh, B. Balachandran, *Applied Nonlinear Dynamics: Analytical, Computational, and Experimental Methods* (Wiley-VCH Verlag, GmbH, Weinheim, 1995).

[86] R.S. Burington, *Handbook of Mathematical Tables and Formulas* (McGraw-Hill Inc., New York, 1973).

[87] W. Szemplińska-Stupnicka, J. Rudowski, Steady states in the twin-well potential oscillator: computer simulations and approximate analytical studies, *Chaos* **3**, 375–385 (1993).

[88] W. Szemplińska-Stupnicka, The analytical predictive criteria for chaos and escape in nonlinear oscillators: a survey, *Nonlinear Dynamics* **7**, 129–147 (1995).

[89] I. Kovacic, M.J. Brennan, editors, *The Duffing Equation: Nonlinear Oscillators and their Behaviour* 1st ed. (John Wiley & Sons, Ltd, Chichester, 2011).

[90] A.H. Nayfeh, D.T. Mook, *Nonlinear Oscillations* (Wiley-VCH Verlag, GmbH, Weinheim, 1995).

[91] Y.C. Shu, I.C. Lien, W.J. Wu, An improved analysis of the SSHI interface in piezoelectric energy harvesting, *Smart Materials and Structures* **16**, 2253–2264 (2007).

[92] A. Badel, D. Guyomar, E. Lefeuvre, *et al.*, Piezoelectric energy harveting using a synchronized switch technique, *Journal of Intelligent Material Systems and Structures* **17**, 831–839 (2006).

[93] D. Guyomar, A. Badel, E. Lefeuvre, *et al.*, Toward energy harvesting using active materials and conversion improvement by nonlinear processing, *IEEE Transactions on Ultrasonics, Ferroelectrics, and Frequency Control* **52**, 584–595 (2005).

[94] W.Y. Tseng, J. Dugundji, Nonlinear vibrations of a buckled beam under harmonic excitation, *Journal of Applied Mechanics* **38**(2), 467–476 (1971).

[95] D. Wagg, S. Neild, *Nonlinear Vibration with Control: For Flexible and Adaptive Structures* (Springer, Dordrecht, 2010).

[96] J. Nocedal, S.J. Wright, *Numerical Optimization* (Springer, New York, 2006).

[97] H.A. DeSmidt, 2010 Analysis of a dual gearbox/shaft system with nonlinear dynamic mesh phase interactions. In *Proceedings of the 51st AIAA/ASME/ASCE/AHS/ASC Structures, Structural Dynamics and Materials Conference*; Orlando. AIAA-2010-2893.

[98] C. Padmanabhan, R. Singh, Analysis of periodically excited non-linear systems by a parametric continuation technique, *Journal of Sound and Vibration* **184**, 35–58 (1995).

[99] M.W. Sracic, S. Yang, M.S. Allen. Comparing measured and computed nonlinear frequency responses to calibrate nonlinear system models. In D. Adams, G. Kerschen, A. Carrella. eds *Topics in Nonlinear Dynamics, Volume 3: Proceedings of the 30th IMAC, A Conference on Structural Dynamics 2012*, pp. 255–267 (Springer, New York, 2012).

[100] R. Masana, M.F. Daqaq, Electromechanical modeling and nonlinear analysis of axially loaded energy harvesters, *Journal of Vibration and Acoustics* **133**, 011007 (2011).

[101] Y. Liao, H.A. Sodano, Structural effects and energy conversion efficiency of power harvesting, *Journal of Intelligent Material Systems and Structures* **20**, 505–514 (2009).

[102] W. Szemplińska-Stupnicka, Higher harmonic oscillations in heteronomous non-linear systems with one degree of freedom, *International Journal of Non-Linear Mechanics* **3**, 17–30 (1968).

[103] R.A. Ibrahim, Excitation-induced stability and phase transition: a review, *Journal of Vibration and Control* **12**(10), 1093–1170 (2006).

[104] Z. Wu, R.L. Harne, K.W. Wang, Excitation-induced stability in a bistable Duffing oscillator: analysis and experiment, *Journal of Computational and Nonlinear Dynamics* **10**, 011016 (2015).

[105] S. Wiggins, *Introduction to Applied Nonlinear Dynamical Systems and Chaos* (Springer, New York, 2003).

[106] A.A. Shama, J.B. Mander, S.S. Chen *et al.*, Ambient vibration and seismic evaluation of a cantilever truss bridge, *Engineering Structures* **23**, 1281–1292 (2001).

[107] W.F. Albers, O. Hag-Elsafi, S. Alampalli. Dynamic analysis of the Bentley Creek bridge with FRP deck. Albany, NY: U.S. Department of Transportation Federal Highway Administration; 2007. Report No.: SR-07/150.

[108] S. Kumarasena, N.P. Jones, P. Irwin, *et al.*, Wind-induced vibration of stay cables. McLean, VA: U.S. Department of Transportation Federal Highway Administration; 2007. Report No.: FHWA-RD-05-083.

[109] M. Dykman, K. Lindenberg. Fluctuations in nonlinear systems driven by colored noise. In G.H. Weiss, ed. *Contemporary Problems in Statistical Physics* (SIAM, Philadelphia, 1994).

[110] S.P. Timoshenko, *Theory of Elastic Stability* (McGraw-Hill, New York, 1936).

[111] D.D. Quinn, A.L. Triplett, L.A. Bergman, *et al.*, Comparing linear and essentially nonlinear vibration-based energy harvesting, *Journal of Vibration and Acoustics* **133**, 011001 (2011).

[112] D.D. Quinn, A.L. Triplett, A.F. Vakakis, *et al.*, Energy harvesting from impulsive loads using intentional essential nonlinearities, *Journal of Vibration and Acoustics* **133**, 011004 (2011).

[113] D.K. Kondepudi, I. Prigogine, G. Nelson, Sensitivity of branch selection in nonequilibrium systems, *Physics Letters A* **111**, 29–32 (1985).

[114] D.K. Kondepudi, F. Moss, P.V.E. McClintock, Branch selection in the presence of coloured noise, *Physics Letters A* **114**, 68–74 (1986).

[115] H. Risken, *The Fokker-Planck Equation: Methods of Solution and Applications* 2nd edn (Springer-Verlag, Berlin, 1989).

[116] F. Moss, P.V.E. McClintock, eds, *Noise in Nonlinear Dynamical Systems, Volume 1: Theory of Continuous Fokker-Planck Systems* (Cambridge University Press, Cambridge, 1989).

[117] P. Hänggi, P. Jung. Colored noise in dynamical systems. In I. Prigogine, S.A. Rice, eds *Advances in Chemical Physics, Vol. 89*, pp. 239–326 (John Wiley & Sons, Inc., New York 1995).

[118] P. Hänggi, Correlation functions and master equations of generalized (non-Markovian) Langevin equations, *Zeitschrift für Physik B* **31**, 407–416 (1978).

[119] P. Hänggi, T.J. Mroczkowski, F. Moss, *et al.*, Bistability driven by colored noise: theory and experiment, *Physical Review A* **32**, 695–698 (1985).

[120] L. Fronzoni, P. Grigolini, P. Hänggi, *et al.*, Bistable oscillator dynamics driven by nonwhite noise, *Physical Review A* **33**, 3320–3327 (1986).

[121] F. Moss, P.V.E. McClintock, eds, *Noise in Nonlinear Dynamical Systems, Volume 3: Experiments and Simulations* (Cambridge University Press, Cambridge, 1989).

[122] L. Gammaitoni, P. Hänggi, P. Jung, *et al.*, Stochasic resonance, *Reviews of Modern Physics* **70**(1), 223–287 (1998).

[123] H. Liu, Z. Huang, T. Xu, *et al.*, Enhancing output power of a piezoelectric cantilever energy harvesting using an oscillator, *Smart Materials and Structures* **21**, 065004 (2012).

[124] L. Tang, Y. Yang, H. Wu, Modeling and experiment of a multiple-DOF piezoelectric energy harvester. In *Proceedings of Active and Passive Smart Structures and Integrated Systems 2012* vol. 83411E (*SPIE*, 2012).

[125] L. Tang, Y. Yang, A multiple-degree-of-freedom piezoelectric energy harvester model, *Journal of Intelligent Material Systems and Structures* **23**, 1631–1647 (2012).

[126] S. Qi, R. Shuttleworth, S. Olutunde Oyadiji, *et al.*, Design of a multiresonant beam for broadband piezoelectric energy harvesting, *Smart Materials and Structures* **19**, 094009 (2010).

[127] R.L. Harne, K.W. Wang, On the fundamental and superharmonic effects in bistable energy harvesting, *Journal of Intelligent Material Systems and Structures* **25**, 937–950 (2014).

[128] R.L. Harne, A. Sun, K.W. Wang, Leveraging nonlinear saturation-based phenomena in an L-shaped vibration energy harvesting system, *Journal of Sound and Vibration* **363**, 517–531 (2016).

[129] P.L. Green, E. Papatheou, N.D. Sims, Energy harvesting from human motion and bridge vibrations: an evaluation of current nonlinear energy harvesting solutions, *Journal of Intelligent Material Systems and Structures* **24**, 1494–1505 (2013).

[130] A. Erturk, J.M. Renno, D.J. Inman, Modeling of piezoelectric energy harvesting from an L-shaped beam-mass structure with an application to UAVs, *Journal of Intelligent Material Systems and Structures* **20**, 529–544 (2009).

[131] W. Szemplińska-Stupnicka, J. Bajkowski, Multi-harmonic response in the regions of instability of harmonic solution in multi-degree-of-freedom non-linear systems, *International Journal of Non-linear Mechanics* **15**, 1–11 (1980).

[132] P. Malatkar, A.H. Nayfeh, Steady-state dynamics of a linear structure weakly coupled to an essentially nonlinear oscillator, *Nonlinear Dynamics* **47**, 167–179 (2007).

[133] J. Liang, W.H. Liao, Impedance modeling and analysis for piezoelectric energy harvesting systems, *IEEE/ASME Transactions on Mechatronics* **17**, 1145–1157 (2012).

[134] X. Tang, L. Zuo, Enhanced energy harvesting using dual-mass systems, *Journal of Sound and Vibration* **330**, 5199–5209 (2011).
[135] W. Zhou, G.P. Penamalli, L. Zuo, An efficient vibration energy harvester with a multi-mode dynamic magnifier, *Smart Materials and Structures* **21**, 015014 (2012).
[136] O. Aldraihem, A. Baz, Energy harvester with a dynamic magnifier, *Journal of Intelligent Material Systems and Structures* **22**, 521–530 (2011).
[137] A. Aladwani, M. Arafa, O. Aldraihem, *et al.*, Cantilevered piezoelectric energy harvester with a dynamic magnifier, *Journal of Vibration and Acoustics* **134**, 031004 (2012).
[138] R.L. Harne, M. Thota, K.W. Wang, Concise and high-fidelity predictive criteria for maximizing performance and robustness of bistable energy harvesters, *Applied Physics Letters* **102**, 053903 (2013).
[139] I.L. Cassidy, J.T. Scruggs, S. Behrens, *et al.*, Design and experimental characterization of an electromagnetic transducer for large-scale vibratory energy harvesting applications, *Journal of Intelligent Material Systems and Structures* **22**(17), 2009–2024 (2011).
[140] J.P. Den Hartog, *Mechanical Vibrations* (McGraw-Hill, New York, 1947).
[141] Z. Wu, R.L. Harne, K.W. Wang, Energy harvester synthesis via coupled linear-bistable system with multistable dynamics, *Journal of Applied Mechanics* **81**, 061005 (2014).
[142] C. Zhang, R.L. Harne, B. Li, K.W. Wang, Reconstructing the transient, dissipative dynamics of a bistable Duffing oscillator with an enhanced averaging method and Jacobian elliptic functions, *International Journal of Non-Linear Mechanics* **79**, 26–37 (2016).

5

Sensing and Detection

This chapter describes how the dynamics of bistable systems have been recently leveraged to introduce new and important functionality for sensing and detecting changes in structures. On the whole, the *dynamic bifurcations* of bistable systems are the keys to the advanced sensing functionalities, since the drastic variation in qualitative response before and after crossing a bifurcation is far more apparent than tracking traditional dynamic behaviors, like a natural frequency, which exhibit less obvious change for the same monitored parameter shift. First, the chapter surveys the many developments to date regarding the use of dynamic bifurcations of bistable systems, developments which are more numerous on the microscale due to the proliferation of bistable structures in that setting. Then, several sections present recent studies that consider both micro- and macroscale bistable sensing architectures which harness and integrate bifurcation- and resonance-based dynamical phenomena to introduce new detection functionality and to enhance emerging sensing principles via the strategic design and operation of the sensors.

5.1 Topic Review

Although nearly all life forms possess some sensory means, humans have the distinguishing trait of *fabricating* sensors to evaluate changes in the environment which may not be detectable or easily quantified using the five, in-born senses. The obelisks constructed by the ancient Egyptians are potentially the first man-made sensors; the structures provided a measure of the sun's passage across the sky and thus a means to determine the incremental amount of a day which had elapsed [1]. The need for more accurate navigation across the seas and oceans drove Chinese inventors to develop the magnetic compass, potentially around the turn of the first millennium [2], to improve upon global positioning strategies based on tracking celestial bodies.

Sensor development has come a long way from the needs to know the time and one's position (although these are no less important data now than thousands of years ago!). Of more pertinent interest to this book are sensor systems designed to monitor the changing dynamics of structures. This is because, according to

Harnessing Bistable Structural Dynamics: For Vibration Control, Energy Harvesting and Sensing, First Edition. Ryan L. Harne and K.W. Wang.

classical mechanics, the change of a structural dynamic indicates that one or more of the structural or environmental properties have correspondingly varied. Regardless of scale, this is a fundamental axiom of structural monitoring [3]. Thus, whether tracking the accumulation of biological matter upon (or in) an oscillating microcantilever [4,5] or detecting the multiscale dynamic changes of civil infrastructures like highway bridges [6], the deviation of a vibrational response from a previous behavior serves as the common indicator that *something* is different about the system. Therefore, by monitoring the deviation over time and relating it to a model or nominal characteristics of the system, one could quantify the variation in terms of a meaningful feature or property that is anticipated to change, whether due to internal or external influences.

The motivations for sensing changing structural dynamics are numerous. The application areas of particular interest in this book are structural health monitoring (SHM) and tracking the adsorption of miniscule mass or actions of force fields on micro-/nanosystems. In the framework of SHM, because mechanical or civil infrastructures are some of the most costly assets in modern civilization (including systems such as grid-scale electrical power generators, aircraft, bridges, buildings, etc.), enormous commercial and government interests unite to ensure that after construction and deployment the systems operate with a maximum of integrity and performance as intended [7]. To these ends, the massive structures are monitored in real-time or in managed intervals, where the measured structural dynamics are probed for deviations that might indicate the onset of damage or an undesired state of operation. The commonly employed vibration-based SHM methods track the changing natural frequencies and modal behaviors of the structures (typically low frequency phenomena) which are readily evaluated via distributed sensor systems [8,9]. Although it is challenging for these methods to detect small, initialized damage sites due to the inverse correlation between defect size and vibrational frequency [3,10], the vibration-based approaches have maintained their popularity in practice due to their straightforward implementation and because they readily provide valuable indicators of the "global" health of a structure.

The extensive proliferation of micro-/nanosystems as (electro)mechanical sensors is in part driven by the corresponding advancement in microprocessor technologies, because such small-scale sensors are fabricated using many of the same techniques as the ever-shrinking computer chips [11–13]. As a result, micro-/nanosystems can be economically fabricated in large batches and are being produced on smaller and smaller scales [14]. Due to their size, these sensors are able to detect extremely small quantities of matter or applied loads with the many of the same frequency-shift-based strategies that are employed in the SHM field. On the other hand, because of the small scales involved, other forces act on the sensors to meaningful and *measurable* extents [15], such as the Casimir force, van der Waals, and optical fields, to name a few. As relates to mass and force monitoring, by tracking the change of dynamic features of the small-scale sensors, scientists and engineers can identify changing airborne concentrations of noxious gases and chemicals [16], environmental cues like humidity inside packaged electronics [17], quantum-scale force fields [18], and biological materials [19], among other important characteristics. Thus, as sensitive laboratory transducers or as devices integrated with conventional electronics, micro-/nanoscale sensors systems have broken enormous ground to enhance our measurement capabilities for a large variety of natural and synthetic phenomena, thereby advancing our knowledge of their underlying principles.

Until very recently, bistable systems have been unexplored in the context of sensing for large-scale structures. Later sections of this chapter will describe the recent advances made by exploiting bistability in sensing strategies useful for SHM practices. In the other sensing venue of interest here, bistable devices have a long history as actuators, sensors, and transducers, due in part to their relatively straightforward fabrication on micro- and nanoscales. For consistency, hereafter such small-scale devices will be referred to as *micro*scale systems/structures/and so on, unless it is specifically intended to denote a *nano*scale structure.

5.1.1 Bistable Microsystems

A survey of the historical development of bistable microstructures requires a wider view than only towards those investigations having sensing as the proposed application basis. Take for instance the fact that many actuators can be used in "reverse" as sensors, and vice versa, although the performance may be significantly better for one purpose than the other. As a result, a fully-informed evaluation of the overall historical development of bistable microsystems must account for a broader range of explorations than only those based around sensing, since innovations made in the design, fabrication, processing, and operational understanding of bistable microsystems (as actuators, switches, relays, valves, resonators, etc.) may lend important insight into the development of bistable microsensors specifically. To yield a more encompassing examination of the scientific and engineering developments in this field, this broader view is the strategy adopted in the following review. Moreover, an abbreviated summary is necessary in order to highlight the developments particularly relevant to this book since the field encompasses a vast community of investigators and an abundant number of published works. Greater details can therefore be found in other references [11,15,20].

Some of the earliest investigations of bistable microsystems examined them as the unintended consequences of microbeam fabrication when both ends of the beams were kept attached to the original substrate after the beam was released (i.e., etched free from below to give conventional "clamped–clamped" boundary conditions). Due to residual stresses imposed by the fabrication process, the beams would buckle out-of-plane upon release from the underlying sacrificial substrate layer, like that illustrated in Figure 5.1. Fang and Wickert [21] characterized this phenomenon analytically and experimentally early in the timeline of bistable microsystem development. The curling or buckling of microstructures due to the fabrication procedures is a well-known and often-encountered reality in the field [11,22]. In fact, this phenomenon has been deliberately exploited to produce mechanically buckled microbeams [23] although the strategy may present challenges for large batch production if consistent design parameters are required.

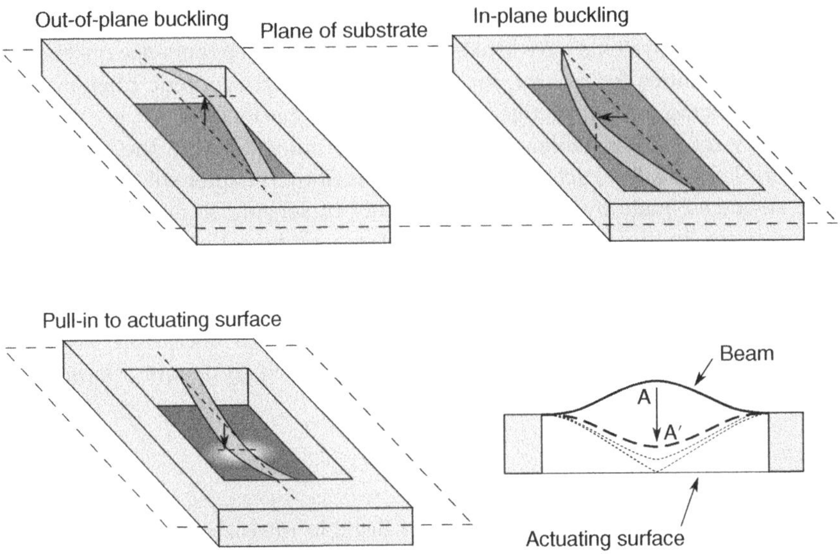

Figure 5.1 Distinction between out-of-plane and in-plane buckling of bistable microstructures, and illustration of the pull-in phenomenon.

The use of design geometry constraints is one means to fabricate bistable microstructures having more reproducible mechanical properties. Howell and colleagues have made particular advances in the study and development of such systems [24]. This research team has examined a myriad of architectural forms which empower bistability (often, bistable in-plane) [25–29]. By incorporating a considerable compliance in the structural members, many of the geometries that have been explored allow for a low stress distribution in the system as it rests in one of the two statically-stable equilibria. This is an important advantage if the structure undergoes many transitions between the stable states. Without such attention, long-term fatiguing of the system could cause the resting configurations to progressively deviate and reduce the effectiveness and performance of the device. The geometrically bistable architectures introduce novel functionality for electromechanical microsystems, wherein the bistable constituent may store and release elastic energy via the switching of stable states. The use of such compliant bistable microsystem designs also enhances fabrication repeatability as compared to the reliance on residual stress phenomena to effect bistability. The group has examined these devices for their utility as accelerometers, shock sensors, thermal actuators, switches, positioners, and more [30–34]. In short, the research team has identified and elucidated a wide variety of opportunities afforded by the implementation of inherently bistable microstructures.

Many researchers have investigated innovative concepts for bistable microstructures used as switches, actuators, or positioners, which are also regularly fabricated as in-plane systems. Han *et al.* [35] developed a nonlinear finite element model to compute the electromagnetic actuation force necessary to operate an in-plane bistable microswitch; the actuation efforts are important to quantify to ensure the successful long-term deployment of the microsystem.

The advantages of low actuation power requirements for an electromagnetically controlled bistable beam were examined by Park and Hah [36], and the theoretical predictions were validated through experimentation on a batch of fabricated devices. Wang *et al.* [37] showed that oscillatory electromagnetic actuation on an in-plane bistable micromechanism could be performed at almost half of the required current as a static actuation strategy. These works demonstrate that electromagnetically actuated bistable microsystems may provide non-trivial power savings if the switches are employed repeatedly, such as for conventional signal relays, mixers, switches, and valves. On the other hand, magnet handling and deposition during bistable microsystem fabrication can be challenging [38], and many other multifield phenomena have been examined to exert more easily controlled forces upon the devices.

Thermal activation of a mechanically bistable microbeam system was proposed by Qiu *et al.* [39] in the context of realizing a fail-safe switch for the prevention of power supply overheating. The device employed two parallel beams joined together by a central clamp to inhibit asymmetric buckling motions which could compromise the intended functionality of the structure. Saif and collaborators [40–42] developed a thorough foundation in the understanding of bistable microstructures buckled in-plane and actuated via electrical or optical fields. The theoretical insight was verified with experimental studies that even helped to explain anomalous dynamic behaviors related to seemingly inconsistent requirements for effecting switches between the stable states [42]. Kumar and Rhoads [43] extended the framework to consider the transient dynamics of optically actuated bistable microswitches, and examined the light pulse times required to induce a transition from one stable state to the other. Casals-Terré *et al.* [44] demonstrated analytically and experimentally that oscillatory actuation of mechanically bistable microbeams could reduce the voltage potential amplitude needed to snap the device from one stable state to the other, which agrees with the observations made in regards to electromagnetic actuation of bistable micromechanisms [37].

The potential applications for inherently bistable micromechanisms are truly numerous, particularly due to the proliferation of microtechnologies which could benefit from (electro)mechanical devices to replace all-electrical components (resistors, inductors, capacitors, transistors, etc.) that have less desirable performance on such small scales. In this spirit, bistable microsystems have been explored as non-volatile memory units, compared to the conventional type of random-access memory which is volatile and hence loses the stored data if power is removed. Because common binary data is resolved using two states (0s and 1s), the two stable configurations of bistable microsystems appear to be opportune binary mechanical realizations around which to base a non-volatile data storage strategy. Roodenburg *et al.* [45] used electrostatic fields to switch a buckled microbeam from one stable state to the other and found highly repeatable data storage results could be obtained once a "burn-in" operation period had been conducted. Intaraprasonk and Fan [46] proposed transmitting optical pulses into a parallel pair of bistable waveguide beams to switch between the odd and even modes and demonstrated a data read-out strategy using the simulated results. Bagheri *et al.* [47] coupled an optical cavity and buckled nanobeam system and experimentally identified the state of the beam based on the refractive index through the coupled system; laser pulses were the mechanism used to switch the non-volatile memory element from one of the stable states to the other. Indeed, due to the proliferation of fiber-optic and laser technologies, the optical actuation of bistable microswitches, relays, actuators, or memory units provides appreciable benefits in terms of a straightforward integration and packaging with other, common electronics.

Researchers have also proposed valve designs which leverage the two statically stable configurations of bistable microstructures to maintain a pressure gradient between different volumes. Wagner *et al.* [48] introduced a novel valve design composed of two bistable, coupled microhemispheres that were actuated electrostatically to switch between stable hemisphere curvatures. Schomburg and Goll [49] used thermo-pneumatic effects to actuate a bistable microvalve and demonstrated that the design could withstand high pressure loads for such a small scale platform [50]. Capanu *et al.* [51], Bintoro *et al.* [52], and Sutanto *et al.* [53] exploited magneto-elastic interactions in the fabrication and operation of bistable microvalve architectures; the research teams examined the benefits of electromagnetic actuation in particular, for low voltage, and hence low power, switching performance. In the ever-expanding field of microfluidics for laboratory and medicinal purposes [54,55], the utilization of inherently bistable microstructures to govern flow characteristics may provide advantages that alternative valve mechanisms do not so effectively achieve.

As the examples above demonstrate, there are numerous multifield phenomena available on the microscale to induce bistability and/or to switch between the stable states of the microstructure: in conjunction with elastic restoring forces, researchers have used electrical fields, optical momentum, pressure or thermal gradients, electromagnetic forces, and other means to enable the transition of state. On the other hand, *electrical* actuation of microsystems is by far the most common method of microstructure interrogation [11]. The fast actuation performance, high energy density, and ease of fabrication and implementation make electrostatic actuation a popular choice for the static and dynamic control of microscale devices. For such microelectromechanical systems (MEMS), a voltage difference is imposed between a suspended structural surface and a fixed surface which results in an electrostatic force between them. Because the fixed surface is sufficiently anchored, the structure is deflected from its original configuration, providing a practical means to control the static and dynamic displacement of the MEMS.

With the numerous advantages of electrostatic actuation come costs. For instance, the phenomenon is strongly nonlinear – the exerted force is inversely proportional to the square of the distance between the capacitive surfaces – and therefore requires care to deploy whether statically and dynamically (using DC and AC signals, respectively). In fact, this feature has been the motivation for a vast amount of research. As pertains to bistable microsystems, electrostatics introduces the opportunity to simultaneously effect bistability and govern the device dynamics. Thus, if the structure is not bistable due to geometric design, residual stresses, or so on, then a static voltage potential difference may be applied to effect the characteristic double-well potential energy profile of a bistable structure. Thereafter, voltages above or below this level induce the desired dynamics for whatever the application may be.

There is a particular concern encountered by employing electrostatic fields to control MEMS, particularly when the devices deflect out-of-plane in an axis normal to the actuating surface. In such a scenario, a large electrostatic force could overcome the elastic restoring force of the microstructure and result in a phenomenon known as "pull-in", when the microstructure adheres to the underlying actuating surface, Figure 5.1. One may consider an analogous behavior occurs when two attractive magnet faces come into contact. Since pull-in essentially leads to the failure of the system, a host of studies have sought to characterize its onset and have investigated appropriate methods for static and dynamic actuation that avoid the catastrophic consequences and ensure the reliable operation of bistable MEMS.

The desire to use bistable microsystems as switches effectively means that the devices are regularly acted upon by step forces for the fastest operation. For electrostatically actuated MEMS, this poses concern if the structure is pulled towards the stable configuration that is closer to the actuating surface: too great an attractive force could suddenly lead to pull-in. This is illustrated in Figure 5.1 when the target switch is from state A to A', but the nearness of the actuating surface may inadvertently lead to pull-in of the beam to the actuating surface. Krylov [56] explored the use of Lyapunov exponents to determine the dynamic stability of bistable MEMS actuated out-of-plane in this fashion; the computational efficiency of the approach was apparent when higher-order modal responses represented a non-trivial contribution to the overall behaviors. A theoretical and experimental investigation of statically and dynamically actuated bistable MEMS was undertaken by Krylov *et al.* [57], providing a rigorous framework with which to evaluate the vulnerability of a given electrostatic actuation strategy to the pull-in effect.

Das and Batra [58,59] developed a model of bistable microstructures using exact continuum balance laws and Maxwell's equations which led to the observation that bistable microarches are less susceptible to pull-in than microbeams. The reports also indicated that shallow arches would undergo symmetric snap-through motions and that the symmetric behaviors had greater resistance to pull-in than asymmetric buckling motions. When reduced model fidelity is desired for the sake of alleviating computational cost, Gutschmidt [60] discovered that lower-order models could in fact be more accurate. The author presented a prescription for this interesting result by providing analytical modeling guidelines based upon the microbeam design under study and the overall goals of the analysis. Ouakad and Younis [61] focused on the dynamic behavior of electrostatically actuated microarches and showed that the perturbation strategy (the method of multiple scales) used to approximately solve the nonlinear governing equations was insufficient to predict the existence or activation of large amplitude interwell vibrations (i.e., continuous snap-through dynamics). Another analytical study by Younis *et al.* [62] explored how damping and the AC actuating voltage magnitude could be tailored to induce bistability as compared to directly triggering pull-in of the MEMS arch, although the corresponding experimental system possessed too small a gap below the suspended arch to realize bistability in the laboratory examinations.

Alkharabsheh and Younis [63] found that unintentional under-etching of bistable arches could substantially change the qualitative dynamic behaviors of the microsystems by lending an extra flexibility at the beam boundary conditions that compromised the integrity of the device to pull-in. The authors also examined how axial forces, characteristic of the residual stresses imposed by fabrication or temperature change, would influence the static arch shapes and consequent dynamics [64]. Extending from [57], Medina *et al.* [65] observed that the nonlinearity of the electrostatic force plays a significant role in the analytically derived criteria which govern the onset of symmetric and asymmetric snap-through. By including the influences of pre-stress into the investigation [66], the authors discovered that bistability may no longer be possible to be effected via the electrostatic fields since the voltage required to initiate a snap-through to the other (ideally stable) equilibrium exceeded that of the critical pull-in voltage. Krylov *et al.* [67] showed that such concerns of pull-in were remediated by the employment of fringing electrostatic fields because the actuating surfaces were no longer positioned in the deflection plane of the bistable microbeam, and hence afforded greater bistable beam deflections using a variety of actuation strategies. It is clear that bistable microsystems, and in particular arches, have been extensively probed in recent years, providing numerous guidelines for their effective implementation.

5.1.2 Bifurcation-based Microsystem Applications

The investigations surveyed above give broad, and in many cases application-independent, insights into the deployment of bistable microstructures. Thus, the conclusions drawn, for example, regarding the dynamic pull-in sensitivities of electrostatically actuated bistable microarches, are just as relevant for utilizing the systems as accelerometers as they are for operating the devices as mass sensors. However, there are numerous microsystem applications that necessarily focus upon key dynamic behaviors for their functional purpose. The strong static and dynamic nonlinearities of bistable microstructures introduce novel opportunities for a number of common microtechnology applications.

The exploitation of the dynamic *bifurcations* of bistable microsystems has recently attracted attention due to the substantial difference in qualitative structural behaviors before and after activation. This could be, for example, the difference between an initial low amplitude intrawell oscillation around one of the stable equilibria that is suddenly destabilized to trigger the large amplitude interwell snap-through dynamic. The unmistakable changes in bistable microsystem response provide several practical benefits: numerous bifurcations may be exploited; the relative change in response amplitude from intra- to interwell motions (or vice versa) is much greater than other saddle-node bifurcation phenomena; and the capability to induce such dramatic transitions via an all-mechanical system (i.e., bistability via pre-stress) means the microstructures may operate without need for actuating power (e.g., a sensor that detects threshold accelerations via the resulting switch of state).

Researchers have recently uncovered the advantages of harnessing the bifurcations of bistable microsystems for key applications that have been historically explored using a variety of more traditional device designs and operational approaches. One such classic application of microsystems is to realize electromechanical filters (electromechanical mass-spring-dampers) as replacements for such devices that would otherwise employ all-electrical counterpart components (resistors, inductors, capacitors, etc.). The benefits to such replacement are that MEMS may exhibit substantially less damping, provide sharper filtering cut-off characteristics, and may take up less volume than the electrical counterparts [68]. Thus, in this context a MEMS is developed that exhibits a certain frequency sensitivity (passband) due to its resonance features; therefore, the input harmonic response signal is allowed to pass (or not) based upon the input signal similarity to the MEMS sensitivity [69].

The exploitation of nonlinearity to provide novel functionality towards these ends is not new, and indeed many research teams have investigated the opportunities [68,70–73]. However, the use of triggering the continuous snap-through dynamics of bistable MEMS as an exceptionally sharp electromechanical signal filter is an idea proposed and examined only very recently. Ouakad and Younis [74] initiated these studies with numerical and experimental explorations that determined the excitation frequency and AC voltage requirements to trigger the snap-through motions of bistable microarches. It was evident that the amplitude of the AC voltage governed the width of the filter passband represented by the existence of snap-through. The experimental work verified these opportunities and also indicated that attempts to trigger snap-through at frequencies close to higher-order modes could lead to failure of the device because of the large actuation voltages required to trigger the bifurcation in response amplitude. Overall the investigation provides important first insights towards realizing high performance signal filtering based upon the dynamic bifurcations in a bistable MEMS [74].

Leveraging the dynamic bifurcations of nonlinear MEMS has another recently explored avenue of application. The idea is based around the sensing and detection of bifurcations as a means to monitor changing characteristics of the microsensor as compared to the traditional approach of tracking resonant frequency peaks and phases which may be less obvious indicators of change. Bifurcation-based (BB) sensing has origins in the classical physics community where the voltage responses of a microwave-driven, superconducting Josephson tunnel junction are exploited for their strongly nonlinear sensitivity. When a shift in the weak, harmonic input signal modulates the critical current, the saddle-node bifurcation of the junction output voltage from low to dynamic amplitude may be activated which enables a novel means to amplify (i.e., detect) the small change in the input signal [75]. Because the monitoring strategy is based upon activating one of two dynamic response regimes, the amplification or detection methodology is effectively *digital* and hence able to detect (in theory) individual quantum events exhibited in the input signal [76].

The strongly nonlinear dynamics of MEMS are likewise prime tools around which to develop BB sensing approaches. Zhang and Turner [77–80] exploited the sub-critical pitchfork bifurcation in parametrically excited microresonators actuated by non-interdigitated comb drives to detect mass accumulation upon the sensor. When mass adsorption shifted the bifurcation frequency such that it coincided with the constant frequency of excitation, the large amplitude parametric resonant dynamics were triggered, indicating a threshold amount of mass accumulation had been exceeded. The examinations demonstrated the reduced vulnerability of the BB mass sensing method to noise and damping effects as compared to tracking the conventional resonant behavior of the sensor [81]. The investigators also showed that the mass adsorption sensitivity of the BB sensing strategy was two orders of magnitude greater than the conventional approach. Li *et al.* [82] examined a counterpart BB sensing approach for parametrically excited microresonators based upon a significant reduction of the phase noise variance. When the sensors are excited near to the parametric resonance bifurcation, the phase of the sensor response correlates strongly with the phase of the excitation such that the variance between the two signals effectively collapses (is "squeezed" to a consistent value from a previous and comparably random state). Using this phenomenon as a basis, the researchers proposed a sensing strategy that tracks the bifurcation event without actually activating the large amplitude responses, thereby reducing the dynamic motions which must be undertaken and hence prolonging the service life of the sensor [82].

Younis and Alsaleem [83] investigated the opportunities enabled by triggering the saddle-node bifurcation (or "fold" bifurcation) of an electrostatically actuated microbeam to detect miniscule changes of mass on the device. Because this bifurcation behavior is evident for all resonance frequencies so long as the level of excitation is sufficiently large, the authors considered actuation strategies at frequencies around the super- and subharmonic resonances in addition to the fundamental resonance. The experimentation corroborated the numerical results and showed that the saddle-node can be readily exploited for microscale mass sensing. Kumar *et al.* [84] examined a piezoelectrically actuated, softening MEMS for BB sensing and evaluated the repeatability of the strategy when the devices were functionalized to bond to airborne methanol. The research team also devised an analytical model formulation from which the quantity of mass accumulation on the sensor could be derived in a closed form expression [85]. Experimental studies provided further insight by suggesting the technique is not without flaw since premature bifurcation events were inadvertently triggered [85].

The works described above all share the application context that the sensing objective is to detect small changes in the mass of the sensor; these are induced because the functionalized surface of the device is conducive to bonding to target analyte materials such as NO_2, methanol, water vapor, and so on. Like electromechanical signal filtering, microscale mass sensing is a classic application for micromechanical and microelectromechanical oscillators [16]. Until recently, *bistable* devices had not been examined for their exploitation in a BB microscale mass sensing strategy. Because bistable systems have several unique bifurcation characteristics, some of which were leveraged in the recent examination regarding signal filtering [74], their lack of exploitation in the context of microscale mass sensing suggests prime opportunities for investigation are outstanding.

5.1.3 Summary

As evidenced in the surveys above, researchers have developed a broad and considerably application independent understanding of the dynamics of microscale bistable systems. Although the insights are useful towards the development of bistable MEMS sensors, the findings do not specifically suggest or introduce opportunities for which the exploitation of the bistable microsystem dynamics may be particularly beneficial. Only recent examinations have begun to explore the possibility of *leveraging* the bifurcations of microscale bistable systems, for example to provide enhanced electromechanical signal filtering performance. As a result, much remains to be understood and explored in regards to *harnessing* the dynamics of bistable microsystems for the purposes of sensing, detection, filtering, and many other objectives. Moreover, until very recently, there have been no investigations on the exploitation of bistable system dynamics for monitoring the health and/or overall behaviors of *large*-scale structures. To fill in these gaps in knowledge and understanding, the following sections present several studies which explore new ideas for the effective implementation of bistable system dynamics for sensing and detection purposes on micro- and macroscales.

5.2 Detecting Changes in Structures by Harnessing the Dynamics of Bistable Circuits[1]

A common vibration-based sensing strategy is to translate an observed shift in system natural frequency to an influence upon or within the system due to an identifiable parameter change. The simplicity and ease of deploying this strategy has led to it becoming a broadly established approach for structural monitoring, maintaining applicability and viability all the way from the nano- [14] to the macroscales [8]. On the other hand, such an approach has clear limitations. For example, there are challenges in the confident detection of frequency and phase shifts when damping and noise distort the clarity of the resonance features [86] and strong nonlinearities may be readily induced on smaller scales that consequently make monitoring a "linear" resonance frequency shift an ill-defined objective [87].

The recent exploration of using bifurcations has led to promising opportunities to surmount these challenges, particularly for micro- and nanoscale mass sensing because of the ease of activating such behaviors in the microsystem architectures [79,81,83,84]. While demonstrations of BB sensing for MEMS mass sensing are so far promising, there are limitations to be resolved. Chief among these are the reusability of some of the microsensors following bifurcation dynamics [83]; the repeatability of bifurcation frequency if crossed numerous times [84] that may be influenced by mass adsorption rate (or sweep rate during characterization [88]); the potential for fatigue failure when operating a structure so as to continually transition between large and small vibrations; and limited bifurcation utility when employing softening Duffing systems

1 Material from this section has been reproduced with permission from R.L. Harne and K.W. Wang, Robust sensing methodology for detecting change with bistable circuitry dynamics tailoring, Applied Physics Letters, 102, 203506 (2013); doi: 10.1063/1.4807772. Copyright 2013, AIP Publishing LLC.

(i.e., one-time activations as mass uniformly accumulates on the sensor). Since bifurcation phenomena require considerable nonlinearity to provide enhanced detection sensitivity, the approaches have thus far been applied only for micro- and smaller-scale systems where nonlinear dynamics are prevalent [68,87].

These concerns motivate innovative ideas that promote wholly new approaches to exploit bifurcations to detect the changes in structures. The aim of this study is to introduce and investigate one such idea and demonstrate the effectiveness and versatility of the approach. Indeed, the proposed BB sensing strategy created here is as applicable to MEMS mass sensing as it is to SHM, thus providing an avenue to extend the robustness of BB sensing to large-scale structures for the first time. The following sections describe the proposed sensing architecture and detection methodology and investigate the utility of the approach to provide new and exciting opportunities for both large- and small-scale structural monitoring.

5.2.1 Bifurcation-based Sensing Platform Based on Bistable Circuitry

The sensing system proposed and examined in this work is shown in the schematic of Figure 5.2. The sensing componentry requires, at the most fundamental level, a bistable electric circuit connected to an electromechanical element which itself attaches to or fully represents the structure intended for monitoring. The bifurcations in the bistable circuit output voltage response are ultimately employed as the indicators of change in the monitored structure. As a result, the proposed approach does not require the monitored structure to undergo bifurcating dynamics, and may indeed utilize a structure that exhibits only linear, smoothly-varying behaviors. By the selection of a statically *and* dynamically

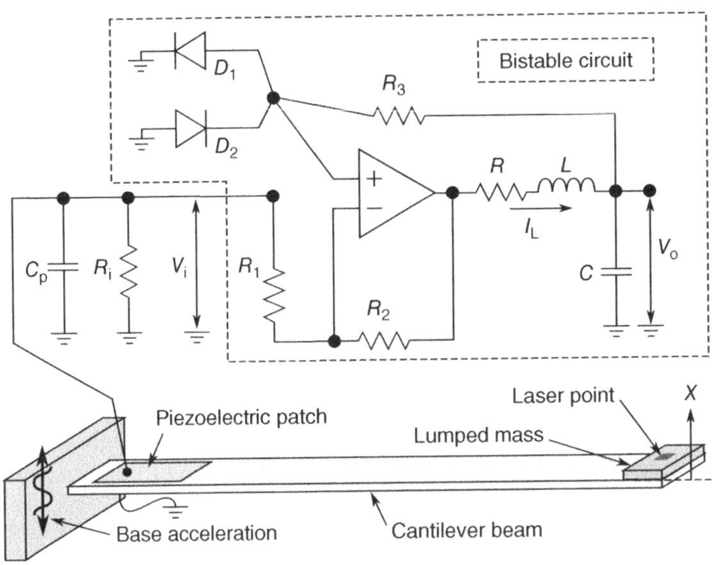

Figure 5.2 Configuration of the sensing platform. In experiments, a cantilevered beam is excited at its base. A piezoelectric patch attached near the cantilever clamped end provides a transduced voltage to a bistable circuit.

bistable element – unlike the past examinations which have used only dynamically bistable devices such as softening Duffing oscillators [83,84] or parametrically excited microresonators [79,81] – new bifurcation phenomena are introduced which provide the opportunity to detect numerous, discrete, and tailored thresholds of parameter change.

The fundamental sensing platform component requirements suggest a broad applicability of the proposed BB detection strategy. For example, the monitored structure and transducer components may be realized in practice by a piezoelectric MEMS resonator or by applying a piezoelectric patch to a mechanical or civil structure, such as that illustrated in Figure 5.2; of course, these systems are both common platforms for microscale mass sensing and SHM [84,89]. The output voltage from the transducer is the input voltage for the bistable circuit. This is illustrated in Figure 5.2 for a system where a monitored cantilever beam has a piezoelectric patch near the clamped end of the beam. The transduced voltage from the piezoelectric patch is provided to the bistable circuit as input voltage V_i following a gain adjustment determined by the value of the resistance R_i. Note that in Figure 5.2, the capacitance C_p represents the capacitance of the piezoelectric material and not a discrete circuit component.

Because of the leading op-amp at the input into the bistable circuitry, which ideally prevents outward flow of current, the strongly nonlinear electrical oscillator does not backwards-couple with the electromechanical system to be monitored. Thus, the monitored structure may be harmonically excited at a single frequency and the dynamics of the structure are tracked via the changes in the transduced voltage V_i as acquired from the piezoelectric patch. Consequently, a change in the monitored structural parameters that results in a shift in the amplitude of the structural dynamics will yield a variation of input voltage level to the circuit.

Figure 5.3 plots a representative case of a bistable system excited at constant frequency but varying input excitation level to exemplify the dynamic saddle-node bifurcations in displacement amplitude which may be induced by a change in excitation level. The results of the plot are generated using the generalized, analytical model formulation presented in [90] which is applicable to the broad class of bistable systems traditionally modeled as possessing the double-well potential energy profile [91]. In Figure 5.3a, the low amplitude intrawell oscillations near point D are destabilized due to the minute increase in excitation level, thus causing the sudden transition to the large amplitude interwell (snap-through) vibrations having amplitude denoted by point E. In the framework of the proposed sensing architecture, the excitation level variation is realized because a change of an important structural parameter results in a shift in the structural dynamic response amplitude. The transduced input voltage to the bistable circuit consequently shifts in level, and leads to the crossing of the saddle-node bifurcation at point D. This causes the snap-through dynamics of the circuit to be suddenly triggered due to the small deviation in structural parameters.

Additionally, an operation near the bifurcation at point G in Figure 5.3a may trigger the sudden drop down to the bistable circuit output response amplitude denoted by point H. This work demonstrates the *sequential* utilization of the bifurcations that induce the dramatic changes in response amplitude represented by points D to E and G to H. New opportunities for BB sensing are

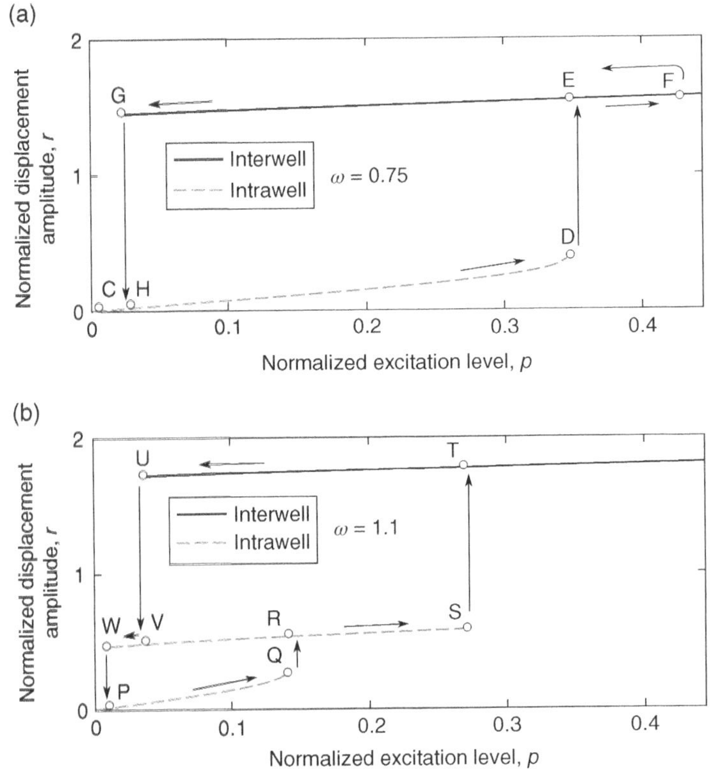

Figure 5.3 Bifurcations in the normalized displacement amplitude r response, as a function of excitation level p. Results are computed using the Eq. (3) and (4) in [90]. In (a) $\omega = 0.75$ and (b) $\omega = 1.1$. $(\gamma, \beta) = (0.02, 1)$. Arrows indicate the trajectories followed in consequence to increasing and decreasing level of excitation.

introduced by this sensing strategy. It enables means to detect small threshold amounts of structural parameter deviation in addition to a means to quantify the amount of change over time, merely by allowing the two bifurcations to be activated in sequence. The passive determination of net parameter variation via successive bifurcations has yet to be enabled. The strategic integration of the monitored electromechanical structure and the bistable circuit dynamics uniquely offers this versatile opportunity for parameter detection and monitoring. Extending further than that which is explored here, by the operation at other excitation frequencies, such as the example illustrated in Figure 5.3b, three bifurcations in the response amplitude of the bistable system may be exploited in sequence, providing still further means for progressive parameter quantification. It is clear that by employing a statically and dynamically bistable element as integrated with the monitored structure, the sequential bifurcations which may be harnessed for passive detection of gradual parameter changes significantly advances the versatility of BB sensing from the existing benchmarks [79,81,83,84,88].

5.2.2 Bistable Circuit Model Formulation and Validation

The sequential activation of bifurcations in the circuit output voltage is used in this study. The bistable circuit depicted in Figure 5.2 was recently synthesized [92] and was shown to exhibit chaotic responses characteristic of the Duffing–Holmes oscillator [93] when excited by a sinusoidal input voltage. This indicates the dynamic similarity between the circuit and a generalized bistable system such as that formulated by the model in [90] and exemplified by the responses in Figure 5.3. Assuming a single-mode approximation of the structural response [94] and ideal op-amp and diode characteristics, the governing equations of the system shown in Figure 5.2 may be derived.

$$\dot{\boldsymbol{x}} = \begin{bmatrix} 0 & 1 & 0 & 0 & 0 \\ -k/m & -b/m & -\theta/m & 0 & 0 \\ 0 & \theta/C_p & -1/R_iC_p & 0 & 0 \\ 0 & 0 & 0 & 0 & 1/C \\ 0 & 0 & 1/L & 0 & -R/L \end{bmatrix} \boldsymbol{x} - \begin{bmatrix} 0 \\ p\sin\omega t \\ 0 \\ 0 \\ F^*/L \end{bmatrix} \tag{5.1}$$

Here $\boldsymbol{x} = [x_1, x_2, \ldots, x_5]^T = [X, \dot{X}, V_i, V_o, I_L]^T$. Parameters m, b, and k represent structure modal mass, damping, and stiffness, respectively; p is the amplitude of base acceleration (or the negative mass-normalized direct forcing amplitude); $\omega/2\pi$ is the excitation frequency in Hz; θ is the electromechanical coupling coefficient; C_p is the piezoelectric capacitance; R_i is the input load resistance; R, L, and C are the resistance, inductance, and capacitance of the bistable circuit, respectively; and the overdot represents a time derivative. It is additionally assumed that $R_3 >> \sqrt{L/C}$ which is borne out in all examinations conducted here. The voltage function F^* is piecewise linear with respect to the output voltage $x_4 = V_o$ across the capacitor C:

$$F^*(x_4) = \begin{cases} F_1^* = x_4 + gV_d; & x_4 < -V_d \\ F_2^* = (1-g)x_4; & |x_4| \le V_d \\ F_3^* = x_4 - gV_d; & x_4 > V_d \end{cases} \tag{5.2}$$

Assuming that the op-amp and diodes are ideal, the feedback gain is determined by $g = R_2/R_1 + 1$. The voltage drop over an open diode is V_d.

The piecewise linear voltage function F^* is plotted in Figure 5.4 using the identified experimental system parameters provided in Table 5.1. As indicated in Figure 5.4, the circuit is statically bistable due to the two stable equilibria output voltages, $V_o^* \approx 1.18$ V, that may be maintained according to $F^*(V_o^*) = 0$. The non-zero, statically stable voltages are of course the consequence of the powered op-amp. Because the voltage function is composed of piecewise linear components, the restoring potential energy $U = \int F^* dx_4$ is the conjoining of three parabolas, which therefore yields the traditional double-well potential energy profile characteristic of bistable oscillators [91]. For completeness, it is also noted that the bistable circuit examined here exhibits a symmetric double-well potential energy profile around $V_o = 0$ V which is the unstable equilibrium.

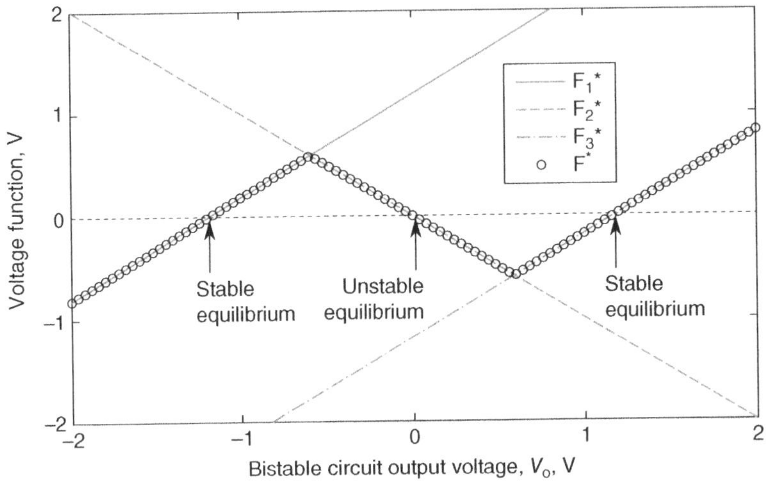

Figure 5.4 Voltage function of the bistable circuit.

Table 5.1 Experimental system parameters.

m, g	b, Nsm^{-1}	k, Nm^{-1}	θ, mN V^{-1}	C_p, nF	R_i, kΩ
65.1	0.192	252	0.135	1.65	9.71
R, Ω	L, H	C, μF	R_1, kΩ	$R_{2,3}$, kΩ	V_d, V
46.0	0.50	52.9	9.64	9.68	0.59

Experiments are performed with the full system depicted in Figure 5.2 to validate the governing Eq. (5.1) such that a favorable employment of the circuit may be determined for the sensing examinations and to validate the model. The aluminum beam has length 412 mm, width 31.6 mm, and thickness 3.15 mm. The tip mass is 38.2 g and its velocity is measured via laser interferometer. The piezoelectric patch is an MFC M-2814-P1 by Smart Material Corp. The circuit components include 1N4148 diodes and a LM741CN op-amp. The remaining parameters are given in Table 5.1. The structural mass and stiffness are the equivalent parameters for a fundamental mode approximation [95]; the damping b is determined from open-circuit ring-down responses of the beam; and the electromechanical coupling coefficient θ is experimentally identified by disconnecting the electromechanical structure from the bistable circuit and measuring the voltage across the resistor R_i.

In two evaluations, the system is excited at 10 Hz with different levels of base acceleration p. The theoretical and experimental beam velocity $\dot{X}_i$ and output voltage V_o resulting from the two input base acceleration levels are plotted in Figure 5.5. The bistable circuit output voltages are normalized with respect to the static equilibria voltage amplitude $|V_o^*|$ such that the black dotted lines at ±1 in the figure indicate the stable equilibria configurations of the circuit. The theoretical responses are predicted by numerically integrating Eq. (5.1). Figure 5.5a shows the beam and circuit responses due to low input acceleration $p = 0.438$ m s^{-2}.

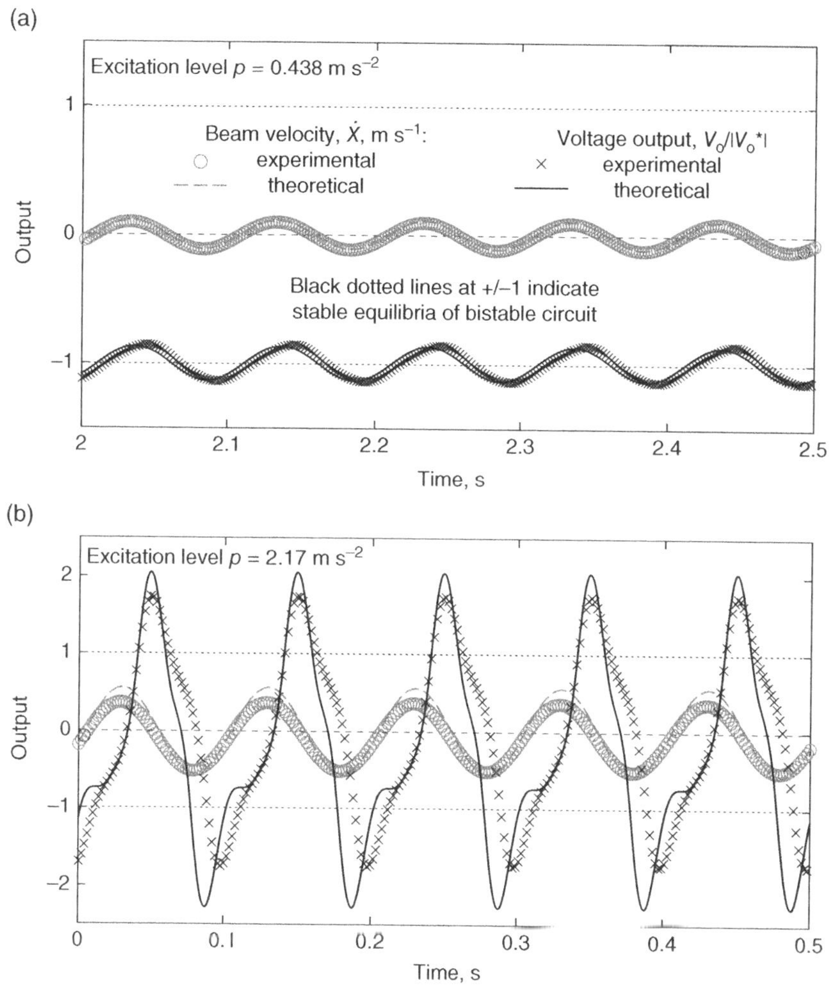

Figure 5.5 Theoretical and experimental responses of the cantilever beam tip velocity and the normalized bistable circuit output voltage. The base acceleration on the beam is (a) 0.438 and (b) 2.17 m s^{-2}.

In this scenario, the bistable circuit output voltage settles into a low amplitude intrawell response, evidenced by the steady-state oscillations around one of the stable equilibria. Increasing the base acceleration level to $p = 2.17$ m s^{-2} induces the interwell, snap-through circuit response; in this event, the circuit output voltage crosses the unstable central equilibrium $V_o^* = 0$ V twice per excitation period, Figure 5.5b. Overall, the theoretical results are in good quantitative agreement with the measured data. This verifies the assumption that there is negligible backwards-coupling from the circuit to the electromechanical system, such as is evident in the coupling terms of Eq. (5.1). These results and observations allow the proposed system to be employed in ways that trigger the bifurcations in output voltage response, exemplified in Figure 5.3a, without a concern that two-way coupling influences may distort the overall dynamic framework of BB sensor operation.

5.2.3 Investigation of Operational Parameters Suited for Bifurcation-based Sensing

The sensitivities of the system to changes in harmonic input voltage level V_i and frequency ω are evaluated to determine operating conditions which are well-suited for the sensing objectives. Due to the above validation that there is no adverse backwards-coupling from the circuit behaviors to the electromechanical structural responses, the circuit may be examined on its own to fully characterize the sensitivities of the circuit to triggering dynamic bifurcations. Thus, in the following experimental study, the bistable circuit is disconnected from the piezoelectric transducer and is excited by a harmonic input voltage provided by a function generator.

Figure 5.6 plots the magnitude of the measured bistable circuit output voltage frequency response function (frf) which is the ratio of the bistable output voltage V_o cross-spectrum to the input voltage V_i autospectrum. To acquire the data, a function generator supplies a slowly forward-swept input voltage at a given frequency of excitation. After the sweep, the input voltage amplitude is incrementally increased to repeat the procedure. In the figure, lines of constant phase lag with respect to the input excitation voltage are indicated by the thin solid curves and numbers in degrees. The thin dashed line in Figure 5.6 indicates an

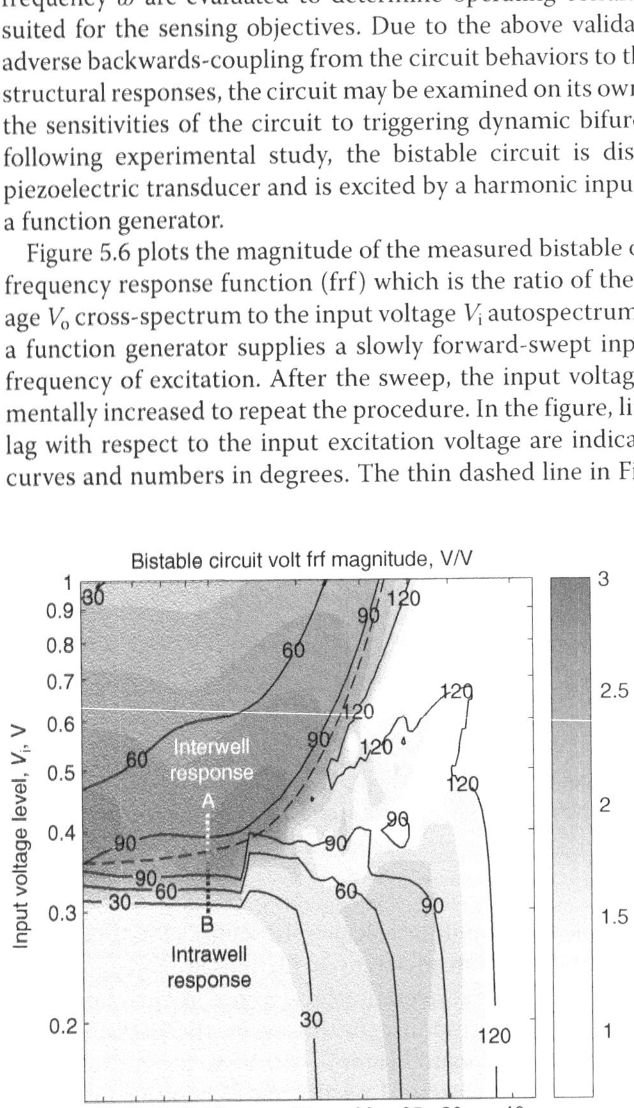

Figure 5.6 Experimentally measured bistable circuit output voltage frf magnitude. Lines of constant frf phase lag in degrees are plotted as the thin solid curves. The thin dashed curve represents the approximate transition region from the intra- to the interwell dynamic regimes.

approximate demarcation between the intrawell and interwell dynamic regimes. The increase of darkness in the contour represents a greater magnitude of output voltage frf.

At the transition between the two, steady-state dynamic regimes, an abrupt change in circuit response amplitude is observed, shown by the sudden increase in frf magnitude (the abrupt darkening of the contour shading). A key operational area is observed in Figure 5.6 denoted by the dotted line connecting the two points labeled A and B. Operated in this area, the bistable circuit may be excited at a constant frequency around $\omega/2\pi=10$ Hz and shifts in the input voltage level may force the circuit through the saddle-node bifurcation which separates the intra- and interwell response regimes, leading to a dramatic change in the output voltage amplitude. In the light of the prior validation of Eq. (5.1), such changes in input voltage V_i level may be representative of constant base acceleration (or direct forcing) conditions but varying structural parameters, such as m, b, or k, that one may wish to monitor.

5.2.4 Experimental Study of the Proposed Bifurcation-based Sensing Approach

Having identified favorable operating conditions for the bistable circuitry to detect changes in the monitored electromechanical system via the BB sensing strategy, the complete experimental set-up of Figure 5.2 is then employed. The investigations probe how the change in structural mass added to the beam tip may yield a corresponding change in input voltage level and induce the successive bifurcation phenomena. In the following experiments, the system is excited at 9.82 Hz which is less than the baseline (no added mass) structural natural frequency identified to be 9.90 Hz. Thus, the increasing mass at the beam tip decreases the natural frequency and amplifies beam response X at the excitation frequency 9.82 Hz; hence the input voltage level V_i is likewise amplified. After a certain quantity of mass is accumulated upon the excited cantilever beam, the structural natural frequency is reduced to the extent that the base acceleration at 9.82 Hz is exciting the structure *above* resonance such that the beam response amplitudes will decrease when further mass is added. Comparably, the input voltage levels decrease for these conditions. From start to finish, the bistable circuit input voltage levels increase and then decrease, first inducing the bifurcation upwards in amplitude, like that of the D to E jump in Figure 5.3a, and ultimately followed by the corresponding drop downwards in output voltage amplitude, like G to H in Figure 5.3a. This operational strategy is qualitatively similar to following a trajectory from B to A to B in Figure 5.6.

The beam is then excited at $\omega/2\pi=9.82$ Hz with an arbitrary (but low) level of base acceleration; the frequency and amplitude of base acceleration do not change throughout the course of the experiments. The input resistance R_i is strategically selected to tailor the level of input voltage V_i such that the circuit output voltage initially exhibits intrawell dynamics close to the saddle-node bifurcation. In other words, the circuit operating state, as coupled to the structure via the piezoelectric transducer, is initialized close to that indicated by point B in Figure 5.6. Adjustment of R_i is a primary means to tailor the detection sensitivity, that is, the initial nearness to a bifurcation. Once initially selected, the resistance R_i, given in Table 5.1, does not change over the course of experimentation. Wax masses are weighed with a microgram scale and incrementally applied to the beam tip. Following each mass addition, the structural natural frequency and damping ratio are determined from a long-time natural response of the system when excited by low level white Gaussian noise. Then, the bistable circuit output voltage V_o is measured when the structure is excited harmonically at $\omega/2\pi$=9.82 Hz. Next, the bistable circuit is disconnected from the piezoelectric transducer to measure the directly transduced voltage across the resistor R_i, here referred to as the direct output. In other words, the direct output is what would be considered in a traditional frequency-shift-based structural monitoring approach. The autospectra of the bistable circuit output and direct output voltages at 9.82 Hz are normalized with respect to the data acquired without added mass on the cantilever beam.

Figure 5.7 presents the changing response characteristics as functions of added mass ratio which is the ratio of the progressive mass accumulation to the baseline

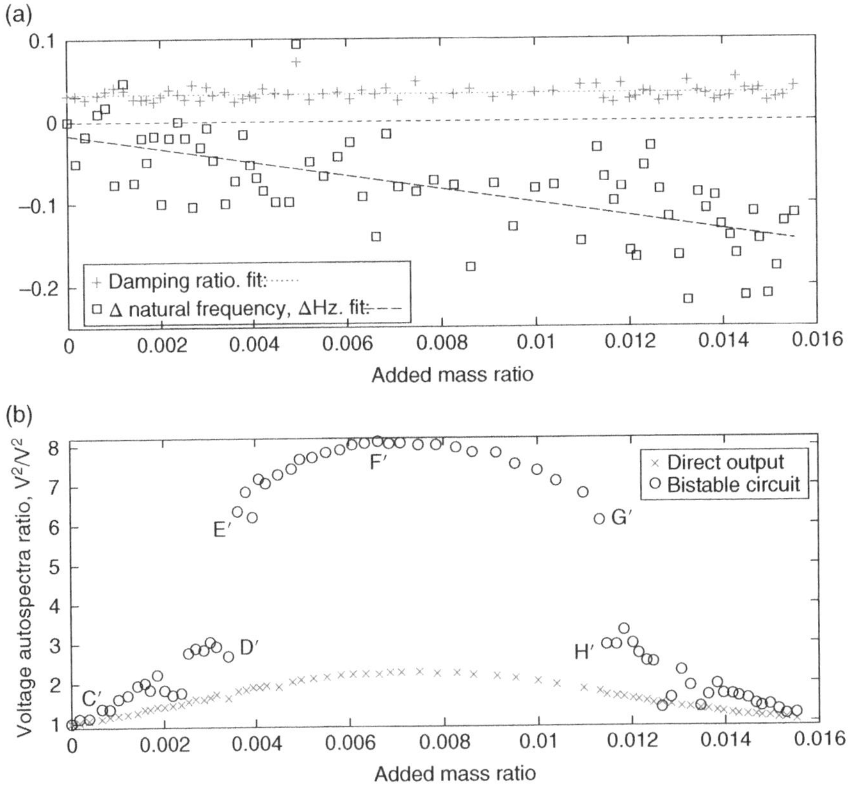

Figure 5.7 (a) Damping ratio and change in natural frequency of the monitored structure as mass is added to the host monitored structure: cantilever beam. (b) Output voltage autospectra normalized to the case of no added mass. The letter locations correspond to similar response features as exhibited in the bifurcation curve presented in Figure 5.3a.

structural mass. Figure 5.7a shows that the damping ratio does not change as mass is added to the structure; however, the natural frequency decreases with accumulated mass but does not do so uniformly. In Figure 5.7b, both the direct and bistable circuit output autospectra amplitudes initially increase due to mass accumulation. However, once an added mass ratio of 0.0034 is reached, the constant frequency excitation at 9.82 Hz is now much closer to the reduced natural frequency of the electromechanical structure. This causes the response level of the beam to amplify which likewise increases the input voltage level V_i. Consequently, with an increase in mass ratio to 0.0036, the bistable circuit crosses a saddle-node bifurcation to trigger the large amplitude snap-through dynamics observed in the output voltage because the level of input voltage is too great to sustain the low amplitude intrawell circuit behaviors. Using such a small deviation in the monitored parameter to induce a large qualitative difference in the sensor response is the characteristic feature of all BB sensing approaches. On the other hand, the proposed sensing architecture and detection strategy enable additional possibilities.

Following a continued increase of mass upon the structure, which eventually leads to the result that the excitations at 9.82 Hz are *above* the progressively-reducing structural resonance frequency, the bistable circuit output undergoes a second bifurcation for the mass ratio of 0.0114 since the input voltage level has reduced to the extent that the interwell dynamic regime can no longer be sustained. Thus, net mass accumulation of ratio 0.0078 is *passively detected* via the sequential activation of the two critical events. In contrast, other BB sensing strategies necessitate active, feedback-loop-based tracking hardware to determine the progressive change of structural features. With careful design and deployment of the bistable circuit, the specific quantity of accumulated mass which induces the sequential bifurcations, denoted by labels E′ and G′ in Figure 5.7b, may be modified. It is clear then that the proposed BB sensing architecture and detection methodology offer highly sensitive parameter change detection alongside means to track progressive deviations in the monitored structural features. Finally, Figure 5.7 also plainly shows that the proposed approach is far more apparent in terms of its confident determination of parameter change when compared to the ambiguous variations in the structural natural frequency that the traditional frequency-shift-based method would use.

5.2.5 Summary

There are several advances and opportunities provided by the proposed BB sensing methodology. The approach uses the integration of a bistable electric circuitry to enable highly-sensitive detection of structural change, while the sensitivity of monitoring frequency shift suffers due to the uncertain identification of the natural frequency. By using electrically-induced bifurcation events, a microscale implementation of the method may employ a conventional MEMS resonator in contrast to exploiting the induced strongly nonlinear dynamics of such microsystems. This simplifies a microscale deployment of the proposed system by isolating the nonlinear dynamics to an easily tailored electrical system which is moreover not subject to fatigue. Importantly, the proposed sensing architecture and detection strategy are well-suited to be applied for SHM, thus introducing the sensitivity and robustness that is characteristic of BB sensing principles to large-scale structural contexts. Since the influence of the structure-transducer upon the circuit is ideally a pure input voltage, the sensitivity of the bistable circuit sensor is easily tailored by adjusting the input resistor R_i; this allows one to eliminate false positives caused by initial operation too close to the saddle-node bifurcation. When compared to past BB MEMS sensing approaches which exploit a single bifurcation event for sensitive detection of threshold mass accumulation, the proposed protocol provides multiple bifurcations that may be used in sequence to denote pre-determined quantities of change. This is a degree of robustness and versatility not evidenced by the existing bifurcation-based sensing methods. Therefore, by exploiting the dynamics of an all-electrical bistable system, the proposed methodology greatly enhances the versatility of BB sensing and extends its range of applicability to fields previously not considered.

5.3 Improving Damage Identification Robustness to Noise and Damping Using an Integrated Bistable and Adaptive Piezoelectric Circuit[2]

In the previous study of Section 5.2, a novel opportunity to introduce bifurcation-based sensing into the context of large-scale SHM was made possible by leveraging the dynamics of a bistable electrical circuit. On the microscale, the investigations of BB mass sensing have shown that such approaches are far less prone to deteriorating performance due to the influences of damping [79] and measurement noise [81] when compared to the long-standing strategy based upon tracking the resonance frequency peak (or 90° displacement phase lag with respect to the excitation force) [96]. Yet, to be utilized in a damage identification approach for SHM, the framework presented in Section 5.2 is incomplete. For instance, the experimental study considered mass accumulation on a cantilevered beam and not the introduction of damage to the beam. Additionally, a useful damage identification protocol is needed to identify both the *severity* and *location* of the structural defect in order to provide a timely remediation.

To address these issues and assess one promising implementation for SHM, this study investigates a vibration- and bifurcation-based sensing framework for large structures centered around an integrated bistable and adaptive piezoelectric circuitry system. The following sections first broadly describe the motivations towards the development of the proposed platform, and then introduce the specific elements and the functionality each constituent provides. A model formulation for the integrated sensing system is created and rigorously examined to

2 Material from this section has been reprinted by permission of ASME, from J. Kim, R.L. Harne and K.W. Wang, Enhancing structural damage identification robustness to noise and damping with integrated bistable and adaptive piezoelectric circuitry. Journal of Vibration and Acoustics 137:011005, 2015. doi: 10.1115/1.4028308. Copyright © 2015 by ASME.

evaluate the effectiveness and robustness of the proposed solution. The performance of the proposed system is compared with that of the commonly adopted, vibration-based damage identification strategy which uses frequency shifts from the peak responses of resonance features to detect structural properties deviation.

5.3.1 An Integrated Bistable and Adaptive Piezoelectric Circuitry for Bifurcation-based SHM

For large-scale structures, global damage identification methods, which include the vibration-based approaches [97,98], evaluate damage-induced changes of the structural dynamics, often the deviations in natural frequencies and/or mode shapes. As a result, these methods may assess the structural health over a relatively large area or volume, provide a means for damage detection even when the affected sites are inaccessible or not in the proximity of a transducer location, and are often more straightforward to implement than localized approaches. These advantages strongly justify the popularity of the global damage identification approaches [8,9]. Among the vibration-based methods, damage identification utilizing resonance frequency change, termed frequency shift-based methods [99,100], are regularly deployed because resonance frequencies may be more easily measured even with a single sensor and are usually less contaminated by noise when compared to the other vibration-based dynamic property of mode shapes [86]. Yet, frequency shift-based methods have two notable limitations: (i) a deficiency of data for accurate damage identification, and (ii) an inhibited capability to detect frequency shifts due to damage effects when resonances are distorted by noise and damping.

The issue of data deficiency is related to the number of practically measurable and identifiable resonance frequencies which are typically far fewer than the number of DOF required for characterizing the damage via a corresponding structural model, such as a finite element model [86]. As a result, because the modal system of equations is underdetermined, the damage identification model solution is not unique. Jiang *et al.* [101,102] proposed a remedy to this concern by coupling an adaptive piezoelectric circuitry with the structure. The method introduces additional resonant frequencies and vibration modes to the complete electromechanical structural system to enrich the complete set of modal frequency measurements. By favorably tuning the inductance values in the circuitry, a greater number of frequency responses with an increased number of resonance frequency shifts are obtained. The study demonstrated that the enriched data set of resonance frequency shifts improved the ability to solve the inverse problem and recovered the accuracy of damage location and severity identification [89,101,102]. Moreover, the method is straightforward to implement due to the ease of integrating piezoelectric transducers and circuitry with the structural system.

The second concern of frequency shift-based approaches is related to accurately determining the damage-induced resonant frequency shifts [103]. Since noise and damping are inevitable, the distorted resonance features may make a confident and correct identification of the frequency shifts challenging if one directly detects the peak response (or 90° phase lag with respect to the excitation). When erroneously identified frequency shifts are combined with the deficiency of the measurement data, damage prediction error can become adversely amplified [104]. To overcome this shortcoming, sensitivity enhancing control techniques have been investigated [105–107]. The approaches manipulate closed-loop eigenvalues and -vectors by an active feedback control system integrated with the structure, such that structural response changes due to damage are enhanced. On the other hand, the benefits of the sensitivity enhancing control concepts are curbed since they require multiple transducers for implementation and are cumbersome due to the active implementation.

To surmount the challenges encountered in the accurate determination of damage-induced resonant frequency shifts and in light of the findings of the previous study of Section 5.2, one may envision that a bistable circuitry may be configured with the electromechanical structure in such a way as to determine the shift in resonance frequency by using bifurcation-based detection principles. This study proposes one such operational procedure, illustrated in Figure 5.8.

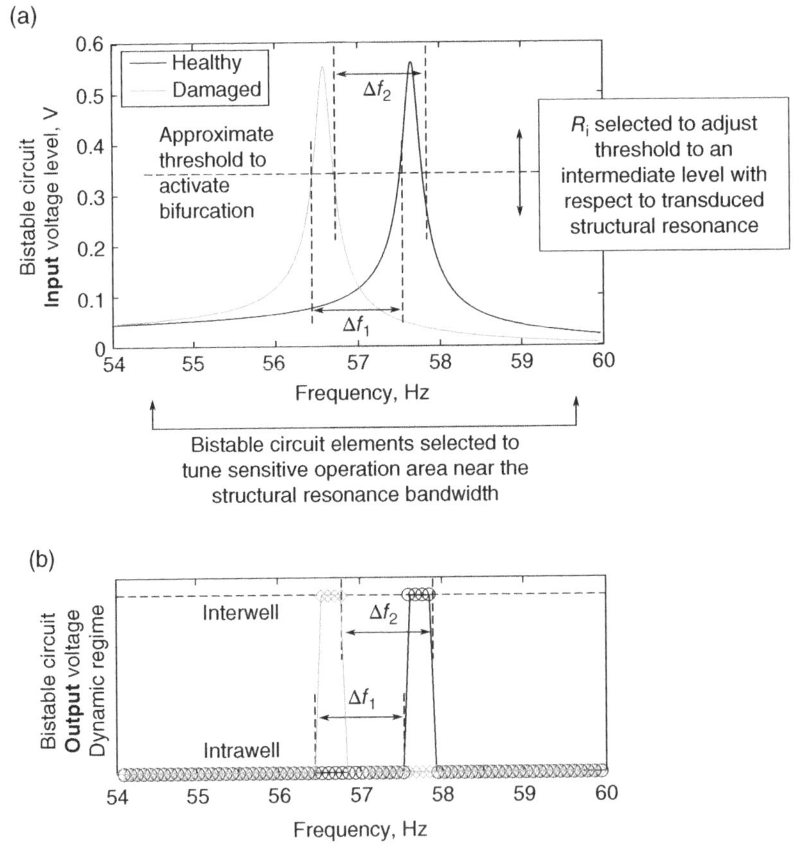

Figure 5.8 (a) Bistable circuit input voltage level. (b) Representative output voltage dynamic regime triggering as observed from the bistable circuit.

As was shown in Section 5.2, through the strategic integration of the structure and bistable circuitry, the structural responses effectively become the input voltage response for the circuit. Therefore, the response amplitudes of the host structural vibrations before and after damage, measured via the piezoelectric transducer, may be provided at discrete frequencies of input voltage to the bistable circuit following gain and frequency adjustments useful to tailor the system sensitivity. Specifically, the bistable circuit elements are designed so as to favorably tune the saddle-node bifurcation frequency sensitivity (such as the 10 Hz operational area identified in Figure 5.6) to the frequency of the structural resonance feature to track. Then, the resistance R_i (see Figure 5.2) is adjusted so that the amplitude of the structural response measurements align the approximate bifurcation threshold input voltage level to an intermediate amplitude of the transduced resonance feature. This is the threshold denoted by the thin dashed curve in Figure 5.6 and likewise exemplified with respect to the structural resonance feature as transduced using the piezoelectric patch in Figure 5.8a. Configured in this way, the integrated bistable circuit and electromechanical structural system is capable of identifying the damage-induced resonant frequency shifts using the following method.

The method to determine the frequency shifts with the bistable circuitry may be conducted on- or offline. When employing the offline implementation, the spectral amplitudes of host structural response (whether excited naturally or by direct methods) are computed from measurements taken without connection to the bistable circuit. Then, the corresponding voltage level for each frequency after the gain adjustment, as described above, is used as the input signal to the bistable circuit. In the online execution of the approach, the structure is excited at a single frequency and the response from the piezoelectric transducer is fed directly to the bistable circuit following gain adjustment; such a direct integration of the electrical and mechanical domains is precisely that which was explored in Section 5.2 and is shown in the schematic of Figure 5.2. The steady-state circuit output voltage response is then evaluated for whether it obtained intra- or interwell oscillations. This transforms the complete steady-state response information into one binary datum: in other words, at this excitation frequency is the bistable circuit output voltage dynamic regime an intrawell or interwell behavior? Repeating this process across the spectrum containing the resonance frequency will yield two sudden changes in the qualitative nature of the transformed data set: one frequency at which the amplitude increases from intra- to interwell behavior, and another for the opposite transition between dynamic regimes. Repeating this process again for the damaged structural responses yields two more bifurcations at certain frequencies.

Figure 5.8b presents example output results from this overall procedure, indicating that over certain bandwidths the interwell responses are activated whereas across other frequencies the intrawell responses are obtained. Thus, from healthy to damaged states, two shifts in frequency are acquired, shown in Figure 5.8b and marked as Δf: one corresponding to the jumps up in output voltage level, and one corresponding to the drops down in level. In this study, all of the following results utilize the mean of the two frequency shifts which is considered to be the damage-induced frequency shift of the resonance. Moreover, repeating this process using different gain adjustments or different parameters of bistable circuit elements leads to still further frequency shifts, all of which may be taken into consideration when determining a mean frequency shift from healthy to damaged structural states. By these procedures, the frequency shifts induced by damage may be quantified and employed in the damage identification routine.

5.3.2 Overview of Damage Identification Using Integrated Adaptive Piezoelectric Circuitry

On the other hand, determining the damage-induced frequency shifts of the few readily measured natural frequencies is insufficient in and of itself to identify the structural damage location and severity [89]. For this reason, the bistable circuitry must be integrated with an additional architecture to resolve both concerns related to the otherwise favorable frequency shift-based damage identification approaches. The developments by Jiang *et al.* [101,102] have shown that a frequency shift-based structural monitoring method using integrated adaptive piezoelectric circuitry for data enrichment surmounts the concerns related to mathematically problematic inverse equations to identify the damage. This is the additional architecture that is integrated with the bistable circuitry in this study. Together the two sub-systems - the bistable circuit and the adaptive piezoelectric circuitry - empower the proposed damage identification system and strategy.

To understand how the adaptive piezoelectric circuitry architecture [101,102] is capable of resolving the concern of underdetermined inverse problems in the damage identification strategy, a brief review of the mathematical foundation of the idea is required. Figure 5.9 shows a schematic of the adaptive piezoelectric circuitry as integrated with the monitored structure in one representative approach derived from [101]. Based on a finite element model, the equations of motion for the system shown in Figure 5.9 can be written as [108]

$$\begin{bmatrix} \boldsymbol{M}_s & 0 \\ 0 & \boldsymbol{L} \end{bmatrix} \begin{Bmatrix} \ddot{q} \\ \ddot{Q} \end{Bmatrix} + \begin{bmatrix} C & 0 \\ 0 & \boldsymbol{R} \end{bmatrix} \begin{Bmatrix} \dot{q} \\ \dot{Q} \end{Bmatrix} + \begin{bmatrix} \boldsymbol{K}_s & \boldsymbol{K}_c \\ \boldsymbol{K}_c^T & \boldsymbol{K}_p \end{bmatrix} \begin{Bmatrix} q \\ Q \end{Bmatrix} = \begin{Bmatrix} 0 \\ V_e \end{Bmatrix} \tag{5.3}$$

where $\mathbf{q}$, $\boldsymbol{M}_s$, $\boldsymbol{C}$, and $\boldsymbol{K}_s$ are the displacement vector, mass, damping, and stiffness matrices of the mechanical structure, respectively; and $\boldsymbol{Q}$, $\boldsymbol{L}$, $\boldsymbol{R}$, and $\boldsymbol{K}_p$ are the

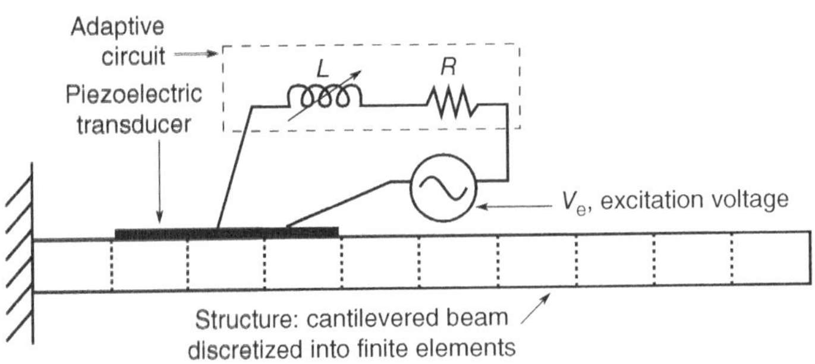

Figure 5.9 Illustration of monitored structure with piezoelectric transducer and adaptive circuit.

electrical charge flow vector, inductance, resistance, and inverse capacitance matrices of the circuit, respectively. K_c is the coupling term between mechanical and electrical domains, and V_e is the excitation voltage input. From Eq. (5.3), the following eigenvalue problem for the healthy structure may be obtained,

$$\left(-\lambda_i \tilde{M} + \tilde{K}\right)\{\phi_i\} = 0, \text{ where } i = 1, 2, \ldots, N_{el} \tag{5.4a}$$

$$\tilde{M} = \begin{bmatrix} \mathbf{M}_s & \mathbf{0} \\ \mathbf{0} & \mathbf{L} \end{bmatrix}, \tilde{K} = \begin{bmatrix} K_s & K_c \\ K_c^T & K_p \end{bmatrix} \tag{5.4b}$$

where λ_i and ϕ_i are the ith eigenvalue and -vector, respectively; $\tilde{M}$ and $\tilde{K}$ are the generalized mass and stiffness matrices of the undamaged system with boundary conditions applied, respectively; and N_{el} is the number of discretized elements in the finite element model. By a similar routine, the eigenvalue problem for the damaged structure is formulated:

$$\left(-\lambda_i^{dam} \tilde{M}^{dam} + \tilde{K}^{dam}\right)\left\{\phi_i^{dam}\right\} = 0, \text{ where } i = 1, 2, \ldots, N_{el} \tag{5.5}$$

where λ_i^{dam} and ϕ_i^{dam} are the ith eigenvalue and -vector of the damaged system, respectively.

By manipulating Eq. (5.4a), it is possible to derive the sensitivity matrix $\boldsymbol{S}$ of the eigenvalue change $\delta\boldsymbol{\lambda}$ with respect to variation in elemental stiffness parameters $\delta\boldsymbol{d}$ [101].

$$-\{\delta\boldsymbol{\lambda}\} = \boldsymbol{S}\{\delta\boldsymbol{d}\} \tag{5.6a}$$

$$\{\delta\boldsymbol{\lambda}\} = \begin{bmatrix} \delta\lambda_1 & \delta\lambda_2 & \ldots & \delta\lambda_m \end{bmatrix}^T \tag{5.6b}$$

$$\{\delta\boldsymbol{d}\} - \begin{bmatrix} \delta d_1 & \delta d_2 & \ldots & \delta d_{N_{el}}{}^T \end{bmatrix} \tag{5.6c}$$

$$S_{ij} = \{\phi_i\}^T \tilde{K}_j^T \{\phi_i\}, i = 1, 2, \ldots, m; \; j = 1, 2, \ldots, N_{el} \tag{5.6d}$$

where m is the number of measured resonant frequencies of the integrated system.

Eq. (5.6a) is usually significantly underdetermined ($m < N_{el}$), which results in erroneous damage identification. However, Jiang *et al.* [101,102] demonstrated that by selectively tuning the inductance, it is possible to alter the dynamics of the integrated system to obtain a suite of resonance frequency shifts due to the same structural damage. In this way, a set of altered inverse equations are obtained

$$\begin{Bmatrix} -\delta\boldsymbol{\lambda}(L_1) \\ -\delta\boldsymbol{\lambda}(L_2) \\ \vdots \\ -\delta\boldsymbol{\lambda}(L_n) \end{Bmatrix} = \begin{bmatrix} \boldsymbol{S}(L_1) \\ \boldsymbol{S}(L_2) \\ \vdots \\ \boldsymbol{S}(L_n) \end{bmatrix} \{\delta\boldsymbol{d}\} \tag{5.7}$$

The merit of employing the tunable adaptive piezoelectric circuitry may be observed by considering Eq. (5.7). Since n inductance values are applied to the circuitry, the number of resonance frequency measurements can be increased to $n \times m$, and thus the inverse problem becomes much less underdetermined. Given correct frequency shift determination throughout the enriched data set, related to the eigenvalue changes $\delta\boldsymbol{\lambda}$, an accurate identification of the damage is more readily achieved. Additionally, due to the global nature of the vibration-based detection method, the past studies observed an insensitivity of the method's accuracy related to the piezoelectric transducer position with respect to the damage site [101,102], which highlights the versatility of the approach.

5.3.3 Verification of Bifurcation-based Detection of Frequency Shifts

The data enrichment routine by inductance tuning with the adaptive piezoelectric circuitry therefore requires the determination of a suite of frequency shifts. The traditional and straightforward approach of direct peak detection (DPD) is limited by the accuracy of the approach which, in fact, substantially degrades when noise and damping distort the response signals. As will be shown in examinations presented in the following sections, even mild damping and a low level of noise can lead to significant error for damage identification if employing DPD for the determination of frequency shifts. As a result, this study proposes the incorporation of the bistable circuit for a *BB* determination of the frequency shifts required by the damage identification model formulation described in Section 5.3.2. Together, the sensing system is an *integrated bistable and adaptive piezoelectric circuitry*. By this union, the performance advantages of both elements may resolve the two concerns which have historically plagued the practical deployment of frequency shift-based structural damage identification.

The examinations by Jiang *et al.* [101,102] verified the benefits of using the adaptive piezoelectric circuitry to facilitate the more accurate identification of structural damage, assuming that damage-induced frequency shifts are appropriately determined. However, the viability of the bistable circuitry as a means to determine these frequency shifts has not yet been established. Therefore, using the offline strategy as described in Section 5.3.1, this section explores the accuracy of the BB method based around bistable circuitry as compared to the traditional approach of DPD, when these two approaches are used to provide the necessary frequency shifts to the damage identification model.

The following investigations use the bistable circuit model formulation that was experimentally validated in Section 5.2.2. Now, however, the integrated configuration that is considered consists of a host structure with the bistable *and* adaptive piezoelectric circuit network as shown in Figure 5.10a. Specifically, the schematic illustrates the online implementation of the frequency shift determination strategy based upon the BB method. The example host structure is a cantilevered beam evenly discretized into ten finite elements and the piezoelectric transducer is considered to be perfectly bonded to the top surface of the beam from the second to the fourth element. The transducer is connected to a tunable inductive circuit with a resistor R. Without loss of generality, damage is represented by 15% structural stiffness reduction on the second element in the finite

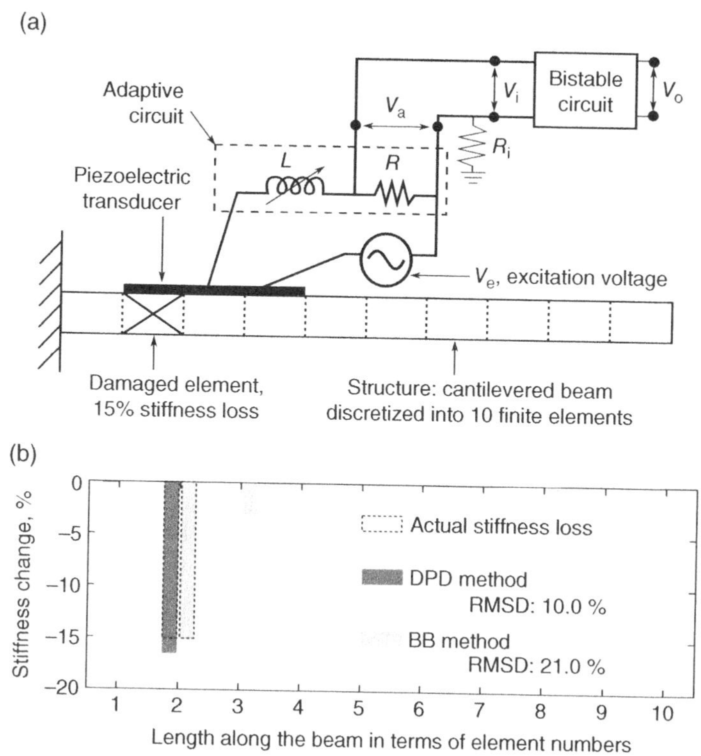

Figure 5.10 (a) Configuration of the cantilever beam integrated with bistable and adaptive piezoelectric circuitry. Damage is simulated as a 15% stiffness loss on the second beam element. (b) Results of the damage identification model.

Table 5.2 System parameters.

Cantilever beam	length, mm	thickness, mm	width, mm	Young's modulus, GPa	loss factor	density, kg m^{-3}
	209.2	3.175	38.1	71	0.003	2700
Piezoelectric transducer	length, mm	thickness, mm	Young's modulus, GPa	density, kg m^{-3}	dielectric constant, β_{33}, V m C^{-1}	piezoelectric constant, h_{31}, N C^{-1}
	62.8	0.191	66	7800	7.1445	1.0707

element model of the beam; note that this element position concurrently indicates the modeled physical position and extent of the structural damage. Relevant parameters used in the damage identification model are listed in Table 5.2.

The proposed damage identification approach is summarized as follows. The healthy structure is first evaluated as a baseline. The excitation voltage V_e is swept over the chosen frequency range, and the response levels of the electromechanical system are measured as voltage drops V_a across resistor R. Around each natural frequency of interest, the amplitudes of voltage V_a are initially scaled to a suitable level by tuning the resistor R_i, and this signal is the excitation for the bistable circuit V_i. The activation of bistable circuit bifurcations is then evaluated by analyzing the output voltage level V_o of the bistable circuit as described in Section 5.3.1. Then, the inductance of the adaptive circuit is selectively tuned following the procedure of Jiang *et al.* [101,102], and the process is repeated for a desired number of inductances to enrich the data set. Once completed, this forms the data set of bifurcation frequencies for the healthy, baseline structure. Following damage, the process is repeated and the gain adjustment[s] employed for the healthy structure are used for the counterpart assessments following damage. This method yields a second data set of bifurcation frequencies. The difference between the data sets of bifurcation frequencies is therefore the collection of frequency shifts to be used in the damage identification routine described in Section 5.3.2. To comment on practical implementation of the on- and offline implementations of the BB method, for an ideal op-amp junction in the bistable circuit shown in Figure 5.2 no backwards coupling of electrical response will occur to influence the voltage V_a across resistor R [109]. In practice, very small coupling will exist for an online implementation of the BB method. To eliminate this concern, offline BB frequency shift detection can be implemented, since the offline method employs structural dynamics measurements taken without connection to the bistable circuit as detailed in Section 5.3.1.

In this initial evaluation, the response signals are free of additive noise. The data set of the lowest three resonance frequency shifts is enriched by selecting 14 inductances which are used in the tuning procedure, yielding a total of 42 frequency shifts. The resonance frequency shifts are determined by conventional DPD and by the proposed BB approach. In the following studies, the BB method was implemented with numerical simulations of the bistable circuit responses using the governing Eq. (5.1) and using random initial conditions (hence, an offline implementation). The DPD frequency shift results were generated by direct identification of the shifts. Then, damage features are determined from model evaluations and the prediction results are compared.

The resonance frequency shifts as determined by BB and DPD approaches are given in Table 5.3. The frequency shifts obtained by DPD are considered to be accurate because no noise is included in this evaluation. Due to the selection of inductances, the first and second resonance frequency *shifts* show great sensitivity when the inductance is varied around 40.0 H. On the other hand, when the inductance is tuned around 1.04 H, the second and third resonance frequency *shifts* exhibit greater change, and hence greater sensitivity. The multiple frequency shifts acquired by either DPD or BB approaches indicate that information about damage is enriched, which reduces the underdetermined formulation of Eq. (5.7) for damage identification. Table 5.3 shows that the frequency shifts detected by BB and DPD methods exhibit little absolute difference. The mean and standard deviation of the differences are −0.01 and 0.05 Hz, respectively, while the maximum difference is 0.15 Hz.

Figure 5.10b presents the damage location and severity predictions following BB or DPD frequency shift detection. It is seen that the results from the DPD and BB methods both predict close to the actual 15% stiffness loss at the second element. Root-mean-square deviation (RMSD) between the predicted stiffness

Table 5.3 Damage-induced resonance frequency shifts measured by DPD and BB methods under inductance tuning.

	First resonance frequency shift, Hz			Second resonance frequency shift, Hz			Third resonance frequency shift, Hz		
Inductance, H	DPD	BB	difference	DPD	BB	difference	DPD	BB	difference
0.89	1.32	1.34	−0.02	1.22	1.37	−0.15	0.14	0.10	0.04
0.94	1.32	1.33	−0.01	1.16	1.25	−0.09	0.19	0.25	−0.06
0.99	1.32	1.31	0.01	1.01	1.13	−0.12	0.38	0.52	−0.13
1.04	1.32	1.31	0.01	0.44	0.52	−0.08	0.97	1.02	−0.05
1.09	1.33	1.32	0.01	0.05	0.13	−0.08	1.28	1.31	−0.03
1.14	1.30	1.29	0.01	0.03	0.06	−0.03	1.33	1.33	0.00
1.19	1.31	1.31	0.00	0.02	0.06	−0.04	1.33	1.32	0.01
32.5	1.20	1.20	0.00	0.12	0.09	0.03	1.31	1.18	0.13
35.0	1.10	1.10	0.00	0.19	0.21	−0.02	1.31	1.31	0.00
37.5	1.00	1.00	0.00	0.29	0.30	−0.01	1.32	1.31	0.01
40.0	0.87	0.87	0.00	0.45	0.43	0.02	1.31	1.30	0.01
42.5	0.73	0.73	0.00	0.58	0.61	−0.03	1.31	1.28	0.03
45.0	0.59	0.60	−0.01	0.72	0.71	0.01	1.31	1.22	0.09
47.5	0.48	0.48	0.00	0.82	0.82	0.00	1.31	1.23	0.08

reduction $\delta \boldsymbol{d}^{\text{predict}}$ and the actual damage $\delta \boldsymbol{d}^{\text{actual}}$ is employed as a metric to quantify prediction error. Note that RMSD is a metric that quantifies the error of both damage location and damage severity predictions.

$$\text{RMSD} = \sqrt{\frac{\sum_{i=1}^{N_{el}} \left(\delta d_i^{\text{predict}} - \delta d_i^{\text{actual}}\right)^2}{\sum_{i=1}^{N_{el}} \left(\delta d_i^{\text{actual}}\right)^2}} \tag{5.8}$$

RMSD by the DPD and BB methods lead to relatively small error levels: 10.0 and 21.0%, respectively. From these results, it can be concluded that the BB method provides a comparable level of performance to the traditional DPD strategy with respect to frequency shift determination and the ultimate damage identification when the response signals are free of noise.

5.3.4 Improving the Accuracy of BB Frequency Shift Detection Through a Greater Number of Evaluations

The prior section employed a single run of the BB frequency shift detection procedure, with randomly selected initial conditions in the bistable circuit simulation, for use in the damage identification routine. It was found that the identification results following noise-free BB frequency shift determination differed slightly from the DPD approach. Because the dynamics associated with the saddle-node bifurcation used in this research are sensitive to initial conditions [91,110,111], and thus activation of the bifurcations may be slightly inconsistent from one time series circuit simulation to the next, it may be anticipated that an increased number of runs using randomly selected initial conditions to determine a greater number of frequency shifts may improve confidence in the BB damage identification result. With a greater number of runs, the final mean value of the detected frequency shift may then be used in the damage identification routine.

Therefore, using the electromechanical structural model described in Section 5.3.3, the BB frequency shift detection procedure was carried out for 40 runs with randomly selected initial conditions covering a sufficiently sampled area of the initial condition parameter space. The mean and first standard deviation of the resulting frequency shifts determined by the BB method are plotted in Figure 5.11a (circles and error bars) as compared to the actual resonance frequency shifts (squares) which are the DPD method values. The horizontal axis indicates the inductance values used for the data enrichment routine. The individual shifts determined from all 40 runs for each inductance are plotted as crosses. For conciseness, results from the first resonance peak are plotted; results

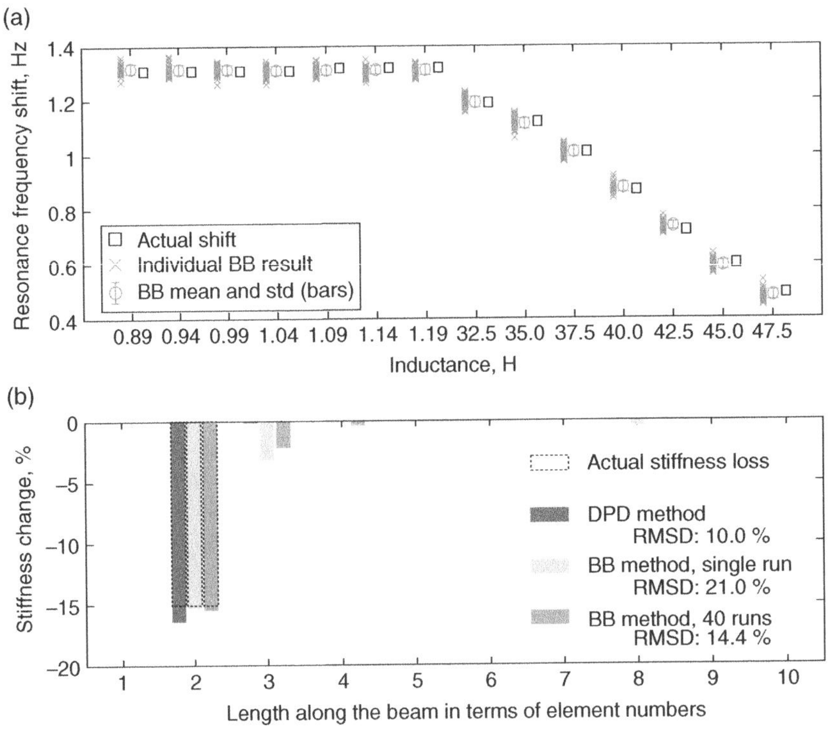

Figure 5.11 (a) Frequency shifts determined by 40 runs of BB method each having random initial conditions for each inductance. (b) Damage identification using the mean results from the 40 resonance frequency shifts.

from the remaining two resonances are comparable. As shown in Figure 5.11a, the actual resonance frequency shifts are within 1 standard deviation of the BB method mean shift, and the BB method accurately tracks the shifts as inductance is tuned to enrich the data set for the damage identification routine.

The means of the 40 frequency shifts for each of the three shifting resonance frequencies were then used for damage identification; the results are provided in Figure 5.11b. The accuracy of the damage features predicted by the mean of 40 runs is improved from that computed from a single run: the resulting RMSD is reduced from 21.0 to 14.4%. The findings indicate that a greater number of runs using the BB method may enhance the accuracy of frequency shift determination. Practically, carrying out an increased number of runs using the online implementation approach of the BB method may be a time-consuming procedure. However, the additional time necessary for the increased number of runs is less of a concern in the offline approach since the method may be conducted numerically, as done in this research, and take advantage of parallel computing architectures to expedite the process.

5.3.5 Investigation of Noise Influences for a Mildly Damped Structure

Sections 5.3.3 and 5.3.4 evaluated BB frequency shift determination when the response signals from the structure and adaptive piezoelectric circuitry network were free of noise. However, measurement noise is often unavoidable due to factors including piezoelectric transducer bonding, fidelity of wires and electrical components, electromagnetic radiation, and ground loops. Therefore, this section investigates the addition of low-pass filtered noise, ranging from 44 to 32 dB signal-to-noise ratio (SNR), with the structural response signals considered in Sections 5.3.3 and 5.3.4 to assess how the detection of frequency shifts by DPD and BB methods are affected and how the final damage identification results are influenced. Here, the SNR is selected according to the definition of "signal" to be the peak response amplitude at the transduced structural response peak when uncorrupted by noise. To appreciate the low level of noises employed, Figure 5.12 compares an example response resonance used in the following investigation for the noise-free case to that including 32 dB SNR additive low-pass filtered noise, which is the highest level used in the following exemplary investigations. In practice, such low-level noise could be caused by conventional sensor sensitivity constraints [112].

The investigation of Section 5.3.3 was repeated but a low-pass filtered white noise was now added to the response signals with each case of inductance tuning. Thus, for each of the 14 inductance tuning cases, different frequency shifts for the three resonance peaks may be determined following both DPD and BB methods should noise influence the results. For each scenario, a single run of the BB frequency shift detection method is employed and, likewise, the DPD approach is utilized with a structural response spectra distorted by the same randomly generated additive noise. Figure 5.13 plots the frequency shift error showing the 42 individual results (crosses) for each case of noise and detection method, while the mean values are shown as circles with first standard deviations indicated by error bars. The numerical values of first standard deviation are provided alongside each case.

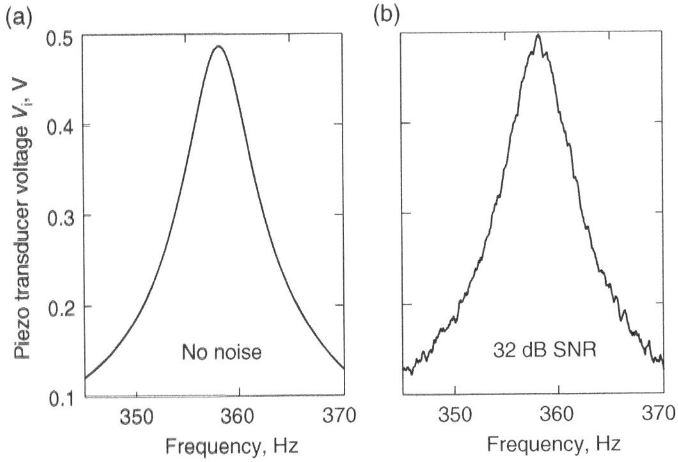

Figure 5.12 Example of (a) noise-free integrated system resonance and (b) that with 32 dB SNR additive noise.

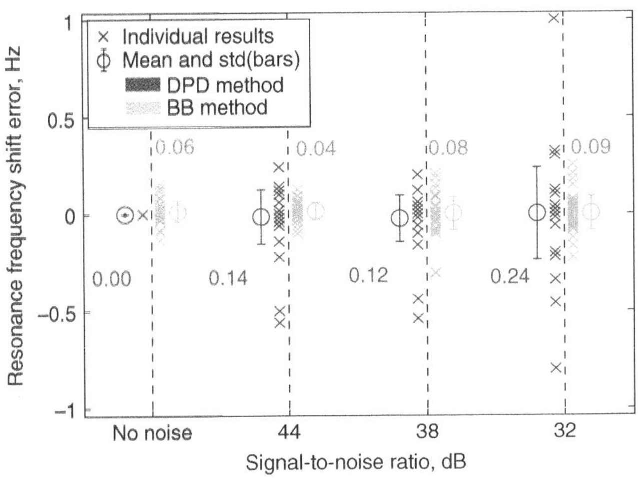

Figure 5.13 Frequency shifts determined by DPD or BB approaches and their standard deviations as signal-to-noise ratio increases. Values of standard deviations for each case of SNR and detection method are provided in the figure alongside the data points.

As detailed in Sections 5.3.3 and 5.3.4, the frequency shifts determined by the BB approach when the responses are noise-free exhibit a small but finite spread of values and thus a small range of error. However, as noise is added to the response signals, the deviations of the frequency shifts by the BB method remain small while those determined by DPD grow significantly, even for the lowest level of noise. Figure 5.14 presents the corresponding damage identification results. Figure 5.14a shows that damage location results by DPD alter dramatically with the addition of the lowest level of noise. Although the DPD method still correctly predicts approximately 15% stiffness reduction at the second element,

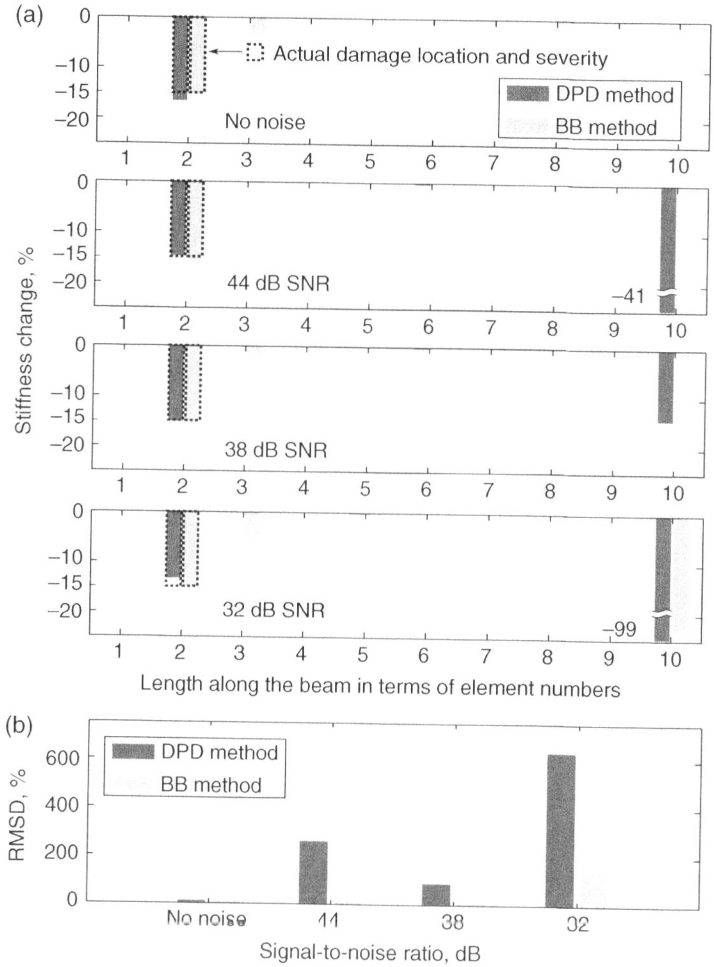

Figure 5.14 (a) Damage identification comparison for structure with mild damping and with varying degrees of low level additive noise. (b) RMSD of damage identification.

now a 41% stiffness reduction is estimated at the tenth element. In this case, the incorrect DPD predictions may mislead corrective measures. Figure 5.14b presents the RMSD which indicates that the BB method is substantially less influenced by the noise addition and its damage identification maintains a higher degree of confidence for correctness.

5.3.6 Investigation of Noise Influences for a More Highly Damped Structure

The prior investigation considered a lightly damped structure with loss factor 0.3%. While some structural systems may have such low damping in an ideal configuration, the realistic influences of boundary conditions, imperfections in mounting, and even structural-acoustic factors will often contribute in ways that increase damping levels in civil and mechanical structures. Thus, this section repeats the noise study of Section 5.3.5 but increases the structural loss factor from 0.3 to 0.5%.

Figure 5.15a presents the resulting damage identification comparing DPD and BB frequency shift detection approaches as noise increases; Figure 5.15b provides the corresponding RMSD. In tandem with increased structural damping levels, Figure 5.15 shows that DPD-based damage identification becomes unreliable and provides a justification for the common suspicion that the approach has clear disadvantages despite its appealing ease of implementation. In contrast, the BB method for frequency shift determination provides improved robustness to noise and damping and the more accurate damage predictions attest to its performance benefits. While the BB method ultimately yields some predictive error,

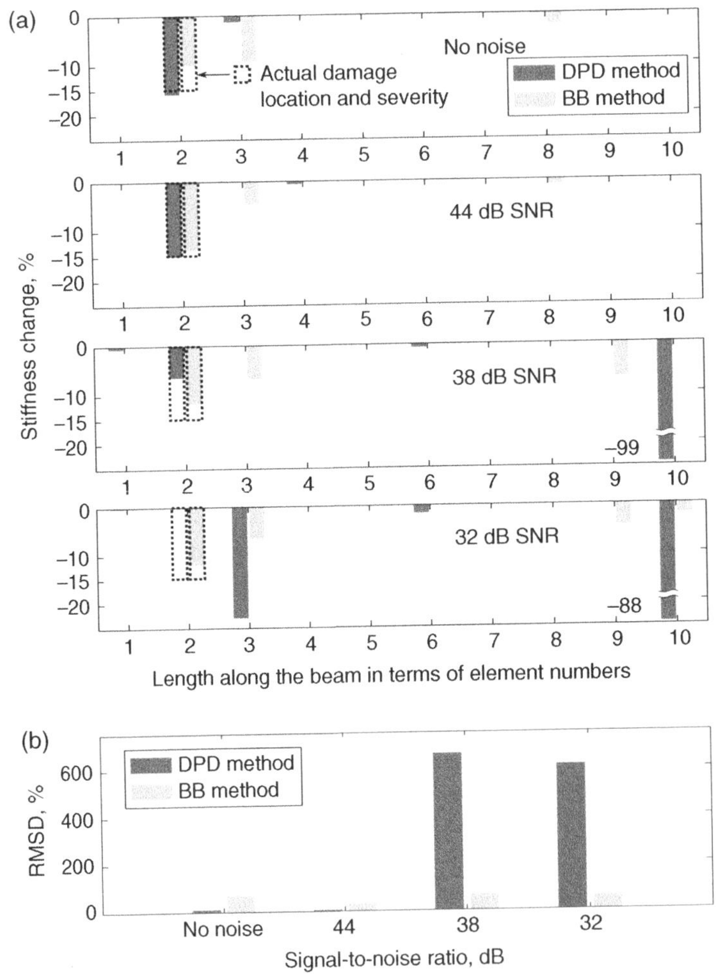

Figure 5.15 (a) Damage identification comparison for structure with increased damping and for varying degrees of low level additive noise. (b) RMSD of damage identification.

as shown in Sections 5.3.3 and 5.3.4, increasing the number of runs, using a greater number of gain adjustments, or tailoring the bistable circuit parameters could provide means to heighten confidence in its predictive capability and further enhance its performance for accurate damage identification.

5.3.7 Summary

This study investigated an integration of bistable and adaptive piezoelectric circuitry to enhance the robustness and accuracy of structural damage identification which are otherwise compromised due to noise and damping. Detecting frequency shifts via a dynamic BB strategy provides an alternative to traditional direct peak detection. In agreement with findings previously observed in microscale mass sensing studies [79,81], the BB method as leveraged in this SHM context is found to be significantly less affected by the addition of noise and increased damping in the response signals and therefore provides a higher level of confidence in its final damage identification results. With the BB method, increasing the number of runs, using a greater number of gain adjustments, or modifying bistable circuit parameters could provide ways to determine a mean frequency shift value that may further enhance the accuracy of the damage identification. In summary, by harnessing the collective dynamics of bistable and adaptive piezoelectric circuitry, a new pathway for accurate and robust structural damage identification may be forged.

5.4 Passive Microscale Mass Detection and Progressive Quantification by Exploiting the Bifurcations and Resonant Dynamics of a Two DOF Bistable Sensor[3]

Although the intelligent use of bistable system dynamics as a means to detect the changing features of large-scale structures is a recently investigated idea, there is considerably greater history to the development and understanding of bistable *micro*system dynamics, as described in the review of Section 5.1. On the other hand, a considerable proportion of this established research area draws observations and conclusions which are application independent. In other words, the findings are generalized insights, such as the critical loading conditions to buckle the device or the properties of the various dynamical regimes, features which are as important in a passive sensing application as

---

3 Material from this section has been reproduced with permission from R.L. Harne and K.W. Wang, A bifurcation-based coupled linear-bistable system for microscale mass sensing. Journal of Sound and Vibration 333 pp. 2241–2252, copyright 2013 Elsevier Ltd. Reproduced with permission. All rights reserved. doi: 10.1016/j.jsv.2013.12.017. Material from this section has been reprinted by permission of SAGE Publications, Inc., from R.L. Harne and K.W. Wang, Passive measurement of progressive mass change via bifurcation sensing with a multistable microelectromechanical system. Journal of Intelligent Material Systems and Structures 26(13) pp. 1622–1632, copyright 2014 by The Authors. doi: 10.1177/1045389X14546781.

they are in an active use of a bistable microsystem. Consequently, there is considerably less understanding as to how one may *exploit* the dynamics of bistable microsystems for specific applications and thus leverage the knowledge for a variety of practical purposes.

The recent and growing interest in the use of dynamic bifurcations as indicators of change in microsystems has given evidence that BB microscale sensing shows notable promise for the accurate, sensitive, and robust detection of chemicals, shocks, loads, and more [79,83,85,113]. While the sensor architectures investigated in these works are each unique, the bifurcations which they exploit for detection purposes share fundamental analogues. This is because, for one-dimensional oscillators, the dynamics near bifurcations are reducible to motions that are modeled by one of four response equations (the *slow* dynamics). These are the equations that define the saddle-node (also termed "fold"), super- and subcritical pitchfork, and transcritical bifurcations [110,114]. For example, the dynamics of the parametrically excited systems examined in Refs. [70,79,81] are effectively modeled using the subcritical pitchfork bifurcation equation of motion expressed by $\dot{x} = \mu x - x^3$, where initially $\mu < 0$ such that the original response is the trivial motion ($x = 0$). Then the sensors are acted upon in ways that cause $\mu \geq 0$ thereby activating a large change in the qualitative dynamics of the system [115]. Similarly, the sensor architectures investigated in Refs. [83,85,109,113] exploit the saddle-node bifurcation which reduces the two-dimensional dynamics to motions governed by the one-dimensional equation of motion $\dot{x} = \mu - x^2$ where $\mu > 0$ in the initialized state and $\mu \leq 0$ upon triggering and thereafter.

The dynamical reduction that occurs when various structures are propelled towards the crossing of a bifurcation places a variety of sensor architectures upon equal planes of performance. On the other hand, approaching the bifurcation (i.e., inducing parametric resonance, causing a pull-in event, etc.) is strongly dependent on the actuation means, triggering mechanisms, system integration, and so forth. Together, these factors indicate that in some fundamental respects BB microsensors have significant similarities – specifically in terms of ultimate performance limits – but differ in design, operation, and other properties that are important for understanding the sensor behaviors close to and removed from the bifurcation.

Thinking in this reducible dynamic framework, bistable systems undergo three distinct dynamic bifurcations. Figure 5.16 presents these bifurcations as they appear in terms of the normalized displacement amplitude response – the saddle-node (SN), period-doubling (PD), and supercritical pitchfork (PF) bifurcations – as a given level of excitation and frequency is varied. The parameters used to generate the results are provided in the figure caption. Although more refined analytical tools are needed to explicitly predict the PD bifurcation than the fundamental harmonic balance that is employed to produce the analytical results in Figure 5.16 [116,117], the open circle with PD label in the figure denotes the approximate frequency at which this bifurcation occurs. From Figure 5.16a–c, the excitation level is increased many-fold. As a result, it may be deduced that exploiting the onset of the pitchfork bifurcation as a detection mechanism requires the system to undergo very high levels of excitation, potentially leading to very large amplitude snap-through behaviors once the lower

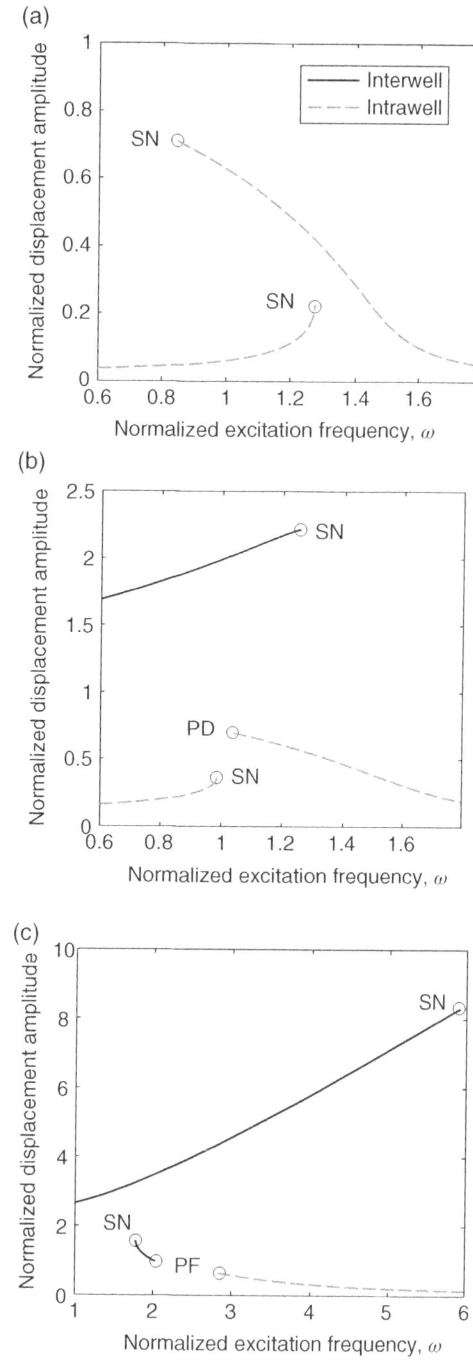

Figure 5.16 Types of dynamic bifurcations in a bistable system. SN: saddle-node. PD: period-doubling. PF: pitchfork. Responses predicted using a constant and one-term Fourier series expansion as the steady-state solution to the equation of motion $x'' + \gamma x' - x + \beta x^3 = p\cos\omega\tau$. Parameters (p, γ, β) are (a) (0.06, 0.09, 0.7), (b) (0.25, 0.09, 0.7), and (c) (4.5, 0.09, 0.7).

amplitude interwell response is destabilized. Such an operating state could easily lead to the early failure of the bistable microsystem. Additionally, the PD bifurcation is well-known to occur next to a dynamic regime characterized by chaos [118,119]. Thus, the PD bifurcation provides a less consistent means to detect a sudden change of state in a bistable system. The net result is that the *SN bifurcations* from one intrawell response amplitude to another (Figure 5.16a) or from an intra- to an interwell vibration dynamic regime (and vice versa) (Figure 5.16b) are the most useful bifurcations to exploit in a bistable structure. These may also be harnessed for levels of excitation which are relatively small and hence cause reasonable amplitudes of displacement.

The exemplary plots of Figure 5.3 in Section 5.2 illustrated the concept that a bistable sensor system may facilitate *sequential SN bifurcation activation* if employed in such a way as to result in excitation amplitude changes that correlate to the important system parameter to monitor. If this capability may be leveraged in a BB microscale sensor platform, it would lead to the ability to *passively* detect *and* quantify changing amounts of mass (or load, force, etc.) on the sensor. To date, the quantification of change in the monitored parameter via BB sensing platforms has only been achieved using active controls and sweeping strategies with parametric resonance-based sensors [79,82]. Thus, the exploitation of bistable microsystems in the context of BB sensing may provide leaps forward in terms of performance and versatility. The challenges to realize such enhanced sensing capability are to effectively design and deploy a bistable system-based sensor wherein excitation amplitude changes occur due to the variation of the monitored system parameter.

The following sections present the development and study of a bistable microsystem architecture that incorporates innovative ideas to meet these challenges. The purpose of the sensor and the context of the study is for microscale *mass detection* although the concepts presented below may readily apply to other sensing objectives, such as forces, humidity, temperature, etc. Following a generalized platform overview, a mesoscale proof-of-concept demonstration sensor is presented and examined. A model of this sensor is verified and then investigated in detail as to the sensitivities and capabilities of the correspondingly proposed sensing strategy to detect and quantify mass adsorption on the sensor as time elapses. It is shown that harnessing the dynamic bifurcations of a bistable microsystem provides great opportunity for microscale mass detection.

5.4.1 Sensor Architecture and Operational Principle Overview

In contrast to the past BB sensor platforms, the sensor developed in this study uses a statically *and* dynamically bistable element. This manifestation of bistability empowers particularly unique and advanced sensing capabilities when employed within the framework of the proposed sensor architecture. Section 5.2 described one platform that induced an effective sweep of excitation level on a bistable circuit to trigger two bifurcations in sequence. Such a sweeping style resulted due to the excited host structure passing through a resonance feature caused by structural mass accumulation and a transduction element which directly transmitted the measured responses as an input voltage to the electrical circuit. To leverage an analogous idea of sequential bifurcation activation in a microscale sensor, the architecture must be carefully devised. While the measured response of a linear, resonant microsensor may be passed to a bistable circuit to monitor change of mass on the small device, such an overall design defeats the purpose of realizing a small scale sensor since the circuit elements are likely to be substantially larger than the size of the mechanical resonator. Consequently, as shown in the following paragraphs, a practical achievement of the sequential activation of bifurcation in a bistable microsystem inevitably requires a fully-coupled two DOF sensor architecture. Moreover, it will be seen from the evidence presented below that the proposed sensing concept is based upon the successful integration and leveraging of resonant *and* BB dynamic principles, thus creating a unique link from the established vibration-based sensing strategies to the emerging concepts that have exploited nonlinear dynamics.

A schematic of such a two DOF sensor platform is presented in Figure 5.17a, showing an excited, host microcantilever with an inset, buckled beam at the free end. The buckling of the inset beam relative to the configuration of the host microcantilever may be realized via magnetic, electrostatic, or other force fields. A variety of specific sensor embodiments will be discussed in Section 5.4.9. More generally, the necessary architecture constitutes a larger host microsystem and a bistable inclusion which is inertially small with respect to the host. The bistable inclusion must also be appropriately positioned and coupled to the primary, generalized coordinate of motion of the host structure. Figure 5.17c shows the more generic model of the two DOF sensor architecture, where the host is excited by a base acceleration, which could also be realized in practice as a

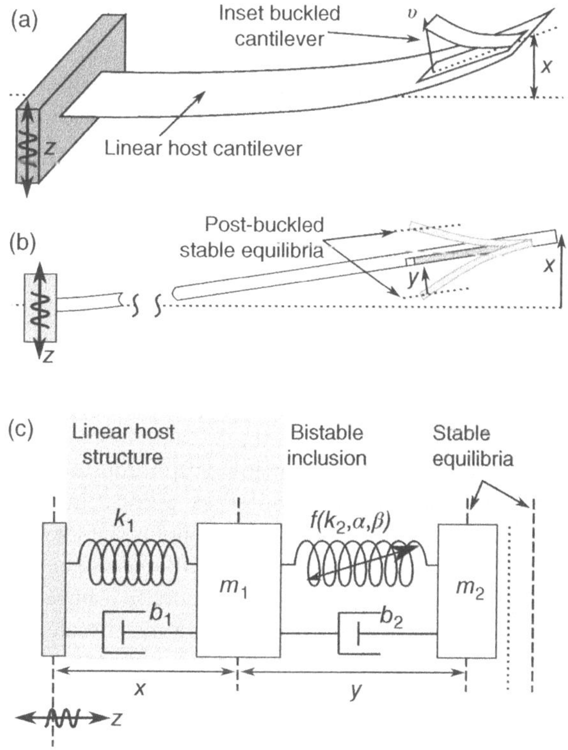

Figure 5.17 Schematic of coupled linear–bistable micromechanical sensor architecture. (a) Realization via host microcantilever and inset buckled cantilever. (b) Schematic showing coordinate convention used in model. (c) Lumped parameter representation of (b).

generalized force on the primary, generalized coordinate of motion via comb drive or piezoelectric actuation, among other possibilities.

There are also dynamic design and operational strategies required to leverage the advanced operational characteristics of the proposed architecture. The linear natural frequency of the bistable inclusion should be much greater than the host structure natural frequency, such that the second linear mode natural frequency of the coupled system is much greater than the fundamental mode resonance. Additionally, according to the sensing strategy devised below, the sensor is only employed with excitations at frequencies around the fundamental mode resonance. Thus, if the two DOF system is operated such that the bistable inclusion initially oscillates with small amplitude intrawell vibrations around one of the stable equilibria, then the low amplitude dynamics of the sensor architecture are effectively modeled via traditional, linear modal analysis. The linear dynamics of the sensor platform around its fundamental mode may then be equivalently expressed by the displacement response-to-excitation transfer function magnitude of a driven, damped harmonic oscillator [120]

$$H = \left\{ \left[k - \Omega^2 m(t) \right]^2 + \left[\Omega b \right]^2 \right\}^{-1/2} \tag{5.9}$$

where the excitation frequency is Ω and the progressively increasing modal mass is $m(t) = m + \delta m(t)$, which depends on the accumulating quantity of mass $\delta m(t)$. Figure 5.18a illustrates the trends in the linear modal response as mass adsorbs on the larger host resonator. By one sensor operating procedure, a constant level of excitation/base acceleration is used at a frequency which is less than the fundamental mode resonance prior to mass accumulation, shown as point 1 in Figure 5.18a. Thus, as mass adsorbs on the host, the effect is to reduce the fundamental natural frequency, thereby gradually amplifying sensor response levels because the resonance draws nearer to the constant frequency that excites the sensor. Figure 5.18b illustrates how the changing amplitude of the fundamental mode response leads to a varying amplitude of vibration for the bistable inclusion, thus creating a conceptual link between the fundamental mode response amplitude H and the excitation level "passed" to the bistable inclusion.

Eventually, as enough mass adsorbs on the host, a time is reached at which the amplified level of the fundamental mode response is great enough to trigger the dynamic, SN bifurcation in the bistable inclusion displacement, which induces the sudden transition from low amplitude intrawell to large amplitude snap-through vibrations, represented by point 2 in Figure 5.18. The fine degree of mass adsorption necessary to push the bistable element across the critical point is the trademark of high sensitivity BB mass sensing. But unlike the other passive sensing strategies for BB microsensors that have been explored to date, further mass accumulation on the two DOF microscale sensor may still be monitored and quantified. Continuing with the example, mass adsorption will ultimately reach a level at which the reduced fundamental natural frequency coincides with the excitation frequency, and the effective amplification gain to the bistable inclusion is maximized, as in point 3 in Figure 5.18. Thereafter, any continued

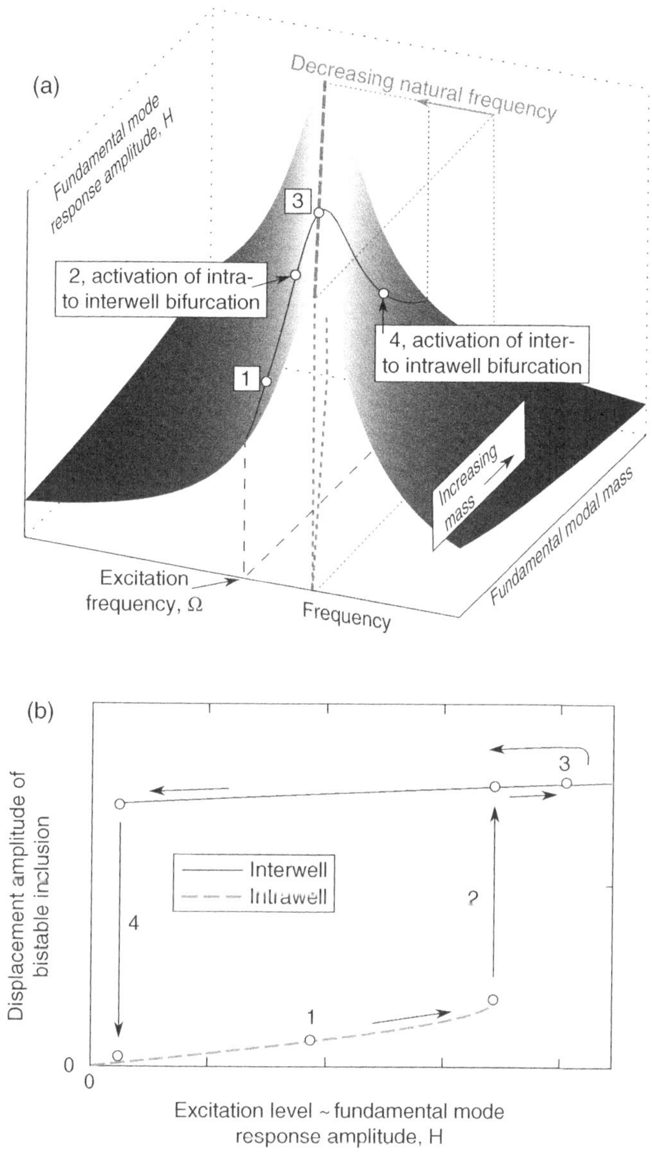

Figure 5.18 (a) Fundamental mode displacement response transfer function amplitude, *H*, presented as a function of the excitation frequency and increasing fundamental modal mass. This amplitude is equivalent to the excitation level which is passed to the bistable inclusion of the system so as to induce the bifurcation featured in (b). The numbers in (a,b) are correlated between phenomena happening with respect to modal dynamics and those with respect to bifurcation dynamics of the bistable inclusion.

mass accumulation results in an above-resonance excitation for the system and thus reduces the effective excitation level which is passed to the bistable element, finally leading to the bifurcation drop down in response amplitude (inter- to intrawell oscillations), as in point 4.

Conceptually, the sequential activation of bifurcations is straightforward when considering the influence of the varying fundamental mode response amplitude as an effective excitation level sweep up and then down for the bistable inclusion [121]. Such dynamic influence is comparable to recent vibration energy harvesting studies that investigated the addition of "dynamic magnifier" resonators to amplify external excitation levels to enhance energy harvesting performance for the attached electromechanical oscillators [122–124]. In the sensing context, two bifurcation events observed in sequence are needed to quantify mass over time in a passive manner. Because each bifurcation may be related to a critical mass level through a bifurcation analysis, the activation of two bifurcations in sequence indicates a measure of mass adsorption related to the passage of time from the first to the second. This ability notably contrasts previous BB micro-scale mass sensing studies which used actively controlled strategies to measure the progressive accumulation of mass [79] or employed passive methods that only indicate a single threshold amount of mass was adsorbed [83,84]. Nevertheless, the examinations of the proposed sensor architecture to be presented below examine the full range of operational procedures: individual detection events activated either via controlled frequency sweeps (for constant mass per sensing trial) or progressive mass accumulation (for constant excitation frequency), and sequential activation of bifurcations triggered via frequency sweeps or progressive mass accumulation. In this way, a thorough characterization of the sensor architecture and dynamical principles may be obtained, yielding a rich assessment of the various potentials enabled by the integrated resonant and BB dynamics.

5.4.2 Experimental Proof-of-concept Sensor Architecture

Proof-of-concept two DOF sensors are fabricated in order to verify the proposed combination of resonant and BB sensing principles and validate a model formulation that will be explored in greater detail in the sections below. The experimental sensor architecture is shown in Figure 5.19 and is fabricated according to the general schematic provided in Figure 5.17a. In the proof-of-concept sample, a small inset beam is located at the primary beam free end and the inset is buckled through the repulsive interaction of magnets. To ensure that only the inset beam is deflected by the magnetic interaction, the inset is machined with reduced thickness, approximately 480 μm as compared to host beam thickness 1180 μm. This fabrication detail is evident in Figure 5.19c,d which also demonstrate the two post-buckled stable states of the small inset beam. The proof-of-concept sample includes a recessed track that allows tailoring of the bistability by modifying the gap between the repulsive magnets which is related to the axial compressive load that governs the specific post-buckled inset beam configurations. As shown in Figure 5.19b, the host beam is cantilevered from an excitation source. In this way, to realize the required sensor architecture components described in Section 5.4.1 and shown in the schematic of Figure 5.17c, the primary structure is a cantilevered beam while the small bistable inclusion is the inset beam that is buckled via repulsive magnet interactions and is cantilevered from the end of the host beam that is opposite to the excitation.

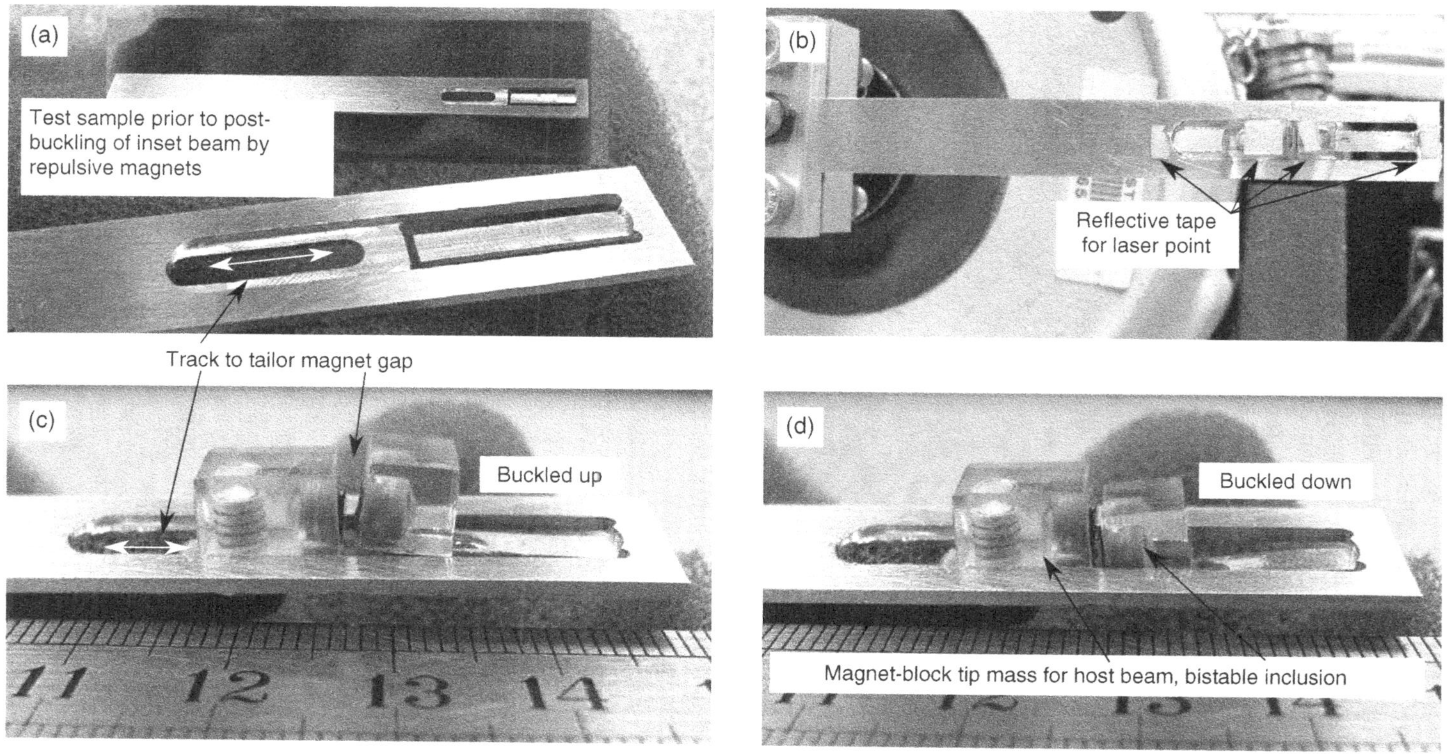

Figure 5.19 Proof-of-concept experimental sensor architecture. (a) Host beam with inset beam, showing fabrication details prior to magnet attachment. (b) System cantilevered from shaker for testing. (c) and (d) Buckled states of inset bistable beam due to magnetic repulsion.

5.4.3 Model Formulation of the Sensor Architecture

A model of the two DOF sensor is formulated to devise design and operational guidelines for the effective implementation of the resonant and BB sensor and sensing strategies. The governing equations are derived according to lumped parameter assumptions which are appropriate for many continuously-distributed microscale systems, particularly when the motions are representative of only those related to the fundamental mode (whether pre- or post-buckled) [11,87,125]. Due to the specific experimental system, the structure is modeled as undergoing base excitation, although a similar equation system may be formulated for direct forcing of the host structure (e.g., a piezoceramic actuator on the clamped host beam end). The generalized coordinates as denoted in Figure 5.17a are x for the host structure displacement relative to base motion, and v for the relative displacement between the bistable element and host structure. The governing equations of the lumped parameter system are derived using Newton's second law to be

$$m_1(\ddot{x}+\ddot{z})+b_1\dot{x}+k_1x-b_2\dot{v}-\Lambda'(v)=0 \tag{5.10a}$$

$$m_2(\ddot{v}+\ddot{x}+\ddot{z})+b_2\dot{v}+\Lambda'(v)=0 \tag{5.10b}$$

where m_i, b_i, and k_i are the effective mass, damping, and linear stiffnesses of the sub-systems, respectively, and $i=1, 2$ indicate the host structure and bistable inclusion; $\Lambda(v)$ is the potential energy of the bistable element expressed as a function of the relative displacement coordinate v; and the overdot and apostrophe operators denote time and spatial derivatives, respectively. In this study, the symmetric, double-well potential energy profile is used for the bistable inclusion which represents a variety of microscale bistable system potentials when taking into account a predominant fundamental buckling mode dynamic behavior [67,126]:

$$\Lambda(v)=-\frac{1}{2}k_{2L}v^2+\frac{1}{4}k_{2NL}v^4 \tag{5.11}$$

The spatial derivative of the potential energy, $\Lambda'=\partial\Lambda/\partial v$, is the restoring force from which may be computed the fixed points of the bistable sub-system: $\Lambda'(v^*)=0 \rightarrow v^*=0, \pm\sqrt{k_{2L}/k_{2NL}}$. Here, v^* are the static equilibria which include two stable configurations that are symmetric with respect to a central unstable configuration, $v^*=0$. A coordinate transformation is applied by defining $y=v-v^*$ such that the bistable inclusion response coordinate y is zero at one of the stable equilibria, as indicated in the schematics of Figure 5.17b,c. Substitution of the transformation into Eq. (5.10) and simplification leads to

$$m_1\ddot{x}+b_1\dot{x}+k_1x-b_2\dot{y}-k_2y\left[1+\alpha y+\beta y^2\right]=-m_1\ddot{z} \tag{5.12a}$$

$$m_2\ddot{y}+\left[1+(m_2/m_1)\right]\left[b_2\dot{y}+k_2y\left(1+\alpha y+\beta y^2\right)\right]-(m_2/m_1)(b_1\dot{x}+k_1x)=0 \tag{5.12b}$$

Here, $k_2 = 2k_{2L}$, while $\alpha = 3\sqrt{k_{2NL}/k_{2L}}$ and $\beta = k_{2NL}/k_{2L}$ are stiffness proportionality constants, and $-\ddot{z} = Z_a \sin \Omega t$ is the harmonic base acceleration of amplitude Z_a and angular frequency Ω. In the event of asymmetries in the restoring force of the bistable inclusion [67,126], the proportionality constants α and β may be appropriately modified from the prior definitions which then indicates that the equilibria are more generally positioned at $v^* = \left[-\alpha \pm \sqrt{\alpha^2 - 4\beta}\right]/2\beta$ and $v^* = 0$, where the latter is a stable equilibrium and the larger absolute value of the two prior locations is the second stable equilibrium.

5.4.4 Experimental Validation of the Model Formulation and Numerical Examinations of System Operation

One set of identified system parameters from an experimental proof-of-concept sensor is provided in Table 5.4; the values are used in the following model validations. The bistable inclusion is seen to possess asymmetry between the stable equilibria around the unstable position and this is reflected in the identified proportionality constants α and β. The fundamental mode natural frequency of the system is $\Omega_1/2\pi$=61.3 Hz while the second mode occurs at $\Omega_2/2\pi$=184 Hz.

The following experimental study adopts a greater focus on verifying the key mass sensing principles which uniquely characterize the proposed sensor architecture and is less focused on highlighting improved detection accuracy. In the following experimental trials, the mass on the host cantilever beam remains constant during each examination. For a constant level of excitation, the excitation frequency is slowly swept from 55 Hz at a rate of +0.05 Hz s^{-1} until the point in time at which the SN bifurcation is crossed, thus causing the transition from low amplitude intrawell to snap-through motions of the bistable inclusion. Each trial is stopped once the critical event is triggered; this is because the inset bistable beams were found to fatigue rapidly (exacerbated by the thin aluminum design) and thus invalidate the complete response trajectories that are measured after the system passed through the fundamental mode resonance frequency. Consequently, the experimental investigations examine a controlled frequency sweeping strategy to activate the first bifurcation. The sections which follow this experimental study will examine the sequential activation of the critical events via a variety of operational protocols. Once the SN is crossed in the experimentation, the excitation frequency is recorded. Since this frequency may be monitored over time via repeated frequency sweeps, as has been employed with the parametric resonance sensors [79], the progressive shift of this critical frequency may be used to compute the accumulated mass quantity following suitable bifurcation analysis.

Table 5.4 System parameters identified from experimental sensor architectures and operation and used in simulation.

m_1, g	m_2, g	b_1, N s m^{-1}	b_2, N s m^{-1}	Z_a, m s^{-2}
3.52	0.705	0.0673	0.0228	15
k_1, N m^{-1}	k_2, N m^{-1}	α, m^{-1}	β, m^{-2}	
637	766	1.27×10^3	4.64×10^5	

Figure 5.20a presents the time series of an example frequency sweep experiment where no wax mass is added to the host cantilever beam end. As the excitation frequency gradually sweeps toward the fundamental mode of the system, the vibrational amplitudes of the coupled system response are magnified. The simulated results, Figure 5.20b,c, computed by numerically integrating Eq. (5.12) using the experimental system parameters in Table 5.4, are in very good agreement with the measurements, Figure 5.20a. In fact, during the sweep when the instantaneous frequency of excitation is approximately 59 Hz, an order-3 harmonic is induced within the asymmetric bistable inclusion; this is observed via the host beam as a small spike in velocity (around time 71 s in the experiment). The experimental observation of this feature is closely replicated in numerical simulation, Figure 5.20b around time 60 s. It is emphasized that this phenomenon is not inherent in the use of the coupled system design of the linear and bistable elements; this claim will be further verified by the absence of such phenomena in later investigations using other representative sensor architecture design parameters. The phenomenon was conclusively determined to be a coincidence of the exact parameters of the fabricated samples and therefore the parameters used in simulations presented in Figure 5.20b,c. As indicated above, the second mode linear natural frequency of the system is 184 Hz, which is coincidentally almost exactly three times the fundamental 61.3 Hz. In other words, the two DOF platform was inadvertently designed with a 1:3 ratio of lowest order natural frequencies. It is known that for a nonlinear two DOF system having a 1:3 ratio of natural frequencies and a cubic nonlinearity, excitation close to the fundamental natural frequency will induce a strong higher harmonic response which may suddenly become manifest, known as the internal resonance phenomenon [127]. This is an undesirable coincidence of the experimental sample fabrication, partially due to the exact degree of magnetic repulsion-buckling employed which partially governs this frequency. As a general rule-of-thumb, fractional or integer multiples of the natural frequencies should be avoided in sample fabrication.

Considering again the frequency sweep shown in Figure 5.20, the bistable inclusion response is pushed continually closer to the bifurcation point by the steady amplification of the host beam response as the harmonic excitation sweeps closer to the fundamental mode natural frequency. In the experiment, the bifurcation is suddenly activated and evident according to a surge and spike in host beam velocity, Figure 5.20a around a time of 93 s. In the simulation around a time of 92 s, the bifurcation is also accompanied by a spike in beam velocity, Figure 5.20b, while the bistable inclusion jumps to the other potential well, Figure 5.20c, since the steady-state mean of the displacement amplitude undergoes a constant shift after about 92 s. The overall good agreement between the measured and simulated results helps to validate the modeling assumptions, equation system formulation, and the identified experimental system parameters. In addition, the results exemplify that the successful integration of resonant and bistable bifurcation-based dynamics empowers novel sensing functionality in this particular sensor architecture.

Figure 5.20 Slow frequency sweep into the SN bifurcation. (a) Experiment: host beam velocity. (b) Simulation: host beam velocity. (c) Simulation: bistable inclusion displacement. Inset images denote zoomed views of bifurcation activation. The reduced-darkness portions of the curves in the simulated results (b,c), for times approximately greater than 92 s, are responses which are predicted to occur after the activation of the jump but which were not measured experimentally because the trial was finished.

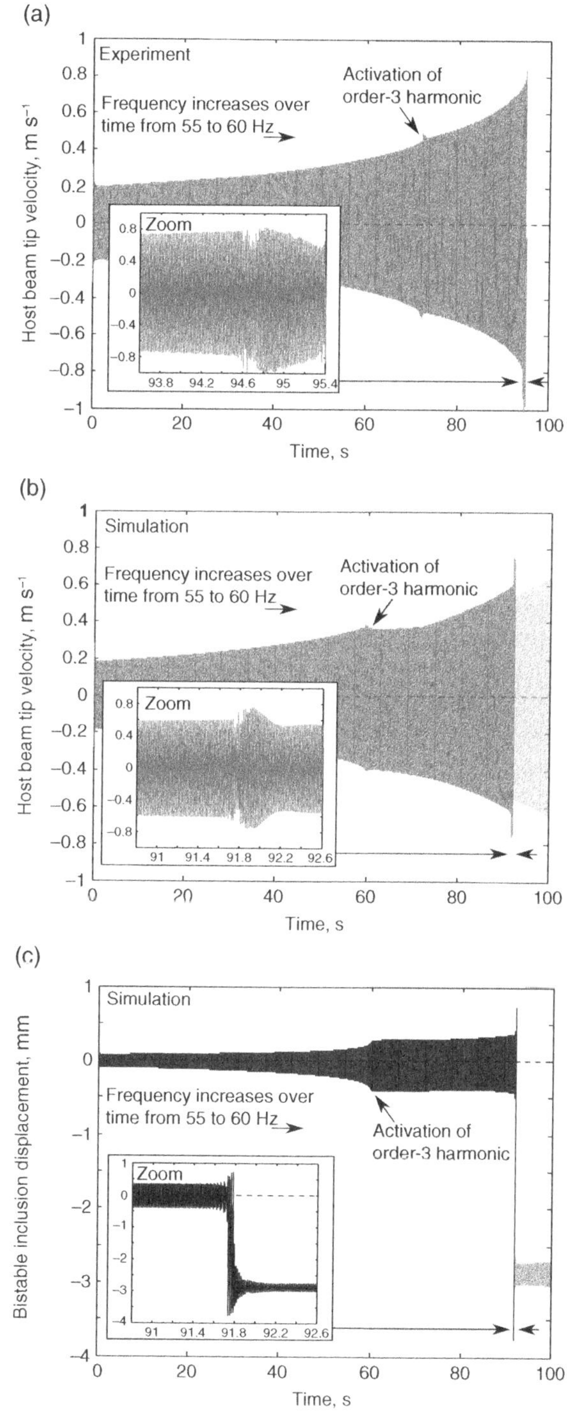

5.4.5 Examination of Passive Quantification of Mass Adsorption via Sequential Activation of Bifurcations

The model is used in order to explore the ability of the sensor and one particular sensing strategy to activate the SN bifurcations of the architecture in sequence. Because the critical system parameters (e.g., the host structure mass) that induce each SN event may be related to a closed form, approximated analytical solution to the governing Eq. (5.12) [90,116], the direct observation of two bifurcations sequentially serve to indicate that a particular quantity of mass has adsorbed upon the sensor over time, assuming all other design and operational parameters are otherwise constant. Thus, the sensing strategy examined in this section is that which was illustrated via the conceptually linked plots of Figure 5.18a,b, where a constant level of excitation (here, base acceleration) at a constant frequency drives the sensor and the progressive accumulation of mass on the architecture leads to a variably-amplified fundamental mode response as time elapses, successively triggering the two bifurcations of response amplitude for the small bistable inclusion.

First, the parameter set of the sensor architecture is scaled down to microscale dimensions, including the asymmetric design aspects which are particularly common on the microscale [45,47,128], particularly in consequence to residual fabrication stresses. The parameter set is provided in Table 5.5. In the following simulation, for simplicity it is assumed that mass is adsorbed only upon the larger host structure such that the time-varying mass of the host is $\bar{m}(t) = m_1 + \delta m(t)$. A mass ratio, mr, is defined as the ratio of adsorbed mass $\delta m(t)$ to the host linear structure mass m_1. In this and all following simulations of progressive accumulation of mass upon the sensor, the linear mass ratio *rate* is constant during the simulation trial, reflective of a target analyte steadily diffusing in the environment during a MEMS mass sensing application. The host structure is excited at a constant frequency that is 99.5% of the fundamental mode natural frequency prior to mass accumulation, $\Omega = 0.995\ \Omega_1$, and at a constant level of excitation provided in Table 5.5.

The results of the simulation, presented in Figure 5.21, are in excellent agreement with the conceptual description of this passive sensing strategy presented in Section 5.4.1. The unique elements and dynamic sensing events described conceptually are all evident: the progressive amplification and then reduction of sensor response due to the fundamental resonance shifting downwards in frequency because of mass adsorption; the activation of a low-to-high amplitude (intra- to interwell) response bifurcation for the bistable inclusion; and the

Table 5.5 System parameters used in simulation for sequential activation of bifurcations.

m_1, ng	m_2, ng	b_1, nN s m^{-1}	b_2, nN s m^{-1}	Z_a, m s^{-2}
5	0.1	471	8.2	7.0
k_1, N m^{-1}	k_2, N m^{-1}	α, m^{-1}	β, m^{-2}	
7.11	0.299	4.57×10^5	4.64×10^{10}	

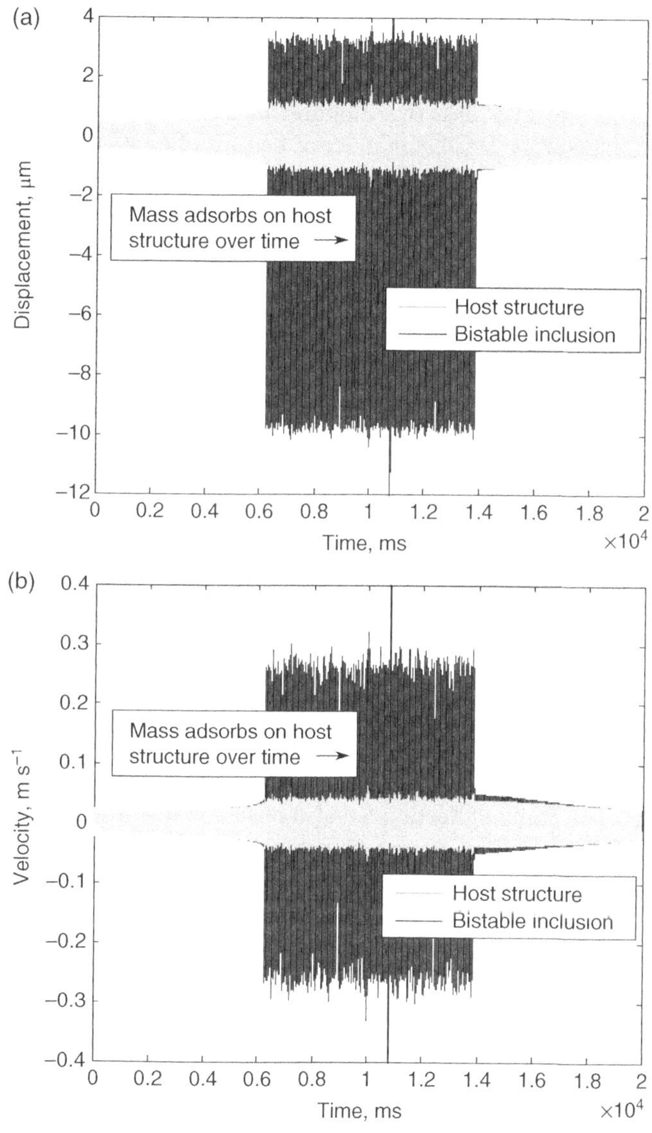

Figure 5.21 Exemplary model simulation of sequential bifurcation activation due to mass accumulation upon host structure. Sensor sub-system (a) response displacements and (b) velocities.

later activation of an inter-to-intrawell bifurcation down in amplitude. When the high amplitude response is activated, there is a sudden and apparent increase in bistable inclusion velocity, Figure 5.21b, and the displacement of the inclusion clearly undergoes snap-through motions, as evidenced in Figure 5.21a via the bias in the mean displacement response during the period of larger amplitude vibrations. While the energetic oscillations are activated, both DOF of the sensor exhibit highly perturbed responses due to the strongly nonlinear interwell vibrations of the bistable inclusion and the more pronounced elastic coupling between the two dynamic bodies.

With knowledge of the time elapsed between sequential bifurcations and the rate of mass adsorption, one may directly compute the mass accumulation. However, this method may not be feasible in practice because the rate of accumulation could be unknown. Thus, as described above, closed form approximate analytical solutions to the sensor architecture governing equations are required to identify the critical conditions which lead to the onset of both bifurcations. With derived expressions for the critical mass values to induce the bifurcations, the occurrence of both jumps in sequence *passively* indicates an amount of mass was adsorbed in the span of time which elapsed such that both bifurcations were triggered. This sensing and mass quantification strategy sharply contrasts the more limited threshold-only detection of mass currently enabled via the softening Duffing-based micro-oscillators [83,84] or the quantification of mass that is provided in the parametric microresonator studies only with the addition of active controls to constantly vary the excitation conditions [79,82,129].

5.4.6 Experimental Comparison of Bifurcation Triggering and Fundamental Mode Natural Frequency Reduction as Consistent Detection Metrics

The investigations of BB change detection for large-scale structures in Section 5.2 found that triggering the SN bifurcation in a bistable circuit was a more reliable indicator of the uniform increase in mass on the monitored cantilever beam than using the shift in the peak structural response around the natural frequency as the measure of change. The study of Section 5.3 also observed this trend: even mild damping and very small additive noise which could obscure a sharp resonant peak response at a natural frequency significantly degrade the usefulness of the classical, direct peak detection approach.

To investigate the consistency of these detection metrics for the proof-of-concept two DOF sensor architectures are explored here, experimental trials are performed using the controlled frequency sweeping strategy used in Section 5.4.4. Now, however, after each activation of the SN upwards from the low amplitude intrawell vibrations of the bistable inclusion to the interwell snap-through dynamics, the trial is stopped and wax mass is added to the free end of the host cantilever beam. Next, the two DOF sensor is excited with very low level white Gaussian noise for 60 s to identify the fundamental mode resonance frequency from the frf of the measurements. Then, the trial is restarted from the beginning sweeping frequency and allowed to proceed until the SN is crossed once more. This process is repeated for many iterations of steadily and incrementally increasing mass on the host beam.

The results of the experimental frequency sweeps for a range of added mass up to 0.012 mass ratio are collected in Figure 5.22. Figure 5.22a demonstrates how the measured frfs of the host beam are affected as mass accumulates upon the free beam end. The peaks of the frfs, identified as the fundamental mode natural frequencies, are plotted as circles. Clearly, the combined influences of measurement noise, damping, and finite time duration for sampling inhibit desirable

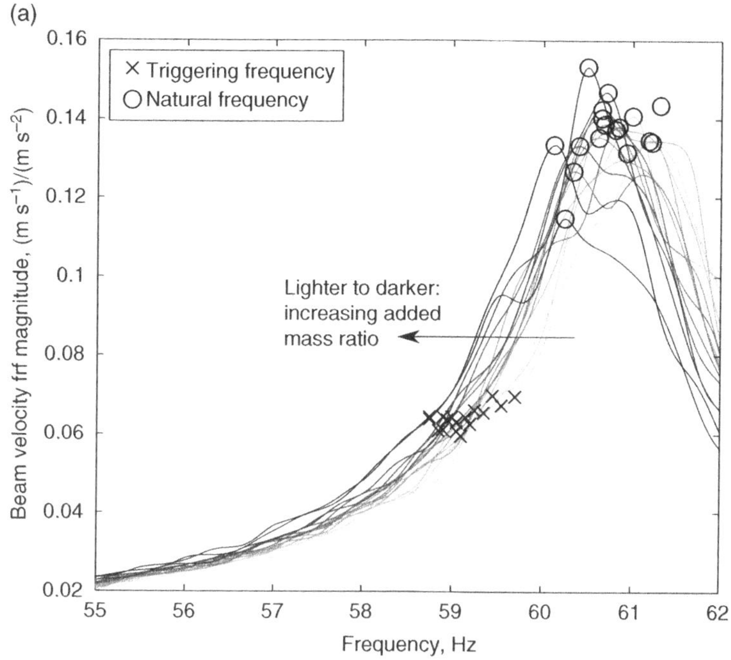

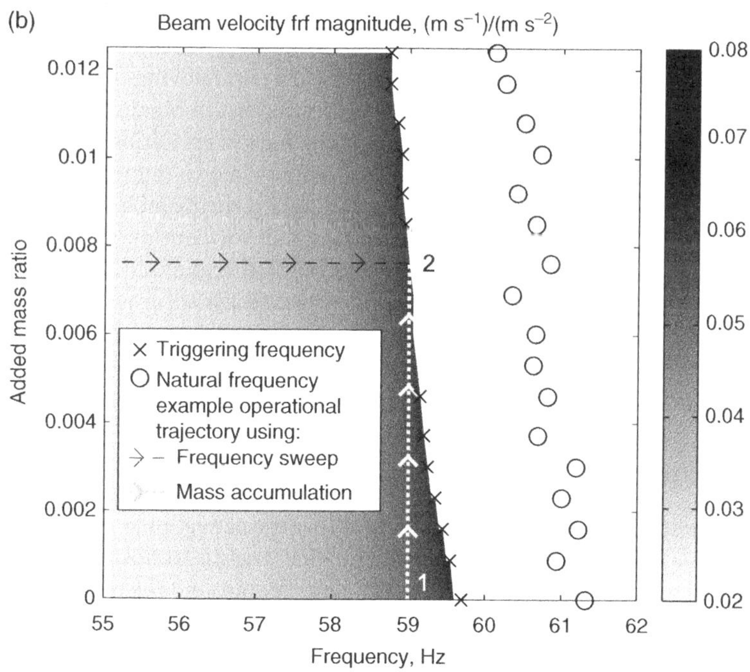

Figure 5.22 (a) Host beam frf due to low-level white noise input denoting natural frequencies and the corresponding frequencies at which the SN bifurcation is triggered for cases of mass addition. (b) Measured frf magnitudes for all cases of added mass ratio and triggering frequencies.

sharp resonant peaks, thus making traditional frequency shift-based mass sensing based upon a direct peak detection method a less reliable indicator of accumulating mass. It is noted that the limitation imposed by a finite time duration of the sampling is comparably negligible since 60 s of data are recorded for a response which resonates around 60 Hz. Thus, a substantial amount of data is available with which to very accurately identify the resonance, should the unavoidable measurement noise and damping not inhibit the result. Therefore, these latter two phenomena are the weaker links in the consistent identification of the fundamental mode natural frequency.

Figure 5.22b plots frf results for the trials conducted for each case of mass addition. The contour represents the instantaneous frf amplitudes measured during frequency sweeps; the transition from grayscale contour to the unshaded region denotes the initial triggering of interwell oscillations. The bifurcation frequencies are shown as crosses which define the boundary between the intrawell to interwell oscillations, similar to the point labeled as 2 in Figure 5.18. In Figure 5.22b, while the natural frequency (circle data points) is observed to reduce non-uniformly with mass accumulation, the frequency at which the bifurcation is activated consistently reduces upon each case of mass addition. Consequently, there is no mistaking mass accumulation using BB detection as compared to tracking the fundamental natural frequency.

In Figure 5.22b, the dashed and dotted lines, respectively, depict operational trajectories the sensor follows using either the controlled frequency sweep method (and constant adsorbed mass) or the passive strategy with constant excitation conditions (and increasing adsorbed mass). When employing the passive strategy, it is shown that the constant excitation frequency is a means by which to tailor the quantity of mass that initially triggers the sensor: excitation frequencies initially closer to the bifurcation require less added mass to induce the first jump. Although each experimental trial is stopped when the bistable inclusion snap-through dynamics are activated, it is recalled that for continued sweeps, either by frequency or accumulating mass, the second bifurcation downward in response is eventually encountered. For the time-varying mass accumulation trajectory plotted in Figure 5.22b as the dotted line, the labels 1 and 2 correspond to the comparable points as those indicated in the conceptual plot in Figure 5.18, to link the experimental results closely to the previous conceptual illustration.

5.4.7 Stochastic Modeling and Noise Sensitivities

By the results shown in Figure 5.22, it is apparent that the activation of the initial SN bifurcation is a more consistent determinant that mass is uniformly added from trial to trial than the use of the peak sensor response around the fundamental mode (identified as the natural frequency). This section seeks to investigate the consistency of activating this bifurcation for a varied range of sweeping rates and additive noise levels. Both sweeping strategies are considered: sweeping excitation frequency with constant mass adsorption, and time-increasing adsorbed mass with constant excitation frequency.

The motivation behind this examination is two-fold. First, bifurcation activation is particularly sensitive to stochastic processes since an otherwise stable, fixed static or dynamic operating state near a critical point may be inadvertently destabilized via a small stochastic perturbation [75,76,130]. And second, mass sensing is necessarily a non-stationary process – that is, mass accumulates at a finite rate over time or the sensor is persistently swept via the excitation conditions towards the bifurcation point – and triggering bifurcations is strongly dependent upon non-stationary processes [118,129–131]. Thus, the link between *stochastic* influences in bifurcation-based sensing and the *rate of bifurcation activation* is pivotal to elucidate in order to effectively harness the proposed detection strategy based on exploiting the combined resonant and BB dynamics.

Therefore, for insight into the stochastic sensitivities of the proposed sensing method and two DOF sensor, Eq. (5.12) is expressed in Itô stochastic differential equation (SDE) form for the following investigation. In addition to the harmonic base excitation, a stochastic process $\zeta(t)$ acts upon the host structure:

$$
\mathrm{d}\boldsymbol{x} = \begin{bmatrix} x_2 \\ \left\{-k_1x_1 - b_1x_2 + k_2x_3\left(1+\alpha x_3+\beta x_3^2\right)+b_2x_4\right\}/m_1 + Z_\mathrm{a}\sin\omega t \\ x_4 \\ \left\{\begin{matrix}\left(m_2/m_1\right)\left(k_1x_1+b_1/x_2\right)-\left[1+\left(m_2/m_1\right)\right] \\ \left[k_2x_3\left(1+\alpha x_3+\beta x_3^2\right)+b_2x_4\right]\end{matrix}\right\}/m_2 \end{bmatrix}\mathrm{d}t + \begin{bmatrix} 0 \\ S \\ 0 \\ 0 \end{bmatrix}\mathrm{d}B \tag{5.13a}
$$

$$
\boldsymbol{x} = \begin{bmatrix} x_1 & x_2 & x_3 & x_4 \end{bmatrix}^\mathrm{T} = \begin{bmatrix} x & \dot{x} & y & \dot{y} \end{bmatrix}^\mathrm{T}; \langle \zeta(t) \rangle = 0; \quad \langle \zeta(t)\zeta(t_0) \rangle = S^2\delta(t-t_0) \tag{5.13b}
$$

where $-\ddot{z} = Z_\mathrm{a}\sin\omega t + \zeta(t)$. The stochastic process $\zeta(t)$ is zero-mean, white Gaussian noise of intensity S^2, and $\mathrm{d}B/\mathrm{d}t = \zeta(t)$. In microscale applications, additive white noise processes $\zeta(t)$ represent commonly encountered uncertainties and or parameter variations induced by phenomena such as thermomechanical fluctuations, Brownian motion, and self-heating [96,132,133].

Prior to exploring the noise and non-stationary characteristic dependences, Eq. (5.13) is experimentally validated for responses near the fundamental mode. Figure 5.23a,b present measured and simulated beam velocity to base acceleration frf curves, respectively, of the two DOF sensor platform when excited by low-level white noise, $S = 0.1$, as mass is incrementally added to the host beam free end from one experiment/simulation to the next. For each case of added mass ratio, a 15 s time series of system response to white noise is measured or simulated. For the numerical simulations, base acceleration is not included, $Z_\mathrm{a} = 0$, and Eq. (5.13) is integrated using the SDE toolbox in MATLAB with the Euler–Maruyama algorithm. The frf results are normalized to the peak of the case without added mass. As mass is added to the end of the cantilevered host beam, the resonance frequencies – identified as the peaks of the plots in Figure 5.23a,b – shift non-uniformly downwards in frequency. The change in

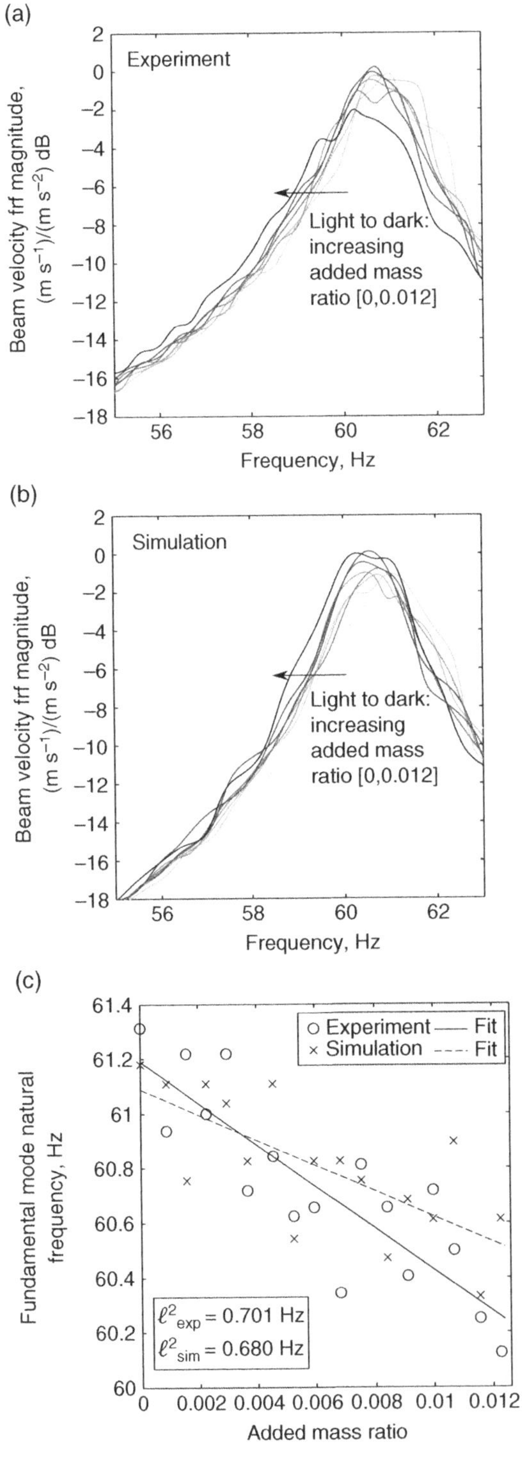

Figure 5.23 (a) Experimentally measured frf around the fundamental mode as added mass increases (light to dark). (b) Comparable simulated frf responses due to stochastic excitation. (c) Measured and numerically simulated fundamental mode natural frequencies as mass increases, with linear fits and residual norms (in inset).

natural frequency due to the mass accumulation is shown in Figure 5.23c. Collectively, noise interferences, damping, and the finite duration of the time series clearly make the resonances less sharply defined than ideal. Good agreement is seen comparing experimental and simulated trends, and Figure 5.23c shows that residual norms of the shifting natural frequency exhibit a similar deviation against linear fits. These findings verify that stochastic and damping influences are appropriately modeled with respect to the system parameters in Table 5.4, and support the further use of the model to evaluate sensitivities of the proposed system architecture and sensing method related to additive stochastic processes and non-stationary operations.

Then, the responses of Eq. (5.13) having additive noise and periodic base excitations are simulated using the sweeping styles explored earlier: controlled frequency sweeps using constant adsorbed mass or the passive approach of constant excitation frequency using progressive mass accumulation over time. Each sweeping style employs a constant level of the harmonic excitation. The important parameter distributions to consider are the values of the measure (frequency or mass accumulation) that occur at the jump event. In this work, excitation-to-noise ratio (ENR) is used to denote the dominance of the harmonic base excitation relative to additive stochastic effects. This is comparable but contrasted against SNR because in this examination the characteristics of the excitation are particularly important. To more clearly understand the ENR relationship, for example a total excitation having 0 dB ENR means the variance of the additive stochastic process is equal to the periodic base acceleration amplitude. Thus, an excitation with 0 dB ENR is significantly more dominated by noise than a 40 dB ENR process which is primarily driven by the harmonic base excitation. In microscale mass sensing applications, a low ENR may be surmounted by applying higher harmonic drive levels at the cost of inducing greater nonlinearities in response and potentially fatiguing the sensor [87].

Figure 5.24 presents the results of the investigations for the two sweeping styles as functions of ENR level and at various rates of sweeping. The bar heights represent mean values of the monitored parameter: frequency of first crossing in Figure 5.24a and added mass at first crossing in Figure 5.24b. The bar height values are determined from 40 trials, while the associated error bars are one standard deviation from the mean value. It is noted that trials conducted with slower sweep rate and high ENR (low additive noise) induce the most accurate triggering values because such cases approach the limiting condition of stationary and deterministic bifurcation activation [130].

Figure 5.24a plots results of the frequency sweep studies. In these simulations, a frequency sweep begins at 55 Hz, and forward, linear sweep rates are employed from 0.005 to 0.2 $\mathrm{Hz\,s^{-1}}$. It is seen that for high ENR the individual cases of frequency sweep rate converge to distinct values of the first crossing parameter, which is here given by the frequency of the bifurcation activation. It is also observed that faster sweeping rates lead to greater deviation and mean values of the bifurcation frequency. This indicates that faster sweeps induce delayed bifurcations; in other words, the events are activated for critical values of the excitation frequency which are in excess of the stationary and deterministic result. In contrast, for low ENR (high noise), premature bifurcations are activated for all sweeping

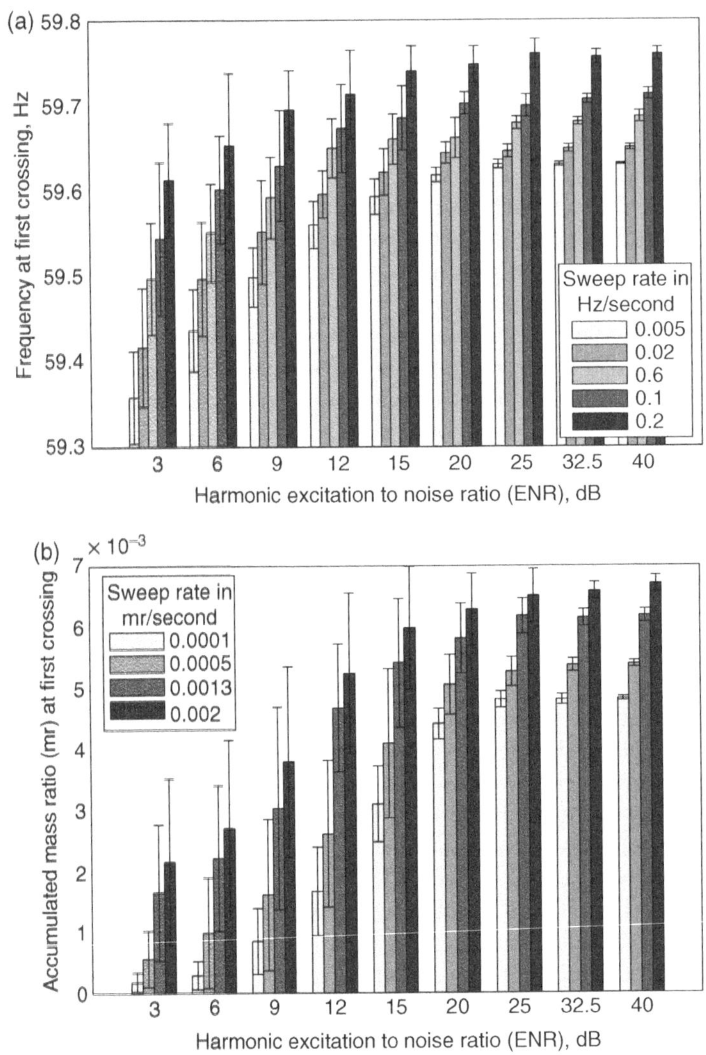

Figure 5.24 Dependence on additive noise excitation contribution for (a) controlled excitation frequency sweeps into the bifurcation and (b) passive, mass accumulation sweeps into the bifurcation (mr = mass ratio).

rates. This may seem counterintuitive but it reflects the fact that noise may induce a bifurcation in advance of the system reaching the deterministic critical condition even in the absence of non-stationary parametric influences [130].

Figure 5.24b shows results for the passive, mass accumulation sweeps where the base acceleration is provided at a constant frequency of 59 Hz while mass gradually accumulates on the host beam end. Several linear rates of accumulating mass ratio (mr) are used. Similar to the frequency sweeping method, low noise (high ENR) and slow mass accumulation rates converge to specific mr values to trigger (approximately 0.0048) and exhibit narrow standard deviations around this mean. This value of added mass to induce the bifurcation is approximately the stationary and deterministic critical mr value that would be determined from a stationary bifurcation analysis of Eq. (5.12). As the ENR is gradually reduced such that the noise becomes more influential, the additive noise induces premature jumps with respect to the critical mr values determined for high ENR. The influence becomes apparent as ENR reduces from approximately 20 to 12 dB and is exacerbated still more for smaller ENR. Under these conditions, the accumulated mass necessary to cross the bifurcation is reduced (premature bifurcation). By comparing the range of critical activation parameters between Figure 5.24a,b and the sweeping rates considered, it may be concluded that bifurcation activation is substantially more sensitive to change in mass accumulation rate than the frequency sweeping rate. For example at 15 dB ENR, Figure 5.24a shows that the change in mean critical jumping frequency is about 0.5% for a two order-of-magnitude variation in frequency sweeping, whereas Figure 5.24ab shows that the mean critical absorbed mass changes by almost 100% for a two order-of-magnitude change in mass accumulation rate. These observations are not intended to be a comprehensive evaluation of the sensor and sensing methods' overall sensitivities, but represent first efforts to examine the noise dependences exhibited by the two forms of sweeping strategies that may most usefully trigger bifurcations in the proposed sensor platform.

5.4.8 Operational Parameter Influences for Passive Sensing Strategy

There is considerable merit to the sensing strategy that employs constant excitation conditions and allows the progressive accumulation of mass on the sensor to induce the two SN bifurcations in sequence, which translates to the ability to *passively* quantify a measure of adsorbed mass over a time interval. Indeed, this capacity is particularly unique for the proposed two DOF coupled linear and bistable sensor platform and sets it apart from the previously explored BB sensors [79,83,84,113]. The prior section examined the stochastic and non-stationary characteristics of this sensing protocol in regards to the initial SN activation. To develop well-supported conclusions regarding the performance and reliability of this passive mass sensing strategy, this section seeks to probe the approach in greater detail and removes the prior focus upon the first bifurcation alone. The following studies employ simulations of the sensor model to assess the critical influences of excitation conditions and monitored mass accumulation rate and to determine useful design and operational guidelines for the passive mass sensing strategy.

Figure 5.25a presents the results of changing the constant harmonic excitation frequency in terms of the final amount of added mass ratio that is measured, for a wide range of sensor excitation levels and fixed mr adsorption rate of 0.001. Since all of the mass adsorption rates considered in this section are linear in time, and the initial and final times of bifurcation activation are detected in the simulations, the amount of mass that is quantified from the first to second bifurcations is computed directly. Also, the excitation upon the two DOF is purely harmonic: additive noise is omitted. The minimum excitation level is selected to ensure that sequential bifurcations are activated in all simulated

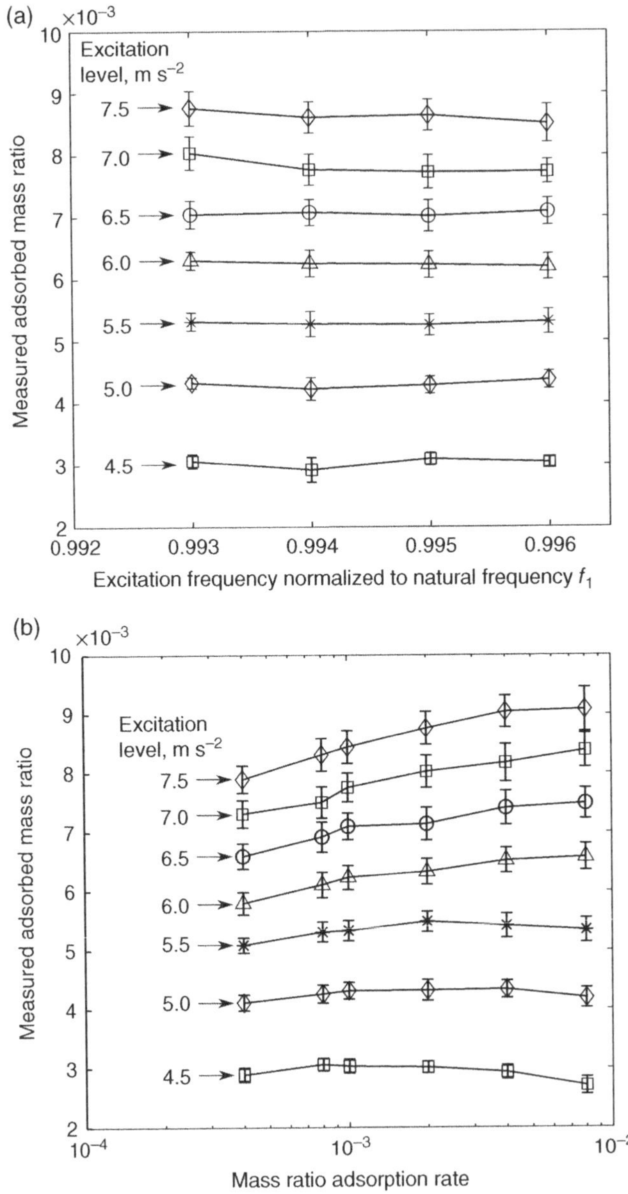

Figure 5.25 (a) Measured adsorbed mr as a function of constant excitation frequency and level. Mass ratio accumulation rate for all trials is 0.001. Data points are mean values across 40 simulated trials, bars are first standard deviations. (b) Measured adsorbed mr as a function of mr adsorption rate and excitation level. Data points are mean values across 5 sets of excitation frequency each using 40 simulated trials, bars are first standard deviations.

trials. Each data point represents the mean absorbed mass ratio value determined from 40 simulated trials, while bars are first standard deviations. Random initial conditions are employed for each trial. From the findings reported in

Figure 5.25a, it is seen that the change of excitation frequency has negligible influence upon the final amount of added mass which is measured. In contrast, the level of harmonic excitation plays a clear role in adjusting the measured mass value. However, the results also show that there is deviation around the mean value, and the adsorbed mass amount is not entirely insensitive to change in excitation frequency.

These findings reflect the initial condition dependence of strongly nonlinear systems and the non-stationarity of the system (mass changing over time) which lead to non-uniform activation of the bifurcations from trial to trial. In Figure 5.25a, the deviation around mean adsorbed mass values is plainly seen to increase as the excitation level increases. This result is intuitive when considering the conceptual framework of activating the bifurcations using the two DOF sensor architecture. As illustrated in Figure 5.18a, the shifting fundamental resonance leads to amplification and then reduction of the sensor response at a fixed excitation frequency. Since mass accumulates at a linear rate in the simulated trials, every time instant corresponds to a respective value of $\partial H/\partial m$ according to the slope of the fundamental mode resonance, as shown in Figure 5.18a. By increasing sensor excitation levels (here, the base acceleration amplitude), the representative bifurcation points 2 and 4 in Figure 5.18a effectively move down the surface contour, which means that the critical events activated at slower rates $\partial H/\partial m$. Because prolonged steady-state oscillation close to a bifurcation may lead to spurious switching events due to the initial condition sensitivity of strongly nonlinear systems [118,130,131] and because the SN bifurcation here utilized may exhibit fractal erosion of the basin of attraction as the system passes through the critical set of parameters [134], the "shallower" slopes $\partial H/\partial m$ caused by too high excitation levels are not conducive to repeatable triggering. It is for these reasons that Figure 5.25a indicates that an increasing excitation level produces larger deviations of quantified mass value; in comparison, smaller excitation levels effectively leverage the "steeper" slopes of the shifting fundamental resonance to more reliably activate bifurcations from the same increment of adsorbed mass over time. This finding indicates a clear means to directly ensure that the sensing strategy maintains viability by modification of the sensor excitation level. This is a versatility for passive BB microscale mass sensing otherwise achieved only by employing actively controlled approaches to date [129].

Because mass accumulates upon the sensor platform at a finite rate in practical mass sensing contexts, the sensor response is non-stationary. This factor is important because the specific parameters of periodic non-stationary excitations have a strong influence on the outcomes of both nonlinear *and* linear system responses [118,135,136]. Consequently, sensor sensitivity to change in adsorption rate must be characterized for important design and operating guidelines. Change of adsorption rate represents practical microscale mass sensing scenarios, such as rapid diffusion of a noxious gas or slow particle accumulation in a controlled laboratory environment, to identify the ultimate sensitivity performance for a given sensor. Thus, a wide range of rates should be considered to thoroughly assess the proposed sensor architecture and sensing strategy.

Figure 5.25b plots results of the measured adsorbed mass ratio as a function of the mass accumulation rate. Each data point is the mean value for a given excitation level computed from 40 simulated trials across a range of 5 constant excitation frequencies. Thus, each data point in Figure 5.25b is the mean value from 200 simulations each begun with random initial conditions. The bars are first standard deviations of the results from the mean. Ideally, the variation of adsorption rate should not influence the ultimate amount of mass that accumulates on the sensor spanning the sequential bifurcations. However, Figure 5.25b indicates that across the range of excitation levels, there is some sensitivity in the measurements to changes in adsorption rate. In general, the faster the rates, the greater the deviation of results. These trends become more apparent for increased excitation levels. Non-stationary passage through a bifurcation is a process prone to variability, but as higher excitation levels drive the sensor to greater degrees of strongly nonlinear response, the sensitivity to the ultimate point at which the bifurcations occur is likewise heightened. The trend for higher excitation levels and faster adsorption rates is an increase in the adsorbed mass that is measured.

Recall that from the perspective of the small bistable inclusion, the effect of mass accumulation is like an effective excitation level sweep up and down to sequentially induce the bistable element bifurcations, illustrated via Figure 5.18. The prior trend of increasing measured mass as excitation level and absorption rate both increase is explained because the SN bifurcation that induces the sudden drop downward from inter- to intra-well oscillation regimes (label 4 in Figure 5.18b) is more liable to cause delayed bifurcations during non-stationary excitation level sweeps than is the opposite SN bifurcation upwards in response amplitude (label 2 in Figure 5.18b) [118]. Thus, results of Figure 5.25b reveal that for the highest excitation levels, the sensor remains activated in the interwell response for too long and consequently quantifies an erroneous, increasing amount of accumulated mass. Uncovering these important trends helps to determine sensor excitation levels desirable to minimize adverse influence of non-stationarity on the reliability of the passive mass measurement strategy.

A final evaluation of the model is motivated by practical constraints which may be encountered in obtaining data read-out from the two DOF sensor platform. Depending on the fabrication process selected for the microscale sensor, it could be challenging to electrically isolate the bistable inclusion from the host so as to strictly measure the response of the smaller inclusion. Thus, it may be necessary to rely on response measurements acquired solely from transduction which is proportional to motions of the host resonator. Thus, this examination evaluates the feasibility of monitoring only the host linear resonator velocity response, which, it is assumed, is measured via one of many common transduction methods proportional to velocity, for example, optical, piezoresistivity, and so on. Specifically, the metric of performance used here is the ratio of variance of the host structure squared velocity, defined using Eq. (5.14)

$$\text{variance ratio} = \text{var}\left(|\dot{x}|^2_{\text{during}}\right) / \left\{\left[\text{var}\left(|\dot{x}|^2_{\text{before}}\right) + \text{var}\left(|\dot{x}|^2_{\text{after}}\right)\right] / 2\right\} \tag{5.14}$$

The variance ratio is computed as the host structure velocity squared variance *during* the activation of the snap-through dynamics of the bistable inclusion divided by the mean of the variances of the respective intrawell dynamics *before*

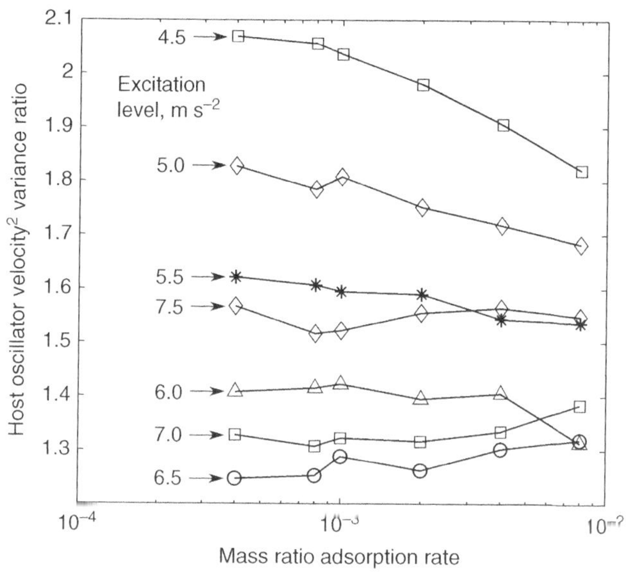

Figure 5.26 Ratio of host oscillator velocity^2 $|\dot{x}|^2$ variance during energetic response period to before/after as a function of mass ratio adsorption rate and excitation level. Data points are mean values across 5 sets of excitation frequency each using 40 simulated trials.

and *after* the triggering of the bifurcation. Put briefly in words, larger values of this metric indicate that detection of the bifurcation events is more readily observed using only the data read-out from the host linear structure.

Figure 5.26 presents results of the mean values of variance ratio determined from 5 constant excitation frequencies using 40 simulated trials each, across a range of excitation levels and mass adsorption rates. The results show that the variance ratio is interestingly minimized for excitation level 6.5 m s^{-2}. An explanation for this feature is not immediately apparent but is the focus of future study; however, low variance ratio is undesirable and thus attention is directed to the other findings evident in Figure 5.26. Importantly, the smaller levels of excitation yield the greatest variance ratios beneficial for bifurcation detection, particularly for the slower mass ratio adsorption rates. The variance ratios for the smaller excitation levels reduce as mass adsorption rate increases but still remain greater than that provided by higher excitation levels.

Recalling that the smallest excitation level in the simulated studies was selected because it ensured that both bifurcations were activated at some time over the course of the sensing trials, the results of Figure 5.25 and Figure 5.26 collectively suggest that the smallest level of excitation able to activate both jump events is the most beneficial for the proposed sensor architecture and passive sensing strategy. Through the parametric investigations, several important conclusions may be drawn. It is seen that the smallest excitation level (i) leads to the least deviation around the mean value for the amount of adsorbed mass that is measured, (ii) shows reduced sensitivity to adverse change in measured mass as accumulation rate changes, and moreover (iii) leads to the most useful host linear resonator variance ratio for mass detection from the more easily acquired measurements of the larger host resonator. The results, therefore, set valuable guidelines for implementation of the proposed sensor to ensure its operation remains effective and reliable.

5.4.9 Sensor Embodiments and Fabrication Strategies

Figure 5.17c illustrated the most generic realization of the proposed two DOF sensor architecture. Section 5.4.1 also described the dynamic design and operational principles which must be followed to effectively leverage the combined resonant and BB dynamics for new microscale mass sensing performance and functionality. Here, additional remarks are provided as pertain to specific, potential microscale sensor embodiments which meet the requirements for the platform as described above.

There are a suite of opportunities to realize the coupled linear–bistable sensor platform shown in Figure 5.17a using coupled microcantilevers. With the aims of enhancing the sensitivity and topographical reconstruction resolution for atomic force microscopes, several variations of microcantilever-in-cantilever sensors have been fabricated where the small oscillator inclusions have been machined facing, opposing, and orthogonal to the free end of the host cantilever [137–139]. Such designs would meet the elastic and geometric requirements of the linear–bistable sensor illustrated in Figure 5.17a, but magnet deposition would also be needed to induce bistability. Recent studies in vibration energy harvesting using energetic bistable systems have provided detailed fabrication protocols and demonstrated successful deposition of repulsive permanent magnets on microcantilever ends [38,140]. Other developments in microrobotics have introduced novel ways to pattern and deposit alternating polarities of magnetic layers for the purposes of translating and rotating microscale robots when acted upon by externally controlled fields [141,142]. Together, these sensor designs and fabrication methods meet the architecture requirements of the proposed microscale sensor in Figure 5.17a.

Alternative sensor platforms may be envisioned which satisfy the coupled linear–bistable design framework. For example, Figure 5.27 is a schematic for coupled in-plane oscillators, where an electrostatically actuated host oscillator contains an inset curved beam. Thus, bistability is realized by the geometry itself rather than electrostatic, magnetic, or otherwise multifield interactions. Such configuration would employ comparable etching procedures as a recently realized large-stroke actuator which utilized serially-connected bistable curved beams with an inset, actuated mass [143]. The success of fabricating microscale, serially-connected linear and bistable in-plane oscillators supports the feasibility of the proposed schematic in Figure 5.27. Moreover, by utilizing a geometrically bistable inclusion – that is, bistability by mechanics – one may alleviate intricate coupling concerns related to achieving bistability via electromechanical factors [113,144] or other multifield methods generally. These examples emphasize that successfully realizing the passive mass sensing strategy on the microscale is not constrained to one specific sensor architecture and may be employed according to fabrication expertise and packaging preferences, be they for out-of-plane cantilever vibrations or in-plane shuttle oscillator dynamics.

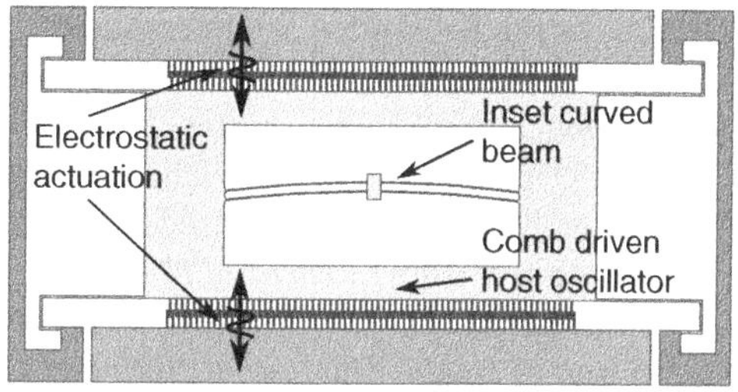

Figure 5.27 Schematic of coupled linear–bistable sensor architecture employing in-plane dynamics of electrostatically-actuated, host comb drive shuttle oscillator and inset curved beam.

5.4.10 Summary

By strategically integrating and harnessing the bifurcations and resonant dynamics of a two DOF bistable sensor, this study has shown that new performance potential and functionality may be invested in a microscale mass sensing platform. The utility of exploiting sensitive bifurcation phenomena has recently been recognized as a prime tool for microscale mass detection and quantification. This study developed a novel sensor architecture and investigated several sensing strategies that advance the possibilities for this sensing context. Using a host of experimental and numerical examinations, design methodologies and operational guidelines were derived to elucidate the connections between detection sensitivity and consistency with the operating states of the sensor. A means to quantify the progressive accumulation of mass on the sensor is also empowered via the strategic architecture composition and operational protocol, thus eliminating the need for active controls to be incorporated to accomplish the same end. It is also exemplified that the sensor architecture requirements are met by a myriad of platform designs, which indicates the potential to realize the performance and functionality advances is amenable to various sensing objectives, scales, and fabrication means. The ongoing interest in BB microscale sensing [82,145] shows that this is an area of growing interest and potential. The new opportunities generated via exploiting the bifurcation and resonant dynamics of the proposed two DOF bistable sensor may become a firm foundation for future developments in this area.

References

[1] E. Burton, *The History of Clocks and Watches* (Crescent Books, New York, 1979).

[2] J. Needham, *The Shorter Science and Civilisation in China* (Cambridge University Press, Cambridge, 1986).

[3] K. Worden, C.R. Farrar, G. Manson, *et al.*, The fundamental axioms of structural health monitoring, *Proceedings of the Royal Society A* **463**, 1639–1664 (2007).

[4] N.V. Lavrik, M.J. Sepaniak, P.G. Datskos, Cantilever transducers as a platform for chemical and biological sensors, *Review of Scientific Instruments* **75**, 2229–2253 (2004).

[5] T.P. Burg, *et al.*, Weighing of biomolecules, single cells and single nanoparticles in fluid, *Nature* **446**, 1066–1069 (2007).

[6] F. Ansari, *Sensing Issues in Civil Structural Health Monitoring* (Springer, Dordrecht, The Netherlands, 2005).

[7] C.R. Farrar, K. Worden, An introduction to structural health monitoring, *Philosophical Transactions of the Royal Society A* **365**, 303–315 (2007).
[8] H. Sohn, *et al.*, A review of structural health monitoring literature: 1996–2001. LA-13976-MS. Los Alamos National Laboratory; **2004**.
[9] W. Fan, P. Qiao, Vibration-based damage identification methods: a review and comparative study, *Structural Health Monitoring* **10**, 83–111 (2011).
[10] S. Gopalakrishnan, M. Ruzzene, S. Hanagud, *Computational Techniques for Structural Health Monitoring* (Springer, London, 2011).
[11] M.I. Younis, *MEMS Linear and Nonlinear Statics and Dynamics* (Springer, New York, 2011).
[12] M. Gad-el-Hak, *MEMS: Design and Fabrication* (Taylor & Francis, Boca Raton, 2006).
[13] M.J. Madou, *Fundamentals of Microfabrication: The Science of Miniaturization* (CRC Press, Boca Raton, 2002).
[14] K.C. Schwab, M.L. Roukes, Putting mechanics into quantum mechanics, *Physics Today* **58**, 36–42 (2005).
[15] W.M. Zhang, H. Yan, Z.K. Peng, Electrostatic pull-in instability in MEMS/NEMS: a review, *Sensors and Actuators A: Physical* **214**, 187–218 (2014).
[16] T. Thundat, E.A. Wachter, S.L. Sharp, *et al.*, Detection of mercury vapor using resonating microcantilevers, *Applied Physics Letters* **66**, 1695–1697 (1995).
[17] C. Zhang, L. Guo, L.F. Wang, *et al.*, Passive wireless integrated humidity sensor based on dual-layer spiral inductors, *Electronics Letters* **50**, 1287–1289 (2014).
[18] H. Miao, K. Srinivasan, V. Aksyuk, A microelectromechanically controlled cavity optomechanical sensing system, *New Journal of Physics* **14**, 075015 (2012).
[19] C. Lissandrello, F. Inci, M. Francom, *et al.*, Nanomechanical motion of Escherichia coli adhered to a surface, *Applied Physics Letters* **105**, 113701 (2014).
[20] R.C. Batra, M. Porfiri, D. Spinello, Review of modeling electrostatically actuated microelectromechanical systems, *Smart Materials and Structures* **16**, R23–R31 (2007).
[21] W. Fang, J.A. Wickert, Post buckling of micromachined beams, *Journal of Micromechanics and Microengineering* **4**, 115–122 (1994).
[22] H. Guckel, T. Randazzo, D.W. Burns, A simple technique for the determination of mechanical strain in thin films with applications to polysilicon, *Journal of Applied Physics* **57**, 1671–1675 (1985).
[23] M. Vangbo, Y. Bäcklund, A lateral symmetrically bistable buckled beam, *Journal of Micromechanics and Microengineering* **8**, 29–32 (1998).
[24] L.L. Howell, *Compliant Mechanisms* (John Wiley & Sons, Inc., New York, 2001).
[25] B.D. Jensen, L.L. Howell, L.G. Salmon, Design of two-link, in-plane, bistable compliant micro-mechanisms, *Journal of Mechanical Design* **121**, 416–423 (1999).
[26] M.S. Baker, L.L. Howell, On-chip actuation of an in-plane compliant bistable micromechanism, *Journal of Microelectromechanical Systems* **11**, 566–573 (2002).
[27] J.W. Wittwer, M.S. Baker, L.L. Howell, Robust design and model validation of nonlinear compliant micromechanisms, *Journal of Microelectromechanical Systems* **15**, 33–41 (2006).

[28] J.W. Wittwer, M.S. Baker, L.L. Howell, Simulation, measurement, and asymmetric buckling of thermal microactuators, *Sensors and Actuators A* **128**, 395–401 (2006).
[29] N.B. Hubbard, L.L. Howell, Design and characterization of a dual-stage, thermally actuated nanopositioner, *Journal of Micromechanics and Microengineering* **15**, 1482–1493 (2005).
[30] B.J. Hansen, C.J. Carron, B.D. Jensen, *et al.*, Plastic latching accelerometer based on bisatble compliant mechanisms, *Smart Materials and Structures* **16**, 1967–1972 (2007).
[31] B. Todd, M. Phillips, S.M. Schultz, *et al.*, Low-cost RFID threshold shock sensors, *IEEE Sensors Journal* **9**, 464–469 (2009).
[32] J.K. Anderson, L.L. Howell, J.W. Wittwer, *et al.*, Piezoresistive sensing of bistable micro mechanism state, *Journal of Micromechanics and Microengineering* **16**, 943–950 (2006).
[33] D.L. Wilcox, L.L. Howell, Fully compliant tensural bistable micromechanisms (FTBM), *Journal of Microelectromechanical Systems* **14**, 1223–1235 (2005).
[34] N.D. Masters, L.L. Howell, A self-retracting fully compliant bistable micromechanism, *Journal of Microelectromechanical Systems* **12**, 273–280 (2003).
[35] J.S. Han, J.S. Ko, Y.T. Kim, *et al.*, Parametric study and optimization of a micro-optical switch with a laterally driven electromagnetic microactuator, *Journal of Micromechanics and Microengineering* **12**, 939–947 (2002).
[36] S. Park, D. Hah, Pre-shaped buckled-beam actuators: theory and experiments, *Sensors and Actuators A: Physical* **148**, 186–192 (2008).
[37] D.A. Wang, H.T. Pham, Y.H. Hsieh, Dynamical switching of an electromagnetically driven compliant bistable mechanism, *Sensors and Actuators A: Physical* **149**, 143–151 (2009).
[38] T. Emery. Fabrication of bistable MEMS systems for energy harvesting. In V. In, A. Palacios, P. Longhini, eds, International Conference on Theory and Applications in Nonlinear Dynamics (ICAND 2012) pp. 123–128 (Springer, New York, 2014).
[39] J. Qiu, J.H. Lang, A.H. Slocum, A curved-beam bistable mechanism, *Journal of Microelectromechanical Systems* **13**, 137–146 (2004).
[40] M.T.A. Saif, On a tunable bistable MEMS – theory and experiment, *Journal of Microelectromechanical Systems* **9**, 157–170 (2000).
[41] M. Sulfridge, T. Saif, N. Miller, *et al.*, Optical actuation of a bistable MEMS, *Journal of Microelectromechanical Systems* **11**, 574–583 (2002).
[42] M. Sulfridge, T. Saif, N. Miller, *et al.*, Nonlinear dynamic study of a bistable MEMS: model and experiment, *Journal of Microelectromechanical Systems* **13**, 725–731 (2004).
[43] V. Kumar, J.F. Rhoads, Modeling and analysis of an optically-actuated, bistable MEMS device, *Journal of Computational and Nonlinear Dynamics* 7, 021007 (2012).
[44] J. Casals-Terré, A. Fargas-Marques, A.M. Shkel, Snap-action bistable micromechanisms actuated by nonlinear resonance, *Journal of Microelectromechanical Systems* **17**, 1082–1093 (2008).

[45] D. Roodenburg, J.W. Spronck, H.S.J. van der Zant, *et al.*, Buckling beam micromechanical memory with on-chip readout, *Applied Physics Letters* **94**, 183501 (2009).
[46] V. Intaraprasonk, S. Fan, Nonvolatile bistable all-optical switch from mechanical buckling, *Applied Physics Letters* **98**, 241104 (2011).
[47] M. Bagheri, M. Poot, L. Mo, *et al.*, Dynamic manipulation of nanomechanical resonators in the high-amplitude regime and non-volatile mechanical memory operation, *Nature Nanotechnology* **6**, 726–732 (2011).
[48] B. Wagner, H.J. Quenzer, S. Hoerschelmann, T. Lisec, M. Juerss, 1996 Bistable microvalve with pneumatically coupled membranes. In *Proceedings of MEMS 1996*. pp. 384–388.
[49] W.K. Schomburg, C. Goll, Design optimization of bistable microdiaphragm valves, *Sensors and Actuators A* **64**, 259–264 (1998).
[50] C. Goll, W. Bacher, B. Büstgens, *et al.*, Microvalves with bistable buckled polymer diaphragms, *Journal of Micromechanics and Microengineering* **6**, 77–79 (1996).
[51] M. Capanu, J.G. Boyd, P.J. Hesketh, Design, fabrication, and testing of a bistable electromagnetically actuated microvalve, *Journal of Microelectromechanical Systems* **9**, 181–189 (2000).
[52] J.S. Bintoro, P.J. Hesketh, Y.H. Berthelot, CMOS compatible bistable electromagnetic microvalve on a single wafer, *Microelectronics Journal* **36**, 667–672 (2005).
[53] J. Sutanto, P.J. Hesketh, Y.H. Berthelot, Design, microfabrication and testing of a CMOS compatible bistable electromagnetic microvalve with latching/unlatching mechanism on a single wafer, *Journal of Micromechanics and Microengineering* **16**, 266–275 (2006).
[54] N.T. Nguyen, S.T. Wereley, *Fundamentals and Applications of Microfluidics* (Artech House, Boston, 2006).
[55] C. Kleinstreuer, *Microfluidics and Nanofluidics: Theory and Selected Applications* (John Wiley & Sons, Inc., Hoboken, 2014).
[56] S. Krylov, Lyapunov exponents as a criterion for the dynamic pull-in instability of electrostatically actuated microstructure, *International Journal of Non-Linear Mechanics* **42**, 626–642 (2007).
[57] S. Krylov, B.R. Ilic, D. Schreiber, *et al.*, The pull-in behavior of electrostatically actuated bistable microstructures, *Journal of Micromechanics and Microengineering* **18**, 055026 (2008).
[58] K. Das, R.C. Batra, Pull-in and snap-through instabilities in transient deformations of microelectromechanical systems, *Journal of Micromechanics and Microengineering* **19**, 035008 (2009).
[59] K. Das, R.C. Batra, Symmetry breaking, snap-through and pull-in instabilities under dynamic loading of microelectromechanical shallow arches, *Smart Materials and Structures* **18**, 115008 (2009).
[60] S. Gutschmidt, The influence of higher-order mode shapes for reduced-order models of electrostatically actuated microbeams, *Journal of Applied Mechanics* 77, 041007 (2010).
[61] H.M. Ouakad, M.I. Younis, The dynamic behavior of MEMS arch resonators actuated electrically, *International Journal of Non-Linear Mechanics* **45**, 704–713 (2010).

[62] M.I. Younis, H.M. Ouakad, F.M. Alsaleem, *et al.*, Nonlinear dynamics of MEMS arches under harmonic electrostatic actuation, *Journal of Microelectromechanical Systems* **19**, 647–656 (2010).
[63] S.A. Alkharabsheh, M.I. Younis, Dynamics of MEMS arches of flexible supports, *Journal of Microelectromechanical Systems* **22**, 216–224 (2013).
[64] S.A. Alkharabsheh, M.I. Younis, Statics and dynamics of MEMS arches under axial forces, *Journal of Vibration and Acoustics* **135**, 021007 (2013).
[65] L. Medina, R. Gilat, S. Krylov, Symmetry breaking in an initially curved micro beam loaded by a distributed electrostatic force, *International Journal of Solids and Structures* **49**, 1864–1876 (2012).
[66] L. Medina, R. Gilat, S. Krylov, Symmetry breaking in an initially curved pre-stressed micro beam loaded by a distributed electrostatic force, *International Journal of Solids and Structures* **51**, 2047–2061 (2014).
[67] S. Krylov, B.R. Ilic, S. Lulinsky, Bistability of curved microbeams actuated by fringing electrostatic fields, *Nonlinear Dynamics* **66**, 403–426 (2011).
[68] J.F. Rhoads, S.W. Shaw, K.L. Turner, Nonlinear dynamics and its applications in micro- and nanoresonators, *Journal of Dynamic Systems, Measurement, and Control* **132**, 034001 (2010).
[69] F.D. Bannon, J.R. Clark, C.T.C. Nguyen, High-Q HF micromechanical filters, *IEEE Journal of Solid-State Circuits* **35**, 512–526 (2000).
[70] J.F. Rhoads, S.W. Shaw, K.L. Turner, Nonlinear response of resonant microbeam systems with purely-parametric electrostatic actuation, *Journal of Micromechanics and Microengineering* **16**, 890–899 (2006).
[71] A. Erbe, H. Krommer, A. Kraus, *et al.*, Mechanical mixing in nonlinear nanomechanical resonators, *Applied Physics Letters* 77, 3102–3104 (2000).
[72] A.T. Alastalo, V. Kaajakari, Intermodulation in capacitively coupled microelectromechanical filters, *IEEE Electron Device Letters* **26**, 289–291 (2005).
[73] M. Koskenvuori, I. Tittonen, Towards micromechanical radio: overtone excitations of a microresonator through the nonlinearities of the second and third order, *Journal of Microelectromechanical Systems* **17**, 363–369 (2008).
[74] H.M. Ouakad, M.I. Younis, On using the dynamic snap-through motion of MEMS initially curved microbeams for filtering applications, *Journal of Sound and Vibration* **333**, 555–568 (2014).
[75] R. Vijay, M.H. Devoret, I. Siddiqi, Invited review article: the Josephson bifurcation amplifier, *Review of Scientific Instruments* **80**, 111101 (2009).
[76] A.B. Zorin, Quantum-limited electrometer based on single cooper pair tunneling, *Physical Review Letters* **76**, 4408–4411 (1996).
[77] W. Zhang, K.L. Turner, 2004 A mass sensor based on parametric resonance. In *Proceedings of Solid-State Sensor, Actuator and Microsystems Workshop*; Hilton Head Island, South Carolina. pp. 49–52.
[78] W. Zhang, K.L. Turner, 2004 Noise analysis in parametric resonance based mass sensing. In *Proceedings of IMECE 2004 ASME International Mechanical Engineering Congress and Exposition*; Anaheim, California. IMECE2004-61412.
[79] W. Zhang, K.L. Turner, Application of parametric resonance amplification in a single-crystal silicon micro-oscillator based mass sensor, *Sensors and Actuators A* **122**, 23–30 (2005).

[80] K.L. Turner, W. Zhang, 2001 Design and analysis of a dynamic MEM chemical sensor. In *Proceedings of the 2001 American Control Conference*; Arlington, VA. 1214–1218.

[81] Z. Yie, M.A. Zielke, C.B. Burgner, *et al.*, Comparison of parametric and linear mass detection in the presence of detection noise, *Journal of Micromechanics and Microengineering* **21**, 025027 (2011).

[82] Li. L.L., E.L. Holthoff, L.A. Shaw, *et al.*, Noise squeezing controlled parametric bifurcation tracking of MIP-coated microbeam MEMS sensor for TNT explosive gas sensing, *Journal of Microelectromechanical Systems* **23**, [1228]–1236 (2014).

[83] M.I. Younis, F. Alsaleem, Exploration of new concepts for mass detection in electrostatically-actuated structures based on nonlinear phenomena, *Journal of Computational and Nonlinear Dynamics* **4**, 021010 (2009).

[84] V. Kumar, *et al.*, Bifurcation-based mass sensing using piezoelectrically-actuated microcantilevers, *Applied Physics Letters* **98**, 153510 (2011).

[85] V. Kumar, Y. Yang, J.W. Boley, *et al.*, Modeling, analysis, and experimental validation of a bifurcation-based microsensor, *Journal of Microelectromechanical Systems* **21**, 549–558 (2012).

[86] M.I. Friswell, Damage identification using inverse methods, *Philosophical Transactions of the Royal Society A* **365**, 393–410 (2007).

[87] R. Lifshitz, M.C. Cross. Nonlinear dynamics of nanomechanical and micromechanical resonators. In H.G. Schuster, ed. Reviews of Nonlinear Dynamics and Complexity: Volume **1**, pp. 1–52 (Wiley-VCH Verlag, GmbH).

[88] K.L. Turner, C. Burgner, Z. Yie, *et al.*, Nonlinear dynamics of MEMS systems, *AIP Conference Proceedings* **1339**, 111 (2011).

[89] K.W. Wang, J. Tang, *Adaptive Structural Systems with Piezoelectric Transducer Circuitry* (Springer, New York, 2008).

[90] R.L. Harne, M. Thota, K.W. Wang, Concise and high-fidelity predictive criteria for maximizing performance and robustness of bistable energy harvesters, *Applied Physics Letters* **102**, 053903 (2013).

[91] I. Kovacic, M.J. Brennan, editors, *The Duffing Equation: Nonlinear Oscillators and their Behaviour* 1st ed. (John Wiley & Sons, Ltd, Chichester, 2011).

[92] E. Tamaševičiūtė, A. Tamaševičius, G. Mykolaitis, *et al.*, Analogue electrical circuit for simulation of the Duffing-Holmes equation, *Nonlinear Analysis: Modelling and Control* **13**, 241–252 (2008).

[93] P. Holmes, A nonlinear oscillator with a strange attractor, *Philosophical Transactions of the Royal Society of London A* **292**(1394), 419–448 (1979).

[94] N.E. DuToit, B.L. Wardle, S.G. Kim, Design considerations for MEMS-scale piezoelectric mechanical vibration energy harvesters, *Integrated Ferroelectrics* **71**, 121–160 (2005).

[95] S.S. Rao, *Mechanical Vibrations* 4th edn (Pearson Prentice Hall, Upper Saddle River, New Jersey, 2004).

[96] K.L. Ekinci, Y.T. Yang, M.L. Roukes, Ultimate limits to inertial mass sensing based upon nanoelectromechanical systems, *Journal of Applied Physics* **95**, 2682–2689 (2004).

[97] E.P. Carden, P. Fanning, Vibration based condition monitoring: a review, *Structural Health Monitoring* **3**, 355–377 (2004).

[98] E. Figueiredo, G. Park, K.M. Farinholt, *et al.*, Use of time-series predictive models for piezoelectric active-sensing in structural health monitoring applications, *Journal of Vibration and Acoustics* **134**, 041014 (2012).
[99] O.S. Salawu, Detection of structural damage through changes in frequency: a review, *Engineering Structures* **19**, 718–723 (1997).
[100] A. Majumdar, D.K. Maiti, D. Maity, Damage assessment of truss structures from changes in natural frequency using ant colony optimization, *Applied Mathematics and Computation* **218**, 9759–9772 (2012).
[101] L.J. Jiang, J Tang, K.W. Wang, An enhanced frequency-shift-based damage identification method using tunable piezoelectric transducer circuitry, *Smart Materials and Structures* **15**, 799–808 (2006).
[102] L. Jiang, J. Tang, K.W. Wang, On the tuning of variable piezoelectric transducer circuitry network for structural damage identification, *Journal of Sound and Vibration* **309**, 695–717 (2008).
[103] H.L. Chen, C.C. Spyrakos, G. Venkatesh, Evaluating structural deterioration by dynamic response, *Journal of Structural Engineering* **121**, 1197–1204 (1995).
[104] B.H. Koh, S.J. Dyke, Structural health monitoring for flexible bridge structures using correlation and sensitivity of modal data, *Computers & Structures* **85**, 117–130 (2007).
[105] L.J. Jiang, J. Tang, K.W. Wang, An optimal sensitivity-enhancing feedback control approach via eigenstructure assignment for structural damage identification, *Journal of Vibration and Acoustics* **129**, 771–783 (2007).
[106] L. Jiang, K.W. Wang, An experimental-based frequency sensitivity enhancing control approach for structural damage detection, *Smart Materials and Structures* **18**, 065005 (2009).
[107] B.H. Koh, L.R. Ray, Feedback controller design for sensitivity-based damage localization, *Journal of Sound and Vibration* **273**, 317–335 (2004).
[108] J. Tang, K.W. Wang, Vibration confinement via optimal eigenvector assignment and piezoelectric networks, *Journal of Vibration and Acoustics* **126**, 27–36 (2004).
[109] R.L. Harne, K.W. Wang, Robust sensing methodology for detecting change with bistable circuitry dynamics tailoring, *Applied Physics Letters* **102**, 203506 (2013).
[110] L.N. Virgin, *Introduction to Experimental Nonlinear Dynamics* (Cambridge University Press, Cambridge, 2000).
[111] R. Wiebe, L.N. Virgin, I. Stanciulescu, *et al.*, Characterizing dynamic transitions associated with snap-through: a discrete system, *Journal of Computational and Nonlinear Dynamics* **8**, 011010 (2013).
[112] J.P. Lynch, K. Loh, A summary review of wireless sensors and sensor networks for structural health monitoring, *Shock and Vibration Digest* **38**, 91–130 (2006).
[113] P.A. Hassanpour, P.M. Nieva, A. Khajepour, Stochastic analysis of a novel force sensor based on bifurcation of a micro-structure, *Journal of Sound and Vibration* **330**, 5753–5768 (2011).
[114] J.M.T. Thompson, G.W. Hunt, *Elastic Instability Phenomena* (John Wiley & Sons, Ltd, Chichester, 1984).

[115] P. Glendinning, *Stability, Instability and Chaos: an Introduction to the Theory of Nonlinear Differential Equations* (Cambridge University Press, Cambridge, 1994).
[116] W. Szemplińska-Stupnicka, J. Rudowski, Steady states in the twin-well potential oscillator: computer simulations and approximate analytical studies, *Chaos* **3**, 375–385 (1993).
[117] S.Y. Kim, Y. Kim, Dynamic stabilization in the double-well Duffing oscillator, *Physical Review E* **61**, 6517–6520 (2000).
[118] C.H. Lu, R.M. Evan-Iwanowski, Period doubling bifurcation problems in the softening Duffing oscillator with nonstationary excitation, *Nonlinear Dynamics* **5**, 401–420 (1994).
[119] M. Panyam, R. Masana, M.F. Daqaq, On approximating the effective bandwidth of bi-stable energy harvesters, *International Journal of Non-Linear Mechanics* **67**, 153–163 (2014).
[120] D.J. Inman, *Engineering Vibration* (Prentice Hall, Saddle River, New Jersey, 2001).
[121] R.L. Harne, K.W. Wang, Passive measurement of progressive mass change via bifurcation sensing with a multistable micromechanical system, *Journal of Intelligent Material Systems and Structures* **26**, 1622–1632 (2015).
[122] O. Aldraihem, A. Baz, Energy harvester with a dynamic magnifier, *Journal of Intelligent Material Systems and Structures* **22**, 521–530 (2011).
[123] W. Zhou, G.P. Penamalli, L. Zuo, An efficient vibration energy harvester with a multi-mode dynamic magnifier, *Smart Materials and Structures* **21**, 015014 (2012).
[124] Z. Wu, R.L. Harne, K.W. Wang, Energy harvester synthesis via coupled linear-bistable system with multistable dynamics, *Journal of Applied Mechanics* **81**, 061005 (2014).
[125] M. Sato, B.E. Hubbard, A.J. Sievers, Colloquium: nonlinear energy localization and its manipulation in micromechanical oscillator arrays, *Reviews of Modern Physics* **78**, 137–157 (2006).
[126] L. Ruzziconi, S. Lenci, M.I. Younis, An imperfect microbeam under an axial load and electric excitation: nonlinear phenomena and dynamical integrity, *International Journal of Bifurcation and Chaos* **23**, 1350026 (2013).
[127] P.R. Sethna, Steady-state undamped vibrations of a class of nonlinear discrete systems, *Journal of Applied Mechanics* **27**, 187–195 (1960).
[128] D.J. Joe, *et al.*, Stress-based resonant volatile gas microsensor operated near the critically buckled state, *Journal of Applied Physics* **111**, 104517 (2012).
[129] C.B. Burgner, K.L. Turner, N.J. Miller, *et al.*, Parameter sweep strategies for sensing using bifurcations in MEMS. In *Solid-State Sensors, Actuators, and Microsystems Workshop*, Hilton Head Island, South Carolina, pp. 130–133.
[130] N. Berglund, B. Gentz, *Noise-Induced Phenomena in Slow-Fast Dynamical Systems: A Sample-Paths Approach* (Springer, London, 2000).
[131] C. Baesens, Slow sweep through a period-doubling cascade: delayed bifurcations and renormalisation, *Physica D* **53**, 319–375 (1991).
[132] J.R. Vig, Y. Kim, Noise in microelectromechanical system resonators, *IEEE Transactions on Ultrasonics, Ferroelectrics, and Frequency Control* **46**, 1558–1565 (1999).

[133] A.N. Cleland, Thermomechanical noise limits on parametric sensing with nanomechanical resonators, *New Journal of Physics* **7**, 235 (2005).
[134] M.S. Soliman, J.M.T. Thompson, Integrity measures quantifying the erosion of smooth and fractal basins of attraction, *Journal of Sound and Vibration* **135**, 453–475 (1989).
[135] F.M. Lewis, Vibration during acceleration through a critical speed, *Transactions of the American Society of Mechanical Engineers* **54**, 253–261 (1932).
[136] D.L. Cronin. Response of linear, viscous damped systems to excitations having time-varying frequency. Ph.D. thesis. Pasadena, California: California Institute of Technology; 1966.
[137] A.F. Sarioglu, S. Magonov, O. Solgaard, Tapping-mode force spectroscopy using cantilevers with interferometric high-bandwidth force sensors, *Applied Physics Letters* **100**, 053109 (2012).
[138] M. Loganathan, D.A. Bristow, Bi-harmonic cantilever design for improved measurement sensitivity in tapping-mode atomic force microscopy, *Review of Scientific Instruments* **85**, 043703 (2014).
[139] N.H. Shaik, R.G. Reifenberger, A. Raman, Microcantilevers with embedded accelerometers for dynamic atomic force microscopy, *Applied Physics Letters* **104**, 083109 (2014).
[140] B. Andò, S. Baglio, C. Trigona, *et al.*, Nonlinear mechanism in MEMS devices for energy harvesting applications, *Journal of Micromechanics and Microengineering* **20**, 125020 (2010).
[141] W. Jing, N. Pagano, D.J. Cappelleri, A novel micro-scale magnetic tumbling microrobot, *Journal of Micro-Bio Robotics* **8**, 1–12 (2013).
[142] W. Jing, D.J. Cappelleri. Towards functional mobile magnetic microrobotics. In I. Paprotny, S. Bergbreiter, eds *Small Scale Robotics: From Nano-to-Millimeter-Sized Robotic Systems and Applications* (Springer, Heidelberg, 2014).
[143] Y. Gerson, S. Krylov, B. Ilic, *et al.*, Design considerations of a large-displacement multistable micro actuator with serially connected bistable elements, *Finite Elements in Analysis and Design* **49**, 58–69 (2012).
[144] S. Abu-Salih, D. Elata, Experimental validation of electromechanical buckling, *Journal of Microelectromechanical Systems* **15**, 1656–1662 (2006).
[145] V.N. Nguyen, S. Baguet, C.H. Lamarque, *et al.*, Bifurcation-based micro-/nanoelectromechanical mass detection, *Nonlinear Dynamics*, doi 10.1007/s11071-014-1692-7 (2014).

6

Emerging Themes and Future Directions

This chapter reflects upon the recent, flourishing research developments and outcomes of harnessing the dynamics of bistable structures as reported throughout this book, and projects how these insights will be extended into future investigations where further enhancements for the areas of vibration control, vibration energy harvesting, and sensing and detection will be realized. Through the first sections of this chapter, the research pursuits of emerging importance and increasing attention are described, detailing the new themes that are found to cultivate great promise for the respective topic area. Then, the chapter concludes with a consolidated summary of challenges that remain to be resolved to reap the full benefits of bistable structural dynamics in practice and provides an outlook of how the new research campaigns are on track to resolve these challenges and create further opportunities for numerous applications of vibration control, energy harvesting, and sensing.

6.1 Emerging Themes

The case study examples of Chapters 3 through 5 detailed research efforts motivated to elucidate fundamental principles that underly effective methods to use the dynamics of bistable structures. This section describes how these research themes continue to evolve, in large and small ways, towards a more comprehensive construction of the methods. In addition, this section describes other emerging research themes, being formulated in diverse scientific communities, that share a foundation of concepts with those elaborated upon throughout this book, and thus may benefit in their ongoing pursuits by implementing the investigative strategies described herein.

6.1.1 Vibration Control

The need to dissipate, isolate, or absorb vibration energies is a timeless requirement to realize high-performance, high-quality, safe, and robust operation of engineered systems and environments. As reported by the case studies presented in Chapter 3, substantial insights have been acquired regarding the effective development and deployment of bistable structures wherein the dynamic

Harnessing Bistable Structural Dynamics: For Vibration Control, Energy Harvesting and Sensing, First Edition. Ryan L. Harne and K.W. Wang.

responses provide the characteristics needed to substantially enhance the vibration-relevant qualities of the applications. The common theme of these investigations is that the negative stiffness associated with transitions between the two statically-stable equilibria of the bistable structure facilitate exceptional damping properties and vibration isolation and absorption features, when strategically leveraged for the specific need.

Negative stiffness therefore introduces performance and functionality that do not have equals in other realizations of passive-adaptive structures. A significant development upon this basis continues to transpire due to high promise for post-buckled structures to enable novel characteristics for vibration and wave energy damping, isolation, and control [1]. One avenue of particular attention in the engineering communities is the use of soft materials, such as elastomers, that employ periodic grids of voids through their cross-sections such that combinations of geometric constraints and applied stresses lead to sudden collapse of the architecture into another volumetric state. Effectively, a bistable or multistable structure is realized according to the loading conditions, as observed in Figure 6.1 showing an elastomer stressed between two configurational states of equilibrium for the same applied load. Transitions between these states thus stores and releases elastic energy, in analogy to the two states of equilibrium for a post-buckled beam. By use of elastomeric or polymer materials that are found throughout many vibration control applications, these metamaterial concepts continue to be aggressively studied for their potential implementation as absorbers of vibration, impact, and wave energies. The term metamaterial refers to a system with macroscopically observable properties unlike those associated with its constituents or associated bulk materials [2–5]. For instance, the elastomeric metamaterial shown in Figure 6.1 exhibits a negative Poisson's ratio (lateral contraction under transverse compression) which is not the property anticipated for

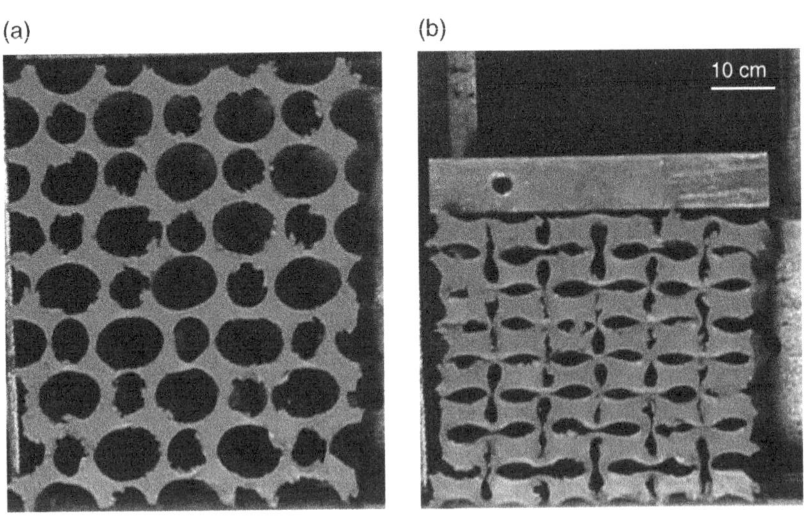

Figure 6.1 (a) Before and (b) after collapse of the cellular metamaterial internal topology between two states of equilibrium for the same load condition.

a bulk sample of the same elastomer. Thus, by the architecture of the elastomer, innovative system-level properties are achieved, which justifies the distinguishing classification of metamaterial.

Original efforts by Mullin *et al.* [6] characterized these potentials in the context of transforming the phononic properties of material systems, with potential photonic applications at micro and nano-scales. Enormous stiffness change was observed following the buckling and collapse of the periodic grid of circular voids into compacted ellipses, like that seen in Figure 6.1. Motivated by this opportunity, Bertoldi and colleagues [7–9] have since shown that such buckling instabilities can be leveraged in elastomeric metamaterials for remarkable adaptation of vibration and wave energy propagation properties, and for novel energy trapping and isolation performance. Through extensive computational model efforts and experimental validations, the research team has demonstrated that particular architectural features give rise to these adaptive characteristics, such that fundamental principles may be elucidated to extend the concepts from meso to microscales [9]. Thus, from such foundation, vibration and wave energy control properties and functionality are shown to be switched on and off according to the design and deformation states [10].

Willshaw and Mullin [11] have also reported that these characteristics are not limited to two-dimensional metamaterial platforms, like that in Figure 6.1, but may be extended to three-dimensional buckling behaviors. However, in the experimental realization considered in the study the buckling modes were nevertheless dominated by two-dimensional deformation trends. According to the design extensions envisioned by Shan *et al.* [9], such factors may not be a general limitation. Yet, while Shim *et al.* [12] investigated a pressure-induced, post-buckled elastomeric metamaterial in three-dimensions (in other words, the stable states lead to deformations in all three dimensions), the experimental realization of statically bi/multistable three-dimensional elastomeric platforms has yet to be reported under pure mechanical loads or stresses. Because most forces, stresses, and waves are applied or transmitted in three dimensions through structures and materials, the realization of a three-dimensional bi/multistable elastomeric metamaterial would be a genuine breakthrough for putting these novel concepts into practice.

The interest in periodic architectures for vibration and wave energy control functions is, in part, driven by the simplicity of repeated cells, modules, or units that facilitate manufacture or an easy assembly into greater platforms. In this spirit, 3D printed (i.e., rapid prototype) concepts have become a recent attraction for applications of post-buckled structures for vibration-based applications. Fulcher *et al.* [13] explored 3D printed negative stiffness vibration isolators and characterized the tunability of the isolation performance according to the change in design variables, application of buckling-inducing constraints, and presence of imperfections. This work provides a thorough assessment of the use of 3D printed, bistable polymer platforms for vibration and shock isolation applications. Using selective laser sintering (SLS) as the fabrication means in the case studies, the performance evaluations discussed the precision of manufacturing, presence of viscoelasticity in the underlying polymer properties, and concerns regarding potential snap-back if the system is ineffectively leveraged under shock loads. Building from this knowledge, the research team later created and explored a SLS-based honeycomb that contracts and expands according to the collapse and recovery of periodic grids of buckled beams [14]. A contemporary study by Restrepo *et al.* [15] explored a variant on this design. Due to triggering a series of buckling events according to a chain-like architecture of buckled beams, a multitude of stable equilibria are uncovered while substantial energy damping is realized under periodic loading conditions. A consistent theme in the two reports [14,15] is the challenge of modeling the dynamics of the periodic, 3D printed multistable structures, which exhibit substantial variation in snap-through sensitivities from one post-buckled 'unit cell' to the next. Ironically, it is ordinarily envisioned that 3D printing provides greater repeatability in manufacturing. Nevertheless, the conditions that promote buckling and the static and dynamic properties of structures beyond such critical limits are very sensitive to changing design parameters [13]. Despite such difficulty in obtaining close correspondence between model results (often from finite element simulations) and measured data, these research teams effectively demonstrated the opportunities for shock energy absorption by collapse of the buckled structures at an extent appreciably greater than the collapse of other periodic architectures, such as a hexagonal honeycomb [13–15].

Florijn *et al.* [16] elucidated how this property of cascading snap-through events may be realized via the soft, cellular elastomeric metamaterials. By use of spatially-graded lateral strains, the metamaterials may be designed in a manner to cultivate successive buckling behaviors that enhance the net energy dissipation from cyclic, transverse loading. Indeed, the group [16] reported a phenomenological model that identified key parameters governing the onset of the bifurcation that yields such topological reconfiguration. Alongside compelling experiments, the study effectively characterized the dramatic and reproducible energy dissipation performance that may be realized under periodic loading conditions of the multistable elastomeric metamaterials.

It is evident that as researchers establish numerous experimental observations of the promising vibration damping, isolation, and absorption properties of multistable structures, composed from bistable constituents, the development of effective and accurate models must keep pace in order to identify the appropriate design strategies by which such advanced features may be reliably cultivated. On the other hand, these built-up architectures realize a remarkable complexity in the dynamic response, which is apparent considering the analytical rigor involved in dissecting the characteristics of one-dimensional bistable structures, see Chapter 3. Despite these difficulties, researchers continue to spearhead novel theoretic methods by which to elucidate the dynamics of multistable structures. Nadkarni *et al.* [17] used a continuum analogy to assess the propagation of waves through chains of bistable elements, which is similar to a discretization of a phase-changing material. The analysis revealed that linear dispersive, soliton, and kink solitary waves are all supported in the 'mechanical metamaterial' of series-connected and grounded bistable constituents. Putting this in an alternative context, the findings show that the presence of negative stiffness can be leveraged as an effective rectifier of input wave energy, thus passing an input wave in unique ways according to the waveform shape and net wave energy. The research team then generalized these conclusions to the transport of diffusive and dissipative phase transitions, whether propagating through discrete or continuous media [18]. The extensive analytical development found that passing multiple phase transitions (i.e., snap-through events in the discrete case) is modeled according to a superposition of each transition behavior. This is an intriguing discovery, verified by the numerical simulations, because the equations of motion are strongly nonlinear and one would not ordinarily anticipate that superposition would hold under such circumstances. The findings therefore indicate that wave energy transport in built-up multistable materials and structures exhibits certain scale-free principles of behavior that may be leveraged in future development of energy-absorbing engineered systems, including multistable carbon nanotube foams [19] and mesoscale 'dissipation ratchets' [20].

In the investigations on one- or two-dimensional bistable structures described throughout this book, it is clear that elastic and magnetic forces can be strategically combined to induce bistability. In the framework of periodic architectures, Schaeffer and Ruzzene [21] have examined the means for adaptive vibration and wave energy transfer through multistable magnetoelastic lattices. The unit cells of the lattices are realized by elastic and magnetic interactions that govern the motions of a mass particle; from such multiphysics interactions, the particle realizes two states of static equilibrium according to the relative deformations of the mass from one side of the unit cell to the other. Using Bloch wave analysis for each state of equilibrium throughout the lattice, the authors reveal the enormous collection of wave energy propagation behaviors supported by the multistable lattices, giving remarkable tunability in vibration control performance or wave guiding [22]. Similar to the low-dimensional bistable analogs, the multistable magnetoelastic lattices also realize dynamic stiffness tuning based on the configuration of the equilibrium supported at a given time. Thus, methods to modulate the properties were discovered using a novel homogenization routine and continuum system approximation [23]. Using the homogenization approach, the properties adaptation of the lattices by way of dynamic reconfiguration among states may be compared side-by-side with respect to other adaptive structures and materials. It also enables the equivalent characteristics of the lattices to be modeled as members of greater engineered systems, such as the lattices serving as members within sandwich structures [23].

Considering these emerging directions in the vibration control context, it is clear that new ideas based on the assembly of bistable structures, or unit cells, into multistable systems are gaining traction for the adaptive and exceptional vibration control performance that is found to be engendered in such architectures. As the experimental realizations demonstrate the enticing promise of such platforms, the theoretical developments are attempting to keep pace, although the complexity of the formulations is reaching a state that computational (simulation) methods are becoming a more common solution.

6.1.2 Vibration Energy Harvesting

Of course the energy dissipated, trapped, or redirected in vibration control contexts is better localized, captured, and electromechanically-converted for applications of energy harvesting. To realize self-powered or -charging small-scale electronics, such vibration energies must be concentrated within the aperture of the energy harvesting device and, ideally, amplified according to the dynamic characteristics of the device response, whether they are resonance-based or are the result of nonlinear phenomena. Thus, in contrast to leveraging the various dynamic regimes of bistable structures to adapt the vibration control properties of structural systems, effective harvesting of ambient vibrations using electromechanical bistable devices entails sustaining the large amplitude snap-through dynamic regime as regularly as possible. The survey in Section 4.1 shows that early numerical and experimental evidence motivated a flurry of research efforts that have thus far produced valuable conclusions on the best means to design and deploy bistable energy harvesters so that snap-through vibrations are more assured. Theoretical developments, such as those within the case studies throughout Chapter 4, have identified many of the fundamental means by which scalable design concepts may be harnessed for energy harvesting enhancements, including the critical frequency sensitivities of steady-state snap-through dynamics and amplitude-dependent performance characteristics. Such insights have continued to drive bistable energy harvester development, as researchers now better understand the principles that constitute overall good practices for design and operation.

As a result, the research considering bistable structures for energy harvesting purposes still flourishes and investigators continue to propose innovative new designs that leverage snap-through dynamics for high performance. Several themes of investigation have emerged towards uncovering best device solutions and implementation methods. One is the honest assessment that practical working conditions rarely if ever involve the ideal designs and excitation scenarios presumed in the early generation of models and studies, whether they considered perfectly symmetric bistable device architectures, pure harmonic vibration input energies, or other simplifications. While the ideal scenarios facilitate useful insights derived from relatively straightforward model investigations of the complex, nonlinear electromechanical dynamics (see Chapter 4 for examples), the predicted performance outcomes may overstate the achievable electrical energy capture with respect to actual working conditions. When the bistable energy harvesting technology is tested alongside other benchmark methods, such overprediction of performance may be a discouragement for adoption of the proposed technology if the advances in energy capture above and beyond the benchmarks are unrealized.

Thus, an emerging theme of the recent investigations is to study bistable vibration energy harvesting platforms under working scenarios or design constraints that are more reflective of a real-world application. Collectively, these efforts seek knowledge of the more nuanced sensitivities of bistable vibration energy harvesting systems, which better rounds out the understanding of the platforms towards the successful transition of technology to practice. For example, Cao *et al.* [24] examined the influence of time-varying restoring force characteristics in a magnetoelastic bistable energy harvester, where the transient properties resulted from application of the device on a moving person so that gravitational influences varied over the course of each stride. Using a proof-of-concept device, the experimentation quantified the relative merits of leveraging bistability for converting human-induced vibrations of the harvester into energy when compared to the same device with the bistability-inducing magnets repositioned so as to either result in monostable or linear cantilevered harvesters. In particular, the bistable platform was shown to be more effective in the application due to the compatibility between the low frequency sensitivities introduced by the magnetoelastic bistability and the frequencies associated with human walking and running. Although its influence was not quantified explicitly, the practical, time-variation in the restoring force did not deter the bistable harvester from achieving the better performance among the compared platforms [24].

Using the same beam architecture based on repulsive magnet effects to induce bistability, Leng *et al.* [25] studied an alternative method to incorporate time-varying restoring forces to enhance the energy capture capability of the bistable energy harvester. The device used an elastic magnet support to promote the snap-through oscillations under stochastic excitations. The measurements show that the approach consistently improved the capability for snap-through to be maintained compared to the device with rigid supports that had inhibited such large amplitude oscillations. Recently, the authors of this book proposed, studied, and quantified the impact of axially-compliant supports on a compressed, simply-supported energy harvesting beam, Figure 6.2, and found conclusively that snap-through oscillations were more readily achievable and sustainable when the axial compliance was increased [26]. This conclusion suggests an easy design means whereby axial support compliance may be explicitly incorporated to enhance the device robustness to changing excitation conditions that otherwise might suppress the favorable snap-through vibrations. Hosseinloo and Turitsyn [27] investigated direct means to control the restoring force (equivalently, potential energy profile) of a bistable energy harvester and articulated a control strategy that injected small amounts of power into the bistable device so as to effectively maintain the snap-through oscillations, even under random excitation conditions such as human walking-induced vibrations. Together, these studies [24–27] show that certain practical working environments and realistic design factors are helpful to the energy capture performance. Indeed, when guided by effective model predictions, these practical conditions may

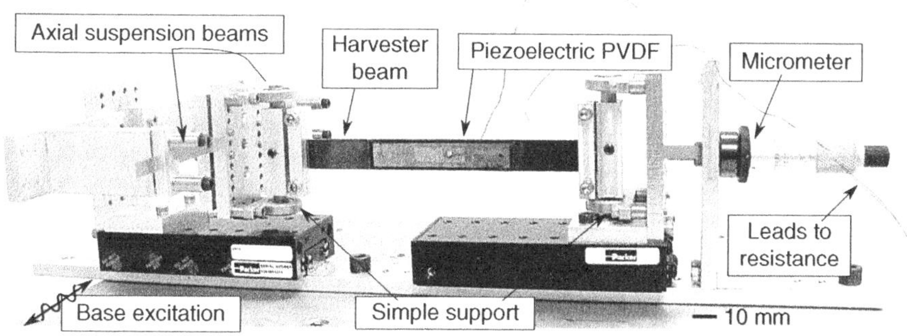

Figure 6.2 Axial compliance integrated into a compressed nonlinear energy harvesting beam system. Photograph of proof-of-concept platform with axial compliance in the end support realized by two cantilevers attached between the base excitation source and one simple support of the beam.

enhance the energy harvesting promise due to an informed method to harness the unique nonlinear dynamics of the bistable vibration energy harvesters.

Interdigitated piezoelectrics have been a mainstay of vibration control [28,29] and energy harvesting [30,31] applications for years due to the enhanced electromechanical authority enabled by leveraging the d_{33} coupling coefficient. Recently, Van Blarigan *et al.* [32] and Ando *et al.* [33] explored bistable vibration energy harvesting clamped-beam systems that incorporated such electrode patterning. The use of interdigitated electrodes for clamped post-buckled energy harvesters is a logical step due to the large axial strains realized in such structures. The researchers characterized these performance benefits and worked to develop accurate models based on computational and empirical characterizations of the respective harvester designs. With such models, implementation of the proposed platforms may be better informed towards achieving optimized performance outcomes.

Lajnef *et al.* [34,35] have made great strides to develop post-buckled energy harvesting systems intended to power embedded condition-monitoring sensors. The researchers have proposed and characterized energy harvesting beams that leverage controlled buckling events induced by quasi-static deformations for significant energy concentration and capture. By taking advantage of the non-resonant qualities of snap-through dynamics, the buckling modes of constrained beams trigger the higher frequency oscillations of attached piezoelectric cantilevers which result in the frequency up-conversion mechanism that was described in Chapter 4 as being ideal for energy harvesting from exceptionally low frequency input energies. Indeed, the studies demonstrated experimentally [34] that mHz vibrations of the environment, for instance temperature-induced deformations upon civil infrastructures, can enable self-powering of a condition-monitoring sensor collocated with the energy harvesting system. These results motivate continued research development to advance the technology to a state of readiness such that the potential of self-powered civil infrastructure monitoring platforms in difficult-to-access, embedded environments may be fully realized.

Surveying the recent investigations of vibration energy harvesting with bistable structures, it is apparent that the pathway to technological commercialization is being aggressively pursued. To reach such a goal, understanding feasibility is required, and identifying the best device solutions and implementation methods are clear emerging themes of the efforts being undertaken to achieve successful technological transition.

6.1.3 Sensing and Detection

As shown through the literature survey and case studies of Chapter 5, the two states of stable equilibrium in bistable structures introduce novel opportunities for efficient actuation mechanisms while the dynamics among the steady-states opens up the potential for high-sensitivity sensing and detection according to strategically triggering the saddle-node bifurcations. To leverage these characteristics with reliability appropriate for engineering applications, it is essential to acquire the understanding of the heightened sensitivities driving such behaviors. As researchers continue their pursuits of harnessing the dynamics of bistable structures in sensing and detection contexts, two research trajectories are emerging: one to better assess the properties and characteristics of ever-smaller-scale bistable systems, and another to leverage folding bi-/multistable structures for engineering applications of reversible switches, detectors, and so forth, particularly at small scales. The following paragraphs survey these two trajectories of research and highlight key outcomes that have thus far illustrated promise and potential for continued investigation.

Structures composed on the micro- and nanoscales are susceptible to many forces and fields which play negligible roles on the meso- and human-scales. For instance, as described in Section 5.1, thermal noise, van der Waals force, film damping, and other phenomena are distinctly important if the largest characteristic dimension of the structure is on the order of microns or smaller. In particular, thermal noise may be high enough such that molecular 'jitters', that is the random oscillations of the structure due to thermal energies, may be so large that it may be necessary to drive post-buckled microsystems with such high levels of excitations that the systems always vibrate in a nonlinear regime [36]. As a result, the working life of the devices may be compromised due to the large periodic stresses. This is but one example: the various small-scale phenomena surveyed above each play nuanced yet critical roles toward determining the total dynamic response of bistable micro/nanosystems. Responding to this concern, researchers have recently been taking focused steps in their study of new generations of bistable micro/nanoscale sensor systems with a greater fidelity suitable for practical application by accounting for such scale-relevant factors.

Huang *et al.* [37] characterized a microbeam resonator that triggered a dynamic response bifurcation in consequence to a tuned Lorentz force and chemical bond potential. This integration of influences from the multiple physics was found to dramatically enhance the device sensitivity and enabled bifurcation transitions at extremely small thermal noise levels, thus opening the door to sense quantum events according to the bifurcation in the bistable microbeam dynamics. Since bifurcation-based sensing originated with emphasis on quantum-level detection ability [38,39], it is fitting that the heightened sensitivities of bistable devices have recently been applied to see this potential borne out using a structural sensor system. Tajaddodianfar *et al.* [40,41] have given evidence that scale is critical towards the snap-through sensitivities of bistable nanobeams and arches, since the local gradient of strain plays an important role in determining loads at which transitions occur from one locally stable state to another, a feature otherwise lost at larger scales where the classical Euler–Bernoulli beam theory is often appropriate as an analytical basis. The reports indicated that asymmetries in arch topology, such as those induced by static electrostatic fields interposed with dynamic actuation fields, will result in differing likelihoods for snap-through events from one state to the other, depending on the starting state. This characteristic is of course articulated elsewhere throughout this book, but the particular impacts of the applied electrostatic fields gain unique relevance for nano systems because misinformed utilization of a bistable nanobeam could result in catastrophic pull-in (Figure 5.1) if the forces exceed critical values, which is an uncommon result if using the meso- or macroscale platforms investigated in the case studies provided here.

The fabrication and actuation of small-scale devices are shown experimentally to be essential factors that govern the repeatability of device response (and hence, functional use) from one specimen to the next [42]. Mohammad and Ouakad [43] addressed the situation that fringing fields, that is, electrostatic fields acting from the sides of the microsystems, may be practically necessary to drive the microbeam according to the scale and fabrication methods. While this feature was previously considered by Krylov *et al.* [44], the more recent investigation characterized the dependences of the symmetric and asymmetric post-buckled modes towards deriving effective reduced-order models of the dynamic system response and hence bifurcation sensitivities to be leveraged for sensing and detection applications [43]. Using reduced-order models is indeed important since predicting the dynamics of distributed-parameter structures, including micro/nanobeams, ordinarily requires the adoption of Galerkin-type assumed solutions to the partial differential equations of motion. A reduced-order model that employs only those needed modes of response that are relevant to a given dynamic behavior will therefore dramatically expedite the nonlinear response prediction (through more computationally efficient studies) and hence propel the system characterization and design. To this end, Medina *et al.* [45] built upon their prior reduced-order modeling foundations [46,47] to elucidate how suddenly-applied electrostatic loads may launch a bistable microbeam from one stable state to the other, or back again. This property was termed 'dynamic trapping' since it could realize a configurational state that statically-applied electrostatic force could otherwise not achieve according to the critical limits associated with pull-in. The authors used the model to show that discrete boundaries of applied load amplitudes and durations could repeatedly trigger the transitions, while certain working conditions exhibited fractal-like sensitivities that might compromise the consistency of practical operation [45]. The new knowledge can therefore guide implementation of electrostatically-actuated bistable microbeams in applications such as sensing switches or mechanical memories. Building upon a well-informed understanding of micro- and nanoscale bistable structures, scientists and engineers are more able to put these platforms to practical use for small-scale sensing and detection purposes since the nuanced physics involved in the devices' operations are becoming more and more clear via the continuing research discoveries.

While conventional engineering structures such as beams, membranes, or plates have their merits in applications on small scales – merits such as their straightforward design, modeling, and fabrication – the functionality of these devices is often limited to serving as one-dimensional dynamic resonators, switches, actuators, and pumps, to name a few of the representative applications of bistable structures surveyed in Section 5.1. The ability to use multidimensional deformations and motions is consequently lost by virtue of the structural units employed. Yet, leveraging two- or three-dimensional structures at small scales can open the opportunity for deployable systems, such as stents, encapsulation methods for drug delivery devices, sensors of biological matter, or machines that perform mechanical work in nanoscale operations [48]. As a result, researchers have recently sought ways by which multiple dimensions of motion may be traversed using ever-smaller structural systems and are finding how bistability can be an enabler for such promising functionality.

To this end, systems that fold open and closed according to only a very few degrees-of-freedom are found to provide remarkable versatility for structural devices composed on small length scales. Origami, the ancient Japanese art of paper folding, has therefore offered inspiration for architectures based on the idea of foldable engineering systems that engender useful functionality and purposes, such as supporting loads, tuning mechanical properties, and, in the case of bistable origami, large shape adaptation [49]. In consequence, origami-inspired engineering design is emerging as a clear vehicle for remarkable system performance tunability. Particular applications of such technology are found in micro- and nanoscales where it is otherwise difficult to fabricate three-dimensional mechanical structures to perform engineering functions. Thus, an emerging theme of recent research is to create a two-dimensional planar structure, which is more readily fabricated, and then to fold it into the three-dimensional form in which it will serve its engineering purpose.

Among the discoveries, it is found that several origami tessellations possess bistability. This introduces the opportunity to leverage the dynamics and tuning associated with stable state transitions of bistable folding structures for small-scale applications where it is less straightforward to achieve comparable characteristics unless a system is actively controlled (which may not be possible on ever-smaller scales). For example, the tessellation termed the "square twist" possesses bistability between open and closed states according to the elongation of the unit, which is the result of natural deformation in the planar facets during the folding process. Silverberg *et al.* [50] used this new understanding to create a square twist at the microscale that was actuated between its two states of stable equilibria using swelling-induced stresses. Alongside a rigorous theoretical development, the innovative experiments demonstrated that bistable origami-based structures cultivate remarkable functionality at small (micro) scales where it is otherwise difficult to deliver useful mechanical work in multiple dimensions or so dramatically change the system configuration, such as for actuation or sensing objectives.

The waterbomb tessellation, Figure 6.3, has also been found to exhibit bistability. Hanna *et al.* [51] characterized these properties of the waterbomb, which result in two stable states that are more visually compatible to the buckling of a

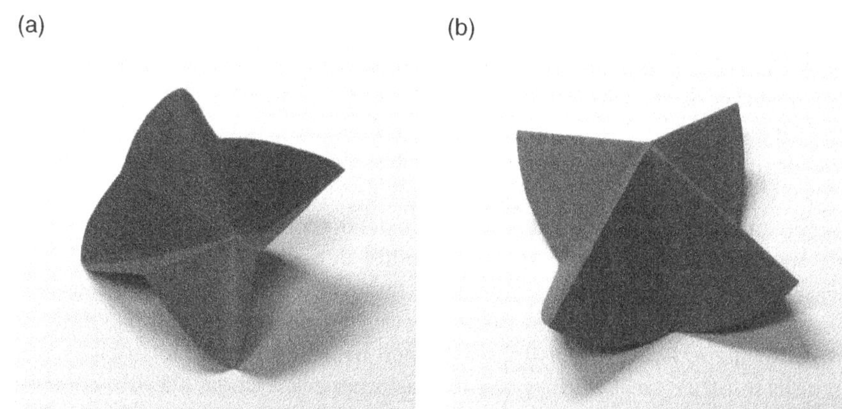

Figure 6.3 Bistable waterbomb origami showing the two stable states of the rotation (a) 'in' and (b) 'out'.

simply supported beam than are the rotational stable states associated with the square twist. The double-well potential energy landscape of the waterbomb was shown to be tunable based on the particular folding stiffness dependences, that were adjusted by the geometric and material features of metallic glass hinges used in the experimental prototypes [51]. By using light- or heat-sensitive materials at fold locations [52], the authors suggested that the experimentally validated model could help inform effective transitioning among the origami waterbomb stable equilibria for diverse engineering applications. Indeed, another research team [53,54] achieved such a method for deliberate transitioning between the two stable states of the bistable waterbomb. Using magneto-active materials bonded to the fold lines of the tessellation, the researchers demonstrated a non-contact actuation of the structure that was guided in design and implementation through corresponding finite element simulations. Thus, by leveraging particular combinations of multiple physics, the transitions among the multidimensional topologies of bistable origami tessellations may be harnessed for engineering purpose.

Using a new concept of 'surrogate folds' which are especially designed hinges that facilitate folding of rigid planar surfaces, Safsten *et al.* [55] studied a multistable folding structure called a kaleidocycle. Unlike other compliant mechanisms, the kaleidocycle may rotate through its multiple stable states continuously without having to reverse direction. This property makes the platform particularly advantageous for applications that necessitate many stable equilibria to transition amongst, including latches, valves, or safety switches. They provided a rigorous derivation of the design parameters that result in various numbers of stable equilibria and the potential energy landscapes associated with them. It was shown that using identical surrogate folds along the torus-like architecture facilitated the device having the least energy differences that must be overcome to transition among the stable equilibria for shape reconfiguration purposes. Such adjustment of static properties corresponds to dynamic characteristics, giving insight for leveraging these behaviors for dynamic performance adaptation.

One research group has made great progress in immediately applying these opportunities at the nanoscale by taking advantage of DNA sequences as a basis to fabricate bistable, DNA origami structures [56]. Zhou *et al.* [57] have demonstrated a method by which DNA sequences may be combined in strands and hinges to create compliant DNA origami mechanisms capable of performing mechanical work, transitioning like nano-latches, and acting as biological sensors based on environmental interactions and bistable configurations. Based on the scale and the background energy associated with thermal noise, these bistable DNA origami cannot be characterized according to maintaining one or another stable equilibrium configuration. Instead, the researchers quantified the state maintenance according to a probability density that was a function of release DNA strands and the temperature (associated with thermal noise) [57]. The study innovatively marries the theoretical mechanics of compliant structures with the enticing opportunities opened by bio- and nanotechnology, and promises to usher great future potentials as other researchers likewise consider DNA origami as a foundation for diverse nanoscale applications of dynamic bi-/multistable mechanical structures [58–60].

Although in 1959 Feynman discussed broadly the desirability of transitioning technology to ever-smaller scales for the advancement of mankind's pursuits [61], only in recent years have the performance and versatility of bistable structures been transitioned to engineering purposes at the small scales of modern times (i.e., microscale and smaller). To achieve such objectives, understanding must drive the effective system development and deployment, which are the primary themes of the recent research. Thus, by better understanding conventional, micro/nanoscale bistable structures like beams and plates and by devising novel small-scale, 3D, bistable structures from folded 2D planes, researchers are establishing firm foundations to translate the advances of bistable dynamics for sensing purposes to length scales well beyond the limits of our human visual inspection.

6.2 Challenges and Future Outlooks

The future is promising for the continued development of structures and materials with bistable characteristics such that innovative performance features are engendered as a result of the unique dynamics of these constituents. Although the authors of this book have extensive research experience in the areas of vibration control, energy harvesting, and sensing wherein the dynamics of bistable structures are capitalized upon, the challenges and subsequent outlooks identified and described in this section are by no means an ultimate authority or final pronouncement: alternative projections may be put forward and supported as confidently as those that follow. Nevertheless, across these three areas, the authors identify two key challenges for the continuing research and development. Surmounting these challenges would substantially promote the potential of the emerging research ventures described in the preceding sections of this chapter and would likely foster new opportunities that may not be predicted at this time. The first challenge pertains to obtaining theoretical grasp and understanding of the dynamic responses of the emerging bi-/multistable structures and materials that suggest high promise for the future, while the second challenge pertains to application relevance and readiness for leveraging the dynamics of bistable structures. These challenges and the outlooks to overcome them are detailed in the following sections to conclude this book.

6.2.1 Theoretical Characterization of the Emerging Bistable and Multistable Structural/Material System Concepts

Throughout the case studies of this book, analytical efforts are complemented by experimental measurements of corresponding scenarios, and sometimes also with numerical simulation methods. A broad strategy for system characterization, such as these numerous case studies exemplify, provides comprehensive support to the conclusions and makes arguments for the high potential of leveraging the dynamics of bistable structures all the more convincing.

Yet, these reported developments have a limitation in that they are focused on low-dimensional architectures having one or two DOF. While such architectures are advantageous in many application contexts, the focus on low-dimensionality also facilitates the tractable implementation of the theoretical methods. At least according to the methods' compositions given here, faithful predictions of the nonlinear responses can be better assured when only a few, nonlinear DOF are present. As detailed in Sections 6.1.1, 6.1.2, and 6.1.3, through exhaustive experimental efforts the emerging research trajectories are uncovering remarkable functionality, operational versatility, and novel performance by leveraging structures composed from many bistable constituents or using multidimensional bistable structures. While the analytical methods employed in the case studies of this book are often applicable to the study of such multistable structures, they become substantially less tractable or of a computational complexity on a par with "brute force" numerical simulation methods if extended from one or two nonlinear DOF to many.

Typically, the researchers who have made such new headway in harnessing the dynamics of multistable structures and materials have corroborated their intriguing experimental findings with finite element (FE) model predictions. This creates two concerns towards attaining generalized understanding and deep insights. First, the accuracy of FE models near points of instability is sometimes questionable and may require one to deliberately introduce asymmetries to prevent unusual computational results [7,9]. Second, the computations are substantially costlier (by orders of magnitude) than even the brute force numerical simulations employed to verify the analytical predictions to a set of nonlinear ODEs. The need for transient FE simulations to characterize buckling-related dynamics means that refined time-steps must be employed approaching every activation of instability and the large number of total DOFs result in substantial computer memory and processing burden, suffered again and again for each time increment [62].

Thus, while researchers are making extraordinary headway in the development and experimental study of multistable structures for novel dynamic performance characteristics, the theoretical grasp and understanding of these systems remain several steps behind. Indeed, such lag in progress between experimental breakthroughs and analytical formulations is wider-reaching than only in regards to efforts in vibration control, energy harvesting, and sensing: it is a delay that concerns the broad context of exploiting multistable structures throughout science and engineering applications. To help these favorable first signs of opportunity prosper and find firm establishment for applications, a new generation of analytical tools must be created for the theoretical characterization of the dynamic behaviors of such multistable platforms.

While some researchers have begun to make such headway, particularly in theoretical studies on wave-based dynamics of multistable structures and materials [18], there are many significant issues to address in new analyses. On the other hand, there are numerous contemporary developments that provide inspiration for scientists and engineers interested in fully grasping the nuanced dynamic responses and sensitivities of the intriguing multistable structures and materials in the emergent research. For example, recent efforts have shown that many-DOF systems may be analyzed via inverse- and receptance-based techniques when there are limited numbers of nonlinear constituents [63]. If such a condition holds, it is shown that the dynamic responses of even hundreds of DOF are readily computed via analysis despite the nonlinearities present [64].

These methods are being derived from techniques previously used for the experimental identification of nonlinear system parameters, for instance to apply such parameters in "full-field" FE models. These methods are relevant to many areas of application, beyond merely their employment for investigations in vibration control, energy harvesting, and sensing [65–67]. Such inverse identification methods, previously employed for experimentation, are finding ways to establish the new analytical tools [63]. Likewise, although not yet suitable for implementation with structures that exhibit dynamics far-from-equilibrium, including many-DOF systems with one or more bistable constituents, these approaches should serve as inspiration for researchers interested in taking the emerging developments in multistable structural systems to new levels of development through the comprehensive characterizations provided by rich analytical investigations.

Considering the reports of the potential for multistable structures and materials to span multiple dimensions and exhibit unprecedented dynamic properties favorable for numerous applications, the next breakthrough required to fuel their application readiness is to gather an effective understanding of these systems' operations and sensitivities. Thus, complemented by the theoretical underpinnings, the eye-catching experimental demonstrations of remarkable energy trapping [9], dynamic topological reconfiguration [50], and wave rectification [68], to name a few examples, will be unstoppable in their progress forward to make a mark in high-profile applications that benefit science, engineering, and society.

6.2.2 Application Relevance and Readiness

Identifying a critical application for a proposed, new technology is essential to better promote its final adoption because the research and development becomes more focused towards genuine problem-solving and towards the identification of working principles that maintain robustness of system performance and functionality. While the recent advances and ideas for harnessing bistable structural dynamics have presented enticing opportunities, many of the developments have been limited to proof-of-concept demonstrators for experimental purposes and have focused upon idealized operational scenarios that may limit the knowledge gained. In addition, few investigations have involved significant industry partnership that could provide the weight to push a novel research idea into a mainstream development and production path.

Among the topic areas considered in this research, the vibration energy harvesting community is perhaps the best positioned to capitalize on industry support to transition the advanced performance of bistable energy harvesters to practical embodiments. As indicated by numerous researchers in the community, the commonly cited end-use for the collected energy from bistable vibration energy harvesters is for microelectronics that require small power for operation but otherwise rely on inconvenient or costly disposable batteries or line transmissions. Indeed, there is significant commercial capacity for self-powered microelectronics. This includes prime applications in the heavy industries where embedded condition-monitoring sensors may be completely inaccessible for line power or battery replacement, for example, pipelines, mining, and drilling [69]. This also includes commercial capacity in the transportation sector where the sheer mass of powering cables for myriad, small sensors in automobiles, aircraft, and spacecraft is now approaching a degree that the plentiful demands for "light-weight" for efficiency will soon begin to require an elimination or significant reduction of cables [70]: hence, an open door for energy harvesting technologies.

To make the biggest "splash", energy harvesting using the enhanced performance capabilities of bistable structural dynamics requires a demonstration of robustness and reliability in practical working environments. Ultimately, the bistable energy harvester must deliver power that is always sufficient for the demand, assuming a reasonable range of vibrations persist to excite the device. As described in Section 6.1.2, researchers are making new headway in this direction by considering more representative situations and inevitable practical circumstances that could otherwise challenge the proposed nonlinear technologies. But, further developments are required, particularly as relate to understanding the susceptibilities of these favorable snap-through dynamics under non-stationary base excitation vibrations, in multiphysics environments where thermal variance or fluid pressures are non-trivial, and pertaining to fatigue and wear of practical device concepts. While researchers have in the past variously addressed these issues for linear or weakly nonlinear energy harvesters [69,71–73], the potential for coexisting steady-state attractors and aperiodic responses in bistable devices complicates the assessment significantly. One alternative approach is to resort to time-consuming and computationally-expensive simulation routines, but generalized insight and understanding are lacking from these purely numerical methods. It is essential that further research be conducted on these questions such that deep, insightful answers may be derived and better design and deployment practices may be put forward. Only by these means will the desirable snap-through dynamics of bistable vibration energy harvesters be most effectively leveraged. Thus, in this way the technology transition will be readily propelled.

For vibration control and sensing, the challenge of application relevance and readiness is more apparent. In contrast to energy harvesting applications, where the unsatisfactory performance of a bistable device may produce only an inconvenient, temporary down-time of signal transmission for condition monitoring (which typically involves slow gradients of relevant change contrasting the fast time scale of harvester oscillation), the ineffective operation of a bistable dynamics-based vibration damper or sensor could lead to immediate, mission-critical consequences. For instance, a temporarily defective bistable structural damper could lead to catastrophic system failure due to excessive system stresses beyond fracture or yield points, while a sensor based on leveraging the bifurcations of a bistable system may be deemed insufficient if it ever fails to detect the target threshold level of a noxious gas for which it was purposed to monitor.

These more serious considerations may be addressed in several ways. One strategy is to adopt more conservative design specifications that yield highly repeatable dynamic response under a set of prescribed conditions, as opposed to an equal likelihood of two, coexisting dynamic regimes occurring, for example, see the results and discussion regarding Figure 3.13. Therefore, for example, to realize the adaptive vibration damping, isolation, or absorption performance desired, one could more confidently rely on the targeted excitation conditions to trigger the sought-after dynamic regime and resulting performance characteristics. A similar approach may be adopted to harness bifurcation-based sensing principles with bistable systems: using sufficient model guidance, one ought to develop the sensor to exhibit a well-defined bifurcation, such as the saddle-node, that is able to repeatedly transition from low- to high-amplitude steady-states based on pre-determined amounts of monitored change in the host structure.

By these means, the impact of perturbations from stochastic disturbances or sudden changes in the excitation regime would be less dramatic since there would be a greater likelihood of sustaining a desired steady-state. While these important themes are discussed and studied in Section 5.4, they largely remain unexplored in the body of reported literature that seeks to investigate the potential for bistable micro/nanosystems to be employed for sensing and detection purposes. Likewise, due to the historical convention of steady-state response analysis for linear vibration control systems [74], researchers have mostly investigated the performance of bistable devices for vibration control applications under the traditional assumptions of harmonic response, without consideration of the roles of imperfections, perturbations, or transients.

Further work must be conducted to characterize these essential factors as they pertain to vibration control and sensing contexts wherein one seeks to beneficially leverage the dynamics of bistable structural systems. As the concepts for bistable dynamics-based vibration control systems are readied for application, it is anticipated that new fundamental challenges will be discovered. These challenges will need to be tackled via the comprehensive approach described in Section 6.2.1, since the corresponding, explicit finite element modeling of the problem may provide immediate resolutions to troubles but may just as easily disguise an important underlying principle that could guide future development. Similarly, the transition of bifurcation-based sensing approaches to large-scale system condition monitoring, in the spirit of the case study examples described in Chapter 5, poses exciting challenges unique to the context of monitoring structural response deviations according to the activation of bifurcations in the dynamics of bistable sensors. Once the sensing strategies are deployed in initial efforts, the difficulties of making "sense" from the triggering of numerous sensor bifurcations will motivate investigations to uncover fundamental behaviors representative of dynamic bifurcations associated more broadly with the phenomenon but specifically relevant to the issue at hand of monitoring structural health. Overall, as further experience is collected in putting the dynamics of bistable structures into practice, a need to obtain deeper understanding of the problem physics will be made clear. Having captured such understanding, one possesses a strong case to forward the technology from concept to prototype to product.

References

[1] N. Hu, R. Burgueño, Buckling-induced smart applications: recent advances and trends, *Smart Materials and Structures* **24**, 063001 (2015).

[2] M.I. Hussein, M.J. Leamy, M. Ruzzene, Dynamics of phononic materials and structures: historical origins, recent progress, and future outlook, *Applied Mechanics Reviews* **66**, 040802 (2014).

[3] N. Kim, Y.J. Yoon, J.B. Allen, Generalized metamaterials: definitions and taxonomy, *The Journal of the Acoustical Society of America* **139**, 3412–3418 (2016).

[4] D. Del Vescovo, I. Giorgio, Dynamic problems for metamaterials: review of existing models and ideas for further research, *International Journal of Engineering Science* **80**, 153–172 (2014).

[5] G. Trainiti, M. Ruzzene, Non-reciprocal wave propagation in spatiotemporal periodic structures, *arXiv*, arXiv:1605.00734v1 (2016).

[6] T. Mullin, S. Deschanel, K. Bertoldi, *et al.*, Pattern transformation triggered by deformation, *Physical Review Letters* **99**, 084301 (2007).

[7] J. Shim, P. Wang, K. Bertoldi, Harnessing instability-induced pattern transformation to design tunable phononic crystals, *International Journal of Solids and Structures* **58**, 52–61 (2015).

[8] S. Shan, *et al.*, Harnessing multiple folding mechanisms in soft periodic structures for tunable control of elastic waves, *Advanced Functional Materials*, doi.org/10.1002/adfm.201400665 (2014).

[9] S. Shan, *et al.*, Multistable architected materials for trapping elastic strain energy, *Advanced Materials* **27**, 4296–4301 (2015).

[10] K. Bertoldi, M.C. Boyce, Mechanically triggered transformations of phononic band gaps in periodic elastomeric structures, *Physical Review B* **77**, 052105 (2008).

[11] S. Willshaw, T. Mullin, Pattern switching in two and three-dimensional soft solids, *Soft Matter* **8**, 1747 (2012).

[12] J. Shim, C. Perdigou, E.R. Chen, *et al.*, Buckling-induced encapsulation of structured elastic shells under pressure, *Proceedings of the National Academy of Sciences* **109**, 5978–5983 (2012).

[13] B.A. Fulcher, D.W. Shahan, M.R. Haberman, *et al.*, Analytical and experimental investigation of buckled beams as negative stiffness elements for passive vibration isolation and shock isolation systems, *Journal of Vibration and Acoustics* **136**, 031009 (2014).

[14] D.M. Correa, T. Klatt, S. Cortes, *et al.*, Negative stiffness honeycombs for recoverable shock isolation, *Rapid Prototyping Journal* **21**, 193–200 (2015).

[15] D. Restrepo, N.D. Mankame, P.D. Zavattieri, Phase transforming cellular materials, *Extreme Mechanics Letters* **4**, 52–60 (2015).

[16] B. Florijn, C. Coulais, M. van Hecke, Programmable mechanical metamaterials, *Physical Review Letters* **113**, 175503 (2014).

[17] N. Nadkarni, C. Daraio, D.M. Kochmann, Dynamics of periodic mechanical structures containing bistable elastic elements: from elastic to solitary wave propagation, *Physical Review E* **90**, 023204 (2014).

[18] N. Nadkarni, C. Daraio, R. Abeyaratne, *et al.*, Universal energy transport law for dissipative and diffusive phase transitions, *Physical Review B* **93**, 104109 (2016).

[19] F. Fraternali, T. Blesgen, A. Amendola, *et al.*, Multiscale mass-spring models of carbon nanotube foams, *Journal of the Mechanics and Physics of Solids* **59**, 89–102 (2011).

[20] I. Benichou, S. Givli, Structures undergoing discrete phase transformation, *Journal of the Mechanics and Physics of Solids* **61**, 94–113 (2013).

[21] M. Schaeffer, M. Ruzzene, Wave propagation in multistable magneto-elastic lattices, *International Journal of Solids and Structures* **56–57**, 78–95 (2015).
[22] M. Schaeffer, M. Ruzzene, Wave propagation in reconfigurable magneto-elastic kagome lattice structures, *Journal of Applied Physics* **117**, 194903 (2015).
[23] M. Schaeffer, M. Ruzzene, Homogenization of 1D and 2D magnetoelastic lattices, *EPJ Applied Metamaterials* **2**, 13 (2015).
[24] J. Cao, W. Wang, S. Zhou, *et al.*, Nonlinear time-varying potential bistable energy harvesting from human motion, *Applied Physics Letters* **107**, 143904 (2015).
[25] Y.G. Leng, Y.J. Gao, D. Tan, *et al.*, An elastic-support model for enhanced bistable piezoelectric energy harvesting from random vibrations, *Journal of Applied Physics* **117**, 064901 (2015).
[26] R.L. Harne, K.W. Wang, Axial suspension compliance and compression for enhancing performance of a nonlinear vibration energy harvesting beam system, *Journal of Vibration and Acoustics* **138**, 011004 (2016).
[27] A.H. Hosseinloo, K. Turitsyn, Non-resonant energy harvesting via an adaptive bistable potential, *Smart Materials and Structures* **25**, 015010 (2016).
[28] N. Hagood, R. Kindel, K. Ghandi, P. Gaudenzi, Improving transverse actuation of piezoceramics using interdigitated surface electrodes. In *SPIE Proceedings Vol. 1917 Smart Structures and Intelligent Systems 1993*, pp. 341–352.
[29] H.A. Sodano, G. Park, D.J. Inman, An investigation into the performance of macro-fiber composites for sensing and structural vibration applications, *Mechanical Systems and Signal Processing* **18**, 683–697 (2004).
[30] H.J. Song, Y.T. Choi, N.M. Wereley, *et al.*, Energy harvesting devices using macro-fiber composite materials, *Journal of Intelligent Material Systems and Structures* **21**, 647–658 (2010).
[31] R.L. Harne, Concurrent attenuation of, and energy harvesting from, surface vibrations: experimental verification and model validation, *Smart Materials and Structures* **21**, 035016 (2012).
[32] L. Van Blarigan, J. Moehlis, R. McMeeking, Low dimensional modeling of a non-uniform, buckled piezoelectric beam for vibrational energy harvesting, *Smart Materials and Structures* **24**, 065012 (2015).
[33] B. Andò, S. Baglio, A.R. Bulsara, *et al.*, A low-cost snap-through-buckling inkjet-printed device for vibrational energy harvesting, *IEEE Sensors Journal* **15**, 3209–3220 (2015).
[34] N. Lajnef, W. Borchani, R. Burgueño, *et al.*, Self-powered piezo-floating-gate smart-gauges based on quasi-static mechanical energy concentrators and triggers, *IEEE Sensors Journal* **15**, 676–683 (2015).
[35] N. Lajnef, R. Burgueño, W. Borchani, *et al.*, A concept for energy harvesting from quasi-static structural deformations through axially loaded bilaterally constrained columns with multiple bifurcation points, *Smart Materials and Structures* **23**, 055005 (2014).
[36] J.R. Vig, Y. Kim, Noise in microelectromechanical system resonators, *IEEE Transactions on Ultrasonics, Ferroelectrics, and Frequency Control* **46**, 1558–1565 (1999).
[37] P. Huang, *et al.*, Generating giant and tunable nonlinearity in a macroscopic mechanical resonator from chemical bonding force, *arxiv*, 1506.07227v1 (2015).

[38] A.B. Zorin, Quantum-limited electrometer based on single cooper pair tunneling, *Physical Review Letters* **76**, 4408–4411 (1996).

[39] R. Vijay, M.H. Devoret, I. Siddiqi, Invited review article: the Josephson bifurcation amplifier, *Review of Scientific Instruments* **80**, 111101 (2009).

[40] F. Tajaddodianfar, H.N. Pishkenari, M.R.H. Yazdi, *et al.*, On the dynamics of bistable micro/nano resonators: analytical solution and nonlinear behavior, *Communications in Nonlinear Science and Numerical Simulation* **20**, 1078–1089 (2015).

[41] F. Tajaddodianfar, H.N. Pishkenari, M.R.H. Yazdi, *et al.*, Size-dependent bistability of an electrostatically actuated arch NEMS based on strain gradient theory, *Journal of Physics D: Applied Physics* **48**, 245503 (2015).

[42] A.M. Bataineh, M.I. Younis, Dynamics of a clamped-clamped microbeam resonator considering fabrication imperfections, *Microsystems Technologies* **21**, 2425–2434 (2015).

[43] T.F. Mohammad, H.M. Ouakad, Static, eigenvalue problem and bifurcation analysis of MEMS arches actuated by electrostatic fringing-fields, *Microsystems Technologies* **22**, 193–206 (2016).

[44] S. Krylov, B.R. Ilic, S. Lulinsky, Bistability of curved microbeams actuated by fringing electrostatic fields, *Nonlinear Dynamics* **66**, 403–426 (2011).

[45] L. Medina, R. Gilat, S. Krylov, Bouncing and dynamic trapping of a bistable curved micro beam actuated by a suddenly applied electrostatic force, *Communications in Nonlinear Science and Numerical Simulation* **36**, 273–284 (2016).

[46] L. Medina, R. Gilat, S. Krylov, Symmetry breaking in an initially curved micro beam loaded by a distributed electrostatic force, *International Journal of Solids and Structures* **49**, 1864–1876 (2012).

[47] L. Medina, R. Gilat, S. Krylov, Symmetry breaking in an initially curved pre-stressed micro beam loaded by a distributed electrostatic force, *International Journal of Solids and Structures* **51**, 2047–2061 (2014).

[48] P.W.K. Rothemund, Folding DNA to create nanoscale shapes and patterns, *Nature* **440**, 297–302 (2006).

[49] K. Miura, T. Kawasaki, T. Tachi, *et al.*, eds, *Origami 6: International Meeting of Origami Science, Mathematics, and Education* (American Mathematical Society, Providence, 2015).

[50] J.L. Silverberg, *et al.*, Origami structures with a critical transition to bistability arising from hidden degrees of freedom, *Nature Materials* **14**, 389–393 (2015).

[51] B.H. Hanna, J.M. Lund, R.J. Lang, *et al.*, Waterbomb base: a symmetric single-vertex bistable origami mechanism, *Smart Materials and Structures* **23**, 094009 (2014).

[52] Y. Liu, J.K. Boyles, J. Genzer, *et al.*, Self-folding of polymer sheets using local light absorption, *Soft Matter* **8**, 1764–1769 (2012).

[53] A. Crivaro, R. Sheridan, M. Frecker, *et al.*, Bistable compliant mechanism using magneto active elastomer actuation, *Journal of Intelligent Material Systems and Structures*, doi 10.1177/1045389X15620037 (2016).

[54] L. Bowen, K. Springsteen, H. Feldstein, *et al.*, Development and validation of a dynamic model of magneto-active elastomer actuation of the origami waterbomb base, *Journal of Mechanisms and Robotics* 7, 011010 (2015).

[55] C. Safsten, T. Fillmore, A. Logan, *et al.*, Analyzing the stability properties of kaleidocycles, *Journal of Applied Mechanics* **83**, 051001 (2016).

[56] L. Zhou, A.E. Marras, H.J. Su, *et al.*, DNA origami compliant nanostructures with tunable mechanical properties, *ACS Nano* **8**, 27–34 (2014).
[57] L. Zhou, A.E. Marras, H.J. Su, *et al.*, Direct design of an energy landscape with bistable DNA origami mechanisms, *Nano Letters* **15**, 1815–1821 (2015).
[58] Z. Zhang, *et al.*, A DNA tile actuator with eleven discrete states, *Angewandte Chemie International Edition* **50**, 3983–3987 (2011).
[59] Y. Liu, J. Genzer, M.D. Dickey, "2D or not 2D": shape-programming polymer sheets, *Progress in Polymer Science* **52**, 79–106 (2016).
[60] P. Ketterer, E.M. Willner, H. Dietz, Nanoscale rotary apparatus formed from tight-fitting 3D DNA components, *Science Advances* **2**, e1501209 (2016).
[61] R.P. Feynman, There's plenty of room at the bottom, *Engineering and Science* **23**, 22–36 (1960).
[62] B.A. Miller. Loosely coupled time integration of fluid-thermal-structural interactions in hypersonic flows. Ph.D. dissertation. Columbus, OH: The Ohio State University; 2015.
[63] X. Wang, X. Guan, G. Zheng, Inverse solution technique of steady-state responses for local nonlinear structures, *Mechanical Systems and Signal Processing* **70**, 1085–1096 (2016).
[64] M.B. Özer, Closed form solutions of weakly nonlinear multi-degree of freedom systems with single nonlinear element, *Journal of Sound and Vibration* **344**, 345–362 (2015).
[65] G. Kerschen, ed., *Nonlinear Dynamics, Volume 1, Proceedings of the 33rd IMAC, A Conference and Exposition on Structural Dynamics* (Springer, New York, 2015).
[66] G. Kerschen, ed., *Nonlinear Dynamics, Volume 2, Proceedings of the 33rd IMAC, A Conference and Exposition on Structural Dynamics* (Springer, New York, 2015).
[67] M. Allen, R.L. Mayes, D.J. Rixen, eds, *Dynamics of Coupled Structures, Volume 4, Proceedings of the 33rd IMAC, A Conference and Exposition on Structural Dynamics* (Springer, New York, 2015).
[68] K. Bertoldi, J. Raney, N. Nadkarni, *et al.*, 2016 Overcoming dissipation with structure: Stable propagation of mechanical signals in soft mechanical metamaterials. In *Proceedings of the APS March Meeting 2016*; Baltimore, MD. K40.00012.
[69] K.A. Cunefare, E.A. Skow, A. Erturk, *et al.*, Energy harvesting from hydraulic pressure fluctuations, *Smart Materials and Structures* **22**, 025036 (2013).
[70] X. Xia, C.P. Lawson, The development of a design methodology for dynamic power distribution management on a civil transport all electric aircraft, *Aerospace Science and Technology* **25**, 125–131 (2013).
[71] T. Seuaciuc-Osório, M.F. Daqaq, Energy harvesting under excitations of time-varying frequency, *Journal of Sound and Vibration* **329**, 2497–2515 (2010).
[72] R. Masana, M.F. Daqaq, Response of duffing-type harvesters to band-limited noise, *Journal of Sound and Vibration* **332**, 6755–6767 (2013).
[73] S.R. Anton, H.A. Sodano, A review of power harvesting using piezoelectric materials (2003-2006), *Smart Materials and Structures* **16**, R1–R21 (2007).
[74] D.J. Inman, *Engineering Vibration* (Prentice Hall, Saddle River, New Jersey, 2001).

Index

Harnessing Bistable Structural Dynamics: For Vibration Control, Energy Harvesting and Sensing, First Edition. Ryan L. Harne and K.W. Wang.

V

W

Printed and bound by CPI Group (UK) Ltd, Croydon, CR0 4YY
16/09/2026
14963479-0006